The Implicit Geno

D0121089

RETURN

The Implicit Genome

Edited by
Lynn Helena Caporale

2006

OXFORD
UNIVERSITY PRESS

Oxford University Press, Inc., publishes works that further
Oxford University's objective of excellence
in research, scholarship, and education.

Oxford   New York
Auckland   Cape Town   Dar es Salaam   Hong Kong   Karachi
Kuala Lumpur   Madrid   Melbourne   Mexico City   Nairobi
New Delhi   Shanghai   Taipei   Toronto

With offices in
Argentina   Austria   Brazil   Chile   Czech Republic   France   Greece
Guatemala   Hungary   Italy   Japan   Poland   Portugal   Singapore
South Korea   Switzerland   Thailand   Turkey   Ukraine   Vietnam

Published by Oxford University Press, Inc.
198 Madison Avenue, New York, New York 10016

www.oup.com

Oxford is a registered trademark of Oxford University Press

Library of Congress Cataloging-in-Publication Data

The implicit genome / edited by Lynn Helena Caporale.
   p. ; cm.
Includes bibliographical references and index.
ISBN-13: 978-0-19-517270-6; 978-0-19-517271-3 (pbk).
ISBN-10: 0-19-517270-1; 0-19-517271-X (pbk).
1. Evolutionary genetics.
[DNLM: 1. Genome.   2. Evolution, Molecular.   3. Immunity—genetics.
4. Repetitive Sequences, Nucleic Acid—physiology. QH 447 I34 2006] I. Caporale, Lynn Helena.
QH390. I47 2006
572.8'38—dc22   2005011590

9 8 7 6 5 4 3 2 1

Printed in the United States of America
on acid-free paper

To Mom and Dad

and to the memory of my Grandparents

Contents

Contributors   ix

An Overview of the Implicit Genome   3
    Lynn Helena Caporale

1. Sequence-Dependent Properties of DNA and
   Their Role in Function   23
       Donald M. Crothers

2. Mutation as a Phenotype   39
       Errol C. Friedberg

3. Repeats and Variation in Pathogen Selection   54
       Christopher D. Bayliss and E. Richard Moxon

4. Tuning Knobs in the Genome: Evolution of
   Simple Sequence Repeats by Indirect Selection   77
       David G. King, Edward N. Trifonov, and Yechezkel Kashi

5. Implicit Information in Eukaryotic Pathogens as the
   Basis of Antigenic Variation   91
       J. David Barry

6. The Role of Repeat Sequences in Bacterial
   Genetic Adaptation to Stress   107
       Eduardo P. C. Rocha

7. The Role of Mobile DNA in the Evolution
   of Prokaryotic Genomes   121
       Garry Myers, Ian Paulsen, and Claire Fraser

8. Eukaryotic Transposable Elements: Teaching
    Old Genomes New Tricks   138
    Susan R. Wessler

9. Immunoglobulin Recombination Signal
    Sequences: Somatic and Evolutionary Functions   163
    Ellen Hsu

10. Somatic Evolution of Antibody Genes   177
    Rupert Beale and Dagmar Iber

11. Regulated and Unregulated Recombination of
    G-rich Genomic Regions   191
    Nancy Maizels

12. The Role of the Genome in the Initiation
    of Meiotic Recombination   208
    Rhona H. Borts and David T. Kirkpatrick

13. Nuclear Duality and the Genesis of Unusual
    Genomes in Ciliated Protozoa
    Carolyn L. Jahn   225

14. Editing Informational Content of Expressed
    DNA Sequences and Their Transcripts   248
    Harold C. Smith

15. Alternative Splicing: One Gene, Many Products   266
    Brenton R. Graveley

16. Imprinting: The Hidden Genome   282
    Alyson Ashe and Emma Whitelaw

    Epilogue: An Engineering Perspective: The Implicit Protocols 294
    John Doyle, Marie Csete, and Lynn Caporale

    References   299

    List of Acronyms   363

    Index   365

Contributors

Lynn Helena Caporale (Editor)
The Judith P. Sulzberger MD
  Genome Center, Columbia
  University, 1150 St Nicholas
  Avenue, New York, NY 10032

Alyson Ashe
School of Molecular and Microbial
  Biosciences, Biochemistry
  Building—G08, University of
  Sydney, New South Wales 2006,
  Australia

J. David Barry
Wellcome Centre for Molecular
  Parasitology, University of
  Glasgow, Anderson College,
  56 Dumbarton Rd,
  Glasgow G11 6NU, United
  Kingdom

Christopher D. Bayliss
Oxford University Molecular
  Infectious Diseases Group,
  Weatherall Institute for Molecular
  Medicine, John Radcliffe Hospital,
  Headington, Oxford OX3 9DU,
  United Kingdom

Rupert Beale
Medical Research Council
  Laboratory of Molecular Biology,
  Hills Road, Cambridge CB2 2QH,
  United Kingdom

Rhona H. Borts
Department of Genetics, University
  of Leicester, University Road,
  Leicester LE1 7RH,
  United Kingdom

Marie Csete
Emory University School of
  Medicine, 1462 Clifton Rd NE,
  Room 420, Atlanta GA 30322

Donald M. Crothers
Department of Chemistry, Yale
  University, 350 Edwards Street,
  PO Box 208107, New Haven,
  CT 06520-8107

John Doyle
California Institute of
  Technology, CDS 107-81,
  1200 E California Blvd,
  Pasadena, CA 91125-8100

Claire Fraser
The Institute for Genome
 Research, 9712 Medical Center
 Drive, Rockville, MD 20850

Errol C. Friedberg
Laboratory of Molecular
 Pathology, Department of
 Pathology, University of Texas
 Southwestern Medical
 Center at Dallas, Dallas,
 TX 75390

Brenton R. Graveley
Department of Genetics and
 Developmental Biology,
 University of Connecticut Health
 Center, 263 Farmington Avenue,
 Farmington, CT 06030-3301

Ellen Hsu
Department of Physiology and
 Pharmacology, State University
 of New York Health Science
 Center, Brooklyn, NY 11203

Dagmar Iber
Mathematical Institute, Centre for
 Mathematical Biology, St John's
 College, University of Oxford,
 24–29 St Giles, Oxford OX1 3LB,
 United Kingdom

Carolyn L. Jahn
Department of Cell and Molecular
 Biology, Feinberg School of
 Medicine, Northwestern
 University, 303 East Chicago
 Ave., Chicago, IL 60611

Yechezkel Kashi
Department of Biotechnology
 and Food Engineering,
 Technicon-Israel
 Institute of Technology, Haifa
 32000, Israel

David G. King
Department of Zoology, Southern
 Illinois University, Carbondale, IL
 62901-6899

David T. Kirkpatrick
Department of Genetics, Cell
 Biology and Development,
 University of Minnesota,
 6–160 Jackson Hall, 321 Church
 St SE, Minneapolis, MN 55455

Nancy Maizels
Departments of Immunology
 and Biochemistry, University of
 Washington Medical School,
 1959 N.E. Pacific Street, Seattle,
 WA 98195-7650

E. Richard Moxon
Oxford University, Molecular
 Infectious Diseases Group,
 Weatherall Institute for
 Molecular Medicine,
 John Radcliffe Hospital,
 Headington, Oxford OX3
 9DU, United Kingdom

Garry S. A. Myers
The Institute for Genome
 Research, 9712 Medical Center
 Drive, Rockville, MD 20850

Ian T. Paulsen
The Institute for Genome
 Research, 9712 Medical Center
 Drive, Rockville, MD 20850

Eduardo P. C. Rocha
Unité Génétique des Génomes
 Bactériens, Institut Pasteur, 28
 rue du Dr. Roux, 75724 Paris
 Cedex 15, France and Atelier de
 Bioinformatique, Université
 Pierre et Marie Curie, 12, Rue
 Cuvier, 75005 Paris, France

Harold C. Smith
Department of Biochemistry and
    Biophysics, Box 712, University
    of Rochester, School of Medicine
    and Dentistry, 601 Elmwood
    Avenue, Rochester, NY 14642

Edward N. Trifonov
Genome Diversity Center, Institute
    of Evolution, University of Haifa,
    Mount Carmel, Haifa 31905,
    Israel

Susan R. Wessler
University of Georgia, Department
    of Plant Biology, Miller Plant
    Sciences Building, Athens,
    GA 30602

Emma Whitelaw
School of Molecular and Microbial
    Biosciences, Biochemistry
    Building—G08, University of
    Sydney, New South Wales 2006,
    Australia

The Implicit Genome

An Overview of the Implicit Genome

Lynn Helena Caporale

I was in a new world, and ... could not help speculating on what my wanderings there ... might bring to light

Alfred Russel Wallace, *The Malay Archipelago* (1869)

Most analyses assume that genomes are to be read as linear text, much as a sequence of nucleotides can be translated into a sequence of amino acids by looking in a table. However, information can evolve in genomes with distinct forms of representation, such as in the structure of DNA or RNA and the relationship between nucleotide sequences. Such information has importance to biology yet is largely unexpected and unexplored. As described in this volume, much of this information, through mechanisms ranging from alternative splicing of RNA to the generation of bacterial coat protein diversity, affects the probability of distinct types of alterations in the nucleic acid sequence. Some genomic DNA sequences affect genome stability, handling, and organization, with implications for the robustness of lineages over evolutionary time. The examples reviewed in this volume, taken from a broad range of biological organisms, both extend our view of the nature of information encoded within genomes and deepen our appreciation of the power of natural selection, through which this information, in its various forms, has emerged.

Introduction

There was a glorious moment, in the 1960s, when, like children first learning to read, we began to perceive meaning in strings of nucleotides in DNA. Suddenly, we could understand that TTT in a protein-coding sequence of DNA meant that the amino acid phenylalanine would be incorporated into a protein. We could, in our minds, and later with computers, directly translate strings of bases,[1] taken as triplets,[2] into strings of amino acids, which then fold up to form three-dimensional proteins.

3

While the challenge remained of perceiving the three-dimensional structure specified by a linear protein sequence, we understood that information was encoded in DNA in a way that was both explicit and linear.

But now, with complete genome sequences before us, instead of triumph comes humility. The strings of nucleotides that we proudly translate into protein amount to less than 2% of the human genome.[3] We know well that additional DNA sequences are involved in the regulation of expression.[4,5] However, fully understanding the information content of genomes will involve expanding our imagination with respect to both what types of information may be there and how information might be represented.

Novel forms of information may be represented in the structure of DNA or in the relationships between sequences, rather than in the sequences themselves. This book focuses on information that leads to editing, splicing, recombination, mobility, and/or which affects the rate, type, and location of mutations of nucleic acid sequences. Such variations in the probability of distinct changes in DNA sequence have implications for evolution. In fact, for many specific examples in this book, certain types of mutation, with an increased probability of being adaptive, are more probable than would be expected from the "average" mutation rate. The reader is challenged to decide whether these examples are just isolated, albeit interesting, cases, specific, for example, to the stress of host–pathogen interactions, which provide no broader insight into genome organization and evolution; alternatively, the reader may decide that the examples of focused genetic variation discussed in this book represent the tip of an iceberg that should command more immediate attention.

One Nucleotide Sequence Can Imply Other Nucleotide Sequences

We recognize information that is encoded in DNA explicitly: in the right context and reading frame, TTTGGG encodes the sequence phenylalanine-glycine. But as the chapters in this book make clear, even information that we have been confident about our ability to interpret, such as protein-coding information, often only is implied in a genome.

Editing and Splicing

Due to splicing and editing of nucleic acid sequences, genomic DNA sequences often do not directly match the sequences of the proteins encoded by the genome.

Alternative Splicing

Although stunning when first observed, we now tend to view the concept of "genes in pieces" as routine—until we attempt to identify the proteins encoded by a newly-sequenced eukaryotic genome. Chapter 15 describes the *Drosophila Dscam* gene, a single gene that, at least theoretically, can, by alternative splicing, generate three times as many proteins as there are protein-coding genes in the genome. For

RNA transcripts to be spliced, they must contain additional information, beyond the base triplets that encode amino acids. This information, which is located both within sequences destined to be exons and those destined to be introns, and across their boundaries, implies the resulting context-dependent pattern of splicing and the "either/or" relationships between certain exons. Exon choice that feeds back to be self-perpetuating can serve as a developmental switch, most famously in sex determination in *Drosophila*.

RNA Editing

When "translating" nucleic acid sequences, we expect to be able to predict the protein sequence, based solely on the table of codons. However, as described in chapter 14, it is not unusual for the primary RNA transcript to be edited as well as spliced before it is translated into protein, so that the RNA sequence no longer is complementary to the DNA sequence from which it was transcribed. Before editing was understood, comparisons between protein and DNA sequences in plant organelles made it appear that chloroplasts use a different triplet code.

In an example cited in chapter 14, at a position that corresponds to an arginine in the protein, the genomic sequence of a human calcium channel is CAG, which bioinformatics software would translate as glutamine; editing to a codon that is translated as arginine depends upon a relationship between DNA sequences: complementarity between the exon that contains the CAG and its neighboring intron.

While "changing" glutamine to arginine (which has an important effect on the properties of the ion channel) requires alteration of a single base in the message, the so-called "cryptogenes" of kinetoplastids undergo far more extensive editing. Some phenylalanines are encoded entirely by Us that are added when the RNA transcript is edited. For software to find in a protein database the protein sequence encoded by that stretch of DNA, it would have to be designed based on an understanding of how the primary transcript is edited upon interaction with separately encoded "guide RNAs."

DNA Gymnastics

As a routine part of their life cycle, ciliates alter their DNA dramatically. As described in chapter 13, although ciliates are unicellular they have two "genomes," each in a separate DNA-containing organelle. The nucleus that undergoes meiosis contains the "germline" genome, while genes are transcribed into mRNA using the other nucleus, which contains the "somatic" genome. In forming their somatic genome, some ciliates eliminate up to 95% of the germline genome sequences, then rearrange and amplify the remainder to generate as many as 1000 copies of each of their genes.

After meiosis and exchange of haploid nuclei with another individual to form the zygote, ciliates must define the sequences that will become part of the new somatic nucleus. Because for some ciliates the order of nucleotides in the germline does not correspond to the order of amino acids in their proteins, germline sequences must also be unscrambled. As described in chapter 13, in the equivalent of a subtractive hybridization experiment, double-stranded RNA moves from the meiotic "germline nucleus" to the old somatic nucleus and then to the developing new somatic nucleus; dsRNA that enters the new somatic nucleus is "interpreted" as a sequence that was not found in the old somatic nucleus, and thus should be deleted. Chapter 13 points out that analysis of how information flow is handled in organisms operating with distinct nuclei could provide a window into DNA sequence "requirements for the divergent nuclear and genomic functions of transcription and partitioning of genetic material."

Because, as described in chapter 13, siRNA and heterochromatin appear to be involved, deletion can be considered an extreme case of "turning off" regions of DNA. In addition, because the sequences that are eliminated between the germline and somatic nuclei have homology with transposable elements, ciliates may represent an extreme example of the potential for transposable elements to define special contexts or behaviors in a genome.

As chapter 13 points out, from telomeres to RNA splicing, to the role of histone acetylation in transcription activation, we have learned a lot by studying ciliates; it appears likely we will learn more, about genome organization and meiosis. The concept that the somatic and germline nuclei communicate via siRNA is intriguing. Our perspective on the biochemical processes that occur as information is transferred between generations likely would be quite different if decades ago molecular geneticists had selected a ciliate, such as *Oxytricha nova,* rather than yeast, as a model unicellular eukaryote organism.[6]

Rules Rather Than Genes

The numbers of different antibodies that can be generated by the vertebrate immune system, as described in chapter 9, and different surface antigens that can be generated by the pathogens that afflict us, as described in chapter 5, are orders of magnitude greater than the number of genes encoded in their respective genomes. Yet if you search the human germline genome for intact genes encoding antibodies, you will not find them there. Both vertebrates and pathogen genomes imply, rather than specify, this diversity of protein sequences, by storing information in gene fragments, along with rules for their assembly. (Alternative splicing as described in chapter 15 provides another way to store information for multiple related protein sequences efficiently.)

The Immunoglobulins

The initial step in construction of a gene encoding an immunoglobulin heavy chain is the recombination of one of a palette of genomic variable regions beside another gene segment. As detailed in chapter 9, information in the flanking "recombination signal sequences" defines which classes of potentially functional gene segments can recombine with each other. In one startling aspect of the formation of an immunoglobulin heavy-chain gene, "diversity" regions may be inserted in any of their three reading frames in a combinatorial assortment next to "variable" regions. In addition, some diversity regions can be inserted in two orientations (i.e. so that either strand can be the "coding" strand), such that the same small patch of DNA may generate six different amino acid sequences. As emphasized in chapter 9, diversity regions also can vary in length, and nontemplated sequences may be added at the junction; this further increases the unpredictable (from the point of view of pathogens) diversity of immunoglobulin binding sites. Thus, the precisely targeted diversity of the immunoglobulins is implied in the genomic DNA.

The Pathogens

Trypanosome surface antigen diversity comes from a repertoire of DNA sequences embedded in cassettes. Flanking sequences facilitate replacement of an antigen that is in an expression site with a copy of another, hopefully (from the pathogen's perspective) not cross-reacting, antigen from the cassette archive.

In addition, as an infection proceeds, mosaic surface antigens may be assembled from multiple partial sequences drawn from an extensive repertoire of "pseudogenes." Chapter 5 points out that generation of mosaic antigens from pseudogenes makes tens of millions, or even billions, of potential surface variants theoretically possible.

As described in chapter 5, diverse pathogens, including *Neisseria, Pneumocystis carinii,* and *Borrelia burgdorferi,* "hide" from their "hosts" by varying their surface antigens rapidly during the course of infection by generating mosaic antigens via gene conversion. Enzymes that catalyze recombination at specific target sequences generate antigenic diversity by such mechanisms as inverting a promoter involved in the expression of surface proteins or replacing a patch of DNA sequence (as described in chapter 7).

Mutable Sequences: Adjustments and Reversible Inactivation

When studying the complex mechanisms that regulate the expression of genes, we traditionally have assumed that a gene's intrinsic activity is arrived at by the action of selection on multiple random point mutations across evolutionary time. However, several chapters in this book, which use terms such as "tuning knobs" (mutable repeats: chapter 4) and "rheostats" (chapter 8), describe biochemical mechanisms that can expedite intergenerational exploration and adjustment of the range of gene activity through less damaging routes than random genome-wide point mutation.

Diversity

Chapters 3, 4, 7, and 11 describe sequences that mutate at frequencies that can be greater than 1000-fold higher than the genome average. Tandem repetitive sequences, such as GGGGGGGGG or CAGCAGCAG, change length frequently; thus such sequences imply a tendency to generate diversity at that locus between generations. Such highly mutable and statistically improbable repetitive sequences are overrepresented in bacterial and eukaryotic genomes but are underrepresented in conserved regions of constitutively-expressed housekeeping genes. Evidence, discussed in chapters 3 and 4, that these mutable sequences are under selection indicates that their mutability affects fitness. Within a bacterial population and also among the descendants of an individual bacterium the rapid rate of mutation at repeat sequences generates diversity at loci, such as those for surface antigens, at which diversity has provided a selective advantages in the past.

Chapter 3 points out that with just 12 genes able to switch between "on" and "off" due to reversibly mutable repeat sequences, a population of bacteria descended from a single individual can explore 4096 (2^{12}) different "versions" of its genome without damage to other loci en route. Thus, when we obtain one complete genome sequence we see only an example of the genomic range of that species. Chapter 3 introduces the concept of a "species genome," which takes into account the "full repertoire of variations in repetitive DNA tracts and locations of insertion elements." The diverse phenotypes of the full set of genomes that are implied in the repetitive sequences of an individual genome extend the potential adaptive capabilities of its descendants to meet classes of challenges that can in effect be anticipated based upon the species' past experience.

As suggested in chapter 4, recent data—ranging from *Drosophila* to cattle—hint that mutable sequences not only can differ among cells in an individual but also are involved in rapid phenotypic adjustments in eukaryotes. Repetitive microsatellites in the human genome can have meiotic mutation rates as high as several percent per generation. These highly mutable loci contribute so much diversity to human populations that some are used for DNA "fingerprinting."

Adjustments

Mutable repeats have the potential not only for generating diversity within a population but also for facilitating the reversible "adjustment" of activity, through selected changes in the frequency of these highly polymorphic alleles, at multiple loci, from generation to generation. In other words, in modeling evolution it is important to consider not only diversity in a contemporary population but also the potential for distinct classes of diversity in the population descended from an individual genome.

Changes in the length of amino acid repeats (encoded by triplet repeats at the DNA level) can have the effect of adjusting biochemical properties such as protein flexibility, affinity for substrate, and strength of protein–protein interactions. As chapter 4 points out, such repeats are markedly overrepresented in transcription factors, protein kinases, and genes encoding developmental regulatory proteins.

As described in chapter 3, changes in length of repetitive sequences located between the right and left sides of the promoter affect the level of transcription in prokaryotes. Variations in the number of repeats can affect transcription activity in eukaryotes as well. Chapter 4 provides the intriguing example of a polymorphic repeat in the first intron of a rate-limiting enzyme in the synthesis of catecholamine neurotransmitters, and estimates that such hypermutable loci may be found in the regulatory regions of as many as several thousand human genes.

As discussed in chapter 6, overrepresentation of closely spaced repeats in mismatch repair genes results in a tendency for recombination to reduce or eliminate the activity of these genes in a proportion of the bacterial population (activity can be recovered through recombination with DNA taken up from the environment). Those bacteria with decreased mismatch repair activity have an increased rate of generation of diversity at repeats elsewhere in the genome that become more unstable when mismatch repair is decreased.

Chapter 8 suggests that the level of transcription also can be affected by the presence of transposable elements in introns. Chapter 14 suggests that the extent of editing of multisubunit ion channel mRNA can serve as a "rheostat", a term also used in chapter 16 with reference to imprinting fo dosage-sensitive genes. As described in chapter 16, epigenetic regulation can adjust, during development, the level of expression of certain alleles through silencing of neither, one, or both alleles in different cells. All may facilitate the combination of robustness and stability that modularity can contribute to development.[7,8]

Forms of Information

Multiple Levels of Messages: Using the Degeneracy of the Genetic Code

Because more than one codon is available to specify most amino acids, additional information may be transmitted along with a protein-coding sequence. Such additional information will constrain the choice between what are otherwise considered to be synonymous codons.[9]

For example, chapter 6 presents two lines of evidence that suggest that selection for increased mutability leads to codon choices that result in the overrepresentation of closely spaced repeats in mismatch repair genes. First, due to the degeneracy of the genetic code, the same sequence of amino acids can be encoded by any one of many less mutable DNA sequences. Second, chapters 4 and 6 present examples in which the presence of a repeated sequence, and thus the property of hypermutability, was conserved, while the nature of the repeat and the amino acid sequence it encoded was not conserved.

In *B. burgdorferi*, the requirement for a conserved nucleotide repeat to enable antigenic variation of a surface protein by gene conversion between two repeats of the five amino acids EGAIK constrains, to a single DNA sequence, what otherwise would be a choice among nearly 200 synonymous sequences that could encode each EGAIK.

As described in chapter 10, information that affects the tendency to hypermutate along a DNA sequence constrains the choice among what are assumed to be "synonymous" codons in immunoglobulin genes. Mutational hotspots in the variable region encode serine using AGY (Y indicates pyrimidine, i.e., C or T), while serine is encoded by TCN (N is any nucleotide) in the constant region.

The choice among synonymous codons also is constrained when a DNA sequence must, in effect, carry a label. As described in chapters 6 and 7, some bacteria "mark" their DNA. For example, in certain bacterial genomes codon choice is constrained by overrepresentation of sequences that facilitate uptake of DNA from closely related bacteria. While only eight copies of a nine-base uptake signal sequence would be expected by chance, there are 1500 copies of that sequence in the *H. influenzae* genome.

Similarly, chapter 15 describes the constraints on codon choice due to exon splice enhancers and information at the exon–intron boundary, which mark the sequence as an exon and define its splice site.

Structural Information

We talk about DNA sequences, and our computer algorithms search them, as if nucleotides were letters that can be read much like those printed on this page. But after reading chapter 1, no reader is likely to view a DNA sequence simply as a one-dimensional string of "letters" again. The distinct tilt and twist of different steps of stacked base pairs along the helix may tend to average out, but for sequences highly enriched in one base, or in which a sequence motif is repeated, there can be substantial deviations from the average coordinates that are used in text-book models of the double helix. The biological consequences of such sequence-dependent variation in DNA structure are discussed in several chapters of this book.

For example, as described in chapter 13, sequences that are as high as 70–80% AT are highly "bendable" and are thought to affect chromatin structure and genome rearrangements in ciliates. As described in chapters 4 and 12, repeats of sequences may favor or inhibit nucleosome assembly, with effects both on gene expression and on the likelihood of genetic exchange between two parental genotypes at that site during meiosis.

As described in chapter 1, protein binding to DNA can be affected by effects of DNA sequence on backbone geometry. In fact, changing the sequence of a four-base "spacer" between two binding sites can change protein affinity by over three orders of magnitude. Similarly, the nature of the spacer sequence affects the efficiency of recognition of immunoglobulin recombination signal sequences, as referenced in chapter 9.

As the number of bases that separate two DNA sequences changes, the two sequences move relative to each other around the helix axis. As shown in chapter 3, changing the relative three-dimensional orientation of the left and right sides of the promoter by changing the distance between them by just 1 or 2 bases can have dramatic effects on interaction with transcription factors and thus on gene expression.

The context of a DNA sequence affects its mutability, which depends not only on the presence and composition of repeats and the direction of replication but also on the nature of flanking sequences. For example, if the sequence between repeats is a palindrome, deletion can be more likely.

Chapter 11 describes unique four-chain structures formed when G-rich sequences separate from the double helix during replication or transcription of the opposite strand. Such G-rich structures can stall replication and increase recombination and cause lethal genome instability, but they have been captured by the ever-creative vertebrate immune system to focus the region-specific recombination that is required for a regulated switch between immunoglobulin heavy-chain classes. Thus, within immunoglobulin heavy-chain introns, G-rich sequences imply a region of recombination that is regulated by transcription.

Different nucleotide sequences may share certain physical chemical characteristics; for example, both A·T and G·C base pairs, but not an A·G base pair, fit the steric requirements of the replicative DNA polymerase. Thus it was possible to be blinded by viewing DNA as "letters," until a palindromic pattern of hydrogen bond donors and acceptors in the major groove was identified as the formerly elusive consensus sequence for P-element insertion in *Drosophila*.[10]

Chapter 1 emphasizes that analyzing the physical chemical properties of DNA sequences is not as straightforward as looking up codons in a table. The natural curvature of runs of As is reinforced when A runs are on the same side of the helix and "cancels out" when they are on opposite sides. Chapter 1 points out, while describing the induction of positive base inclinations at the 3′ end of AAAAAA tracts repeated in phase with the helical screw of DNA, that structural deviation from the coordinates of the standard B-form DNA can be propagated into neighboring sequences; due to the importance of sequence context, the local shape and flexibility of a DNA sequence cannot be predicted simply by adding up the tilt and twist parameters of individual steps. This effect of sequence context can make the identification of nucleotide "consensus" sequences challenging.

Relationships

In examples ranging from mutable repeats to RNA editing, information often is represented in the relationship between sequences, rather than in the specific sequences themselves. Certain RNA strands can form alternative loop structures that can regulate transcription and translation in response to the presence or absence of other molecules in the cell. A classic example is the attenuation loop involved in regulation of the biosynthesis of tryptophan.[11] More recent important examples involve sequences in mRNA at which metabolite binding affects the choice between alternative RNA structures.[12]

The essential, extensive, and evolving biological roles of RNA transcripts that do not encode amino acid sequences have recently become breathtakingly apparent.[13] For example, as discussed in chapters 8, 13, and 14, complementary RNA sequences trigger sequence-specific gene and/or transcript silencing.

Repeats in DNA can be loci of genetic variation, as described in chapters 3, 4, 6, and 7, but certain inverted repeats are sites of stability. During DNA replication on the lagging strand, the tendency of nearly-palindromic sequences to form base-paired hairpins can lead to a high frequency of a specific subset of mutations that result from a tendency to "repair" inexact matches across the hairpin.[14] "Correction" by internal palindromes has been proposed to protect Y chromosome sequences.[15] Relationships between DNA sequences can enable them to interact in ways other than the standard double helix, as discussed in chapter 11. The importance of sequence relationships in RNA in forming biologically-important structures is well appreciated.[16]

Evolutionary Information: Probable Future Genomes

In most discussions of evolution, mutation is described as a random and generally harmful process, with the genome as its hapless victim. However, during evolution, one genome sequence does become other, well-adapted, genome sequences, as a result of multiple mutations.

When genes and mutation were incorporated into evolutionary theory during the first half of the twentieth century (it should be noted that this was not only prior to genome sequencing but also prior to understanding how DNA encodes information or even that DNA is the genetic material), it was assumed that mutation was "random." Of course, those were not Darwin's words, as the concept of genes and the biochemistry of mutations were unknown when he proposed that biological evolution results from selection acting on variation in traits that are inherited.

As described in this book, nucleic acid sequences often change in ways that are not completely random. Intrinsic sequence-dependent variations in DNA sequence context (chapter 1) and in the structure and fidelity of enzymes responsible for replication and repair of DNA molecules (chapters 2, 3, and 6) result in sequence-dependent variations in the types and rate of mutations.

While it is clear that mutation is not random with respect to DNA sequence context, it has been assumed that mutation must be completely random with respect to its potential effects on phenotype because "selection lacks foresight."[17] However, because the world is not completely random, selection can gain a degree of "foresight"—to the extent that a lineage repeatedly must survive the same classes of challenges, such as pathogens surviving attacks by our immune response and our surviving attacks by pathogens.

Indeed, despite the expansive landscape of possible random changes, genomes often find repeated paths to the same solution. Chapter 8 presents a case study of adaptation of yeast to glucose restriction in which multiple clones shared the same breakpoint, at a transposable element. (This is one of many examples of how transposons can be valuable to the genome that hosts them.) Repeated (and reversible) amplification of certain sets of genes can enable stressed bacteria to reach beyond the limit of maximal expression of a single copy of these genes.

Most research in biology focuses on mechanisms that enable an organism to survive through one life cycle, including adaptation to changes in the environment that occur within its lifetime (such as the appearance of lactose). Yet genomes survive through an unbroken chain of living beings across evolutionary timescales. Lineages must survive challenges that may be repeated (such as climate cycles) or extended (such as the ongoing evolution of other organisms in the community). Some, but not all, chapter authors and I suggest that the evolutionary success of certain lineages may result from their genome being more efficient than random at exploring variation that is aligned with the nature of those challenges and opportunities their lineage has faced repeatedly during evolution, which underlies a phenotype of robustness to those classes of challenges[18]; this concept is illustrated by the rapid variation of pathogens' surface antigens (chapters 3 and 5).

Intriguing observations in the literature suggest that selected paths of exploration might exist for systems as diverse as toxin genes for cone snails[19] and, as discussed in chapter 4, skeletal structure in dogs. Such observations should inspire experimental investigation of possible facilitated genomic exploration of lineage-essential traits, such as (pure but irresistible speculation) beaks upon which birds depend for access to available seeds.

As described in chapter 4, natural selection will act indirectly on the mechanisms that generate genome variation, much as it acts directly on beaks and wings. Darwin argued: "Why ... should nature fail in selecting [useful] variations ...? I can see no limit to this power." [20] This power can include selection for (useful) variations of distinct types of mutation along a strand of DNA.

Selection for Mechanisms That Generate Diverse Descendants

The best measure of "fitness" often is the ability to generate diverse descendants. As described in chapter 3, from the perspective of pathogens, hosts are an unsettled landscape. Which bacterium is the "fittest" can be redefined quite suddenly, such as by the appearance of a new antibody.

Reversible mutations generate diversity that enables a lineage to survive when one trait causes a disadvantage, without losing from the lineage a trait that may prove advantageous to descendants under other likely circumstances (chapters 3 and 6). For example, a capsule that shields bacteria from attack by the host's complement system prevents the bacteria from adhering to certain tissues. If the capsule appears and disappears through reversible mutations in repetitive sequences, bacteria will have descendants with and without the capsule, starting from either phenotype. In other words, bacteria resistant to complement will have a significant percentage of progeny that can stick to host tissues, while bacteria that can stick to host tissues will have a significant percentage of progeny that are resistant to complement. Monoallelic expression is another mechanisms that "hides" information (one allele in a diploid organism) that will predictably reappear in a subsequent generation (chapter 16).

As reviewed in chapter 12, two well-known mechanisms that ensure diversity in eukaryotic offspring are independent assortment of maternal and paternal contributions to the genome and recombination between homologous parental and maternal chromosomes during meiosis (with unexpected ratios of offspring phenotypes resulting from unrepaired heteroduplexes).

Frameworks for Exploration

An evolved infrastructure facilitates DNA mobility and recombination at sites where it is more likely to facilitate combination of functional patches of DNA, whether they are genes, regulatory regions, genomic contexts, or exons, rather than damage the recombining sequences.

Frameworks for Gene Movement in Prokaryotes

Chapter 7 describes the "metagenome," comprising potentially valuable information that flows horizontally among bacteria through a well-evolved infrastructure.

Transposable elements, phage, conjugative plasmids, and transposons all participate in the transfer of information among bacteria. Integrons, which carry a promoter and can capture and release gene cassettes, provide an efficient framework for transferring and expressing information; such a set of cassettes traveling together on one integron may carry resistance to all classes of clinically useful antibiotics, as well as to antiseptics and disinfectants. The spread of integrons can be aided by their incorporation into plasmids that have a broad host range.

As described in chapter 7, "genomic islands" combine relatively conserved core regions with other regions that contain a more variable set of genes that is appropriate to the environment of the particular strain or species.

Recombination and Movement of Information in Eukaryote Evolution

As described in chapter 9, the vertebrate immune system, which appears to have emerged through the creative action of a transposable element in the germline, has evolved a genomic infrastructure of gene segments, flanked by specific signal sequences, which are recombined according to specific rules. This genomic framework generates diversity that is aligned with the requirements of the protein; specifically, variation tends to occur where it affects the potential to bind to diverse pathogens without impinging upon effector functions required for pathogen disposal. While the repertoire of antibodies, with distinct pathogen-binding specificities that are many orders of magnitude larger than the currently annotated number of protein-coding genes in the human genome, evolves within the lifetime of each individual, it does so within a framework that was selected over evolutionary time.

Based on the clear evidence of such biochemistry being available to the immunoglobulin genes, it is reasonable to consider that information that aligns the probability of distinct types of variation along a nucleotide sequence with the requirements for protein function might be a characteristic of other "successful" gene families too, facilitating expansion to large numbers of functional members by not depending exclusively on chance and random variation to sculpt each new duplicated copy of the gene.[21] In fact, chapter 9 suggests that hypermutation of the Ig variable regions may have predated in evolution the appearance of their extraordinary somatic recombination.

Chapter 11 suggests that G-rich sequences, which define the sites of exon switching in the immunoglobulin genes, might stabilize four-stranded DNA structures in the germline, which could, through ectopic recombination, facilitate exon shuffling (the idea, first proposed when introns were discovered, that functional domains might move around the genome, thus facilitating an exploration of potential new combinations of information). That G-rich sequences do recombine in the germline is suggested by yeast, in which a meiosis-specific protein that binds G-rich four-chain structures promotes the interaction of two helices. In fact (as described in chapter 12), certain GC-rich sequences create regions distributed over approximately 100 to 500 base pairs that pairs that are hotspots of recombination in meiosis.

Chapter 9 presents several lines of evidence for ongoing activity in germ cells of the transposon-derived enzyme RAG (recombination activating gene), which is involved in immunoglobulin gene segment rearrangement. The continued role of transposable elements in eukaryote genome evolution, including two mechanisms for exon shuffling, is discussed in chapter 8. As transcription of elements that transpose through an RNA intermediate often continues into flanking host DNA, such elements can contribute to exon shuffling due to their propensity to carry host sequences; these transcripts return to the nucleus and can prime reverse transcription into chromosomal DNA at sites nicked by a transposon-encoded endonuclease. Thus a transcribed element potentially can carry, to another place in the genome, any sequence that is its 3′ neighbor. Such exploration of new combinations of functional pieces of DNA can lead, for example, to adding a regulatory domain to an enzyme, which might be one step in linking together a control network, as described in chapter 8.

Even without "jumping," transposable elements can facilitate exon shuffling, by providing homology that enables ectopic recombination between otherwise unrelated sequences in their neighborhood, much as recombination at G-rich sequences described in chapter 11 does not require extensive sequence homology. Whether transposons move around a genome through an RNA intermediate, directly as DNA, or are no longer "jumping," they spend most of their time as DNA, part of the genome. While the probability of ectopic recombination between any two elements may be low, in aggregate, the one million *Alu* elements in the human genome are likely to contribute significantly to genome evolution. Chapter 8 reports that *Alu* elements, dispersed through the genome, also can become new exons.

In addition, both trypanosomes (chapter 5) and chicken immunoglobulins (chapters 9 and 10) point to the important role of pseudogene fragments, which contribute diverse sequences by gene conversion. Pseudogenes are widespread; there is growing evidence for the contribution of information derived from pseudogenes both to gene evolution and to regulation of gene expression.[22]

Genome Organization: The Importance of Neighborhood and Context

Both the mutability and the "meaning" of a nucleotide sequence depends on its context. For example, a T is more likely to be deleted or added in a run of other Ts. And as for that nucleotide run, whether TTTTTTTTT "means" phenylalanine–phenylalanine–phenylalanine or a specific level of gene expression depends upon whether the mononucleotide run is in a "protein coding" region or separates the left and right side of a bacterial promoter. The effect of context can be felt over a range of distances, from neighboring twists of the helix (chapter 1) to chromosomal regions (chapters 8 and 16).

Gene Expression Is Affected by Neighboring Transposons

With the potential to involve that gene in a regulatory hierarchy through RNAi-mediated generation of heterochromatin, Chapter 8 describes two routes by which transposon-derived sequences in the neighborhood of a gene can lead to dsRNA. Transcription initiated in a host sequence can generate dsRNA by reading through an element that contains terminal inverted repeats. Alternatively, transcription initiated in a transposable element can generate dsRNA if its promoter drives transcription into a nearby gene on the antisense strand. (When the neighboring gene is transcribed, sense and antisense transcripts anneal, forming dsRNA.) Thus genes that neighbor the same transposable element yet are scattered through the genome might be coordinately silenced by an environmental trigger that affects the element's promoter.

Regions of Monoallelic Expression

We are used to thinking that as humans we are strictly diploid, with two active copies of every gene (except those on the X and Y chromosomes)—one inherited from our father and one from our mother, but which now are equivalent and both equally "ours." As described in Chapter 16 however, one copy of a parentally imprinted locus is turned off, depending upon the parent of origin. We are, in effect, haploid at that locus, with the information on the silent copy just passing through us to be revealed in a future generation. Imprinting occurs with distinct patterns, depending upon whether the genome is developing in the male or female germline. Imprinted genes "remember" whether they are from Dad or from Mom. An imprint may be removed in some but not all tissues in the progeny, most notably the developing germline, where it then is re-set according to the sex of the future new parent. The "reasons" for this are the subject of active speculation, but, as discussed in chapter 16, parental imprinting may have its biochemical origin in the requirement of homologous chromosomes to carry parental marks to enable appropriate sorting during meiosis and mitosis. Because whether a gene is imprinted depends upon the genomic context in which it is placed, chapter 16 leads into a literature that discusses the effects on gene activity of information that defines chromosome neighborhoods.

As described in chapter 16, many additional loci in diploid genomes that experience monoallelic expression (also termed "allelic exclusion"), including the T-cell receptor, natural killer receptors, and receptors for odors and pheromones, encode proteins that interact with agents from the environment. An "exposed" suddenly haploid allele will experience direct selection. For example, as discussed in chapter 10, antibody-producing cells undergo selection based upon the binding activity of the expressed allele. As discussed in chapter 5, only one specialized expression site for trypanosome surface antigen genes is transcribed at a time; an active site becomes silenced and an inactive site activated through an as yet mechanistically undefined interaction between the two.

Regulated, Region-specific Recombination

While much recombination is site-specific, some recombination is increased along a region of DNA. In some cases it is clear that such "region-specific recombination" is regulated through transcription. In particular, as described in chapter 11, introns that participate in the immunoglobulin class switch are enriched in G-rich regions that attract recombination; recombination occurs in response to extracellular signals that stimulate transcription across the G-rich region.

Chapter 12 describes the requirement of specific transcription factors, but in contrast to the immunoglobulin class switch no evidence to date for transcription itself, at hotspots of meiotic recombination. Thus, many proteins that have been considered "transcription factors" may have as their primary role the regulated opening of chromatin structure, whether for transcription or for other purposes related to genome organization and handling.

As discussed in chapter 5, the probability of recombination is affected by position on a chromosome as sequences in subtelomeric regions have an increased probability of ectopic recombination. While some genomic regions are hotspots of recombination and/or foci of genomic flux, there also can be great local stability. As described in chapter 12, genome sequencing has led to investigations into the nature of recombination hotspots and of cold spots that may enable certain blocks of genes to remain together from generation to generation. Certain gene regions, such as the HOX genes, experience duplications as a block, but maintain their linear relationship to each other within these blocks across long evolutionary timescales.[23] In contrast, there is a particularly high rate of germline recombination at specific loci in the human genome, such as at the HLA locus,[24] at which diversity is especially important.

The precise location of meiotic recombination within a meiotic recombination hotspot depends upon the specificity of Spo11, the endonuclease that makes the initiating double-strand break. The location of sites favored by this and other nucleases, such as the endonuclease involved in priming and inserting sequences following reverse transcription (chapter 8), also can fall under selection.

Contexts Defined by the Timing of Replication

Preliminary work links the boundaries between "early" and "late" replicating regions with the boundaries of regions with a tendency to amplify during tumor formation and boundaries of syntenic regions that have remained together across evolutionary time,[25,26] indicating that these boundaries have biological, or at least biochemical, identities.

In a wide range of species, whether a region replicates "early" or "late" appears to be one way of defining sequence regions with distinct properties. Chapter 16 indicates that the timing of replication at imprinted loci differs depending on the parent of origin. Chapter 13 points out that regions of DNA that are destined to be left out of the somatic nucleus of ciliates replicate during a second phase of DNA synthesis. Chapter 12 observes that many cold spots of meiotic recombination replicate late during pre-meiotic replication when chromatin alterations occur that subsequently influence region-specific double-strand break formation during meiosis. Chapter 7 observes that sequence conservation is much higher near the origin of replication than at the terminus in prokaryotes.

When distinct regions of the genome replicate at different times or in different territories, nucleotide pools and relative levels of mismatch repair proteins (chapter 6), polymerases with distinct fidelity (chapter 2), and other protein activity, all can differ giving these regions their own spectrum of mutations (chapter 2). Combined with sequence-dependent variations in DNA structure (chapter 1), there can be distinct region- and time-dependent effects on the mutability of different classes of sequences, such as mononucleotide or tetranucleotide repeats (chapter 3). This provides a biochemical route by which multiple genes and regulatory pathways can experience selection through effects on the mutation phenotype.[27]

To the extent selective pressure has operated on variations in the probability of mutation along a DNA sequence, not only will classes of sites in a genome have different probabilities of distinct types of mutation, but those classes that have a higher probability of success will tend to become more frequent than those that are more consistently deleterious.

Regulated Variation in Nucleic Acid Sequences

Much like the regulated alterations in nucleic acid sequence involved in alternative splicing (chapter 15) and ciliate genome reorganization (chapter 13), certain classes of mutations in nucleotide sequences are affected by the levels of distinct gene products and therefore can be regulated.

Effects of Gene Products on Mutation: Lessons From the V-region

As described in some detail in chapter 10, changes in the balance of activity of specific proteins can change the outcome of a mutagenic insult. For example, C to U deamination occurs spontaneously at a significant rate, estimated to be 100 times per cell per day in the human genome.[28] The immune system has captured this physical chemical property of the nucleotide C, accelerating it enzymatically at sites in the variable region. While mechanisms that repair deaminated C evolved long ago, and are present in bacteria, enzymatic C deamination in the immunoglobulin variable region results in mutations; regulation of repair enzyme activities can affect the nature of the genetic change that results. Since deamination occurs spontaneously throughout the genome, mutation could be affected at any site in the genome not only by targeted enzymatic deamination but also simply by manipulation of repair.

Effects of Gene Products on Variation at G-rich Sequences

As described in chapter 11, an extracellular signal can induce targeted genome rearrangement in B cells by inducing RNA transcripts that initiate from promoters located within immunoglobulin heavy-chain introns. These transcripts do not encode proteins but rather free a G-rich region from the double helix, initiating recombination between two transcribed regions. Thus, at least theoretically, regulated changes in the activity of proteins that interact with G-rich DNA could change the level of genome instability in the germline as well. Indeed, as described in chapter 11, decreased levels of helicases that unwind four-strand structures formed by G-rich DNA do lead to increased genome instability; inhibition of another helicase-like protein leads to genome-wide destabilization of polyguanine tracts in nematodes. Similarly, deficiency of a protein that, at its normal evolved level, prevents cotranscriptional hybrid formation in yeast leads to hyperrecombination that is associated with actively transcribed regions of DNA.

Effects of Temperature and Nutrient Levels

In yeast, as described in chapter 12, the locations of hotspots of meiotic recombination are altered in different environments due to yeast hotspot dependence on transcription factors (and thus on the metabolic states that regulate them). The observation that temperature determines the rate of site-specific recombination in the locus encoding the *E. coli* fimbrial proteins, as described in chapter 3, indicates that environmental signals may not only inform a bacterium that it has arrived in its host but also, as a response, trigger it to access specific genomic sequences that are available in its implied repertoire ("species genome") and which are appropriate to the host environment.

Genomic Change Mediated by Mobile Elements

Barbara McClintock suggested[29] that stress could activate genome reorganization, which we now might describe in molecular terms by saying, for example, that a transposase that recognizes a class of genomic sequences might be induced by a biochemical signal of stress. Perhaps this explains why during adaptation of rice under the stress of selection by humans for growth in temperate climates there was a rapid increase observed in the number of copies of a transposable element that inserts into regions where it may alter the regulation of rice genes, as described in chapter 8. Similarly, chapter 4 points to a repeated contraction of a $(CT)_n$ repeat that is reported to occur only in wheat exposed to the head blight pathogen.

In bacteria, stress activates the SOS response. Prophage, which can facilitate the movement of genes among bacteria, as described in chapter 7, are induced by stress. As described in chapter 3, slippage rates of repeat tracts, such as those found in the genes encoding *N. meningitidis* lipopolysaccharide biosynthetic enzymes, depend on the activity of distinct proteins and so can be regulated. The rates are affected by transcription, the SOS response, and by environmental signals, as well as by mutations with sequence-context effects, such as those that perturb leading and lagging strand DNA synthesis.

As described in chapter 6, a population-level bacterial survival skill results from the presence, in the sequence of mismatch repair genes, of closely spaced repeats that tend to lose and regain mismatch repair activity through recombination. As described in chapters 3, 6, and 12, the diversity of sequences that can be recombined into a genome is increased when mismatch repair is decreased. Chapters 6 and 7 describe ways in which bacterial competence for, and acceptance of, DNA present in the environment itself can be regulated by environmental signals.

A Genome Is Not a Bag of Letters

The number of genome sequences available for analysis suddenly has become close to overwhelming. In addition, many new high-throughput laboratory techniques will inform issues raised in this volume. For example, as chapter 12 points out, the availability of whole-genome microarrays will overcome the challenge of obtaining statistically significant data regarding the probability of recombination across a genome, including at cold spots.

However, as we ramp up our data gathering and crunching, it is important to note that genomes are not just bags of genes, and genes are not just strings of context-independent letters. A nucleotide's "meaning" and behavior depends on its relationship to neighboring sequences and on the nature and relative levels of proteins expressed under distinct circumstances in that genome. Therefore, we must be cautious in drawing broad conclusions about adaptation from experimental studies that use laboratory constructs rather than genes with their own mix of "synonymous" codons and in the context in which they evolved. As discussed in chapter 6, bacteria that lose mismatch repair activity through recombination between closely spaced repeats may regain activity by recombination at these sites with exogenous DNA. Thus they are able to use the mutator phenotype as a temporary step to adaptation without then having to evolve new mismatch activity from scratch. Simply deleting a mismatch repair gene in the laboratory does not duplicate the endogenous behavior.

Similarly, as pointed out in chapter 5, laboratory-adapted strains of trypanosomes switch surface antigens at rates that are far lower than those observed during infection; and, when grown in the laboratory, *B. burgdorferi* tends to lose the plasmids required for generation of diversity through gene conversion.

Added to these observations is our growing appreciation for the regulatory role of RNA transcripts that do not encode protein, which serves to emphasize the importance in biological regulation and evolution of complementary relationships between sequences. Therefore, in spite of the great success of experiments that have cut and pasted DNA, it is now time to consider, in experimental design, what we might learn by studying sequences in the contexts in which they evolved.

Summary: A Sequence That Implies Other Sequences

For over half a century, we have been in the thrall of the double-helical structure of DNA, which, in an instant, revealed that information can be transferred between generations by a simple rule: A pairs with T, G pairs with C. In its beautiful simplicity, this structure, along with the table of codons worked out in the following decade, had entranced us into believing that we can fully understand the information content of a DNA sequence simply by treating it as text that is read in a linear fashion. While we have learned much based on this assumption, there is much that we have missed.

Now, at the beginning of the twenty-first century, with entire genome sequences appearing before us, biologists can appreciate the feeling experienced by physicists at the beginning of the twentieth century, of the firm ground slipping out from under their feet. We are indeed in a "new world."

Vast unannotated spaces appear in genome sequences, then fill up with transcripts that do not encode proteins. Far from being a passive tape running through a reader, genomes contain information that appears in new forms, and which creates regions with distinct behavior. Some genome regions are "gene rich," some mobile, some full of repeats and duplications, some sticking together across long evolutionary distances, some readily breaking apart in tumor cells. Even protein-coding sequences can carry additional information, taking advantage of the flexible coding options provided by the degeneracy of the genetic code. When viewed at the level of the RNA transcript or the DNA itself, "synonymous" codons are not always synonymous. Even something as familiar as an RNA transcript with alternative splice forms bears information "about" itself: defining exons and introns, and framing out how it will be spliced under different circumstances. There are new concepts to capture from the flood of often surprising new data.

Even the most senior among us has become a student again, working to learn from this rich font of new information; unless we remain open to the possibilities of discovering new types of information, and novel ways of encoding information, we may fail to perceive such information when it appears before us.

The chapters in this volume touch on one or more of three interconnected themes: information can be implied, rather than explicit, in a genome; information can lead to focused and/or regulated changes in nucleotide sequences; information that affects the probability of distinct classes of mutation can have implications for evolutionary theory.

Rather than simply summarize chapter by chapter in this overview, I have worked to integrate these observations into a conceptual framework. Whether or not a reader (or a chapter author!) is intrigued with the extent to which variations in the probability of mutations enable the ability to evolve itself to evolve under selective pressure, what is clear is that there are "deeper" ways to look at and understand nucleotide sequences and sequence contexts than had been obvious from viewing base sequences solely as linear strings of letters.

All of us who have contributed to this volume anticipate that this book will inspire readers to ask their own challenging questions, and to suggest their own syntheses of the unexpected, unexplored, and unexplained information in the rapidly expanding genomic databases. New terms are suggested, such as "implied information," "metagenome," and "species genome." At times even such familiar words as "information" and "code" seem to constrain our ability to describe what is represented in the genome. The Epilogue proposes adapting the concept of "protocol," which is widely used in describing another dynamic, robust, and evolving entity: the internet. Even our view of what can be "inherited" is challenged, as the population descendant from an individual bacterium inherits diversity from one specific manifestation of diversity at each highly-mutable locus, enabling survival in a broader range of niches than the individual parent itself could survive.

As this volume ranges across a broad field of biological problems it cannot be comprehensive, rather is intended provide a unique perspective on, and a doorway into, the literature while inspiring the reader's imagination to expand its appreciate of the types of information that may be represented in a genome. As you journey through the pages of this volume, may your reflections lead you to question assumptions (including those of this editor) ask more challenging questions in your own research, and send you on a path to additional startling discoveries.

Acknowledgments I would like to thank the many readers—some known to the chapter authors, some anonymous, some chapter authors themselves—who contributed their time to review one or more chapters. I also would like to thank those who contributed advice on the selection of chapter authors, and those at Oxford University Press who contributed their hard work "behind the scenes" without whom this book would not be in your hands. Most of all, I would like to thank the chapter authors for their thought-provoking contributions to this volume.

Much of the time devoted to this book was exchanged for time I might have spent with family, including Rockella, Parker, Michael, and Brooks, and so this chapter is dedicated to them.

1

Sequence-Dependent Properties of DNA and Their Role in Function

Donald M. Crothers

Overview

The genome is a complex nuclear organelle whose function is to encode the information needed to maintain the living state as cells grow and divide, and as generations pass from parents to progeny. Much attention has focused on the DNA sequences that encode proteins, the workhorse molecules of biochemical metabolism and biological structures. However, these sequences account for only a fraction of the total human genome. Moreover, because of the degeneracy of the genetic code, there are many ways to encode a specific protein. Local DNA structure depends on sequence, as do the mechanical properties, such as bending and twisting flexibility. As a consequence, much more information is encoded in the genome than is accounted for by protein sequences. Deciphering the complex secondary code[s] has only begun.

Introduction

The B-DNA helix structure proposed by Watson and Crick[1-3] has been the dominant icon of molecular biology for half a century. Only purine–pyrimidine base pairs, A with T and G with C, are tolerated in the confines of the regular helical structure, shown in figure 1.1a,[4] as refined from fiber diffraction studies. Inherent in the structure is the logic of its replication, since each strand can serve as the template for synthesis of its complement. Incorporation of deoxyribose in the alternating sugar–phosphate backbone confers polarity on the strands: each sugar has a phosphodiester linkage on its 5' and 3' oxygens. By convention, DNA sequences are written in the direction that corresponds to the order of biosynthesis—from 5' end to 3' end. A key feature of the structure is that the two strands in the duplex are antiparallel: the 5' end of one chain and the 3' end of the other chain are at the same terminus of the duplex. The simplicity of a uniform base-paired helix, with 10 bp per turn, 3.4 Å rise per base pair, was a key element in the rapid acceptance of the structure, and the explosive growth of molecular biology that followed.

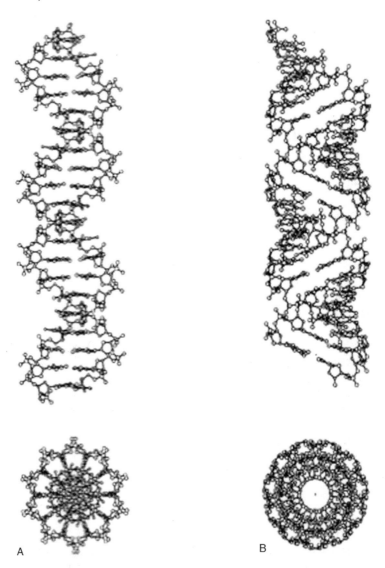

Figure 1.1. DNA structure. Mutually perpendicular views of (A) the "average" calf thymus B-DNA structure, determined from fiber diffraction at high humidity, and (B) A-DNA, determined at low humidity. (Reprinted from reference 4.)

However, 50 years later we recognize that the B-DNA helix shows significant variations in properties, depending on the base sequence. A thermodynamic property that became evident early on[5] is the higher stability conferred against melting or denaturation by a high content of G·C base pairs, readily explained by the greater resistance to breaking base pairs provided by the three hydrogen bonds in a G·C pair compared with the two bonds in an A·T pair (figure 1.2).[6] As a consequence, a DNA molecule that contains 100% G·C pairs has a melting temperature,

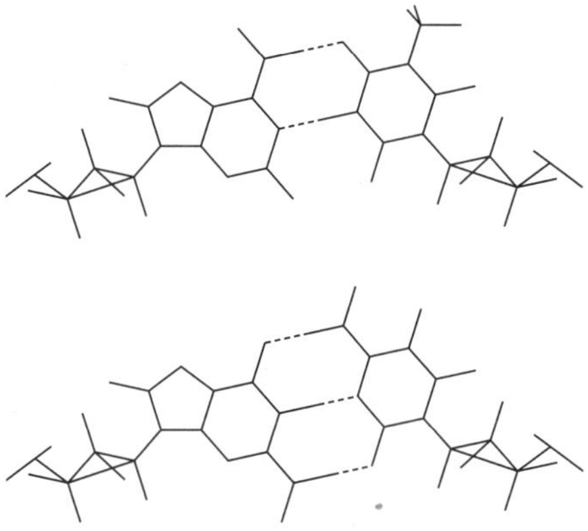

A·T

G·C

Figure 1.2.   Standard base
pairing geometries. (Reprinted
from reference 6.)

or T_m, about 40 °C higher than a molecule containing only A·T pairs (when both are longer than a few hundred base pairs). X-ray and NMR structural studies over many years have revealed substantial variations in local structure from one duplex to another. Further elements of complexity and variety of structure are contributed by superhelical conformations of circular DNAs,[7] dramatic global structural alterations that yield three-stranded[8] and four-stranded structures[9] (see also chapter 11). Formation of these multistranded structures depends on sequence correlations such as runs of pyrimidines on one strand and purines on the other, or on repeated tracts of G residues along one strand in telomeric sequences. Branched nucleic acids also provide important intermediates in recombination and replication.[10]

Another key feature to keep in mind is that only about 5% of the human genome is transcribed into mRNA. Small RNAs with a variety of functions account for an additional percentage. Does the rest of the genome encode essential information? The answer is clearly yes for regulatory regions, where proteins bind sequence-specifically. While the hydrogen bonding functionalities of the DNA bases are clearly important for binding specificity, variations in local DNA structure and in mechanical properties such as bending and twisting stiffness also play important roles. Packaging of DNA from nucleosomes through chromatin and up to mitotic chromosomes also depends in important ways on local structure and mechanical properties. Hence the information content of human DNA is contained not only in the base pair complementarity rules but also in the more subtle sequence-dependent structural and mechanical properties. Deciphering this code is still in its early stages.

Local Structural Variations

How DNA Structure Is Characterized

At a meeting at Cambridge University in 1988, a group of nucleic acid structural biologists agreed on a standard set of parameters to characterize DNA at a local structural level. These are shown diagrammatically in figure 1.3.[6] The top set of

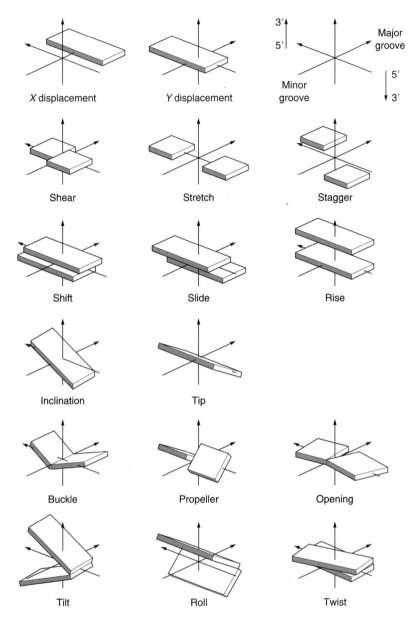

Figure 1.3.   Helical parameters. Translations are shown in the upper part of the diagram and rotations in the lower part. Each section contains base-pair axis, intra-base-pair, and inter-base-pair parameters. (Reprinted from reference 6.)

eight parameters represent distances; they are organized into base-pair movements (top row) relative to the axis, intra-base-pair distances (middle row) and inter-base-pair distances (bottom row). The bottom set of eight parameters represent angles, similarly organized into rows. For reference, most of these quantities are near zero for the classical B-DNA structure, excepting rise (3.38 Å) and twist (36°). More recent refinements of fiber diffraction data[6] for B-DNA have nonnegligible values of propeller twist, −13° to −15°. The classical A-form structure (figure 1.1b), on the other hand, has substantial positive roll (6.3°), positive inclination of the base pairs (12°), substantial x-displacement (4.1 Å), which moves the base pairs away from the helix axis, and slide between base pairs of 1.6 Å. A number of papers on nucleic acid structure that review this field in detail are provided in the *Oxford Handbook of Nucleic Acid Structure*.[11]

Sequence-dependent Variation of Structure in DNA Duplexes

Several approaches have been taken to determine experimentally the sequence-dependent variation of DNA structural parameters. The most direct method is x-ray crystallography of oligonucleotide duplexes, for which there is now a substantial database.[12,13] Gorin et al.[14] provide a useful summary of the major parameters for the 10 independent dinucleotide base-pair steps, and point out some significant correlations between them. (There are 10 independent dinucleotides instead of 4^2 (= 16) because of complementarity; for example, 5′–A–C–3′ is complementary to 5′–G–T–3′ in a duplex context, and hence one implies the other. Four of the dinucleotides, such as 5′–A–T–3′, are self-complementary, leaving 12 dinucleotides that provide six independent base-pair steps.) Averaged over the 10 dinucleotides, for a data set of 195 structures, roll and tilt are both nearly zero (−0.2° and 0.0°, respectively). However, there are significant variations from one dinucleotide to another. For example, the average roll for the G–C dinucleotide is −7.0°, whereas roll for G–G is +6.5°. Tilt generally varies less than roll,[15] by roughly a factor of two. This accords with simple mechanical expectation that it should be easier to rotate about the base pair long axis (roll) than about the short (tilt); see figure 1.3.

Twist is another parameter that varies considerably among dinucleotides, from a low of 30.5° for A–G to a high of 40.0 for T–A. However, it is generally not possible to build a severely underwound helix from these dinucleotide steps because the low or high value of the starting dinucleotide is generally compensated by a correspondingly high or low value of the next dinucleotide. For example, T–A–T has an average twist of 36.7°, much like the average value of 35.8° for A–A–A. Similarly, the average twist for A–G–A is 35.0°.

The structural parameters for DNA duplexes, such as those in the previous paragraph, should not be thought of as rigid values that can be applied to every stretch of DNA. In fact, these parameters vary depending upon on sequence context. In addition, structural differences can arise in the measurement itself, as a consequence of variable crystal packing forces. The root-mean-square (rms) variation, or square root of the variance, around the average value for a particular dinucleotide, averaged again over all dinucleotides, is 5.7° in the Gorin et al.[14] data set. The corresponding value for tilt is 3.6°, in agreement with the greater difficulty of varying tilt from its mechanical equilibrium value. These values accord quite well with the thermal fluctuations in the sum of roll and tilt deduced from the measured bending flexibility of DNA molecules in solution, as discussed below. The observed average variance in twist values in crystals is 5.2°, within the range of values determined from solution studies on the twisting stiffness of DNA.

DNA Curvature

Given the local variations and fluctuations in roll and tilt values, the DNA helix axis is virtually never completely straight, as embodied in the canonical B-DNA model. Most of these variations occur at seemingly random positions along the molecule, except in repeated sequences, and tend to cancel, although the molecule is never quite as straight as it should be given its mechanical rigidity.[16–18] However, there are a few structural motifs that can cause dramatic bending of DNA. In 1982, Marini et al.[19] associated the slow electrophoretic mobility of DNA fragments from the kinetoplast body of *Leishmania tarentole* with DNA curvature. The sequences responsible for bending were subsequently shown to be tracts of oligo(dA):oligo(dT), now generally called A-tracts, each about half a helical turn long, repeated in phase with the DNA helical screw.[20] In this orientation, the relatively small bends, each about 20°, add together to produce a dramatic overall curvature. This curvature can play a significant biological role, for example in transcriptional activation by affecting promoter geometry. Many transcriptional activators are DNA bending proteins, which can either recognize DNA bases (direct recognition) or specific DNA properties, such as flexibility (indirect recognition).[21] In some cases, substitution of an entire promoter region by properly curved DNA can activate in vitro transcription.[22,23]

When DNA is bent, the electrophoretic mobility is reduced because, as a general rule of thumb, the mobility varies with the end-to-end distance in the molecule, which is reduced by curvature. Excess flexibility of A-tracts, rather than or in addition to bending, cannot be the cause of the lower mobility because, in contrast to the decreased mobility observed when the A-tracts are repeated in phase with the DNA helical screw, repetition of A-tracts at 1.5 helical turn spacing yields molecules of normal mobility. A-tracts have the same flexibility no matter what their spacing, but with 1.5 helical turn spacing there is no overall curvature because the bends produce a zig-zag shape.[24] It is, however, very difficult experimentally to distinguish static bending from anisotropic flexibility, in which the helix prefers to bend in one direction rather than the opposite one.[25] The subject of A-tract bending has been reviewed by Crothers and Shakked.[26]

Other motifs that have been implicated as significant sources of DNA bending are the C–A (or T–G) dinucleotide step, which show up enriched in sequences selected for reduced electrophoretic mobility.[27] In this case, flexibility may be the source of the reduced mobility. In addition, sequences of the form GGCC tend to produce curvature toward the major groove,[28,29] particularly in the presence of divalent cations.[30]

Direction and Magnitude of A-tract Bends from Solution Studies

Simple arguments establish the direction of curvature of A-tracts. First, Koo et al.[24] showed that overall curvature is only slightly reduced when A-tracts are alternated with T-tracts on the same strand. In other words, the curvature of an A-tract is unchanged by rotating about a pseudo-dyad axis through its center. This can only be true if the bend is toward the major or the minor groove at the center of the A-tract. Second, Zinkel and Crothers[31] placed the A-tract bend at various phasings relative to the bend of known locus and direction produced by binding of *Escherichia coli* CAP protein to DNA. These observations unambiguously show that the A-tract bend is toward the minor groove at a locus near its center.

Estimates of the magnitude of DNA bending produced by phased A-tracts have varied from about 11° per tract using electrophoresis[32] to about 28° from the rate of cyclization in competition with linear ligation.[33] Measurements of rotational relaxation gave 18°,[34] as did computer simulation of the rate of cyclization versus dimerization of linear DNA fragments containing phased A-tracts at 25 °C.[35] Values near 18° are now generally accepted. Comparative electrophoresis shows that the curvature is modulated by only about ±10% by changes in the nature of the DNA sequence between the A-tracts.[36] However, A-tract curvature is quite temperature-dependent, being greatly reduced at elevated temperature.[37,38]

Fiber Diffraction Studies of Poly(dA):poly(dT)

Fiber diffraction studies of the poly(dA):poly(dT) double helix began with the work of Arnott and collaborators in 1974, [39] and culminated with the work of Alexeev and colleagues in 1987.[40,41] As observed by Crothers and Shakked:[26] "The unique and common features of the [purely] fibre-based structures of poly(dA):poly(dT), which distinguish them most from general sequence B-DNA, are negative inclination of the base pairs with respect to the helix axis (average −6°), high propeller twisting of the base pairs (average −26°), and a very narrow minor groove (average 3.4 Å)." The work of Aymami et al.[42] mixed crystallographic and fiber diffraction data by introducing constraints on the fiber conformation to keep the local structure close to that determined from x-ray crystallographic analysis of a dodecamer containing A_3T_3. The negative roll of the base pairs characteristic of the fiber diffraction structures did not survive these constraints. It should also be noted that in building their model, Aymami et al.[42] also introduced constraints that suppressed buckle of the base pairs, which, as is discussed below, may be a critical factor for understanding the structural basis of DNA bending. The high propeller twist and narrow minor groove are also characteristic of all crystallographic studies of A-tract containing molecules, but negative base pair inclination is not, as discussed below. A minor groove that becomes progressively more narrow toward the 3′ end of the A-tract was also found from hydroxyl radical footprinting studies.[43] It should be kept in mind that crystallography and fiber diffraction examine nucleic acid molecules in a condensed, high-concentration state, in which the structure is subject to perturbation by neighboring molecules. As a consequence, the actual structures in fibers and in the crystal may differ, perhaps accounting for some of the apparently contradictory observations. As an example, the helical repeat of DNA in fibers is 10.0 bp per turn, whereas it is around 10.4 in crystals and solution.

Crystallographic Studies of A-tracts

A number of structures of double-helical molecules containing A-tracts have been published (reviewed by Crothers and Shakked[26]). For this purpose, molecules of sequence $5'-A_nT_n-3'$ are considered to be A-tracts, because their curvature is comparable to A-tracts, but sequences $5'-T_nA_n-3'$ are not, because they are essentially straight;[44] the T–A step has quite different properties than the A–T step. All structures (with an exception discussed below) show that the helix axis is straight and the base pairs are perpendicular to the helix axis. Roll between the base pairs in the A-tract region is very small. Two additional general features characterize A-tracts in all the crystal structures: an exceptionally narrow minor groove and highly propeller-twisted base pairs. Propeller-twisted base pairs are capable of "bifurcated" hydrogen bonds, in which the adenine amino group hydrogen bonds not only to its complement, but also to the T to the 5' side of the complement.[45] This conformation both drives and is stabilized by the very narrow minor groove.

Such bifurcated hydrogen bonds might explain the origin of DNA bending at A-tracts. Shatzky-Schwartz et al.[46] explored this idea by combining electrophoretic measurements of bending with high resolution crystallography on sequences containing $I \cdot C$ or $I \cdot mC$ pairs, which cannot form bifurcated H-bonds with adjacent $A \cdot T$ pairs. In spite of that, propeller twisting persisted, as did curvature. At elevated temperatures, where curvature disappears, so does the propeller twisting, as inferred from resonance Raman spectral studies of hydrogen bonding to the thymine carbonyl group.[47] It should be noted that interrupting A-tracts with $G \cdot C$ pairs, which places an amino group in the minor groove, disrupts curvature,[24] as does introducing a T–A step.[44]

While it seems likely that high propeller twisting of the bases in A-tracts is a likely contributor to curvature, most of the crystallographic evidence is indecisive. Most structures studied (with an exception discussed below) do not show overall curvature directed toward the minor groove at a coordinate frame located at the A-tract center. Rather, the curvature seems to be dictated by crystal packing forces. Based on crystallographic data, Dickerson and colleagues[28,29] have argued that since the A-tracts are straight, the curvature must arise from positive roll in the intervening non-A-tract segments, an argument advanced earlier on general grounds by others.[32,48,49] While this model has the proper direction of curvature and is formally possible, it is not easy to reconcile with the fact that the extent of curvature depends little on the nature of the sequence between the base pairs,[36] while dinucleotide steps differ significantly in roll, as discussed above. Furthermore, the average roll among DNA oligomers that have been studied, which is nearly zero, is clearly not sufficient to give enough positive roll (about 6°) to produce the observed curvature.

Crystal Structure of an A-tract Having the Natural Curvature

Hizver et al.[50] reported the high-resolution structure of a symmetric duplex oligonucleotide of sequence ACCGAATTCGGT, which is of particular interest because it

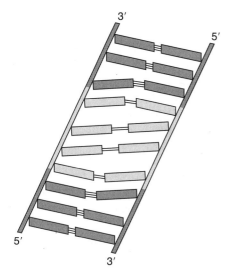

Figure 1.4.   Schematic representation of an unrolled helix viewed at the minor groove. A·T base pairs are in light shading and G·C pairs are in dark shading. The global helix axis is along the vertical direction. The figure illustrates the mediating effect of base-pair buckling between the A-tract region (negative inclination of the base pairs) and the flanking segments (positive inclination of the base pairs). (Reprinted from reference 50.)

is a high-affinity target of the E2 regulatory protein from human papilloma virus. The three independent molecules in the crystal lattice show curvature of about 10° directed toward the minor groove at the site of dyad symmetry in the sequence, in good agreement with expectation from solution studies of AATT-containing sequences.[51] In the crystal structure, curvature results from a combination of negative roll in the A-tract and positive roll in the adjacent segments. These roll angles result in negative inclination of the base pairs in the A-tract and positive inclination in the adjacent sequences. To accommodate the change in inclination, the base pairs buckle at the junctions where the A-tract meets the flanking sequences. Figure 1.4[50] shows these relationships diagrammatically for an "unrolled" or unwound helix, viewed at the minor groove. Note that buckle changes sign across the center of sequence symmetry, as expected for the nearly symmetrical conformation observed in the crystal. The buckle is positive at the 5′ junction with the A–A sequence, and negative at the 3′ junction with the T–T sequence. This view of DNA bending is consistent with the idea that there is not a single source of the A-tract bend,[26,52] but that it is the result of a combination of cooperative effects brought about by inserting an A-tract into "normal" B-DNA.

NMR Structures of A-tracts Having an Appropriately Directed Bend

Two recent NMR structural studies of A-tract-containing molecules, either A_4 or A_6, incorporated residual orientation-dependent dipolar couplings between protons, yielding improved long-range structural fidelity. The resulting structures have the bend directed toward the minor groove at its center.[52,53] In general, there is an appreciable net negative roll in the A-tracts and positive roll in the flanking segments. There is also an appreciable contribution of tilt to the overall bend. Of particular interest is the positive buckle observed throughout the A_4-tract,[52] which causes a clash at the junctions analogous to that shown for the symmetrical structure in figure 1.4. This result reinforces the conclusion from crystallography on ACCGAATTCGGT that a common feature of A-tract-induced DNA bends is base-pair buckle.

The structural model that has emerged bears a resemblance to the original "junction" model that we proposed in 1986.[24] Based on fiber diffraction and optical dichroism studies of poly(dA):poly(dT), the model proposed negative base-pair inclination and positive buckle in the A-tract. What seems to happen in addition is that the base-pair buckle and clash at the junction induce positive base inclination in the adjacent base pairs, which therefore also contribute to the bend, as proposed by Dickerson.[28] This effect transmitted into the adjacent DNA must be relatively independent of sequence in the neighboring DNA in order to account for the observations. Haran and colleagues[54] have shown that the propagation effect is polar, leading to a characteristic induced structure on the 3′ but not the 5′ ends of A-tracts. Curvature is a cooperative phenomenon[55] that requires A-tracts in the length range of about 4 to 8, repeated in phase with the helical screw of DNA. The main lesson here is that local sequence correlations can provide dramatic effects on the global structure of DNA, with important consequences for its looping and folding.

DNA Flexibility

Bending Flexibility and the Persistence Length

Double-helical DNA is a relatively rigid linear polymer, but no such polymer is infinitely stiff, except at absolute zero temperature. One measure of polymer stiffness is the persistence length, P. An intuitive definition is the following. Consider a very long polymer chain containing N segments, whose orientation at one end is along the x axis. Let x_N be the projection on the x axis of the end-to-end vector in the chain. The persistence length is the average value of x_N over all chain configurations in the limit as N becomes infinitely large; it measures the average extent to which the chain "persists" along the starting direction. For DNA an average value is about 150 bp, but persistence length is a sequence-dependent property.

The free energy required to bend DNA can be calculated from the persistence length. Specifically,

$$\Delta G_{bend} = \frac{P}{2L} RT(\Delta\vartheta)^2 \tag{1.1}$$

Where $\Delta_t\vartheta$ is the total bend angle (in radians) in a length L of DNA, and RT is the gas constant time absolute temperature. For example, winding 146 bp of DNA 1.75 times around a core of histones to form a nucleosome requires about $62\,RT$, or about 38 kcal/mol at 37 °C. The energy required for this process comes from establishing electrostatic interactions between DNA phosphates and positively charged or polar amino acid residues on the histones.

Classical methods of polymer physics, such as light scattering and birefringence decay, have been used to determine the persistence length of polymers including

DNA, but the method of cyclization kinetics, originally developed by Baldwin and collaborators[56,57] is particularly apt for DNA. The method relies on measuring the relative rates of cyclization and bimolecular ligation for DNA molecules in the range from about 100 to several hundred base pairs. The ratio of the unimolecular cyclization rate constant to the bimolecular rate constant is called the J factor, in recognition of the original contribution of Jacobson and Stockmayer[58] to the theory of polymer cyclization. J has units of concentration, and can be thought of as the effective concentration of one end of a DNA molecule around another in a ligation-competent configuration. When the concentration of DNA molecules equals J, bimolecular ligation occurs at the same rate as cyclization. Measured values of J cover a very large range, from sub-nM to mM. This subject has been reviewed by Crothers et al.[59]

The cyclization kinetic method is particularly useful for determining the curvature and flexibility of small DNA sequence elements, for example 10 bp segments that make up a single helical turn. Figure 1.5 shows the type of construct used by Roychoudhury et al.[60] to characterize the properties of the nucleosome positioning sequence (NPS) TATAAACGCC, which was found by Widlund et al.[61] by selection/amplification methods starting from mouse genomic DNA to be a motif that is especially favorable for nucleosome formation. The cyclization constructs contain six repeats of an A_5- or A_6-tract, which give curvature to the molecule and determine the rotational setting of the circle, namely with the minor groove of the A-tracts at their centers on the inside of the circle. A variable linker separates the A-tracts from the sequence to be studied, which is typically three repeats of the 10 bp motif to be examined. This linker enables any curvature in the sequence examined to be placed in or out of phase with the A-tract curvature. In-phase molecules have a C-shape, and cyclize more rapidly than S-shaped out-of-phase molecules. The known direction of curvature of the A-tracts enables determination of the direction of curvature of the sequence of unknown properties. It is this feature of the cyclization constructs that enables the effects of curvature and flexibility on the persistence length to be deconvoluted. Finally, there is another variable adapter region of the molecule which enables the overall length of the molecule to be changed systematically; optimal cyclization requires an integral number of helical turns so that no twist or writhe is required for cyclization. When the number of turns is integral, the cyclized molecules are essentially planar, but the molecules have small amounts of writhe and are accordingly nonplanar when the number of helical turns is nonintegral.

Figure 1.6 shows typical results of a cyclization kinetic assay on the set of molecules of (nearly) optimal phasing between the A-tract and NPS curvature. Molecules in which the NPS sequence replaces a normal 30-bp B-DNA sequence cyclize about 100 times faster than the parent molecule. The variation of cyclization efficiency by about 100-fold with a 10 bp periodicity reflects the preference for cyclizing molecules that have an integral number of helical turns.

Theoretical simulation of the cyclization kinetic results, either by Monte Carlo methods[60] or with an analytic theory,[62] assuming a harmonic potential for bending and twisting DNA, allows determination of curvature direction and magnitude, and bending and twisting flexibility or root mean square (rms) variation angular fluctuations. For example, the NPS sequence was found to be bent by about 13°, in a direction either toward the minor groove at the center of the TAA sequence or toward the

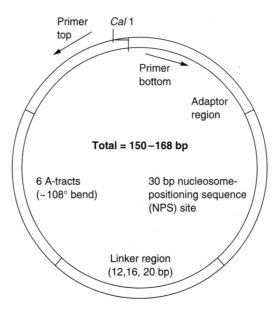

Figure 1.5.   A-tract-NPS minicircles. The six-phased A-tracts and the 30 bp NPS region are separated by linker regions with variable lengths of 12, 16, and 20 bp. The adaptor region enables variation of the overall length of the constructs. (Reprinted from reference 60.)

major groove at the center of the GCC sequence. (Since these two directions give the same overall curvature direction, the method cannot distinguish between them.)

The model assumes that DNA bending by tilt or roll is equally likely; this assumption is needed because simulation shows that the data do not allow one to deconvolute flexibility into separate roll and tilt components. The analysis reveals that the NPS sequence is significantly more flexible than normal B-DNA. The rms angular fluctuations for roll and tilt increase from 4.7° for B-DNA to 6.8° for the NPS sequence. Since the persistence length, and the bending force constant, are proportional to the inverse square of the rms angular fluctuations, they are reduced by nearly twofold in the NPS sequence compared with normal B-DNA. Both the curvature and enhanced flexibility are potential contributors to the increased affinity of NPS-type sequences for core histones to form nucleosomes.

Twist Flexibility

A protein or protein complex frequently interacts with DNA at more than one location. Optimum interaction requires that the helical phasing between the sites be such that the sites are presented in the correct relative geometry. Sequence-dependent variations in twist between the sites can cause misalignment, which can in turn be overcome by energy-requiring fluctuations in local twist. Early measurements of the twist flexibility were based on determining a characteristic elastic property called the Young's modulus, E, and utilizing the assumption that the bending and torsional force constants vary linearly with E.[63] The conclusion was that the force

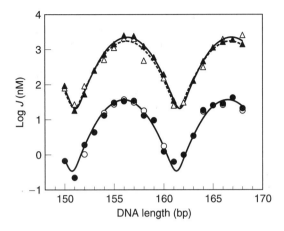

Figure 1.6.   Comparisons of experimental and theoretical log J versus DNA length for constructs with and without the NPS site. The upper sets of curves correspond to constructs having the NPS site with a fixed linker length of 16 bp. The experimental $\log_{10} J$ values are shown as open triangles and the corresponding curve fit is the dashed line. The simulation values are indicated by filled triangles and their curve fit is the solid line. The lower sets of curves correspond to the B-DNA molecules containing a generic 30 bp sequence in place of the NPS site. The experimental $\log_{10} J$ values are the open circles, with the dashed line indicating the curve fit; the simulation values are the filled circles with solid line indicating the curve fit. (Reprinted from reference 60.)

constants or stiffness vary with GC content, and are about four times larger for 100% GC DNA than 100% AT DNA.

More recent direct measurements of the variation in bending and twisting flexibility by the cyclization kinetic method indicate a smaller and more complex variation.[64] Expressed as angular fluctuations in bending, flexibility is greatest for ATAT at 6.4° and smallest for GGCC at 4.45°, which implies a twofold variation in bending force constant. However, there is not a strict correlation with AT content. For example, ACGT at 4.53° is only slightly more flexible than GGCC. Nor does the twist flexibility vary strictly in proportion to the bending flexibility: the ratio of bend to twist angle fluctuations varies from 0.88 for CGCG to 1.25 for ATAT. However, the range of twist fluctuations is similar to that for bending, from 3.9° for GACC to 5.4° for TATA, again corresponding to a nearly twofold variation in the force constant. Accurate measurement of the stiffness of specific DNA sequences clearly requires independent determination of the bending and twisting fluctuations or force constants.

Influence of Curvature and Flexibility on Protein Interactions

Nucleosomes

Nucleosome formation requires that 146 bp of DNA be wrapped 1.75 times around a superhelical axis, with an energy penalty for bending, according to equation (1.1), of about 38 kcal/mol. All indications are that the force constant for DNA bending, which is proportional to the persistence length, can vary by about a factor of two, depending on the sequence (see above). According to equation (1.1), a twofold more flexible DNA would have an advantage over the stiffer molecule by 19 kcal/mol in the bending energy, which would translate into a stronger binding constant by about 10^{14}-fold. However, observed variations, discussed below, are more like one to three orders of magnitude. Clearly, something has been left out of the calculation so far.

Consider the following thought experiment: imagine first a flat surface, to which the DNA binds at all points along its length. In this case, a stiff (straight) molecule has an advantage over a flexible one because the flexible molecule must give up its bending flexibility, with a cost in entropy, when it binds as a strictly linear molecule. The stiff molecule, on the other hand, is already linear and so does not lose as much entropy. At the other extreme, consider a highly curved surface. The flexible molecule then has the advantage because it is more easily bent to adapt to the curved surface. The important implication is that there is an optimum curvature of the surface for which the binding affinity is independent of stiffness. The value of the optimum curvature depends on what one assumes about how the vibrational modes of the molecule are constrained by binding to core histones.[65] For example, if only the bending modes perpendicular to the nucleosome surface are completely constrained by binding, the optimum curvature is 4.5°, whereas if the twisting mode and bending mode parallel to the surface are also constrained, the optimum angle is 7.8°. For comparison, the average curvature of DNA on the nucleosome is 4.3°. The theory predicts that if the surface curvature is less than optimum, more flexible molecules should bind more weakly because of their entropy loss.

Available experimental evidence on this issue comes from study of the variation of nucleosome affinity of DNA sequences with their persistence length.[60,66,67] The experimental evidence indicates in general that higher flexibility enhances binding, meaning that the curvature of the nucleosome is slightly greater than the optimum value. This could be the result, for example, of only partial suppression of the intrinsic bending and twisting modes of DNA when it binds to the nucleosome. The basic conclusion is that the nucleosome is designed to have a curvature that greatly reduces the range of affinities for DNA molecules depending on their stiffness, thus enabling efficient packaging of the bulk of the genome.

It should now be clear what was wrong with the calculation of affinity based only on the bending free energy. The model assumed constraints only on the ends of the DNA molecule, without constraining the path by which it is wound 1.75 turns around the superhelical axis. Thus the calculation leaves out the entropy loss suffered by the flexible chain, and hence overestimates its affinity. However, the calculation would be appropriate for freely looping geometries in which there is no constraint on the DNA molecule between two points of binding, for example to *lac* repressor.

Indirect Readout: The 434 Repressor and E2-DNA Interactions

Recognition of DNA by proteins can be divided into direct and indirect readout effects. Direct readout occurs when the nucleic acid bases interact directly with functional groups on the protein. Common effects are hydrogen bonding to purine N7, the exocyclic amino groups, and carbonyl oxygen. Hydrophobic interaction, for example with the thymine methyl group, also falls in this category. Because they are buried within the core of the helix, the imino nitrogens are not usually involved, unless the helix is broken open by the binding interaction.

Indirect readout falls into two general categories. The first is interaction of a protein functional group with nucleic acid functional group, such as the sugar–phosphate backbone, whose nature is independent of base sequence but whose precise geometric position still depends on sequence. In one example, there are two closely related conformations of DNA, called B(I) and B(II), which differ by coupled rotations about the C3′–O3′ and O3′–P bonds. Because of such sequence-dependent effects on backbone geometry, the binding of a protein that recognizes only one of the conformations can still be sequence-dependent. The preference for one state over another in a particular sequence can be monitored by NMR spectroscopy.[68] The B(I) and B(II) conformations differ in placement of the phosphate relative to the grooves: in the more common B(I) conformer the phosphate is roughly symmetric between the grooves, whereas the phosphate in B(II) swings toward the minor groove, narrowing it locally. Thus the geometry of optimal protein hydrogen-bonding interactions with the phosphate differ in the two conformers. A related phenomenon is narrowing of the minor groove in AT-rich regions, particularly by A-tracts, which can have a profound influence on the strength and geometry of protein–phosphate interactions.

A striking example of indirect readout, in which the protein can distinguish differences in twist at individual dinucleotide steps, is provided by *E. coli* integration host factor (IHF). In spite of the fact that this protein relies almost entirely on indirect readout for recognition,[69] binding is severely reduced when the wild-type sequence TTG in the binding site is replaced by TAG. The total twist over the trinucleotide is constrained by interactions with the first and third phosphates, and is roughly the same for the two trinucleotides, at 82° and 78°, respectively. However, in TTG the twist at the first dinucleotide is 34°, whereas it is 40° in TAG; the second dinucleotide steps twists are 48° and 38°, respectively. These values reflect the trends in average twist seen in crystals of duplex DNA[14]: TT is 36°, TA is 40°, TG is 37°, and AG is 30°. The resistance of the AG step to overwinding places the central base pair in the trinucleotide in an unfavorable position for an interaction with the protein, which is in itself not base-pair specific.[69] There are many other examples in the database of protein–DNA complexes[12,13] of indirect readout resulting from intrinsic sequence dependent variations in DNA structure. Further detailed knowledge of structures of DNA and complexes will enable codes of this kind to be deciphered.

A second kind of indirect readout results when proteins interact with DNA at two or more sites, leaving a stretch of DNA between the contact points that is not touched by the protein. Two examples are the phage 434 repressor[70] and the human papilloma virus regulatory protein E2.[71] In both cases, four noncontacted base pairs separate the protein binding sites. In an early attack on this problem, Hogan and Austin[63] correlated binding strength of the 434 repressor with the ability of the DNA segment to undergo bending and twisting deformations, which were assumed to vary in parallel (see above). A more recent perspective on this problem by Koudelka and colleagues[72] concludes that the GC content, or number of H-bonds between the bases, is not significant, but rather the key factor is the presence of an amino group in the minor groove. Indirect effects on local structure and minor groove width could also be the source of the observations.

In a new approach to the effects of DNA intrinsic properties on protein complex stability, Zhang et al.[64] used the cyclization kinetic method to measure variations in curvature and flexibility of four-base pair spacer sequences placed between the E2 protein contact sites. A statistical mechanical theory was then used to calculate the relative ability of the spacers to begin and end in the proper configuration for binding. Even though experimental measurements showed that the DNA sequences varied in affinity by over three orders of magnitude, the experimental properties of the DNA sequences still were able to predict the relative binding constants within a factor of three for 15 of 16 sequences. (The sole exception was traced to anomalous magnesium ion binding.)

It is now possible to correlate large variations in the protein binding constant with intrinsic properties of DNA sequences. However, it is important to note that the properties of the sequence are collective; that is, they can't be predicted by adding up parameters due to individual dinucleotide steps. Context is all-important. Considerable work remains before the information about local shape and flexibility can be decoded from genomic sequences.

Summary

The structure of distinct DNA sequences can vary dramatically, depending upon both the nature of the sequence and its context. The relative orientation of neighboring sequences in space is affected by both the length and nature of any "spacer" sequences. Computational analyses that treat DNA sequences merely as letters on a page miss much of what is interpreted by molecules that interact with DNA, which do not see "letters," but rather diverse sequence-dependent arrangements of chemical groupings and physical chemical properties in space.

Acknowledgments   This work was supported by grant GM 21966 from the National Institutes of Health, and by support from the National Foundation for Cancer Research for the Yale Center for Protein and Nucleic Acid Chemistry.

2

Mutation as a Phenotype

Errol C. Friedberg

Overview

The stability of DNA sequences across generations is dependent upon a diverse set of biochemical mechanisms that protect the DNA from predictable sources of damage that otherwise would be catastrophic. These biochemical mechanisms range from the repair of damage to nucleotide precursors or to DNA itself, to the replication past any one of a number of lesions, to the proofreading of a new copy of DNA. The spectrum of mutations experienced by a genome will change and their number will increase dramatically when the efficacy of even a single one of these many mechanisms is altered. During sequence comparisons mutations often are treated much like random typographical errors, yet there is a rich biochemistry that underlies the probability of distinct types of errors along a particular genome sequence. This chapter provides an overview of such biochemical mechanisms, with a particular focus on novel polymerases that are able to effect synthesis across distinct classes of lesions in DNA.

Introduction

To enable a cell to survive in the face of otherwise lethal DNA damage, specific biochemical mechanisms have evolved; but these survival mechanisms have a high probability of generating mutations. While mutations can interfere with normal cell functions in both somatic and germline cells, and hence can be deleterious at the level of individual cells, mutations are central to the generation of diversity that is essential for adaptation in Darwinian evolution.

This chapter considers mechanisms that contribute to the genesis of mutations, with an exclusive focus on single nucleotide substitutions. The present discussion underscores the fact that the generation of mutations is primarily a consequence of the failure of cells to remove (repair) DNA-base damage before it is irrevocably fixed as permanent alterations in nucleotide sequence. The importance of effective DNA repair in the prevention of mutations is discussed in chapter 6.

For mutations to be heritable they must of course transpire in the germline. However, most of our current understanding about mechanisms of mutagenesis in multicellular organisms derives from studies on somatic cells. Even in somatic cells, there are differences in the mechanisms by which mutations are generated in different organisms. These will be highlighted only to the extent that they illuminate key aspects of the process of mutagenesis. Chapter 12 discusses generation of genetic variation in meiosis, with a focus on recombination in yeast.

Environmental versus Spontaneous Mutagenesis

It is traditional to distinguish mutagenesis that derives from exposure of the genome to exogenous agents in the environment ("environmental mutagenesis"), from that which derives as a consequence of normal cellular metabolism: so-called "spontaneous mutagenesis."[1,2] This distinction has limited value in mechanistic discussions, however. In particular, many environmental agents that cause damage to DNA are spontaneous in the sense that they are not man-made. For example, ultraviolet radiation from the Sun provided a powerful force for the evolutionary selection of multiple biochemical pathways that prevent, but in certain circumstances (as described below) can generate, mutations.[1,2] Thus the distinction between environmental and spontaneous mutagenesis is useful only when considering cellular responses to genomic insult caused by exposure to DNA-reactive agents that are not normally encountered in the environment, such as synthetic chemicals. Even then, the mechanisms by which such types of genetic insult are repaired, or are converted to permanent alterations in nucleotide sequence, presumably evolved to cope with naturally occurring forms of damage that such agents mimic (as sensed by the repair mechanisms) with respect to their effects on DNA.

Mechanisms of Mutagenesis

With the recently discovered exception of somatic hypermutation in immunoglobulin genes (see chapter 10), most mutations can be considered as consequences of the failure of cells to correct (repair) some form of altered nucleotide sequence or DNA damage, thus prompting them to rely for their survival on mechanisms that can (but do not necessarily) generate permanent alterations in sequence.

Viewed from this conceptual point of view, mutational mechanisms can be conveniently considered under two broad headings. The first are those associated with DNA synthesis by high-fidelity DNA polymerases that are normally employed in DNA replication. Such high-fidelity replication may transpire on completely normal template DNA, or on template DNA that is altered in a manner that does not arrest or impede DNA synthesis by the replicative DNA machinery. The second are those mechanisms associated with DNA synthesis by specialized low-fidelity DNA polymerases which are not used in normal DNA replication and which have a high probability of generating mutations.

Mutations Associated with High-fidelity DNA Replication on Normal Template DNA

Incorporation of Damaged Deoxyribonucleotides

In addition to the nitrogenous bases in the polynucleotide chains of DNA, those in nucleotide precursors, in particular the longer-lived deoxynucleoside triphosphate pools that are immediate substrates for DNA polymerases during DNA replication, are subject to damage. Studies that embrace the full spectrum of known base damage to nucleotides in DNA have not been comprehensively extrapolated to the four deoxynucleoside triphosphates and other DNA precursors. However, oxidative base damage to deoxynucleoside triphosphates has received considerable attention as a potential source of mutations in the genome.[1,2]

In bacteria such as *Escherichia coli*, a gene called *mutT* normally contributes to the avoidance of errors by a mechanism referred to as "nucleotide pool sanitization."[1-3] Endogenous oxidation can convert dGTP in cells to 8-oxo-dGTP. The utilization of 8-oxo-dGTP opposite template A by DNA polymerases during replication generates 8-oxoG·A mispairs. A protein encoded by the *mutT* gene specifically degrades 8-oxo-7,8-dihydro-dGTP (8-oxo-dGTP) and 8-oxo-7,8-dihydro-GTP to the monophosphate form (8-oxo-dGMP), thereby preventing its incorporation into DNA[4] and RNA, respectively. The biological significance of this reaction in *E. coli* is underscored by the observation that inactivation of the *mutT* gene results in the strongest mutator phenotype known in this organism, with as much as a 10,000-fold increase in the level of A·T→C·G transversion mutations.[1,2]

An 18-kDa protein with properties very similar to those of the *E. coli* MutT protein is encoded by the human *MTH1* gene (for *MutT* homolog).[5,6] Expression of the human *MTH1* cDNA in *E. coli mutT* mutant cells partially corrects their mutator phenotype. MutT homologs have also been identified in other organisms.[6] It remains to be determined whether other forms of naturally occurring nucleotide precursor damage can result in mutations during normal DNA replication. Mutator genes with unknown functions are prime candidates for preventing such mutations, and defining their specific substrates may be a useful way of defining novel forms of nucleotide precursor damage.

Replication Errors with Native Template DNA

High-fidelity replicative DNA polymerases catalyze nucleophilic attack of the 3′-hydroxyl of a DNA primer on the α-phosphate of nucleoside 5′-triphosphates, thereby promoting incorporation of a nucleoside monophosphate into the primer strand, with the concomitant release of pyrophosphate.[7] Most DNA polymerases incorporate nucleotides with very low error frequencies (10^{-3} to 10^{-5} errors per incorporation).[8,9] This high fidelity derives from multiple contributing features of DNA replication. First, the specific configuration of the polymerase active site contributes to the geometric selection of nucleotide substrates according to how well the nucleotide template base pair fits into the active site. This represents a good example of an induced fit mechanism by which nonideal substrates impede the rate-limiting conformational change that converts the inactive (open) polymerase into the active (closed) form.[10–12] Second, the nonpolar environment of the active site enhances the favorable energy of A–T and G–C hydrogen bonds and disfavors abnormal base pairings that leave hydrogen bonding groups unsatisfied.[13] Most of the surface of the base pair lying in the active site is shielded from solvent and is extensively contacted through nonpolar interactions. These steric and electrostatic constraints of the polymerase active site selectively specify the shape and hydrogen-bonding arrangement of standard base pairs while strongly disfavoring mispairing with the DNA template.

The replication mechanisms intrinsic to individual DNA polymerases are significantly altered by their interactions with other subunits of the replicative machinery, so-called replication accessory proteins such as proliferating cell nuclear antigen (PCNA), replication factor C (RFC) and replication protein A (RPA). Even then, DNA polymerases occasionally make errors. These are subject to editing by the important $3' \rightarrow 5'$ proofreading exonuclease activity that removes misincorporated nucleotides. The rapid rate of DNA synthesis slows dramatically following nucleotide misincorporation, providing an opportunity for the exonucleolytic removal of nucleotides from the $3'$ end of the primer strand.

The additional proteins just mentioned reduce the mutation frequency associated with normal high-fidelity DNA replication to approximately 1 in 10^9 nucleotides incorporated into the human genome.[1,2] Errors are further reduced by post-replicative mismatch correction, so-called mismatch repair (MMR),[1,2] which requires a series of highly conserved proteins that function in a coordinated manner. In addition to specifically identifying mismatched sites in DNA, the process of MMR must discriminate between the newly synthesized DNA strand carrying the misincorporated nucleotide and the extant template strand, so that it is the incorrectly incorporated nucleotide that is removed. In many bacteria, the mechanism of this strand discrimination is based on differences in the methylation pattern of extant and newly replicated DNA.[1,2] We do not yet know how the template and newly synthesized strands are distinguished in eukaryotes.[1,2]

In summary, it is inevitable that all replicating cells experience some level of spontaneous mutation, in spite of the intrinsic high fidelity of DNA replication and of mismatch repair. Mutagenesis in germline cells presumably contributes to generating the genetic heterogeneity required for Darwinian evolution.

Spontaneous Alterations in Template DNA

Of the five bases normally present in DNA (including 5-methylcytosine), four (cytosine, adenine, guanine, and 5-methylcytosine) contain exocyclic amino groups (figure 2.1). Loss of these groups, a process called deamination, occurs spontaneously under normal physiological conditions,[14] resulting in conversion of the affected bases to uracil, hypoxanthine, xanthine, and thymine, respectively. Some of these conversions result in mutations during high-fidelity DNA replication due to changes in base pairing properties.[15] Cells are able to mitigate against the potentially mutagenic effects of such alterations in nucleotide chemistry by employing specific DNA repair pathways that selectively remove these deaminated nucleotides. The removal of altered bases, for example the base uracil that results

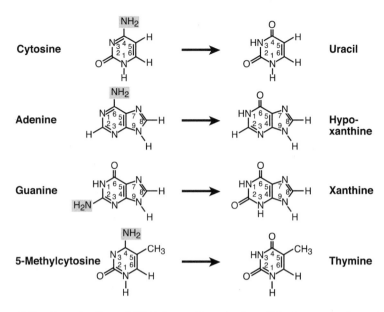

Figure 2.1. Deamination of DNA bases. Deamination of the bases cytosine, adenine, guanine, and 5-methylcytosine transpires under normal physiological conditions and generates nucleotide alterations that can lead to spontaneous mutations.

from deamination of cytosine, is catalyzed by specific DNA glycosylases.[1,2] The resulting abasic site is subsequently repaired, as described in chapter 10.

Primer Template Misalignments During DNA Replication

Mutations during high-fidelity DNA replication also can result from various misalignments of primer and template DNA strands at the primer terminus.[16–18] A common source of such misalignment is a homopolymeric nucleotide run that allows slippage of base pairing and subsequent stabilization by normal base pairing (figure 2.2a). Continued polymerization from such intermediates can result in deletions if the unpaired nucleotide is in the template strand, or in nucleotide addition if the unpaired base is in the primer strand (figure 2.2b). More complex mutations can also result from slipped primer templates. For example, if misincorporation precedes strand slippage but the misincorporated base realigns to generate a properly paired primer terminus, a frameshift mutation will result (figure 2.2b). Another possibility involves initial slippage followed by correct nucleotide incorporation and subsequent slipped alignment, so-called dislocation mutagenesis (figure 2.2c).[16–18] (Some biological consequences of mutations due to insertions and deletions at nucleotide repeats are discussed in chapters 3 and 4.)

The various mechanisms summarized thus far contribute to an undetermined fraction of the spontaneous mutation rate and frequency in living cells through perturbations of high-fidelity DNA replication. But further contributions to spontaneous

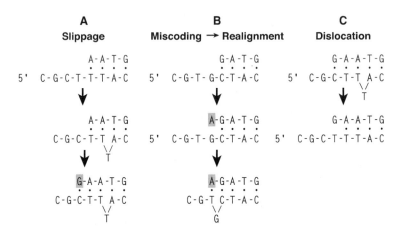

Figure 2.2.   Replication slippage. Various types of abnormal primer-template structures can result from strand slippage at the replication fork and lead to spontaneous mutations.

mutagenesis derive from the replication of DNA templates carrying unusual structures (such as the four-chain structure described in chapter 11) and forms of base damage that frequently arrest normal DNA replication because they cannot be accommodated in the catalytic site of high-fidelity DNA polymerases (see below). In recent years, considerable progress has been made toward understanding how cells cope with such replicative blocks.

Mutagenesis during Repair Synthesis of DNA

As already indicated, when cells suffer genomic insult from exposure to agents that are reactive with DNA they deploy a plethora of responses that evolved to repair the damage by one mechanism or another, thereby restoring the genome to its native configuration and nucleotide sequence.[1,2] Most of these DNA repair mechanisms include a DNA synthesis event to replace normal nucleotides removed during the repair reaction. Although the precise mechanism of repair synthesis is not necessarily identical to that used for DNA replication, some modes of repair synthesis involve the use of high-fidelity polymerases (e.g., during nucleotide excision repair) and hence are subject to a very low frequency of errors.

Mutations Associated with Low-fidelity DNA Replication

If a site of DNA damage is encountered by the replication machinery before it is recognized and repaired, a group of cellular responses is evoked that does not include removal of the offending lesions in DNA. For this reason, this mode of cellular responsiveness is referred to as "DNA damage tolerance." One of the multiple DNA damage tolerance mechanisms that operates in both prokaryotes and eukaryotes is called "translesion DNA synthesis" (TLS).[19–22] Since it is using as a template DNA that is damaged at one or more positions, TLS is highly mutagenic and is, presumably, an important contributor to the frequency of both spontaneous mutagenesis and that caused by exposure to environmental agents that are reactive with DNA and result in arrested DNA replication.

Before discussing TLS in some detail, it is cogent to consider what circumstance determine whether cells deploy DNA damage tolerance as opposed to DNA repair mechanisms when they have sustained DNA damage. Obviously, deficient or defective DNA repair is expected to predispose to the preferential or even exclusive utilization of DNA damage tolerance mechanisms to avoid the lethal consequences of arrested replication. An example of how this situation can lead to a disease state in humans is discussed presently. Another intuitively obvious contribution to the deployment of different cellular responses relates to the stochastic nature of DNA damage. If, for example, base damage occurs very close to the advancing replication machinery, DNA replication may be stalled or arrested before the lesion can be addressed by repair because the offending lesion is inaccessible to DNA repair proteins, for example due to steric constraints. Conceivably, this dynamic is also influenced by regulatory mechanisms that determine relative levels of proteins for damage tolerance and repair. When base damage is unrepaired because of its proximity to the advancing replication apparatus, cells are likely only to need to tolerate unrepaired damage for a single round of DNA replication since repair should have access to the lesion once sites of damage become accessible again after replication.

Translesion DNA Synthesis

As the term implies, TLS is a process by which DNA replication is effected across sites of template base damage that would otherwise arrest normal high-fidelity DNA synthesis, causing cell death.[19–22] Historically, our understanding of the phenomenon derives from efforts to understand the role of specific genes in ultraviolet (UV) radiation induced mutagenesis in *E. coli*. The identification of genes in *E. coli* (such as *umuC/D* and *dinB*) that are required for mutagenesis in cells exposed to UV radiation led to the initial suggestion that the products of these genes somehow relax the fidelity of replicative DNA polymerases, allowing them to bypass templates that cannot be accommodated by high-fidelity polymerases.[23] In recent years, it has been recognized that the products of the *umuC/D* and *dinB* genes are themselves DNA polymerases.[19–22] They, as well as a large number of orthologs and paralogs in both lower and higher eukaryotes, have little overall amino acid sequence identity with members of previously identified families of high-fidelity DNA polymerases. They are thus placed in a new group of proteins called the Y family of polymerases, which is represented at all levels of biological organization (figure 2.3).[19–22]

The unique amino acid sequences outside of the catalytic domain in the Y family members notwithstanding, solution of the crystal structure of several of these polymerases reveals similar palm, thumb, and finger domains to those found in the high-fidelity polymerases.[24–26] But in spite of this structural homology, an analysis of the structural features of these polymerases readily suggests an explanation for their low fidelity for correct template copying.[24–26] As indicated earlier, much of the surface of the base pair lying in the active site of a high-fidelity polymerase is shielded from solvent and is extensively contacted through nonpolar interactions.

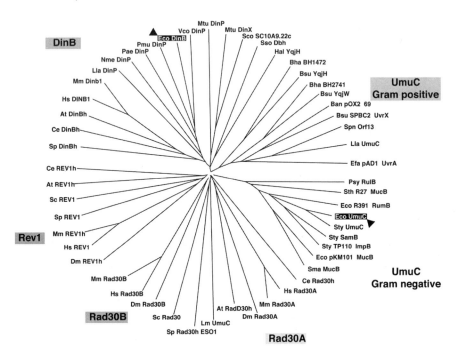

Figure 2.3.   The Y family of specialized DNA polymerases, circa 2001. The *E. coli* DinB and UmuC proteins are prototypes that led to the elucidation of the unrooted tree shown. Four subfamilies, designated DinB, Rad30A, Rad30B, and UmuC, are now recognized.

These steric and electrostatic constraints of the polymerase active site disfavor mispairing with the DNA template. In contrast, structures of several Y family proteins reveal a highly solvent-exposed catalytic site that can accommodate bulky structures, such as pyrimidine dimers and large chemical adducts (figure 2.4).[24–26]

The evolution of DNA polymerases with such relaxed fidelity for pairing nucleotides with the template strand is the essence of translesion synthesis–and of the accompanying misincorporation that so frequently attends this process. Otherwise stated, mutation is the "price" paid for the relaxed fidelity required for translesion synthesis. Indeed, most of these polymerases have a huge error frequency when copying native DNA. In addition to their uniformly reduced fidelity, none of the known specialized DNA polymerases have 3′→5′ proofreading exonuclease activity, and all have markedly reduced processivity, as would be expected if their role is to bypass a lesion and then get out of the way to allow the high-fidelity polymerase to take over again in copying the undamaged region of the template.

Specialized DNA Polymerases

Most of the evidence indicating a biological role of specialized DNA polymerases in TLS comes from in vitro primer extension experiments using purified enzymes and short, naked DNA primer templates. While such systems are clearly highly

A   B

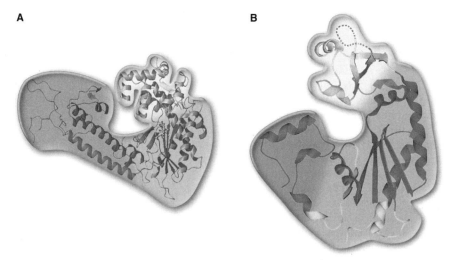

Figure 2.4.   Structures of high- and low-fidelity DNA polymerases. The structure of a high-fidelity DNA polymerase, such as that from bacteriophage T7 (A), reveals little exposure to solvent in the active site pocket. In contrast, low-fidelity DNA polymerases, such as that from the thermophile Dpo4 (B), have a more exposed active site pocket. Note that the fingers and thumb domains of the Dpo4 enzyme are stubbier than those of the T7 enzyme.

reductionist, there are no compelling reasons to question the validity of extrapolating simple qualitative conclusions to living cells, particularly since most high-fidelity enzymes examined under comparable conditions fail to support such replicative bypass. However, it is a general experience that if one biases reaction conditions sufficiently, for example by using a large excess of polymerase, one can demonstrate at least partial TLS of some types of base damage in vitro with almost any DNA polymerase, including even the high-fidelity enzymes. Hence, biological validation of the participation of specialized polymerases in TLS has been eagerly sought, particularly with respect to their lesion specificity.

Data do support the hypothesis that mutants of *E. coli* defective in the *umuC* gene, which encodes a member of the Y family called DNA polymerase V (Pol V), are unable to bypass UV lesions and die rather than mutate. Cells with inactive *UmuC protein*, are nonmutable following exposure to UV radiation.[27,28] Such mutants also suffer increased lethality following UV exposure. These observations, coupled with in vitro TLS by PolV, support the hypothesis that mutations in *umuC*[+] cells derive from error-prone replicative bypass of photoproducts, and that the inability to replicate across such lesions leads to cell death.

Similarly, persuasive biological evidence for the role of specialized polymerases in TLS derives from study of the human skin cancer-prone disease xeroderma pigmentosum (XP).[29,30] XP individuals suffer a markedly increased mutational load in skin cells exposed to sunlight.[1,2] This increase in mutations, which frequently leads to skin cancer, can derive from either of two types of molecular defects. Most XP patients are genetically defective in one of multiple nucleotide excision repair (NER) genes. These encode proteins that assemble into a large complex (repairosome) at

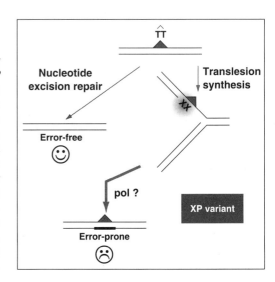

Figure 2.5. Generating mutations in XP variant cells. Unlike most individuals with xeroderma pigmentosum (XP), patients with the variant form of XP are proficient for excision repair but are defective in a particular DNA polymerase (Polη) required for translesion synthesis across certain types of base damage. Consequently, another polymerase is used for this purpose, generating a high frequency of mutations.

sites of helix-distortive base damage, including cyclobutane pyrimidine dimers and [6–4] photoproducts in cells exposed to UV light.[1,2,31,32] A smaller number of XP individuals, who are designated the variant form of XP, are proficient in excision repair of DNA but are defective in a gene designated *XPV* (for XP variant) that encodes a specialized DNA polymerase called DNA polymerase eta (Polη).[28,29]

Biochemical characterization of Polη indicates its ability to support TLS past cyclobutane pyrimidine dimers. Moreover, it does so with remarkable fidelity, incorporating A residues across thymine–thymine dimers (T<>T) most of the time.[28,29] In fact, while retaining the basic property of reduced fidelity, Polη replicates past template T<>T no less accurately than it does past undimerized adjacent TT residues.[28,29] Hence, in living cells, accurate TLS by Polη past thymine dimers reduces the mutational effect of UV radiation to below the threshold required for the development of skin cancer. Since XP variant individuals lack functional Polη it is likely that in these individuals pyrimidine dimers formed by ultraviolet radiation, are bypassed by other specialized polymerases that do not preferentially incorporate A residues, thus increasing the mutational burden (figure 2.5).[28,29]

These data lead to the intriguing hypothesis that individual TLS polymerases evolved to deal with specific types of base damage. In other words Polη evolved specifically to bypass pyrimidine dimers, even though it can also bypass chemically unrelated types of bulky base damage. This hypothesis suggests that when the "correct" polymerase is utilized for TLS across its cognate substrate, the potential for mutation is reduced. Hence, the mechanism of mutations by TLS is by an "incorrect" (noncognate) DNA polymerase incorporating incorrect nucleotides.

As an aside, it is intriguing to ponder the notion that despite the fact that an enzyme such as Polη has an extraordinarily low fidelity, it nonetheless appears to be adapted to incorporate A residues opposite T<>T lesions with remarkable accuracy overall. This may not relate so much to the intrinsic accuracy of Polη as to the fact that during each encounter with a T<>T lesion in the template strand the

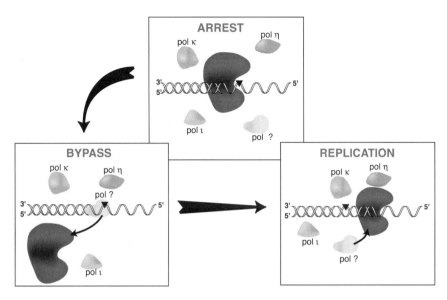

Figure 2.6. DNA polymerase switching. The process of translesion DNA synthesis (TLS) is presumed to require polymerase switching events as shown. The dark kidney-shaped structure represents the DNA replication machinery. In the top panel, this is shown to be arrested at a site of base damage (inverted triangle). For TLS to transpire, one of the multiple specialized polymerases is presumed to occupy the primer-template site (lower left panel). Once TLS is completed, the replication machinery returns to the primer template and normal high-fidelity replication continues (lower right panel).

enzyme is required to incorporate just a few nucleotides. Hence, the error-rate genome wide is small.

Mechanism and Regulation of Translesion DNA Synthesis

The notion that multiple specialized DNA polymerases evolved for the replicative bypass of various types of base damage in DNA raises many interesting questions. What is the nature of the polymerase switching events that temporarily replace the arrested replication machinery at the primer terminus with a low-fidelity poly-merase (figure 2.6)? And once TLS is completed, how is the process reversed, thereby allowing high-fidelity replication to continue past the site of base damage? In light of the reduced fidelity of the specialized DNA polymerases and their poten-tial for generating mutational havoc if they are allowed to copy long stretches of DNA, how is their access to undamaged regions of the genome controlled? Finally, if, as has been suggested above, the redundancy of specialized polymerases in higher organisms (mammalian cells have as many as ten such enzymes, some of which are able to bypass the same types of base damage in vitro) reflects a require-ment for multiple enzymes to effect the bypass of multiple types of naturally occur-ring base damage, how is any particular polymerase selected for its cognate

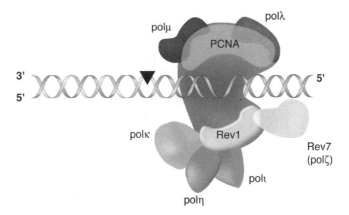

Figure 2.7.   Protein–protein interactions that may be relevant to the DNA poly-merase switching events that transpire during translesion DNA synthesis. Mouse Rev1 protein interacts with multiple other polymerases, including, Polκ, Polη, Polι, and the Rev7 subunit of Polζ. These four proteins all bind to the same region of Rev1 so they presumably do not bind simultaneously. Rev1, Polλ, Polμ, Polκ, Polι and Polη also interact with proliferating cell nuclear antigen (PCNA), and Polη and Polι interact with each other.

substrate(s)? These and other questions are the focus of intensive investigation in a number of laboratories.

At the time of writing this chapter, there is little information about the polymerase switching events just mentioned. In eukaryotes, several specialized polymerases have been shown to interact with several accessory proteins in the replication machinery, notably PCNA (figure 2.7).[21] Additionally, one member of the Y family, Rev1 protein, has been shown to interact with multiple specialized polymerases at the same C-terminal binding site[33] as well as with PCNA (figure 2.7), suggesting a central role of Rev1 in the process of TLS. While Rev1 shares a conserved catalytic domain with the other bone fide DNA polymerases in the Y family, the purified protein has very weak polymerase activity in vitro, incorporating just one or two dCMP residues regardless of the template DNA sequence.[34] For this reason, Rev1 is referred to as a dCMP transferase. Inactivation of this transferase activity does not relieve the requirement of Rev1 for TLS,[35] suggesting that Rev1 has a noncatalytic role(s) in this process. Post-translational modification of various proteins may also be important for TLS. In this regard, recent studies have demonstrated monoubiquitination of PCNA specifically following exposure of cells to UV radiation.[36]

Mutagenic versus Nonmutagenic DNA Damage Tolerance Mechanisms

The recent insights into the mutagenic potential of TLS suggest that this process contributes to some level of spontaneous mutagenesis in cells. But while the phenomenon of TLS has attracted widespread attention as an investigative topic, it is not clear how often this process is actually used in cells. When cells encounter sites of replicational arrest they can employ strategies that essentially skip over the damaged site and continue high-fidelity DNA replication downstream of the damage. Some of the known ways that this is effected involve template switching to the homologous sister DNA duplex.[1,2] A question of considerable interest with respect to spontaneous mutation frequency is: what determines whether such error-free modes of replicative bypass or error-prone TLS are used in any given situation? Intuitively, one might anticipate that in fact error-prone TLS only is employed when nonmutagenic bypass mechanisms fail.

Obligatory Mutagenesis: The Special Case of Somatic Hypermutation

The mechanisms of mutagenesis thus far considered can be viewed as default strategies in the sense that they are incidental rather than obligatory for normal cellular functions. However, mutagenesis is known to operate obligatorily in at least one physiological function: the generation of high affinity antibodies in the lymphoid system. As described in chapter 9, antibody genes are generated by assembling immunoglobulin gene segments. In order to generate a large repertoire of antibodies that can be selected for optimal antigen affinity during an immune response, the antibody coding sequence in immunoglobulin genes and a short region surrounding it are extensively mutated by a process called "somatic hypermutation."[37] The molecular basis of somatic hypermutation has long been a Holy Grail for immunologists, and is considered in detail in chapter 10. This mutagenic process is initiated by a specific cytosine deaminase encoded by a gene called *AID* (activation-induced deaminase), which was originally isolated as a gene expressed specifically in mutating B lymphocytes.[38] (Expression of *AID* converts nonmutating human cells to cells that mutate their antibody genes.[39–41])

The mutation process in B cells is largely targeted to antibody genes, but other highly transcribed genes, such as *BCL6*, are also mutated, albeit at a much lower rate. One wonders whether other, as yet unknown, processes of obligatory physiological mutagenesis occur outside of the immune system. If so, specialized low-fidelity polymerases may have yet other biological functions.

Mutational Spectra

A few words on mutational spectra, a phrase coined to connote the type and precise position of all the mutations in a defined segment of DNA (typically the entire coding region) generated under a defined set of circumstances. Mutational spectra can be highly informative about specific mechanisms of mutagenesis, particularly at so-called "hotspots," locations in the genome where mutations are particularly frequent under a given set of conditions. For example, the observation that most mutations associated with exposure to UV light are located at sites of adjacent pyrimidines suggests that adjacent pyrimidines are targets of UV radiation damage in DNA, as indeed they are.[1,2] In this manner it is sometimes possible to infer the nature of base damage caused by exposure to mutagenic agents.

A detailed discussion about mutational spectra is beyond the scope of this chapter. But it should be noted that the spectrum of mutations generated by any given mutagen is subject to numerous variables, in particular the nucleotide context in which mutagenesis transpires. To embellish the example of adjacent pyrimidines as targets for the formation of cyclobutane pyrimidine dimers and [6–4] photoproducts, both of which can generate mutations when bypassed during TLS, the immediate sequence context in which such dipyrimidines are embedded can significantly influence both the frequency and type of mutations generated.

Other parameters can significantly influence the distribution, frequency, and class of mutations in cells. As pointed out in a recent review, "the leading and lagging strands also can have very different probabilities that, for example, an A will mutate to a G, resulting in different base compositions on the two strands. Through integration of a wide range of cellular activities, including the level of, and balance between, distinct repair, polymerase and proofreading activities encoded and expressed by that genome, and their interaction with different sites in the genome, an overall mutation rates emerges."[42] Mutation rates can also vary as a function of the balance of DNA repair proteins for a particular repair pathway. For example, high levels of expression of a 3-methyladenine-DNA glycosylase relative to apurinic endonuclease increases the mutation rate. 3-methyladenine-DNA glycosylase is an enzyme that recognizes various alkylated bases in DNA and removes the free base during the process of base excision repair. The sites of base loss left are immediately attacked by an endonuclease that specifically attacks sites of base loss in DNA. However, if base removal by the DNA glycosylase greatly exceeds normal rates (as during overexpression), too many sites of base loss are generated and some of these lead to mutations before they can be correctly repaired.[43] Other examples of changes in base composition associated with imbalanced expression of DNA repair enzymes have been documented.[42]

Conclusions

In a retrospective memoir written in 1989, Evelyn Witkin, one of the pioneers of mutagenesis caused by exposure to exogenous agents, notably UV radiation, made the following comment about our knowledge of this topic some 60 years ago: "the prevailing notion until the late nineteen-forties was that mutations were instantaneous events—the mutagen went 'Zap!' and that was that."[44] Our understanding of the molecular mechanisms by which point mutations are generated in DNA, at least in somatic cells, has come a long way since then. We now know that mutations are generated by multiple mechanisms. Some of these are indeed as simple as "zap," in the sense that they arise from alterations in the genome that promote the incorporation of incorrect nucleotides during normal DNA replication. What could be simpler than the incorporation of adenine opposite deaminated cytosine (uracil) to generate a C→T transition mutation? But recent insights into the phenomenon of translesion synthesis across base damage indicate a profound biochemical complexity involving multiple proteins that position different DNA polymerases at the primer-template terminus with exquisite regulation. Aside from deciphering this complexity in precise biochemical terms (a task that may defy simple in vitro experimental models), what remains to be firmly established are the regulatory mechanisms that determine a balance between the mutational burden in the germline genome that generates diversity acted upon by natural selection on the one hand, and avoiding a burden that is deleterious to the descendents of an organism on the other. Another important challenge for the future is to establish experimental techniques that facilitate the investigation of mutational mechanisms in germline cells. As suggested in the body of this article, perhaps some of the processes by which mutations arise in DNA differ fundamentally in germline and somatic cells.

Acknowledgments This article is dedicated to Evelyn Witkin, a pioneer in the study of mutagenesis. The author gratefully acknowledges Paula Fischhaber for discussions and for careful review of the manuscript. Studies in the author's laboratory are supported by the University of Texas Southwestern Medical Center at Dallas and by research grant 5R01 ES011344-09 from the U.S. Public Health Service.

3

Repeats and Variation in Pathogen Selection

Christopher D. Bayliss and E. Richard Moxon

Overview

For a population of bacteria to survive in a fluctuating environment, diversity at certain loci is an essential part of fitness. The potential for diversity at such loci actually can be inherited and selected. A single genome sequence can imply diversity at those loci through the evolution of sequences with a high mutation rate at such loci. We suggest the term "species genome" to encompass the full range of diversity implied by a bacterial genome.

Exploitation of the Implicit Genome for Adaptation and Survival in Hostile Dynamic Environments

The diversity of life forms is extraordinary and it is a matter of profound importance and fascination that the instructions specifying the varieties of shape, size, color, and other attributes of organisms are stored in and transmitted to progeny by sequences of nucleic acids. The availability of complete genome sequences has revolutionized our understanding of the heritable characteristics of organisms, and has provided details of their gene content and organization. For many named bacterial species, at least one, and in some cases more than one, complete genome sequence is now available. This provides an index genomic exemplar of the species, a necessarily selective representation that Caporale has designated the "explicit genome."[1] Nevertheless, one complete genome sequence cannot capture more than a restricted sample of the DNA content (coding and noncoding) and organization of that one strain, let alone the variety within the named species of which that strain is but one member. In this sense, the use of the term "implicit genome" captures the additional information content afforded by the predictable variations in nucleic acid content and organization that more properly describes a "species genome," for example one that takes into account the full repertoire of variations in repetitive DNA tracts or locations of insertion elements. A key issue is that the distinction between explicit and implicit genomes is not merely a matter of descriptive accuracy, but one that has profound implications in considering the evolutionary and adaptive potential of organisms.

Bacterial Pathogens Exhibit Extensive Genetic Variation

Since this chapter will consider in some detail aspects of the adaptive behavior of bacterial pathogens in the context of the dynamic interplay between host and microbe, it is worth extending the concept of the implicit genome to emphasize the potential importance of variations in genome content and organization that are now known to be characteristic of the population biology of different species of certain bacterial pathogens. For example, strains of *Mycobacterium tuberculosis* are considered to display a clonal population structure, but detailed sequencing has shown widespread variation (polymorphisms) from one strain to another[2] and dynamic alterations in the location of insertion elements (IS 6110). Thus, there is considerable species heterogeneity in genome content and organization. Likewise, the complete genomic sequences of several different strains of *Staphylococcus aureus* have revealed an extraordinary diversity between the explicit genomes. The picture here is of a "relatively conserved backbone that is peppered with variation."[3] These within-species variations in DNA content and organization of pathogen genomes have major implications for the emergence of commensal and virulence behavior, drug resistance, and vaccine design.

Bacterial Pathogens Face a Diverse and Evolving Host Environment

Pathogenic bacteria face particularly stringent tests of their adaptive potential because of the dynamic changes in the host environment, the polymorphisms characteristic of innate immunity, and the extraordinary diversity of acquired immune responses (the generation of diversity by the acquired immune system is discussed in chapters 9 and 10). Typically, acute infections of animals occur within a matter of hours following the translocation of the pathogen from one genetically distinct host to another, or from a living vector to a host, or from the inanimate environment to a host. The capacity of bacterial pathogens to negotiate these differing environments is remarkable, as evidenced by the many colorful descriptors that have been coined, including "genetic arms race"[4] and the "Red Queen."[5] The complex and versatile behavior of pathogenic microbes is often depicted metaphorically as 'smart' behavior. This genetic "intelligence" is, of course, not a cognitive process but refers to the concerted interactions of the genetic mechanisms that determine genome stability and genome variation (figure 3.1).[6] The outcome of this interplay, shaped by natural selection, determines the fitness of an organism. In the case of pathogens, the multiple strategies for evading host clearance mechanisms are wrought by the action of natural selection on, for example, genes of genome maintenance (e.g., mismatch repair) and replication (e.g., DNA polymerases that exhibit varying degrees of proneness to error). Concomitantly, the host landscape also is evolving and undergoing rapid modifications, so that as soon as a bacterial pathogen has scaled a fitness peak, the landscape may change and plunge them into a fitness trough (figure 3.2).[7]

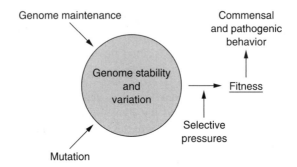

Figure 3.1.   Microbial fitness is driven by a dynamic balance between mutational processes and genome maintenance functions. "Smart behavior" is one metaphor that has been used to describe bacterial pathogens. This metaphor refers not to cognitive processes but to the immense, and apparently purposeful but in fact blind, power of natural selection that acts to generate "fit" organisms. This natural selection acts on the genes responsible for maintaining the balance between the mechanisms mediating genome stability and variation. The outcome of this evolutionary process is fitness, which then dictates the commensal and pathogenic behavior of microbes. (Figure based on ideas originally elaborated by Tone Tonjum and Erling Seeberg.)

Bacterial Pathogens Have Adopted Variable Mutation Rates as a Survival Strategy

A quantitative measure of the fitness of any population of organisms is its basic reproductive rate (R_0), the average number of surviving progeny derived from each founding organism.[8] If R_0 falls below one, then that population and the genome sequence of that lineage of the organism is extinguished. Although the biological determinants of fitness are dynamic and complex, adaptation of an organism to its environment is realized through a combination of gene regulation and gene variation. Thus, bacterial populations can adapt to changes in their environment by classical gene regulation, which facilitates acclimation to frequent environmental changes through a range of coordinated responses, such as catabolite repression, coupled sensor-transducers, and stress response pathways. But, given the diversity and rapidity of environmental changes, what if these stereotypic mechanisms are insufficient? Genetic diversity within populations can provide a stochastic mechanism for generation of variants with heightened fitness in novel environments. The limiting factors are the size, history, and mutation rate of the population. Consequently, selection can act upon the genetic mechanisms determining the rate at which variants arise. But mutations occur without respect for their utility and are more often deleterious or neutral than beneficial. Selection accommodates this trade-off, balancing the fitness gains against the harm caused by mutations (and hence by mechanistic processes that stochastically generate mutations). As discussed below, it has become quite clear that evolution has found ways of enhancing fitness by increasing mutation rates in some selected regions of genomes.

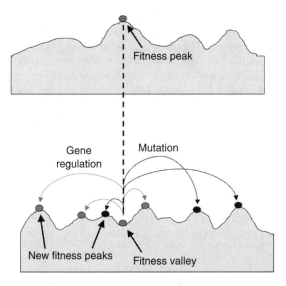

Fitness peak

Gene
regulation

Mutation

New fitness peaks

Fitness valley

Figure 3.2. Rapid and frequent adaptation is critical for survival of bacterial pathogens in the face of a dynamic, hostile, and evolving host landscape. Microbes frequently have to adapt to dramatic changes in their environment. In the top diagram, a bacterial cell has evolved to have a high level of fitness in its current environment. If the environment changes, as depicted in the lower diagram, the cell may find that its fitness level is below the optimal level. The cell adapts to the new environment through alterations in the expression of particular genes, a process achieved either by alterations in classical gene regulatory pathways or by mutation. Notably, mutation also provides an opportunity to explore the adaptive potential of alternate alleles of particular gene products. These pathways are particularly relevant to the stringent and rapid alterations experienced by bacterial pathogens and offer a "short-term" mechanism of adaptation that is not open to other horizontal transfer pathways. (Adapted from a figure previously published by Martin Brookes.)[7]

Hypermutable sequences ("hotspots"), first recognized in 1961 by Seymour Benzer,[9] represent a subset of the total DNA of bacterial genomes. Selection has favored such sequences at specific loci, often called contingency loci,[10] which have evolved to facilitate an efficient exploration and acclimation to unpredictable changes in host environments whilst minimizing the deleterious effects on fitness associated with a generalized increase in mutation rate (see table 3.1). Key features of contingency loci include their propensity to mutate at frequencies strikingly above (>1000-fold) the genome average, a variety of genetic switching mechanism (figure 3.3) and, in many cases, reversible switches, often referred to as phase variation (PV), of phenotypic determinants. Although contingency genes make up a very small fraction of a bacterium's total DNA, they can provide substantial flexibility. For example, as around 12 of the approximately 2000 genes in the *Haemophilus influenzae* strain Rd genome are contingency genes able to switch between "on" and "off," this bacterial strain could display 4096 different phenotypic combinations. Thus multiple contingency genes within a genome engenders a multitude of phenotypic

Table 3.1.   Contrasting effects of global and localized hypermutation on the mutation rates of two functional classes of genes in a genome. This table highlights the fact that global hypermutation (a "mutator" phenotype) increases the mutation rates in all genes in the genome whilst localized hypermutation raises mutation rates only in specific "response" genes in which this mode of mutation has evolved. Response genes are subject to dynamic and stringent selective pressures (e.g., from a host immune system or bacteriophages) that require a stochastic response from the microbe. Core genes mediate intracellular processes and survival in frequently encountered inanimate environments.

| | *Mutation rate* | |
	Response genes	*Core genes*
Nonmutator	Low	Low
Mutator	High (all)	High (all)
Localized hypermutation	High (few)	Low

variants enabling at least one, perhaps even a few, bacteria in a population to adapt to a new environmental challenge. Whilst these intrinsically hypermutable sequences are critically important for adaptation, particularly within hosts, quantum leaps in adaptive potential can also occur, as described in chapter 7, through lateral transfer of DNA, as exemplified by the acquisition of novel pathogenic properties encoded by pathogenicity islands or virulence plasmids.

Although a prominent feature of some, but not all, bacterial pathogens, the strategy of localized hypermutation in contingency locus is so successful that it is a prevalent feature of viruses, fungi, and protozoa.[11] Indeed, as described in chapter 4, evidence will be provided to argue that evolution of hypermutable loci can provide a mechanism for providing quantitative trait variation in eukaryotes, the natural hosts of these various microbes.[12] Furthermore, these mechanisms even play a role in the immune responses of eukaryotes to these microbes (chapter 10). Thus, the principles of localized hypermutation are widespread and, as discussed by Koch,[13] should be considered in the broader context of bacterial acclimation, for example in metabolic processes.[14,15]

Mutational Hotspots Are the Evolutionary Roots of Contingency Loci

The explicit genome highlights the proteins, RNA molecules, and regulatory elements encoded in the DNA sequence as major forces driving evolution of a genome sequence. There can, however, be conflict between these different forces. Thus alterations in coding sequence as a result of external forces may result in loss of a Chi sequence, beneficial for recombinatorial repair in the *Escherichia coli* genome, and increase internal selective forces for evolution of a new Chi sequence in a nearby part of the genome. There are therefore constant adjustments in the DNA sequence as these external and internal selective forces are balanced in the search for the optimal fitness of the organism.

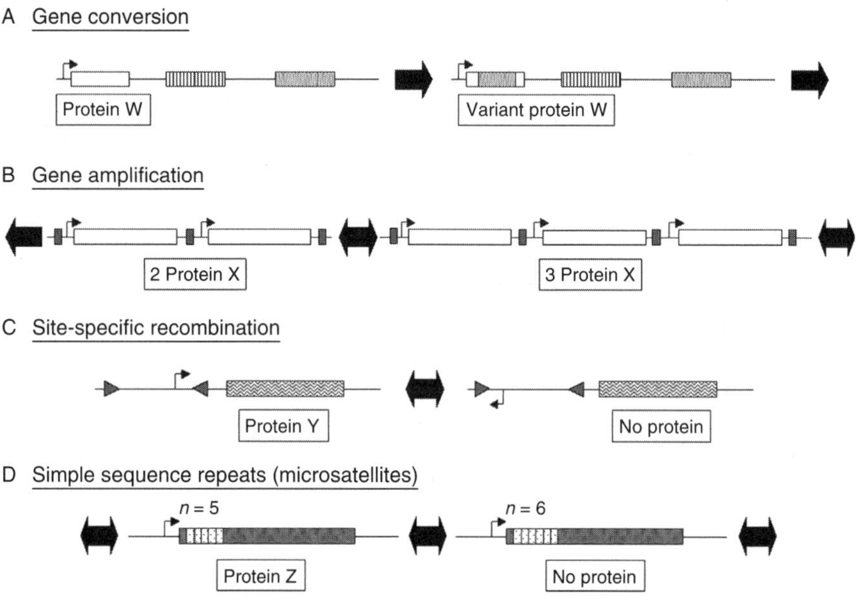

A Gene conversion

Protein W

Variant protein W

B Gene amplification

2 Protein X

3 Protein X

C Site-specific recombination

Protein Y

No protein

D Simple sequence repeats (microsatellites)

$n = 5$

$n = 6$

Protein Z

No protein

Figure 3.3. Features of the major mechanisms of localized hypermutation. Four mechanisms are responsible for elevating mutation rates in the majority of contingency loci. The key features of these mechanisms are highlighted in this figure. Promoters are indicated by small arrows and genes by large rectangles. (A) Gene conversion can occur when multiple nonidentical copies of a gene (shown as differentially patterned rectangles) are present in a genome. One copy templates mutations in another copy but is unaltered in the process. In contingency loci, one copy of the gene is in an expression locus whilst the others are in "silent" loci. This process is not reversible (indicated by unidirectional block arrows) but can realize a vast number of variant genes. Note that this organization also permits for frequent intra- and intergenomic recombination. (B) Gene amplification occurs when multiple, tandemly arranged copies of a locus are present in a genome. Increases or decreases in copy number are generated by intra- or intergenomic recombination between copies of the locus (the ends of which are marked by small oblongs). The genes in a contingency locus of this type have their own promoters such that amplification leads to increases in the amount of gene product. The block arrows indicate that amplification is reversible and can lead to multiple copies of the locus. Note that reduction to a single copy is unidirectional and recovery of this mutational mechanism requires horizontal gene transfer. (C) Site-specific recombination involves inversion of a DNA segment by the action of site-specific recombinases that act at specific DNA sequences (indicated by arrowheads). In the example of a contingency locus, the inverted segment contains a promoter and inversion of the fragment switches gene expression off by altering the orientation of the promoter. The single block arrow indicates that this process is reversible but that the locus can only exist in these two states. (D) Simple sequence repeats are short, tandemly arranged sequences that are prone to "slippage" mutations and hence alterations in the number of repeats present in a particular locus. In contingency loci, repeats are present in the reading frame (as shown) or promoter of a gene and changes in repeat number alter gene expression. The block arrows indicate that this process is reversible and that there is potential for multiple states as the number of repeats is increased or decreased.

Table 3.2.   Number of loci of given repeat type and number in selected bacterial genomes. This table shows the number of loci in each genome that contains a simple sequence repeat tract of the given repeat type and number. The data were derived using the MICAS website. Note that a known contingency locus of *N. meningitidis* (i.e., *siaD*) contains seven Cs and that the number of loci containing seven or more G/C nucleotides in the *N. meningitidis* genome is more than expected based on the frequency of the component parts.[22]

Strain	Genome size, Mbp (%G + C)	G/C mononucleotides, number of repeats per loci					Tetranucleotides, number of repeats per loci			
		5	6	7	8	> 8	3	4	5	> 5
Neisseria meningitidis MC58	2.27 (51)	1416	159	21	8	18	34	1	1	4
Haemophilus influenzae Rd (KW20)	1.83 (38)	509	85	19	6	0	25	0	0	12
Campylobacter jejuni NCTC11168	1.64 (31)	303	19	0	3	26	34	0	0	0
Pseudomonas aeruginosa PA01	6.26 (66)	3122	547	66	7	4	70	0	0	0
Escherichia coli O157:H7	5.53 (50)	2854	449	79	18	6	55	0	0	0

A key selective force of the implicit genome that is both shaped by and sometimes in conflict with the explicit genome is the mutation rate of different DNA sequences. This volume highlights the variations in mutation rate that are inherent in particular DNA sequences. In general, however, there will be selection against sequences with high mutation rates as these sequences are likely to lead to a loss of coding capacity. Thus Gly–Gly–Gly can be encoded by 5′GGG–GGG–GGG3′ but this sequence is highly unstable and is likely to result in loss of one or more nucleotides and inactivation of the protein. Mutational and recombinogenic hotspots in genomes are therefore rare (see the lack of long tetranucleotide repeat tracts in the *E. coli* genome, table 3.2). But, as a result of external and internal selective forces and constant readjustments of the genomic sequence, hotspots can evolve by chance and as a result many hotspots are present and are constantly being generated in genomes.

Mutational hotspots are defined as sequences that have higher mutation rates than the average for the whole genome.[9,16] Hotspots encompass a wide range of mutation rates (see figure 3.4) and may persist in a genome because there is no selection against their presence in a particular locus (see above) or as result of selection for the coding capacity of the mutable sequence. These hotspots, however, provide a substrate for selection, thus in some cases there will be direct selection for the mutations (i.e., the phenotype conferred by the mutation) produced by a particular hotspot. This selection will allow for a significant heightening of the indirect selection for the mutational mechanism in the particular locus that generated variations in the phenotype associated with the locus.[10] These hotspots will then

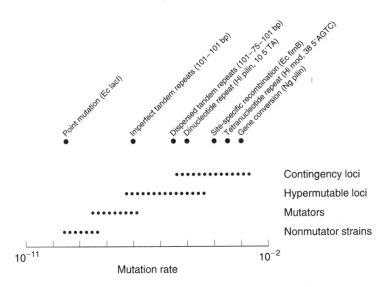

Figure 3.4. Mutation rates of whole genomes and specific loci. Mutation rates vary across genomes and some DNA sequences (termed "hypermutable loci") have mutation rates that are much greater than the average for the genome. These hypermutable loci usually form only a small proportion of a genome. This figure indicates arbitrary ranges for the mutation rates of whole bacterial genomes (nonmutators and mutators) and for hypermutable loci and contingency loci. The ranges for the whole genomes are based on spontaneous mutation rates derived from, for example, assays of antibiotic resistance. Mutators refer to strains of bacteria that as a result of mutations in genome maintenance genes have elevated spontaneous mutation rates. Mutation rates of specific sequences, natural or artificial, are indicated at the top of the figure. (Data are taken from reports by John Drake, Susan Lovett, Ian Blomfield, Steven Seifert, and the authors.)

have the potential to evolve higher mutation rates and become strong hotspots or instantly recognizable examples of hypermutable loci, which provide a contingent response to environmental change. Hypermutable loci in which there is selection for the hypermutable mechanism have been termed "contingency loci." This term places emphasis on the adaptive potential of the locus and has practicable utility for labeling of genomes (see below). It should be recognized, however, that a dynamic relationship exists between hypermutable loci and the subset of these loci that are referred to as contingency loci, and it is difficult to accurately predict, from single genome sequences and our limited knowledge of the selective pressures acting on organisms, into which category particular loci belong.

Contingency loci with high mutation rates are readily identifiable in genomes (e.g., the long tetranucleotide repeat tracts in *H. influenzae*). However, the mutation rates of contingency loci exist in a continuum that overlaps with hypermutable loci (figure 3.4), engendering a practical problem in identifying contingency loci within a genome. This problem is important as contingency loci frequently mark regions of the genome important for pathogenesis of microbes. A consideration of the identification of simple sequence contingency loci (SSCL) further illuminates this problem.

Simple Sequence Repeats and the Genomes of Bacterial Pathogens

Simple sequence repeats are multiple repetitions of a DNA sequence (the unit) in a tandem arrangement. The unit may consist of one to nine nucleotides, termed herein a "microsatellite," or more than nine nucleotides, termed "minisatellites"[17] (see chapter 4 for a fuller discussion of these terms and of eukaryotic microsatellites). Long simple sequence repeat tracts (i.e., exceeding four to ten repeat units) are frequent in the genomes of many eukaryotes but rare in the genomes of prokaryotes.[18,19] These sequences are highly prone to mutation due to their repetitive nature, probably by mechanisms involving "slippage," and this mutability explains their absence from prokaryotic genomes with their high coding densities and sensitivity to frameshifts (see above). However, these sequences are not completely excluded from coding regions, and in fact microsatellites represent a major mechanism responsible for localized hypermutation in contingency loci (i.e., SSCL).[20]

Identification of SSCL in Genome Sequences

The determination of multiple bacterial genome sequences revolutionized the search for SSCL. The *H. influenzae* genome, for example, was found to contain 12 tetranucleotide tracts of greater than five units (see table 3.2), where previously only three were known, and these tracts were all in the coding regions of genes likely to mediate interactions with environmental factors.[21] A similar analysis was performed on the *Neisseria meningitidis* genome sequences, and many novel potential SSCL were identified by searching the genome for simple sequence repeat tracts where the number of repeats was greater than an arbitrary cut-off (e.g., three for tetranucleotide repeats, resulting in 40 loci, see table 3.2). Each of these loci was examined in turn to identify whether the repeat was in the coding region of a gene or in a promoter region (defined as the 100 bp upstream of a start codon). For tetranucleotide repeats, 10 of the loci were deemed to be potential SSCL.[22] For the G/C mononucleotides, the estimates were 27 SSCL and a minimum repeat number of five or more units. Intriguingly, a Markov chain analysis of the occurrence of repeats based on the occurrence of their component parts (e.g., a repeat of seven nucleotides is composed of two repeats of six nucleotides) indicated that there were more tracts of seven G/C than would be expected by chance (note that seven G/C is the lowest repeat number known to be associated with a phase-variable locus in *N. meningitidis*). This analysis indicated that some, but not all, of the tracts with more than six G/C repeats may be maintained in the genome by selection for their mutability. Recently, the observation of polymorphisms between *N. meningitidis* strains in particular repeat tracts suggested that the minimum repeat number for G/C repeats in SSCL may be six units.[23] There are, in the MC58 genome, 159 repeat tracts of this type and length (table 3.2). It is unclear whether these shorter tracts are maintained by positive selection for mutability or represent previously selected tracts that have shortened and are no longer hypermutable. In addition, other factors (such as local DNA context or orientation) may modify mutation rates of repeat tracts such that tracts of equivalent length exhibit significant differences in their mutation rates. Due to these uncertainties, it is currently difficult to calculate a definitive number of SSCL for each genome.

Long Tracts of Inherently Mutable Simple Sequence Repeats Are Typically Found in Obligate Bacterial Pathogens

A number of prokaryotic genomes contain many more repeat tracts than are expected to occur by chance (see tetranucleotide repeats in table 3.2); many of these tracts mark SSCL. To date, bacteria found to have multiple SSCL are obligate host parasites of eukaryotes and tend to have small genome sizes (e.g., *N. meningitidis, N. gonorrhoeae, H. influenzae, Helicobacter pylori* and *Mycoplasmas*), which may indicate that SSCL provide a flexible strategy for survival in the dynamic and variable environments offered by host populations as opposed to inanimate environments. The next three sections highlight the importance of this flexibility, with a focus on variability in antigenicity, phenotype, expression, and mutation rate.

An Extensive Repertoire of Antigenic and Phenotypic Variants Is Generated by Contingency Loci

Contingency loci encode products that mediate interactions of bacterial cells with their external environments. For bacterial pathogens, the cell surface is a major site of interaction with its host, both as a target for the host immune response and as the site that mediates attachment to host cells and acquisition of nutrients. Consequently, contingency loci have evolved to generate two types of variation—antigenic and phenotypic (figure 3.5). The majority of contingency loci fall into one of several groups with distinctive roles in interacting with the environment, namely evasins, adhesins, iron-acquisition proteins, lipopolysaccharide (LPS) biosynthetic enzymes, and restriction-modification (R-M) systems. There are, however, a few contingency loci whose functions do not fit into these groups, and the potential for a contingent alteration in metabolic activity is one that is yet to be fully explored.[14,15]

Rapid Generation of Antigenic Variation by Contingency Loci

Escape of Immune Responses

Antibodies directed against surface components are central effectors of the acquired immune system (as described in chapters 9 and 10) and are readily generated during infections. Contingency loci permit escape as antigenic variants are generated at high frequencies, such that even small populations will contain variants that could escape a response against an antigen present on the majority of the population (see also chapter 5 for a discussion of antigenic variation in eukaryotic pathogens, such as trypanosomes). This process of escape can be accompanied by

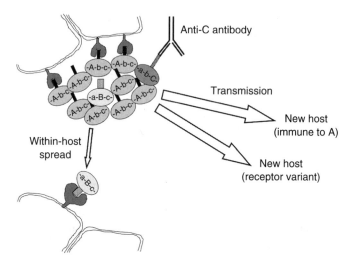

Figure 3.5.   Generation of antigenic and functional variation by hypermutation in contingency loci. A colony of phase variable bacteria attached to host cell tissues is depicted. Attachment is mediated by one of three adhesions (a, b, c) which can undergo phase variable alterations in gene expression (upper and lower cases indicate "on" and "off" phase variants, respectively). The adhesins differ in their antigenic epitopes and tissue tropisms. Thus, variants generated in this colony enable escape of host antibody responses and attachment to alternate tissues either within the same host or during transmission to new hosts. Note that this figure could be adapted to encompass other classes of phase-variable loci, such as iron-acquisition or LPS biosynthetic genes.

retention of function. Gene conversion (see figure 3.3) generates antigenic variants without switching off gene expression (e.g., pilin phase variation in *Neisseria* species). In many other contingency loci, hypermutation results in loss of gene expression (i.e., reversible "on" to "off" switching). In these cases, the presence in a genome of multiple contingency loci with similar functions but different antigenic structures (e.g., three to twelve Opa adhesin proteins in *N. meningitidis* and *N. gonorrhoeae*, three or four hemoglobin-binding proteins in *H. influenzae*, seven capsule loci in *Bacteroides fragilis*, or multiple genes for altering LPS epitopes) permits switching between different antigenic variants with equivalent functional phenotypes.

Antigenic variation generated by contingency loci may also enable transmission to new immune hosts as well as survival of an immune response during the course of an infection in an individual host (figure 3.5). Generation of antigenic variation to escape acquired immune responses is likely to be one of the major functions of contingency loci.

Escape of Bacteriophage Infection

Switching between different antigenic variants through localized hypermutation of genomic regions encoding surface structures could be driven by bacteriophage, which attach to proteins or determinants on the bacterial surface. Some studies have shown that antigen 43, a phase-variable outer membrane protein of *E. coli*, is bound by bacteriophage such that PV of this antigen may mediate resistance to bacteriophage infection.[24] Similarly, it has been proposed that the presence of multiple contingency loci in the genome of *Campylobacter jejuni* enables this species, which is found in extremely high numbers in the intestines of birds and so would be highly susceptible to bacteriophage, to resist infection by generating significant heterogeneity in expression of surface receptors.[25]

Rapid Generation of Phenotypic Variation by Contingency Loci

Contingency loci also provide the potential for generation of extensive variation in function. Although figure 3.5 depicts variation in a particular function (adhesiveness), it also presents a generic view of the way in which contingency loci could generate phenotypic variation. Some specific examples follow.

Evasins

Evasins enable organisms to escape killing by immune effectors. The capsules of many bacterial species have this function. For *N. meningitidis*, the presence of a capsule has been shown to reduce invasion of host cells and thus could inhibit colonization of a host.[26] Noncapsular variants of capsular *N. meningitidis* are frequently found in the upper respiratory tract of humans.[27] These results indicate that phase-variable expression of the capsule may be required to facilitate colonization or survival of an immune response, as dictated by the prevailing conditions.

Adhesins

Adhesins mediate attachment of bacterial cells to host cells. Many contingency loci encode adhesions, and in many bacterial genomes there are multiple such loci. The Opa proteins of *N. meningitidis* are a family of related adhesins that are encoded by contingency loci and are found in multiple copies per genome.[28] Some Opa proteins can be distinguished from others through their ability to bind to distinct host proteins, all of which are members of the CEACAM family of cell-surface proteins. Switching among these Opa proteins may enable *N. meningitidis* to adhere to different cell types within a host and/or to colonize humans who express alternative versions of these host proteins. As described in chapter 5, parasites illustrate the importance of contingency loci that mediate variations in adhesion.[29] *Plasmodium falciparum*, one of the causative agents of malaria, encodes multiple variant copies of a surface adhesin, the Var protein. Switches between *var* genes occurs at high frequencies through a mutational process involving both recombination between the multiple *var* genes within a malaria genome and stochastic switches in expression. This process results in switching between variants with different tissue tropisms, which is vital for the life cycle of the parasite and has a significant impact on its pathogenesis.

Iron-acquisition Proteins

Access to iron is a crucial battleground for bacterial pathogens and their hosts. Iron-acquisition proteins are involved in the uptake of iron or iron-containing compounds (e.g., hemoglobin) by bacteria. Three proteins, LbpA, HpuA, and HmbR, which bind respectively lactoferrin, hemopexin, and hemoglobin, are encoded by contingency loci in *N. meningitidis*.[30] Switching of these loci could enable adaptation to survival in different habitats containing alternative sources of iron. The validity of this theory requires further knowledge of the distribution of various iron sources in the host in relation to niches known to be inhabited by this bacterial species.

LPS Biosynthetic Enzymes

LPS is the major component of the outer membrane of Gram-negative bacteria. This molecule has multiple functions, including provision of an impermeable barrier between the cytosol and the environment, mediating resistance to multiple mediators of the innate immune system and attachment to host surfaces. Many contingency loci encode biosynthetic enzymes that add substituents (e.g., sugars or phosphates) to the outer portions of the LPS molecules.[31,32] For some of these loci, there is evidence of switching between alternate functional phenotypes. One epitope found on LPS molecules is sialic acid. This molecule is negatively charged and has been shown to provide resistance to killing by serum (possibly through inhibiting the action of complement) and also to inhibit or reduce attachment of bacterial cells to host cells.[33] In analogous fashion to the conflicting pressures linked with expression of a capsule, localized hypermutation enables the bacteria to switch between the alternate phenotypes associated with presence or absence of sialic acid on LPS. Similarly, phosphorylcholine epitopes on LPS mediate sensitivity to killing, triggered by C-reactive protein, resistance to bactericidal peptides, and attachment to host cells.[34,35] Thus switching in loci encoding LPS biosynthetic enzymes probably enables bacterial populations to adapt to variations in the prevailing conditions in the host (i.e., presence or absence of an active immune response), to survive in different niches, to move between niches (e.g., into host cells), and possibly to survive in genetically distinct hosts (which may, for example, vary in the activity and specificity of their innate immune effectors). Another intriguing possibility is that switching of some loci may activate the host immune system and increase spread of bacteria within a host.

Restriction-modification Systems

R-M systems modify (usually by methylation) or cleave specific DNA sequences. Contingency loci encoding R-M systems are found in many bacterial genomes; there are often multiple such loci in a genome. All the strong candidate contingency loci of this type encode type I or III systems. These systems are encoded by three and two genetically linked-genes, respectively, with restriction activity requiring a complex of all the gene products such that loss of modification activity (encoded by two or one genes, respectively) also results in loss of restriction activity. Three hypotheses, which are not mutually exclusive, have been proposed to account for the occurrence of localized hypermutation in R-M systems. First, alterations in expression of these systems changes the methylation status of promoters or phase variation rates of other genes, thereby altering the phenotype of the bacterial cell and enabling survival in a new environment.[36] Second, changes in expression alter uptake of external DNA (e.g., by conjugation or transformation), enabling horizontal transfer-mediated alterations in phenotype. Third, changes in expression alter the susceptibility to survival by bacteriophage. In support of the first hypothesis, it is known that methylation of promoters by Dam alters expression of some genes. A second observation is that when mice are infected with a mycoplasma, a switch from "off" to "on" in a phase-variable R-M locus is seen only in mycoplasma recovered from one anatomical site. This switch is frequently accompanied by a switch in a phase-variable surface protein,[37] suggesting that PV of the R-M system stimulates switching in the second gene. The second hypothesis is supported by the observation that "on" and "off" variants of one R-M system in *H. pylori* have different transformation efficiencies.[38] Evidence favoring the third theory is that type I and III R-M systems were originally discovered by their ability to "restrict" growth of bacteriophage and that phase variable sensitivity to bacteriophage infection in *H. influenzae* strains is associated with R-M systems.[39] One modification of this theory that accommodates evolution of phase-variable systems is that the heterogeneity in the population provided by having different combinations of genes "on" and "off" would infringe the ability of a bacteriophage both to spread through a particular clonal population[40] and to develop resistance to an R-M system.

Stochastic Variations in Gene Expression Mediated by Contingency Loci

The majority of contingency loci alter antigenicity or phenotype of variants via changes in expression of the locus. In many cases this involves a simple on–off switch, with additional functional complexity introduced by the presence of multiple loci in a genome (see above) or variations in mutation rate (see below). In a number of cases, however, rather than simple on–off switches, mutations in contingency loci introduce variations in the levels of gene expression. These variations are over and above classical gene regulatory signals to which some contingency loci are still subject. These mutationally induced variations in gene expression may permit greater fine-tuning of the fitness of an organism in response to variations in the environment. Some examples of these variations in gene expression are described below.

Gene Dosage

The capsule locus of *H. influenzae* serotype b strains consists of a tandem duplication of a fragment of approximately 18 kbp encoding all the genes required to synthesize the capsule. Unequal homologous recombination between replicated copies of this duplication occurs at a high frequency, generating variants containing three copies or more of the capsule locus (see figure 3.3). These variants produce more capsule than the parental strains, presumably due to gene dosage (i.e., multiple copies of a gene increase production of the gene product). Further amplification of this locus is associated with higher levels of capsule production, and such variants are frequent among disease isolates.[41]

Use of Initiation Codons in Different Reading Frames

The on–off switching of many SSCL is driven by nontriplet simple sequence repeats located in protein coding regions. Alterations in repeat number shift the reading frame and can lead to truncation of the associated gene product. While truncation operates at the level of translation, for some loci it has been noted that the mRNA molecules bearing stop codons are destabilized in "off" variants. The *lic2A* gene of *H. influenzae* contains a tetranucleotide repeat tract that is preceded by four ATGs, potential start codons (two in one frame and two in another frame). Using reporter gene constructs, three of these ATGs were shown to be functional start codons, but at greatly differing efficiencies (Dixon et al., unpublished data). Thus alterations in the numbers of repeats in the *lic2A* locus could result in switching between three levels of gene expression: high, low, and none. Multiple start codons are present in many SSCL, indicating that variations in gene expression may be a common theme (figure 3.6a).

Modulation of Promoter Activity

In some SSCL, the repeat tract (which can consist of triplet or nontriplet simple sequence repeats) is present in the promoter region. Usually, these tracts are present within the core region of the promoter, which is bound directly by RNA polymerase such that changes in repeat number alter, and even abrogate, gene expression.[42] Phase variation of the *nadA* gene of *N. meningitidis* is mediated by a tetranucleotide repeat upstream of the core promoter[23] in a region bound by two transcriptional regulators, integration host factor (IHF) and Fur. It is thought that alterations in the numbers of repeats alters the spatial positions of these regulators relative to the RNA polymerase holoenzyme that is bound at the core promoter, changing transcriptional activity.[43] A similar model was proposed to explain phase variation of pMGA, a surface lipoprotein, in *Mycoplasma gallisepticum*, where switching is driven by alterations in the length of a 5′GAA tract located upstream of the core promoter.[44] In these cases, alterations in repeat number modulate, rather than abrogate, the levels of gene expression (figure 3.6b). As simple sequence repeat tracts also occur in the promoter regions of eukaryotic DNA, it is possible that repeats may affect expression of eukaryotic genes, as discussed in chapter 4.

Flexibility in Mutation Rates of Contingency Loci

The mutation rate of each contingency locus is determined by the combined effect of a number of *cis*- and *trans*-acting factors. These factors are subject to variation between strains of a species, indicating the potential of a bacterial pathogen to modify its mutation rate in response to alterations in the selective regime of the host population, to competition, or to other factors (figure 3.7). This process may

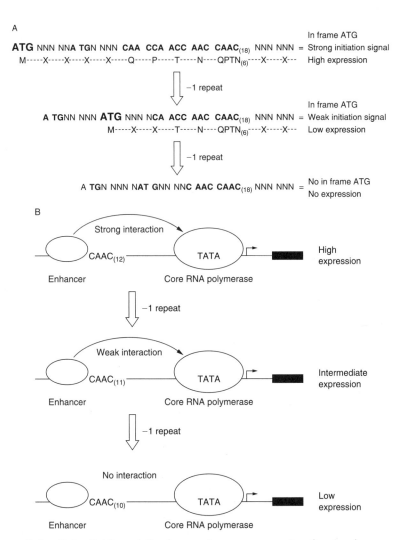

A

In frame ATG

ATG NNN NNA TGN NNN CAA CCA ACC AAC CAAC$_{(18)}$ NNN NNN = Strong initiation signal
M-----X-----X-----X-----X-----Q-----P-----T-----N----QPTN$_{(6)}$----X-----X--- High expression

−1 repeat

In frame ATG

A TGNN NNN ATG NNN NCA ACC AAC CAAC$_{(18)}$ NNN NNN = Weak initiation signal
M-----X-----X-----T-----N----QPTN$_{(6)}$----X-----X--- Low expression

−1 repeat

A TGN NNN NAT GNN NNC AAC CAAC$_{(18)}$ NNN NNN = No in frame ATG
No expression

B

Strong interaction

Enhancer · CAAC$_{(12)}$ —— TATA · Core RNA polymerase · High expression

−1 repeat

Weak interaction

Enhancer · CAAC$_{(11)}$ —— TATA · Core RNA polymerase · Intermediate expression

−1 repeat

No interaction

Enhancer · CAAC$_{(10)}$ —— TATA · Core RNA polymerase · Low expression

Figure 3.6.   Potential for subtle changes in gene expression for simple sequence contingency loci. (A) In-frame simple sequence repeat tracts: in this generic simple sequence contingency locus there are two start codons upstream of a tetranucleotide repeat tract. The two start codons are in different reading frames and are associated with ribosomal binding sites of different strengths. Phase variation between 21, 20, and 19 5′CAAC repeats results in shifting between these two start codons and the third frame, which lacks a start codon. This type of contingency locus has the potential to produce three levels of gene expression, in contrast to other loci that alternate between "on" and "off" gene expression. (B) Promoter-located simple sequence repeat tracts: many tracts of this type are located in regions known to modulate binding of the RNA polymerase holoenzyme. Some tracts are located outside these regions and must, therefore, influence gene expression by altering interactions between the holoenzyme and other transcriptional factors (repressors and enhancers). This figure depicts a contingency locus in which alterations in repeat number change the strength of interactions between an enhancer and holoenzyme located upstream and downstream, respectively, of the repeat tract. This type of contingency locus could produce three or more levels of alterations in gene expression.

69

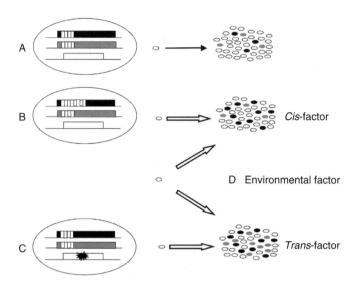

Figure 3.7.   Differential impacts of alterations in cis- *and* trans-*acting factors on the genetic diversity of phase-variable bacterial populations.* The mutation rates of contingency loci can be altered as result of changes in both *cis-* and *trans-*acting factors. These factors have differential impacts on the diversity of the bacterial population. The left-hand part of the figure depicts bacterial cells containing two genes (black and gray rectangles) whose phase-variation rate is determined by repeats (small white rectangles) and a modifier gene (white rectangle). The right-hand part of the figure indicates the amount of genetic diversity that this bacterial cell could generate. In all cases, the genes are "off" (white "cells") in the cell that initiates the population and switches can occur, as a result of mutations in the repeat tracts, during replication, to turn one of the genes "on" (colors of cells and genes are matched). (A) The phase-variable genes have repeat tracts of equivalent length and the modifier gene is intact. (B) The length of one of the repeat tracts is increased, resulting in an elevated mutation rate for one of the phase-variable genes. (C) The modifier gene is inactivated and the mutation rates of both phase-variable genes are increased. (D) This figure highlights the potential role of environmental factors, which could increase the mutation rates of a single locus or multiple loci.

be rapid where, for example, it involves inactivation of a single gene (e.g., generation of mismatch repair MMR mutants, as described in chapter 6) and so could occur during the course of an epidemic. In addition, there is the potential for environmental factors to modify the activities of these *cis-* and *trans-*acting factors. Thus the mutation rates of one or multiple contingency loci could be modified in response to environmental changes (figure 3.7). Progress in this area will be highlighted through the following three examples.

Cis-*acting Factors*

The most common mechanism of localized hypermutation in *H. influenzae* involves tetranucleotide repeat tracts. The rate of PV of these tracts is directly dependent on the number of tetranucleotide repeats in each tract.[45] For example, it has been shown for one of these tracts that the rates of switching increase four-fold as the length of the tract increases from 17 to 38 repeats. These results suggest that the strain-to-strain and locus-to-locus variations in repeat number for *H. influenzae* SSCL will result in differences in PV rate. Variations in repeat number have a similar, but greater, influence on *N. meningitidis* PV rates.[46] While repeat number has a major influence on PV rates of SSCL, further research is needed to assess the possibility that other *cis*-acting factors (e.g., context, transcription, or the nature of the specific sequence that is repeated) also have a significant impact on switching rates.

Trans-*acting Factors*

Studies have also aimed at identifying *trans*-acting factors for *H. influenzae* SSCL. Inactivation of *H. influenzae's dam*-directed MMR system did not increase PV rates of tetranucleotide repeat containing reporter genes.[47] Indeed, to date, mutations in only two *trans*-acting factors, both of which are involved in lagging-strand DNA synthesis, have been found to influence the slippage rates of tetranucleotide repeat tracts: PolI polymerase activity[47] and RnaseHI.[48] Pol-exo mutants lacking the PolI Klenow fragment exhibit a ~30-fold increase in the rate of tetranucleotide repeat mediated PV. RnaseHI mutants increase PV rates less dramatically (approximately fourfold). Thus it seems likely that lagging-strand DNA synthesis is a major source of slippage mutations in tetranucleotide repeat tracts. Moreover, perturbation of lagging-strand DNA synthesis will increase slippage rates. As mutations affecting this process also decrease growth rate, it is likely that only environmental factors that transiently perturb this process could increase slippage rates in natural populations of this species. Because of the small number of studies of repeat tracts with more than three nucleotides per unit in prokaryotes, our current knowledge of the slippage mechanisms is poor, but it seems that, in general, repeat number rather than *trans*-acting or environmental factors has the major influence on mutation rates of these tracts.

In contrast, multiple *trans*-acting factors exert significant effects on the mutation rates of repeat tracts of three or fewer nucleotides. Studies in *N. meningitidis* have demonstrated that mononucleotide repeat-mediated PV rates are elevated by mutations that inactivate MMR[46] or increase expression of Pol IV (encoded by NMB1448; see *Nme dinP*, figure 2.3).[49] Similar studies in *E. coli* indicate that high rates of slippage occur in mutants that perturb leading and/or lagging-strand DNA synthesis.[50] Other factors shown to affect slippage rates include the SOS response, transcription and oxidative stress.[50–52]

The combined effect of sequence context and *trans*-acting factors on PV rate was demonstrated when elevated PV rates were observed in *H. influenzae* MMR-mutants for a reporter construct containing a 5'AT tract but not for the hemagglutinating pilus, whose switching is mediated by a 5'TA repeat tract. On one side of the pilus locus repeat tract are DNA sequences that have the potential to form hairpin structures. It was proposed that these structures activate MMR and, through stimulation of MMR-directed DNA synthesis in this region, increase slippage mutations in the 5'TA-repeat tract of pilus locus. In MMR mutants, this source of slippage mutations would be absent but this decrease would be balanced by an increase in DNA replication-associated slippage.[53]

As inactivation of some *trans*-acting factors can increase slippage rates in repeat tracts without affecting the growth rates of bacteria, there is a potential for variations in the expression or activity of a *trans*-factor to occur between strains. An examination of mononucleotide repeat-mediated PV and spontaneous mutation rates in serogroup A strains of *N. meningitidis* revealed three classes of PV rates (high, medium, and low). Through complementation, many of these variants were shown to have deficiencies in the activities of either *mutS* or *mutL*.[46] As not all the strains with high mutation rates could be complemented with *mutS* or *mutL*, there must be other *trans*-acting factors with the potential to influence the switching rates of mononucleotide repeat tracts in this bacterial species. Notably, strains with high PV rates occurred most frequently among isolates from epidemics. These results suggest that high PV rates could contribute to an increase in transmissibility. It should be noted, however, that many of these strains also exhibited elevated spontaneous mutation rates (i.e., a mutator phenotype) and it cannot be excluded that this phenotype was responsible for the correlation between mutability and transmission.

The dramatic increases in simple sequence repeat-mediated PV rates observed in MMR mutants underlines the fact that a MMR mutator phenotype does not simply increase mutation rates throughout the genome but has differential effects on different DNA sequences. It is possible, therefore, that mutations are targeted in MMR mutants (or other mutator mutants) to cryptic contingency loci (e.g., short, simple sequence repeat tracts) raising the possibility of a much greater range of adaptive phenotypes being explored through localized hypermutation than previously suspected. In addition, there is a potential for *trans*-acting factors to be encoded in contingency loci (see chapter 6 for a description of the frequent inactivation of MMR genes and the role of dispersed repeats in this process, and chapter 4 for a description of the occurrence of repeats in eukaryotic MMR genes). This scenario would necessitate third-order selection and is still speculative.

Environmental Factors

Phase variation of type I fimbriae of *E. coli* is mediated by inversion of a DNA element, containing a promoter, that drives expression of adjacent structural genes. Inversion of this element can be performed by either FimB or FimE, both of which are site-specific recombinases whose activities are influenced by IHF and leucine-responsive regulatory protein (Lrp).[54] Intriguingly, FimE shows a bias for on-to-off switching and is repressed when the element is in the "off" orientation. The rate and direction of fimbrial switching can, therefore, be influenced by the relative levels of expression of FimB, FimE, IHF, and Lrp, and indeed observations indicate that fimbrial PV rates are controlled by environmental signals, such as temperature, branched-chain amino acids, and sialic acid.[55]

Influential Roles for Variations in Mutation Rates and Selection on Population Diversity Generated by Multiple Contingency Loci: Computer Models

The easily recognizable nature of some mechanisms of localized hypermutation at the level of DNA sequence has lead to the conclusion that this process is widespread in prokaryotes (and a number of parasites, see chapter 5) and, based on the overrepresentation of these mechanisms and their location in sites that affect fitness (described above), must have arisen due to selection. The challenge has been to quantify the fitness advantages conferred by the ability to hypermutate, by the variations in mutation rates between loci/strains/species, and by the presence of multiple contingency loci in a genome (figure 3.7). In particular, multiple loci provide a combinatorial explosion in genotypes of potential advantage to the bacteria, and an increasing challenge to the investigator. To date only a few experimental studies have tackled these questions, so most of our understanding of the population dynamics engendered by localized hypermutation has been derived through the use of theoretical and computer models.

While theoretical modeling has aimed to assess the extent to which global hypermutation can increase the rate of adaptation of a microbial population, a definitive answer is in doubt because of hitchhiking and the difficulties in providing reliable quantitative estimates of the number and relative fitness increments of adaptive, neutral, and deleterious mutation sites in a genome. Sniegowski and colleagues[56] presented a simple model of a single locus that could evolve higher or lower mutability and in which the strength of indirect selection on mutability was dependent on the mutation rate. They concluded that, assuming per-locus mutation rates are small ($<10^{-5}$), selection for mutability would be overwhelmed in small populations by drift and in larger, asexual populations by selection at linked loci unless "rescued" by horizontal gene transfer. However, we note that this assumption does not apply for contingency loci, many of which have mutation rates greatly in excess of 10^{-5}; in addition, many of the bacteria that contain contingency loci are also naturally competent for DNA transformation. This model does raise the possibility that there is a definitive lower limit to the mutation rates of contingency loci.

Multiple Contingency Loci Permit Rapid Generation of a Diverse Population in the Absence of Selection

Variations in rates of localized hypermutation may have a major influence on the genetic variability of a bacterial population, something that could be particularly important in populations of limited size (10^6 to 10^9 cells) and when there are multiple contingency loci within a genome. These variables were examined using a computer model of bacterial populations that lacked selection and its associated complications. Populations were initiated by a single "cell," which divided by binary fission and had three sites of localized hypermutation (i.e., nine potential genotypes).[45] In this model, switching rates experimentally observed for tetranucleotide repeat tracts of 10 to 25 repeats always produced populations containing four or more genotypes within 20 generations (i.e., a final population of 10^6 cells).

Interestingly, small differences in PV rates produced proportionally greater impacts on population diversity, thus a threefold difference in PV rates resulted in an 18-fold increase in the number of populations containing nine genotypes after 30 generations. This model demonstrated that heterogeneous populations and complex switches in phenotype can be generated in small populations, but only if the mutation rates are significantly high (i.e., in the range of 10^{-4} to 10^{-5} mutations per division). Small populations are not only relevant to the early stages of infection/adaptation of bacterial pathogens in their eukaryotic hosts but may also be relevant to organisms growing as microcolonies in the highly differentiated environments found in these hosts.

This model emphasizes the influence of high localized mutation rates and the presence of multiple contingency loci within a genome on the amount of diversity present in small bacterial populations. This diversity is, however only a substrate for selection and models presented in the next two sections assess the effects selection exerts on population diversity generated by contingency loci.

Changes in Contingency Loci Derived Populations Are Driven by Selection

While the rate of mutation determines the level of diversity that can be produced in the absence of selection, rapid changes in the composition of a population requires sustained selection to increase the frequency of one subset of the variants generated by mutation. Using a computer model of a single-phase variable locus that can switch between two variants (A and B), in which populations are initiated with only variant A, Saunders et al.[57] found that, in the absence of fitness differences, the number of B variants in a population was proportional to the mutation rate, such that at observed rates ($\sim 10^{-4}$), 2000 generations would be required for B variants to reach 10% of the population. In contrast, a fitness difference of 10% or 50% between these variants resulted in more than 90% of the population being of the alternate phenotype in as few as 90 and 25 generations, respectively. Similarly, Webb and Blaser[58] modeled the evolution of phenotypic variation in *H. pylori*. In this model, Lewis antigen expression was altered incrementally by mutations and each level of expression exhibited differential fitness. When models were run under different selective regimes or in the absence of mutation, it was observed that selection was the major determinant of the rate of evolution, with mutation rates being important only when the initial population was extremely small. These models indicate that selection can rapidly alter the proportions of particular variants in a population, even when these variants exhibit small (i.e., 10%) differences in fitness.

Contingency Loci Permit Survival in Fluctuating Environments

Alternating environments are a particularly stringent challenge that is frequently encountered by bacterial pathogens (e.g., the waxing and waning of an innate immune response). Leigh found that a single locus that could switch between two alleles with differential fitness attributes in two different environments would have an "optimal" mutation rate equal to one over the number of generations between switch of environments.[59] However, this model made several assumptions that are in conflict with observations of the lifestyles of bacterial pathogens; namely, forward and reverse mutation rates would be equal, no involvement of other loci, and "long" deterministic periods between environmental shifts. In particular, most loss of function mutations (e.g., point mutations or frameshifts in nonrepetitive DNA) exhibit significantly different forward and reverse switching rates, and many mutator strains exhibit genome-wide increases in mutation rate (and hence the potential for deleterious mutations in other genes). Michael Palmer and colleagues have shown by analytical arguments and simulation that these factors tend to "depress" selected mutation rates to levels much below those predicted by Leigh's theory (M. Palmer et al., unpublished data). It is of interest to note, however, that contingency loci fit the assumptions of Leigh's models much more closely (i.e., forward and reverse mutation rates are similar and independent of other loci) and therefore should be able to evolve high mutation rates. These models provide a theoretical rationale for the evolution of high mutation rates in contingency loci and underline the extent to which environmental fluctuations may have shaped the variable mutation rates of genomes.

Does Localized Hypermutation Encode the Potential for Accidental Generation of Virulence?

This chapter has emphasized the wide variety of antigenic, phenotypic, expression and mutation rate variations that can be explored through the presence of multiple contingency loci within a genome. Indeed, due to their source in localized variations in mutation rates of DNA sequences, genomes can evolve new contingency loci to respond to new or old challenges. For bacterial pathogens, this means that there is a large amount of genetic variation that can be generated to accommodate variations in the host population (including both genetic and immunological variations) and differences in the rate at which this variation can be generated. One outcome of rapid generation of many genetic variants is the potential for rapid, and even accidental, evolution of virulence. An increase in mutation rate could amplify the range and rate that genetic variants are generated in carriage populations such that there is an elevated risk of "finding" a virulent variant and hence of causing disease. Competition between bacterial variants could also lead to higher mutation rates and hence the potential to cause disease. This situation may be particularly relevant to *N. meningitidis*, where virulence occurs sporadically and is not thought to produce a beneficial outcome in terms of transmission.

Conclusion

Bacteria, and other organisms, survive by transmitting a single genomic DNA sequence, the explicit genome, to their descendants. The presence of multiple contingency loci within such a DNA sequence means that these descendants also acquire the ability to generate high levels of genetic diversity. Transmission of implicit genomes of this type are central to the survival of many host-adapted bacterial commensals and pathogens that are faced by a fluctuating and evolving milieu.

Acknowledgments   We are grateful to Mike Palmer and Marc Lipsitch for providing us with details of their theoretical and computational research on mutation rates and for their thoughtful comments on the manuscript. We would also like to thank Patricia Martin, Kevin Dixon, and Wendy Sweetman for the inclusion of commentary on their unpublished data. Chris Bayliss is supported by a Wellcome Trust Programme grant.

4

Tuning Knobs in the Genome: Evolution of Simple Sequence Repeats by Indirect Selection

David G. King, Edward N. Trifonov, and Yechezkel Kashi

Overview

Simple sequence repeats (SSRs) are highly mutable sites that are distributed throughout eukaryotic genomes. They often occur as functional elements within genes and gene-regulatory regions where mutational changes in repeat number provide extensive variation with minimal genetic load. Implicit in this mutability is the potential for rapid and reversible adjustment of quantitative traits. Although these repetitive elements have been regarded as "junk" or "selfish" DNA, their unique properties are consistent with indirect selection for a "tuning knob" function that facilitates efficient evolutionary adaptation.

Introduction

A simple sequence repeat (SSR) is a DNA tract consisting of a relatively simple base-pair sequence, or "motif," that is repeated over and over in tandem. Although short repeats should be found by chance in any lengthy tract of random DNA, SSRs are longer and far more numerous than expected from chance alone. For example, an estimated 88,000 sites in the human genome include $(AC)_n$ tracts with an average length of over 50 tandem repeats.[1-3] ("$(AC)_n$" refers to n repeats of an adenine–cytosine motif; these repeats are paired with guanine–thymine [GT] repeats on the complementary DNA strand. This sequence might be described equivalently as "$(AC)_n$," "$(GT)_n$," "$(AC)_n \cdot (GT)_n$," or "AC/TG repeats."[4] The probability that a particular sequence of n base pairs will appear at a specified site in a random DNA sequence is approximately $(1/4)^n$ [assuming equal proportions of each nucleotide]. Thus any repeated sequence longer than 20 or so base pairs is unlikely to appear solely by chance, even once, anywhere in the 3×10^9 base pairs of the human genome.)

SSRs based on many different motifs share three common characteristics, with important functional consequences. First, SSRs are found in most regions of most genomes.[1–14] They have been reported within many open reading frames, and they are even more frequent in regulatory regions.[3] The proportion of DNA consisting of SSRs varies substantially among taxa, but it is as high in the compact genome of the pufferfish[6] as in the much larger genomes of many other vertebrates; SSRs make up about 3% of human DNA[2] and over 11% of the *Dictyostelium* genome.[14] In *Dictyostelium*, SSRs within exons occur on average every 724 base pairs.[14] Second, mutations that increase or decrease the number of motif repetitions at a given SSR site are both frequent and reversible. As a result, many SSR sites have several common alleles that differ in the number of repeats. In the human genome, for example, the proportion of AC repeats that are polymorphic is estimated to exceed 90%.[3] Third, and most importantly for this chapter, the number of repeats at an SSR locus can exert a quantitative influence on several aspects of genetic function.[3,15–23]

SSRs are commonly classified into two categories, based on motif length. Those with very short motifs are called "microsatellites"; those with longer motifs, up to several dozen base pairs, are called "minisatellites." These two categories can behave somewhat differently; the motif-length for the boundary between them has been defined at various values between five and ten base pairs.[1,4] In most of the work cited in this chapter, microsatellite motifs consist of six or fewer base pairs, while minisatellites have motif lengths of twelve or more.

Behavior of SSRs

Brief Historical Background

Because most functional elements in the genome are closely linked to one or more polymorphic SSRs, SSRs have been widely used as genetic markers for mapping genes and quantitative trait loci (QTLs). Also, since a dozen or so polymorphic SSR loci may be sufficient to specify a unique genotype for each individual within a population, SSRs are used for DNA "fingerprinting," kinship analysis, and studies of genetic variation within populations.[4,24–26] Until recently, since SSR functionality has not been widely appreciated, most researchers have presumed that extensive SSR polymorphism had no effect on fitness. This view was reinforced by the assumption that the high mutability of SSRs could not be tolerated by natural selection unless SSR alleles did not vary in fitness; otherwise, the population mutational load would be greatly increased.

In the early 1990s, however, investigation of several neurological disorders, including fragile-X and Huntington's disease, provided unexpected but compelling evidence that changes in SSR length could have dramatic consequences. These diseases are characterized by "genetic anticipation," a peculiar pattern of inheritance in which symptoms become more severe and appear at an earlier age as the disease is passed from one generation to the next. The genetic alteration that precipitates the disease is mutational expansion of a repeating triplet, commonly a $(CAG)_n$ or $(CTG)_n$ tract in a protein-coding domain that is translated into a homopolymeric glutamine repeat.[27,28] When the DNA tract exceeds a threshold or "premutation" length of a few dozen repeats, the mutational process enters a pathological mode in which the number of repeats can increase by large increments at each generation. Subsequently, as the sequence expands far beyond its normal size range, symptom severity and age at disease onset appear as functions of tract length and worsen with each new expansion. Relatively small expansions of alanine-encoding GCG or GCC repeats,[29] as well as noncoding repeats,[30] have also been found to cause a number of diseases.

Preliminary commentary on the discovery of triplet repeat diseases expressed astonishment at the existence of such troublesome repetitive sequences,[31] even though the mutability of SSRs had long been known[32] and the potential for repeat number variation to modulate gene function had also been previously recognized.[33,34] Homopolymeric amino acid repeats soon were reported in many critical developmental regulatory proteins, including human androgen receptor, human TATA binding protein, *Drosophila* antennapedia homeotic protein, and mouse homeobox protein *HOX-2.6*.[35-37] Experiments introducing such repeats into the DNA binding domain of a *GAL4* transcription factor demonstrated that variation in amino acid number within the range found in normal (i.e., nonpathogenic) proteins could materially affect transcription activity.[35] Taken together, these early results suggested that quantitative modification of the genetic program produced by mutation in triplet repeat tracts could play a positive evolutionary role.[15,35]

Triplet-repeat diseases brought widespread attention to SSR function, but massive expansion by hundreds of repeats is atypical. Normal allelic variants, including nonpathogenic alleles of the loci involved in triplet repeat diseases, typically differ by only a few repeats. Nevertheless, effects on gene function of such small repeat number differences appear to be commonplace. Research showing effects of repeat number variation on gene activity for SSRs of many different types, not just those with triplet motifs or those located in coding domains, has greatly expanded the potential scope for the role of SSRs in evolution.[3,15-23]

Effects of Repeat Number Variation on Molecular Structure and Function

The mechanisms by which SSRs exert their influence over gene function are as diverse as the functions of DNA. These mechanisms vary by motif and by position relative to regulatory and coding sequences. The effects of repeat number are not necessarily linear and may show maxima or minima at certain values.

At the most fundamental level, an SSR can alter the structural properties of DNA. Guanine tracts and other G-rich repeats, such as $(CGG)_n$, can spontaneously form four-stranded structures (G4 DNA) upon the transient denaturation that accompanies transcription or replication (see chapter 11). Tracts of purine–pyrimidine dinucleotides ($(AC)_n$ or $(AT)_n$) can form Z-DNA.[33,38] Nucleosome assembly is favored by CAG repeats[39] but inhibited, at least in vitro, by TGGA repeats[40] and also by mononucleotide A_n tracts, which have a rigid structure stabilized by additional, unconventional cross-strand bonding (bifurcated hydrogen bonds, see chapter 1) that is excluded from nucleosomes.[41] Mutations that change SSR repeat number have a stronger structural effect than nucleotide substitutions since addition or subtraction of even a single repeat motif, by changing the length of the DNA region, rotates the flanking sequences relative to one another by about 34° per base pair. In regulatory regions, the resulting modification of local and higher-order chromatin structure would be expected to change the spatial disposition of transcription factor binding sites and hence the whole architecture of transcription factor interactions (see chapter 3 for an example).

SSRs in exons are perhaps best understood. Here trinucleotide repeats are the most common SSR type.[8–11,14,42] Even before translation, variation in the number of such repeats can affect transcription activity.[43] However, because trinucleotide motifs are equivalent to codons, changing their number also causes a corresponding change in the number of encoded amino acids. Homopolymeric amino acid repeats encoded by triplet SSRs, including imperfect SSRs incorporating differing but synonymous codons, are found within a substantial proportion of all eukaryotic proteins, estimated at 13% in *Caenorhabditis elegans*, 15% in *Saccharomyces cerevisiae*, 20% in humans and in *Arabidopsis thaliana*, and 27% in *Drosophila melanogaster*.[21] In the genome of *Dictyostelium discoideum*, SSRs (including imperfect SSRs) encoding either homopolymeric amino acid repeats or simple repeats of up to four different amino acids are found in 34% of predicted protein-coding genes.[14] Among those proteins containing at least one homopolymeric repeat, 5–25% contain two or more such repeats, including those proteins in which larger expansions lead to human triplet repeat diseases.[21] Variation in the length of these amino acid repeats can adjust biochemical properties such as protein flexibility, substrate binding, and strength of protein–protein interactions.[36]

SSRs with motif lengths that are not multiples of three also occasionally occur in protein-coding DNA. Repeat-number mutations in these SSRs, in contrast to those in triplet repeats, typically inactivate the protein as they shift the reading frame by adding or subtracting one or two repeats. Such repeat-based inactivation is readily reversible, however, by subsequent mutation (see chapter 3).[44,45]

Variation in repeat number can also affect gene function when SSRs are located outside protein-coding DNA, whether in upstream regulatory regions, 5′ UTRs (untranslated regions), 3′ UTRs, or introns, with many examples of *cis*-regulatory influence by SSRs reported in eukaryotes as well as prokaryotes (see also chapter 3).[3,15–20,22,23] For example, when experimentally inserted into an expression vector plasmid and transfected into cultured mammalian cells, AC repeats can enhance transcription activity by two to ten times, depending on repeat number and distance from the promoter, but not on orientation[33]. The number of AC repeats in a human polymorphism in the first intron of the gene for epidermal growth factor modulates transcription activity both in vivo and in vitro.[46] SSRs can serve as variable strength binding sites for a variety of regulatory proteins. For example, binding of a zinc finger transcription factor (*ZNF191*) to a polymorphic TCAT repeat in the first intron of the human tyrosine hydroxylase gene (the rate-limiting enzyme in the synthesis of catecholamine neurotransmitters) is quantitatively correlated with the number of repeats.[47] RNA transcribed from trinucleotide CTG repeats in the 3′ UTRs of the myotonic dystrophy protein kinase gene can form hairpin folds that bind to and activate the dsRNA-activated protein kinase.[48] Dinucleotide SSRs in 3′ UTRs are frequently associated with membrane-targeted proteins; the fact that the position of such SSRs is frequently conserved among related species suggests that they do have a functional role, presumably regulating translation.[49]

In addition to modifying protein structure and *cis*-regulation of gene expression, SSRs can also serve as sites for replication stalling, an effect that becomes more probable with longer repeat tracts and may be associated with formation of G4 DNA (see chapter 11), triplex DNA,[50] or other structures; as hotspots for recombination, with both motif and repeat number influencing recombination rate (also see chapter 6)[51,52]; and as insertion sites for transposable elements, facilitating the genomic alterations that such elements bring about (see chapter 8).[53,54]

Furthermore, closely spaced repeats are overrepresented in the coding regions of mismatch repair proteins in a variety of species.[55] These repeats implicitly encode variation in the rates at which mutations occur at other sites, since recombination between repeats can disable mismatch repair proteins and thus may permit rapid, reversible adjustment of a population's overall mutation rate (see chapter 6). Activation of error-prone polymerases in stress conditions is well documented.[56] This raises the speculation that SSRs in genes coding for mismatch repair proteins could respond to specific stress-dependent signals by inactivating mismatch repair. In this way, the rate of mutation of SSRs elsewhere in the genome could be increased selectively as part of a metabolically activated stress-response mechanism. Indeed, one research group has recently reported that mutation to particular microsatellite alleles can be reproducibly induced by fungal infection in wheat, suggesting the possible existence of highly specific stress-response mechanisms based on SSRs.[57]

Effects of Repeat-number Variation on Phenotype

Variation in a large and growing number of phenotypic traits has been closely associated with SSR repeat-number polymorphism. The triplet repeat diseases[27–29] represent extreme examples; even ordinary variation in repeat number can yield measurable phenotypic effects. The most extensively studied cases are those in bacteria where repeat number mutations shift a gene's reading frame to produce the reversible gene inactivation that causes antigenic phase shifting (see chapter 3).[44] Similar repeat-based reversible gene inactivation can, through somatic mutation, create black spots on red pigs[58].

Abundant evidence that repeat number can affect phenotype has begun to suggest that QTLs and other functionally polymorphic sites mapped by using SSR alleles may often be the SSRs themselves.[59] Although a correlation between phenotypic and repeat-number variation might result from linkage disequilibrium between SSR alleles and genetic variants at other, undetermined, loci rather than from direct causal relationship, in many cases repeat number itself has been shown to modulate the action of genes that are causally involved in shaping the associated phenotypic traits.

In a survey of specific allelic variants that have been experimentally shown to affect the transcription of 107 human genes, over 20% were repeat-number variants; most of these variants have also been associated with phenotypic consequences affecting biochemical, morphometric, psychometric, or disease susceptibility traits.[3]

Conservative extrapolation from this sample suggests that regulatory regions of several thousand human genes may incorporate functional variation in repeat number, opening wide opportunities for meaningful effects of SSR variation on gene function. Hence a demonstration of tight linkage between SSR alleles and a mapped locus of trait variation should be sufficient to implicate the SSR itself, at least tentatively, as a contributing cause of that variation, worthy of further investigation.

Examples of phenotypic effect for SSR alleles include Holstein dairy cattle, where milk fat production appears to be influenced by repeat-number variation of a minisatellite in the promoter for a key enzyme that regulates synthesis of triglycerides in fat cells. Each 18-base-pair motif of this QTL contains a potential transcription-factor binding site[60]. In Angus beef cattle, length variation in an $(AC)_n$ microsatellite located in the P1 promoter region of the bovine growth hormone receptor gene apparently causes quantitative variation in body weight.[61] Similarly, an $(AC)_n$ polymorphism in a prolactin promoter is associated with variation in gene expression and growth response in the fish tilapia.[62]

The influence of repeat-number variation is especially well-documented for a hexanucleotide repeat in the *period* gene of *D. melanogaster* and for a complex polymorphic repeat in the 5′ region of the gene for a mammalian vasopressin receptor, *avpr1a*. The length of the threonine–glycine tract encoded by the SSR in the *period* gene has experimentally demonstrated effects on temperature compensation in the regulation of the fly's circadian rhythm.[63] In the wild, these effects are apparently acted upon by natural selection, as described below. Differing alleles of the vasopressin receptor (a molecule known to influence male social behavior) yield different levels of expression in cultured rat cells. In prairie voles, variation in the length of this repeat is associated both with differing patterns of gene expression in the male brain and also with corresponding variation in the social behavior of individual males.[64]

Additional evidence for the effect of repeat-number on phenotype has been provided by an extensive study of SSRs, especially those in genes associated with development of body shape, in 92 breeds of domestic dogs.[65] In *Runx-2* (a runt-related transcription factor that regulates osteoblast differentiation and in which inactivating mutations cause craniofacial and other skeletal malformations in humans), the length ratio of two triplet repeats (encoding 18 to 20 glutamines followed by 12 to 17 alanines) is quantitatively correlated with measures of facial shape. The fact that the ratio of the lengths of these two repeats correlates with facial shape much more strongly than does the length of either repeat alone suggests that precise modulation of transcription by the *Runx-2* protein could be facilitated by the pairing of repeats with opposing activities. This same study also found that contraction of a hexanucleotide repeat in the *Alx-4* gene (*Aristaless-like 4*) correlated with the presence of extra toes in Great Pyrenees (polydactyly is an official characteristic for the breed).

In humans, SSRs have been associated with disease susceptibility and with variation in a number of psychological and neurobiological traits.[3,59,66] For example, alleles with exceptionally few CAG repeats in exon 1 of the androgen receptor gene are associated with increased risk of prostate cancer, while those with pathologically expanded repeats cause spino-bulbar muscular atrophy.[43] Variation in a GGC repeat in this same gene has been associated with measures of aggression, impulsivity, and other behaviors.[67] Curiously, both of these repeats are longer in humans than in other primates,[68] a pattern that also has been reported for triplet repeats in other genes.[69] The TCAT repeat described above, which regulates transcription of the tyrosine hydroxylase gene, has been associated with variation in human longevity,[70] as well as with several diseases, including essential hypertension, bipolar disorder, and schizophrenia.[47] The shorter of two alleles of a minisatellite length polymorphism in the promoter of a serotonin transporter gene reduces transcriptional efficiency and increases susceptibility to stress-induced depression in humans[71]; in rhesus monkeys, the shorter allele of this minisatellite is strongly associated with earlier age at the time of a male monkey's dispersal from its natal group.[72]

An especially well-studied repeat polymorphism in humans involves a short run of five or six thymidines in *MMP3*, a matrix metalloproteinase gene associated with heart disease risk.[73] The shorter T_5 allele of this gene binds with lower affinity to a transcription complex and drives higher levels of expression. Allelic differences translate into measurable differences in the elasticity and thickness of arterial walls. The lower expression T_6 allele contributes to the buildup of atherosclerotic plaque and hence to coronary artery disease; the higher expression allele reduces plaque buildup but promotes plaque instability and hence myocardial infarction. Evidence for positive selection of this polymorphism is discussed on page 88.

For the vast majority of SSR sites, the genetic and phenotypic impact of repeat number polymorphism has not yet been measured; in many cases it may well be negligible. Nevertheless, the examples described above, representing only a preliminary sampling of extant SSR variation, establish a role for microsatellites as a store of heritable variation for many different phenotypic traits, and as a source of rapid mutational input for such variation which may be far greater than is commonly realized.

Mutability of SSRs

Mutation rates at SSRs are generally orders of magnitude higher than mutation rates at other sites within the same genome. Estimates range from 10^{-2} (*E. coli*) to 10^{-4} or 10^{-5} (yeast) mutations per locus per cell division, with human rates possibly as high as 10^{-3} per locus per generation.[74] Certain hypervariable SSR loci, including a tetranucleotide repeat in barn swallows[75] and a poly-A tract in intron 2 of the human 3-beta-hydroxysteroid dehydrogenase gene,[76] have meiotic mutation rates as high as several percent; meiotic instability of G-rich repeats can be even higher (see chapter 11).

These elevated rates reflect the occurrence of a specific type of mutation—a stepwise change in the number of repeats. Mutations of this type are facilitated by the repeat structure itself. Their principal cause appears to be slipped-strand mispairing, also called replication slippage, but recombination and gene conversion may also be involved, especially in minisatellites (also see chapter 3).[1,4,29,69,77–81] During slipped-strand mispairing, a replicating DNA strand can shift position, or "slip," by an integral multiple of the motif length while still pairing with a complementary sequence (see figure 2.2, page 44). Although the majority of SSR mutations involve gain or loss of only one or two repeats, larger changes may also occur. The resulting changes in repeat number are not reliably corrected by subsequent DNA proofreading and mismatch repair.[45] However, repeat number mutations are readily reversible: the addition of one or more repeats at any point within an SSR can be undone by subtraction at the same or any other point, and vice versa.

SSR mutation rates vary among species, among loci within a single genome, and for a single locus even among individuals of differing age and sex.[1] The mutation rate at each SSR is influenced by several factors. Those external to the SSR's own base-pair sequence include flanking regions,[77,78] methylation status both within and near the sequence,[82] and polymerase fidelity (see chapters 3 and 6).[45,83] More importantly, however, the specific mutation rate for a given SSR, as well as details of the distribution of mutation sizes and directions, is largely determined by the SSR sequence itself.[1,5,69,78,84] As a general rule, at least in microsatellites, shorter motifs such as mono- or dinucleotides yield higher mutation rates than longer motifs such as tri- or tetranucleotides, as might be expected from the greater precision and degree of DNA "melting" needed for slipped-strand realignment of the longer motifs. The rate at which repeats are added to or subtracted from an SSR generally is greater with higher repeat number. Several reports suggest a bias in favor of mutations that increase the number of repeats, but there are also indications that long repeats may have a bias toward shortening; in any case, some process such as biased mutation or selection must prevent most SSRs from growing without bound.[1,4,78,84,85]

Significantly, SSR mutation rate also is affected by the "purity" of motif repetition, since imperfect repeats can interfere with the ability of the replicating strand to pair with a shifted complementary sequence.[15,47] Thus an SSR's site-specific mutation rate may be modified by local point mutations. This attribute of SSRs illustrates one way in which DNA can simultaneously and independently embody multiple codes.[34,86,87] For example, a triplet-motif SSR that encodes an amino acid repeat can also utilize motif variants in "wobble" codon positions to encode the rate at which slipped-strand mispairings cause changes in the length of that repeat.

Evolution of SSRs

Origin of SSRs

Microsatellite SSRs may originate de novo by ordinary base-pair substitution, which can convert a site with imperfect repeats (sometimes called "cryptic simplicity") into a short microsatellite with perfect tandem repeats.[4,74] A short, hypermutable human poly-A tract, albeit one causing cancer predisposition, has been observed to arise by this mechanism.[88] Additionally, short di-, tri-, and tetranucleotide repeats can arise in sites that had no prior repeat structure, by insertions that duplicate adjacent sequences.[89] Once a few tandem repetitions of a motif exist, the tract has an increased likelihood of mutating by slippage, either disappearing or expanding to higher repeat numbers. Minisatellite origin remains largely unexplored, although with their longer motifs and more complex mutational mechanisms,[80,81] minisatellites presumably require more specialized initiation events than microsatellites.[90,91]

Appropriate sites for microsatellite origin may occur by chance, but such sites are provided in abundance by certain transposable elements that function as "microsatellite generators." For example, flies in genus *Drosophila* contain many copies (over 3000 in *D. melanogaster*) of a "microsatellite initiating mobile element," a transposable element that commonly generates SSRs from internal "proto-microsatellites," either by expansion of short tandem TA repeats or by C-to-T transition mutations that convert imperfect repeats into short SSRs with GTCC or GTCT motifs.[92] In Lepidoptera (butterflies and moths), microsatellites occur in families with very similar flanking sequences, also suggesting an origin from transposable elements.[93] In mammals, microsatellites of several motifs have been associated with *Alu* and other SINEs (short interspersed nuclear elements).[53,94–96] A similar association of transposable elements with dinucleotide SSRs has been reported in barley.[54] Certain microsatellites can in turn facilitate transposition by acting as relatively safe "landing pads"[54] or "retroposition navigators."[53]

Distribution of SSRs

Just as different genomes have different families of transposable elements (see chapter 8), so also do different genomes display a preference for different motif classes of SSRs. For example, $(AC)_n$ is the most common dinucleotide motif in the human and *D. melanogaster* genomes, while in *Caenorhabditis elegans* the predominant dinucleotide is $(AG)_n$, and in *Arabidopsis thaliana* it is $(AT)_n$; $(CG)_n$ motifs are rare in all species examined.[8,11,85] A mutual interaction between microsatellites and transposable elements could help explain such interspecies differences in SSR abundance and motif distribution.

SSR abundance and motif distribution also typically differ by functional region within the genome of a single species. The most striking of such differences occurs between coding and noncoding sequences. Eukaryotic exons and open reading frames are enriched with SSRs having triplet or hexanucleotide motifs whose repeat number may mutate without shifting the codon reading frame; other motif lengths prevail in noncoding regions.[8–11,42] For example, in yeast approximately 83% of trinucleotide repeats occur within open reading frames, while 95% of mono-, di-, and tetranucleotide repeats occur in nontranscribed regions.[9]

For SSRs located in coding regions, the amino acid repeats encoded by triplet repeats also show distinct preferences that may differ from genome-wide codon prevalence, and these codon preferences vary among species.[21] Perhaps most interestingly, trinucleotide repeats are markedly overrepresented in genes encoding developmental regulatory proteins, transcription factors, and protein kinases.[9,14,21,29,35–37] Proteins that function in developmental or transcriptional regulation also commonly include multiple homopolymeric repeats.[21] In contrast, dinucleotide AC repeats, while extremely abundant in noncoding portions of genes, are found in a smaller proportion of constitutively expressed housekeeping genes (43%) than of human genes generally (59%).[97] Genes for eukaryotic structural and cell-surface proteins, especially "intrinsically unstructured proteins," often contain minisatellite SSRs encoding oligopeptide repeats and appear to evolve by repeat expansion.[98,99]

Selection Can Favor Mutability

The fact that SSRs seem to be especially prevalent in functional regions where their characteristic mutability can support viable variation in adaptive traits suggests the hypothesis that natural selection may have preferentially retained SSRs in just such locations. However, the possibility that natural selection might favor SSR loci precisely *because of* their mutability contravenes a deeply entrenched belief that mutations happen simply because replication is imperfect.[100] In his classic text *Adaptation and Natural Selection*, George Williams persuasively argued that natural selection necessarily favors only the highest practical level of fidelity for DNA replication.[101] By this argument, the mutability of SSRs must be an inherent flaw in the genome: it cannot be a function evolved to facilitate adaptation. Nevertheless, although similar arguments remain influential,[102,103] SSRs are excluded from the reach of two principal assumptions upon which these arguments are founded.

The first assumption invalidated by the mutational characteristics of SSRs is that, because most mutations having any phenotypic effect at all will be harmful, any increase in mutation rate must on average be selectively disadvantageous.[101,103] As recognized several decades ago by R. A. Fisher,[104] "mutations of small effect" (i.e., those which yield small changes in fitness-determining traits) may have an almost even chance of being beneficial. On the metaphorical "adaptive landscape" (a rugged topography where peaks represent genotypes of optimal adaptation, see figure 3.2, page 57), mutations of large effect are usually deleterious because random relocations are unlikely to land closer to a peak. However, for any population that is close to but not already on a peak, a mutation of small effect will necessarily shift fitness either uphill or downhill with approximately equal probability. In other words, whenever a prevailing trait value is less than optimal, any small change in that value is just as likely to constitute an improvement as otherwise. Because most populations face variable and/or changing environments, they must often find themselves at some small distance from the nearest adaptive peak. Therefore, mutations of small effect could have been generally beneficial over much of evolutionary time. However, because any single mutation that yields only a small improvement in fitness has a fairly high probability of loss to genetic drift before it can be fixed by selection,[105,106] mutations of small effect are likely to be an important source of genetic variation only if they are also frequent. Mirabile dictu, an abundant and continuing supply of just such mutations is exactly what SSRs are able to provide, assuring plentiful replacement for any potentially beneficial mutations that might be lost to drift. In addition, if conditions change enough to favor a mutation of larger effect, a series of small step-wise SSR mutations can provide a relatively smooth and safe path toward that result.

The second assumption invalidated by the mutational characteristics of SSRs is that replication fidelity is set by genes, such as those encoding DNA polymerases and repair enzymes, whose effects are distributed across the entire genome. Such genes are separable by recombination from the sites at which mutation is occurring. Thus, even if some beneficial mutations were to arise, independent assortment and recombination between the mutant alleles and the replication-fidelity alleles would prevent selection for the beneficial mutations from consistently favoring the increased mutation rate. Nevertheless, although an average genome-wide mutation rate may indeed be determined largely by replication-fidelity genes, local mutation rates also can vary dramatically among individual loci.[32,87] SSRs elegantly demonstrate the dependence of local mutation rate on local DNA sequence. As described above, the high mutation rate at a typical SSR locus is due largely to slippage during replication of the tandem repetitive DNA that comprises the SSR itself. Furthermore, changes in repeat number do not alter the essential character of the site and are readily reversible. Hence, even as an SSR mutates under the influence of its own high mutation rate, it remains an SSR with a high mutation rate; that is, recombination cannot separate mutant SSR alleles from the local process of SSR mutation by which they arise. So selection for a favorable SSR allele also unavoidably favors the mutability by which that allele arose.

SSRs thus provide special circumstances that can foster site-specific high mutation rates. Wherever the characteristic mutability of an SSR contributes beneficial variation, direct selection for favorable mutations will indirectly select the mutability as well. This occurs even while low mutation rates are maintained elsewhere in the genome, where high-fidelity replication remains advantageous. The essential point is that indirect selection for a high mutation rate at a given SSR site can remain consistent even while direct selection driven by fluctuating environmental conditions favors first one allele and then another at that site.

Because SSRs appear to facilitate the adaptive adjustment of quantitative traits, we have described them metaphorically as "evolutionary tuning knobs."[107,108] Others have termed them "agents of evolution."[109] Indeed, this may be their evolved "function," the role for which SSRs are adapted. Evidence that SSRs and the variation they generate have been shaped both directly and indirectly by natural selection, although limited and circumstantial, is beginning to accumulate.

Evidence for Direct Selection Shaping Mutable Sites

"Direct" selection is the familiar process whereby particular alleles increase or decrease in frequency in a population, based on the range of phenotypic effects that each allele confers upon individual organisms. Disentangling the effects of direct selection from those of other evolutionary processes, such as hitchhiking (when neutral or even deleterious alleles are carried to higher frequency by selection favoring a coupled allele at a closely linked locus), is notoriously difficult.[110] Nevertheless, and in spite of only limited efforts to date, a number of studies indicate that selection can act on SSR alleles in natural populations of eukaryotes (prokaryotes are reviewed in chapter 3).

Interspecies differences in mating pattern between prosocial and asocial vole species (prairie and pine voles versus montane and meadow voles) are associated with corresponding variation involving the same complex repeat that, as described on page 82, appears to be responsible for individual differences in male mating behavior.[64] Thus selection for species-specific reproductive strategies may be mediated through variation in such repeats. In the case of the *Drosophila period* gene, also described above, the frequencies of the two most common repeat-number alleles vary across a latitudinal cline, as would be expected if the temperature-related functional effects of these alleles cause fitness differences under differing climatic conditions.[63]

Correlations between SSR alleles and specific environmental variables have also been seen in several other species, although the causal mechanisms relating these SSRs to phenotype and fitness remain unknown. For instance, an analysis of population-genetic structure of acorn barnacles, using six microsatellite loci and two allozyme loci, found that the frequencies of the microsatellite alleles did not appear to behave neutrally but were correlated with latitude and temperature.[111] At "Evolution Canyon," Israel, an analysis of 19 nuclear microsatellite loci and four chloroplast microsatellite loci in seven populations of wild barley distributed across north-facing and south-facing ecological zones revealed striking and significant interslope differentiation of SSR allele distributions; the authors of the study conclude that extremes of microclimatic stress might select for reduced SSR diversity, especially if SSRs function to fine-tune adaptation.[112] Similarly, researchers studying microsatellite variation in 15 wild populations of emmer wheat concluded that correlation of ecogeographical differentiation with climatic and soil factors was based on selection for noncoding SSRs that might previously have been regarded as "junk DNA."[113] (There is additional discussion of sequences that others dismiss as "junk DNA" in chapter 8.)

Analysis of data from several human populations suggests that positive selection in northern Europe has increased the frequency of the shorter of two alleles of a thymidine repeat in the *MMP3* gene (see page 83). This repeat appears to have been derived from a more complex GT repeat in other primates. Human haplotype data indicate that the longer T_6 allele has given rise on several occasions to the shorter T_5 allele. One consequence of the increased frequency of this allele appears to be substantial protection from coronary heart disease; however, the selective agent for the increase remains unknown since *MMP3* has many pleiotropic functions.[73]

The dramatic evolution exemplified by rapid morphological divergence among domestic dog breeds may also involve selection acting upon SSR variation. A recent study showing an association between repeat number and canine body shape (see above) also reported an extraordinary level of SSR polymorphism among dog breeds that contrasts sharply with a lack of evidence for functional single-nucleotide variation in genes known to influence morphological traits.[65] Furthermore, a comparison of orthologous human and dog genes showed dogs with significantly higher levels of repeat purity in 17 repeat-containing genes involved in morphological development but not in a broader sample of 844 orthologous triplet SSRs. By considering the purifying effects of repeat expansion and contraction mutations together with point mutations within the repeats, the authors of this study infer recent selection for altered repeat length in the dog morphogenetic genes. From this evidence, including the correlation of repeat length with body shape described on page 82, they argue that rapid evolution of dog breeds may have depended on a continuing supply of variation that could have been provided only by SSRs.

The hypothesis that an appreciable proportion of SSRs generate adaptively meaningful variation might appear to be inconsistent with the widespread impression that many repeat number polymorphisms are neutral (i.e., not noticeably associated with fitness). However, the SSR alleles that predominate in a population already have been culled by direct selection and are therefore expected to display little apparent difference in fitness. It is only by comparing subpopulations that have recently diverged, as in the examples above of domestic dogs, emmer wheat, barley, and *Drosophila*, that the adaptive effects of SSR loci become evident. Even while extant variation at a particular SSR site is practically neutral, the site retains the potential to generate adaptive variation as its alleles expand and contract.

Indirect Selection for Mutability of SSRs

"Indirect" or "second-order" selection[114,115] is the process that emerges as the mutability characteristics of particular loci are eliminated, preserved, or modified through their consistent association with deleterious or beneficial mutant alleles. The peculiar "tuning-knob" nature of SSRs enables mutational modification not only of each SSR's repeat number but also of its intrinsic mutability. Thus indirect selection should discriminate among SSR alleles with greater or lesser mutability, depending on the advantage or harm conferred by mutations which alter the number of repeats. For example, if a particular SSR has a high mutation rate based on perfect tandem repeats, and if this mutability proves deleterious, then indirect selection will favor descendants of those members of the population in which point mutations have converted the SSR into a more stable tract with imperfect repeats.[15] The human tyrosine hydroxylase gene described on page 80 may illustrate just this process; the more frequent and longer allele contains a single nucleotide deletion, which interrupts the repeat sequence and should discourage further expansion beyond this currently favored length.[47] Conversely, mutability can be restored not only by simple reversal of interrupting point mutations but also by removal of interruptions through replication slippage.[116]

To the extent that indirect selection is an effective evolutionary process, it should eventually eliminate SSRs at sites where variation is consistently detrimental (for example, by eliminating most nontriplet repeats from those protein-coding domains where a change in repeat number would cause deleterious frame-shifting effects)[117] while it relegates to the effects of drift those SSRs whose impact on phenotypic variation is adaptively negligible. But it should also favor the formation and preservation of any SSRs which contribute beneficial variation. Furthermore, given the positive interaction between microsatellite SSRs and transposable elements (page 85), indirect selection for beneficial SSRs could also favor the mechanisms of transposition by which they originated and thus contribute to the prevalence of their associated transposable elements (also see chapter 8).

The clearest evidence for the selective advantage of repeat number mutability comes from bacterial stress-response genes (see chapter 6) and contingency loci where nontriplet SSRs in coding domains permit advantageously reversible gene inactivation by frame shifting (see chapter 3).[44] While haploidy can facilitate the indirect selection of higher mutation rates,[103] the prominence of SSRs in eukaryotic genomes suggests that such advantage is not restricted to haploid organisms. Simple population genetic simulations affirm that indirect selection acting on a diploid population in a fluctuating environment can favor genes with mutability characteristics similar to those of SSRs.[118]

Circumstantial evidence that indirect selection actually does shape the occurrence and mutability of eukaryotic SSRs includes both the distinctly nonrandom patterns of SSR distribution that have been reported in various genomes (above) and the evolutionary conservation of particular SSR loci, even among distantly related species.[69] SSRs that are conserved over evolutionary time tend to be characterized by imperfect repeats,[119] but indirect selection could favor either stability or variability of an SSR, depending on the circumstances. For example, the variable hexanucleotide repeat in the *Drosophila period* gene shows up in homologous human and mouse genes as a shorter and presumably more stable imperfect repeat encoding serine–glycine rather than threonine–glycine.[120] Conservation of SSR function is strongly suggested by a number of SSRs in 3′ UTRs that have homologs in both rat and human even though the specific motif patterns have changed (e.g., from $(GT)_n$ to $(AG)_n$ during evolutionary divergence[49] (see also the discussion of conservation of the presence of closely spaced repeats, rather than the repeat sequences themselves, in chapter 6). Also, a repression domain in the *Drosophila Ubx* protein (a homeobox protein which controls differentiation of body segments) includes an alanine repeat encoded by an imperfect triplet repeat whose location is conserved across several insect orders. The phylogenetic appearance of this SSR corresponds with the evolutionary origin of insects from ancestral arthropods and may have played a role in the emergence of this group.[121]

Conclusion

Evidence that variation in the length of SSRs influences gene expression adds new significance to the familiar idea that gene regulation plays an important role in phenotypic evolution.[122] As general-purpose tuning knobs for gene function,[107,108] SSRs have the qualities of modularity and exploratory behavior that are essential for evolvability.[123,124] SSRs are well suited for incorporation into complex regulatory hierarchies,[122] with transposable elements serving as effective vehicles for their insertion into novel locations on a trial-and-error basis.

By their sheer number and variety, SSRs represent a rich reservoir of information. By their mutability, SSRs provide a prolific source of heritable variation that is renewed with each generation. Many years ago, Sturtevant famously asserted that "mutations are accidents, and accidents will happen."[100] But neither the distribution of SSRs nor the mutations that they produce appear to be accidental. We propose that the abundant presence of SSRs in eukaryotic genome derives, at least in part, from indirect selection for SSRs' role in facilitating the process of evolution. SSRs endow populations with an abundant yet relatively safe supply of raw material for adaptation. It would be astonishing if this great resource of implicit variability had not been turned to advantage by natural selection.

Note Computer programs for SSR analysis of any genome are available at *ATRHunter* (bioinfo.cs.technion.ac.il/atrhunter/) and *Tandem Repeats Finder* (tandem.bu.edu/trf/trf.html).

5

Implicit Information in Eukaryotic Pathogens as the Basis of Antigenic Variation

J. David Barry

Overview

Eukaryotic pathogens, such as trypanosomes, imply far more antigenic variability than is encoded explicitly in their genome, thus enabling them to provide an ever-changing and effective challenge to the host immune system. By the use of gene fragments and pseudogenes combinatorially to generate novel genes, trypanosomes can, in theory, generate a number of surface antigens significantly greater than the total number of genes in the genome.

Introduction

Imagine a system whereby a pathogen, faced during an infection with repeated lethal antibody assaults directed against a particular surface protein, can survive by altering specifically the epitopes (antigenic determinants) of that protein while maintaining the structure of other, functional, parts of the molecule. Imagine further that, at least in theory, the epitopes can be changed many more times than there are genes in the genome. And imagine that this ability is heritable such that, to some extent, the same epitope changes can be predicted to occur in subsequent, distinct infections. This, in a nutshell, is what some pathogens achieve through their antigenic variation systems. Clear evidence of the success of this strategy is the fact that it has evolved repeatedly and independently, involving surface proteins of quite distinct function, in a wide range of organisms. Key to the great success of antigenic variation is the encoding of information implicitly in the genome—novel protein or epitope variants are templated as silent genes or gene fragments that require a recombinational event, or sometimes transcriptional activation, to become expressed.

For simple subcellular or unicellular organisms, ranging from viruses to free-living and parasitic single-cell eukaryotes which do not have the sophisticated homeostatic mechanisms of their multicellular counterparts, a major challenge is responding to a rapidly changing and possibly hazardous environment. An extreme example is living in a mammalian host, where a rapid and intensely antagonistic immune response arises. Evasion of immunity within a single infection or in the broader scope of the host population, some members of which might already have immunological memory or residual antibodies from previous infections, necessitates a strategy by the pathogen that coordinates a number of factors. These include, for example, the manner and rate of its reproduction, maintenance of its fitness in the face of potential competitors, how it is transmitted to new hosts, movement between niches it occupies in its host, and the complexity of its genome. For environmental changes that are perhaps too rapid to allow the microorganism to react through detection and response, such as an exponentially growing immune response, a successful strategy is to have diversity already present in the population, ensuring survival and proliferation. The strategy is based on stochastic, localized hypermutation (or in some systems spontaneous transcriptional switches) acting on a specific subset of genes or gene fragments that encode a phenotype directly related to the environmental challenge, with the outcome of a change of phenotype preemptive of environmental change, rather than responsive to it. Loci that are subject to these events have been termed "contingency" (see chapter 3).[1] Contingency systems are selected because the associated environmental changes recur routinely, such as each time the pathogen infects a new host.

Contingency systems differ from bacterial constitutive mutator systems, where inactivation of DNA mismatch repair confers on the population the ability to present mutants, arising through general rather than localized mutagenesis, to an environment varying in an unpredicted way. Because all their genes, including those essential for survival, are subject to mutation, constitutive and "inducible" mutators suffer a considerable loss of fitness and perform best in the shorter term, where survival of the population is pressing, as described in chapters 3 and 6. Contingency mechanisms are more evolved because they target only the genes whose products are directly related to the environmental demand. Classic contingency systems include phase variation in bacteria, where there is spontaneous toggling between alternative phenotypes, such as presence or absence of the surface capsule, as discussed in more detail in chapter 3. Although these different strategies have distinct adaptive and evolutionary roles, they can sometimes operate together, with a contingency mechanism effecting onset of a constitutive mutator phenotype, as in the case of bacterial mismatch repair genes containing unstable repeat sequences that confer a natural and spontaneous high mutability (see chapter 6).

For eukaryotic pathogens, there has unfortunately been little study of such adaptive evolutionary mechanisms, although there is evidence emerging for the existence of natural constitutive mismatch repair mutator strains of the South American parasite *Trypanosoma cruzi*.[2] Nevertheless, antigenic variation systems are becoming well understood, and the following account focuses on the best-known system, in African trypanosomes, which provides dramatic insight into the workings of what can be termed "eukaryotic contingency systems," and the importance for survival of implicit encoding in the genome.

Antigenic Variation in African Trypanosomes

The *Trypanosoma brucei* antigenic variation system is very extensive, displays typical contingency features, and appears to have evolved beyond similar systems in not only bacteria, but also other simple eukaryotes. It is a clear example of how a daunting informatics problem—how to develop, store, and access selectively the required, very extensive coding information—appears to have been overcome through implicit encoding. Furthermore, the existence of much of the information in partial form may well allow the trypanosome to express it in a structured way during infection, while also providing a secondary benefit, a strategy enabling transmission through a partially immune host population. The following account concentrates on these themes in trypanosomes, with reference as appropriate to other organisms.

Antigenic variation is one form of contingency system. Rather than involving a simple toggle between alternative phenotypes, as in phase variation, it is a constant forward switch for the population of organisms growing within an infected individual host, avoiding detection by the cumulative output of immune responses against the invaders' ever-changing guises. It has been described in a variety of microorganisms, from viruses through bacteria to simple eukaryotes.[3,4] It concerns different types of surface molecule, which in some cases appear to play merely a passive role in immune evasion but in others also have distinct biochemical functions. It is used in different pathogen lifestyles, including intra- and extracellular. These differences at the level of phenotypes nevertheless are outnumbered by the wide range of genetic and molecular mechanisms driving antigenic variation, reflecting the fact that an organism will fix any genetic mechanism that is successful. Attempts to categorize antigenic variation systems, in particular based on genetic mechanisms, generally have been frustrated by their great diversity, as discussed by Borst.[5] Nevertheless, similarities are significant and apparent.

The antigenic variation system of African trypanosomes[6] can be considered to comprise two substrategies, one involved directly in evading immunity during single infections in mammals, and the other probably facilitating transmission through a host population that already has a degree of herd immunity. Trypanosomes meet the mammalian humoral immune system head on. They live extracellularly in the bloodstream and evoke a rapid response of lytic IgM antibodies. Both induction and execution of this response are highly efficient because the trypanosome expresses a protein coat—composed of variant surface glycoprotein (VSG)[7]—that is highly immunogenic, making an excellent inducer and target for specific immunity. The coat is a smokescreen. The only "functions" directly related to the VSG structure are that it forms a properly packed surface coat and it presents epitopes to the host. It is thought also to have a critical passive role, inhibiting innate immune mechanisms and shielding other, necessarily invariant, surface molecules from antibodies. The protein has a hypervariable N-terminal domain of about 450 residues and a more conserved C-terminal domain adjacent to the plasma membrane and hidden from the host.[7] Clonally, as the population expands in the host, trypanosomes switch to distinct VSG coats at a rate of about 10^{-2} switches per cell per population doubling.[8] The new VSG retains the packing and protective properties, while presenting epitopes unrecognized by antibodies against other VSGs. The

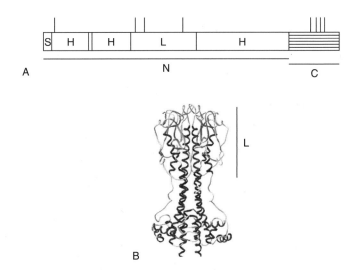

Figure 5.1.   The variant surface glycoprotein (VSG) of T. brucei. (A) The primary structure, with the variable amino (N) and more conserved carboxy (C) terminal domains indicated. S, signal peptide; H, helix; L, loops. Vertical lines denote cysteine residues. (B) The structure of the variable domain homodimer.[66] The first two helical regions of the primary structure form the backbone of each monomer. The loops, shown lighter than helices and denoted "L", decorate the end of the molecule exposed to the host; the "sides" are concealed by dense packing of VSGs. The exposed epitopes are thought to lie in these loops.

variable domain of each monomer of the VSG homodimer has a backbone of two antiparallel α-helices and is decorated with loops exposed to the host (figure 5.1). The loops are thought to contain the variable epitopes, an inference that has not been tested thoroughly due to their physical instability in vitro, which possibly means that they have a conformational nature.[9,10] Thus, each switch effectively retains what is important and changes only what needs to be changed—the epitopes. Although epitopes might be expected to vary more than the rest of the domain, this is not the case, possibly because the sequences encoding the whole domain operate as discrete units in the DNA rearrangements that drive antigenic variation. The wide variation is possible also because many sequence combinations can form α-helical secondary structures.

Population Phenotypes: Ordered Expression in the Mammal

Rather than proceeding through a fixed sequence of single variants, trypanosome antigenic variation presents several variants in each peak of growth. Although such ongoing diversification is important for maximizing evasion possibilities, it must be balanced against uncontrolled expression of the full array of variants early in infection. For example, in a typical host, such as a cow infected with *T. vivax*, the first major wave of parasitemia can include a total of more than 10^{12} parasites,[11] which in one division would yield 10^{10} switched trypanosomes. Random activation from the pool of genes available theoretically might lead to expression of an enormous set of variants in this peak, compromising survival of the host, and consequently transmission of the parasite. Presumably there is selective pressure for a balanced manifestation of diversity, which could be achieved through a degree of order in expression of variants. This is what happens. In a single infection, at least 101 variants, and probably many more, can be produced. In independent hosts, some variants routinely arise early in infection, some arise only late, and there are many with intermediate timings. This progression is specified to a great extent, and in several ways, by the nature of the implicit information—in particular the location and sequence homologies of VSG-encoding genes—and how it is accessed.

Population Phenotypes: Predictability in the Transmission Stage

When the parasite is ingested by a tsetse fly, it dispenses with the VSG coat, regaining it in the salivary glands as the mammal-infective metacyclic parasite stage develops. This population is diverse in VSGs, expressing about 20 different metacyclic VSG (MVSG) coats. Generally, one would expect the infective, transmission stage of a pathogen to be of unpredictable phenotype, to enhance the chances of successful transmission to a host of unknown susceptibility. Surprisingly, however, the set of MVSGs is highly predictable, regardless of which VSG was being expressed by the trypanosomes that originally infected the fly. It has been proposed that this predictability is a trade-off against the diversity in the metacyclic population.[12] Taking the view that antigenic variation coevolved with hosts that are now able to control infections, the challenge for the metacyclic population is to infect such hosts as their antibody levels diminish. The idea has been postulated that the trypanosome has solved this dilemma by evolving a *VSG* gene activation system for this stage that operates via a dedicated subset of *VSGs*, the *MVSGs*, guaranteeing diversity at the expense of predictability (nevertheless, the *MVSG* set evolves, preventing complete predictability). This proposed use of antigenic variation to promote transmission through a partially immune host population is similar to what has been proposed for antigenic variation systems in viruses.[13] Even for RNA viruses, which genetically are more unstable than DNA viruses, the combination of point mutation, recombination arising from copy choice switching during reverse transcription, and the reassortment of genome fragments in multiply infected host cells produces an antigenic variation rate insufficient for evasion of the immune response in the same host.[13,14] Trypanosomes may have developed such a transmission tactic as a refinement of its main antigenic variation system. Furthermore, the existence of most of the information for bloodstream antigenic variation in incomplete form might also contribute to this strategy, as discussed below.

The Genetic Basis of Antigenic Variation

The Silent Gene Archive

Although much is known about the organization and expression of the implicit information for VSG antigenic variation, our knowledge of how switching occurs is still fairly rudimentary. There is an archive of silent *VSG* genes that is so large as apparently to have necessitated specific genome adaptations. The full repertoire of *VSG*s occupies four types of locus, all of which are subtelomeric; primarily two types comprise the silent archive, while the other two contain *VSG* transcription sites (figure 5.2). In *T. brucei*, there are 11 main diploid chromosomes (known as "megabase" chromosomes), some smaller derivatives of uncertain ploidy, and an estimated 100 minichromosomes per nucleus.[15] The largest set of *VSG*s has 700 to 850 genes or pseudogenes, tandemly arrayed at subtelomeres of some of the megabase chromosomes ("arrays") (figure 5.2). Only one of each chromosome pair has been sequenced, but in the single case where sequence is available for both homologues, the allelic arrays have different sets of *VSG*s. As the genome is diploid, it is possible that there are more than 1500 distinct genes or pseudogenes in this set. The second type of locus comprises *VSG*s that are the last, and usually only, genes on minichromosomes, just before the telomere (figure 5.2). VSG-based antigenic variation exists also in *T. congolense* and *T. vivax*, which are, respectively, more primitive than *T. brucei*. As *T. vivax* has generally only possibly one or two minichromosomes[16] (and collateral evolution notwithstanding), one interpretation is that minichromosomes originally were not a component of the antigenic variation system but evolved as an adaptation to it. The adaptation must be important as it is associated with a significant S-phase asynchrony and with modification of the chromosome segregation machinery.[17,18] These small chromosomes are present also in the tsetse midgut stage of the trypanosome and probably are permanent through the life cycle, in contrast to the developmentally regulated generation of functional genome fragments in the macronucleus of ciliates (see chapter 13). The two remaining locus types, at subtelomeres of megabase chromosomes, may account for up to approximately 50 *VSG*s and include the transcription sites for metacyclic and bloodstream *VSG*s (figure 5.2).

Genomic environment, at both gross chromosomal and local levels, is key to the expression of individual *VSG*s and to the strategic use of the archive. Most *VSG*s are flanked upstream by one or more units of a sequence known as the 70-bp repeat (figure 5.2)[19] and have homology also around the 3′ end of their coding sequences.[20] Thus, in general, the sequences encoding *VSG* variable domains are embedded in cassettes, all of which share the 5′ end sequence (70-bp repeats) and the 3′ end sequence. The gene arrays consist mostly of cassettes arranged head-to-tail, with the occasional inversion, usually associated with presence of sequence of the retrotransposon *ingi* (figure 5.3). In these cassettes, the norm is just one 70-bp repeat unit; there are more for minichromosomal cassettes and expression sites have just a few (*MVSG*s) or enormous runs many kilobases long (bloodstream transcription units). The other subtelomeric genes have longer homology stretches in their 3′ environment, containing also shared subtelomeric elements. The *VSG* cassettes in the arrays resemble the dispersed low copy repeats in the human genome (see chapter 8) as, following presumably initial tandem duplication events, they appear to have expanded through segmental duplication and retrotransposon sequence-mediated homologous recombination, leading to *VSG* subfamilies being dispersed widely over different arrays.

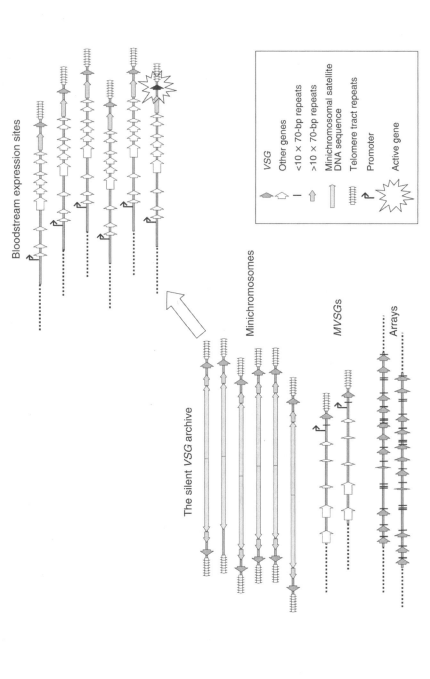

Figure 5.2. *VSG loci in* T. brucei. Only a few examples of each VSG type are shown. The large open arrow depicts the general flow of genetic information associated with expressing implicit information: >90% of events.

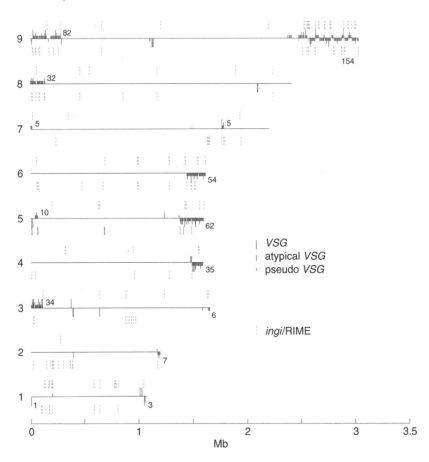

Figure 5.3.   The silent array of VSGs in T. brucei. Vertical lines show the start codon of each (pseudo)gene, with different heights distinguishing pseudogenes, those encoding what might be VSGs missing one or more element for formation of the mature protein, and intact genes. Large numerals denote chromosome number and smaller numerals the number of (pseudo)genes in each array. Dotted lines depict the ends of intact, and fragments of, the retrotransposon *ingi*. Large sections of subtelomeres, including expression sites for *VSGs*, have not been sequenced or assembled and are not included.

With such a large archive, controlled and relatively ordered expression presents a considerable problem. The foremost way each trypanosome achieves expression of just one *VSG* at a time in the bloodstream is that transcription proceeds only from specialized bloodstream expression sites, a collection of 20 or more sub-telomeric loci (figure 5. 2).[21,22] Expression sites have an RNA polymerase I promoter,[23] a set of genes that do not encode VSGs and so far appear not to be associated with VSG antigenic variation, the long run of 70-bp repeats, and then the *VSG*. To be expressed, *VSGs* must gain residence in an expression site. This appears to be a neat solution to one of the problems of controlled use of the archive, but a complication is that there are perhaps 20 or more expression sites in the genome, so there is another layer of control that ensures only one is active in the bloodstream. Transcriptional switching, in which the active site becomes silenced and an inactive one arouses, does occur in bloodstream infection, although in fewer than 10% of switches.[24] There have been a number of theories about this control, which has been likened to allelic exclusion in other systems.[25] (See chapter 16 for additional discussion of allelic exclusion.) The active site is associated with the expression site body, a specialized structure observed by labeling studies at the light-microscopy level,[26] and there appears to be cooperation between the two expression sites involved in the process of transcriptional switching.[27] There are two theories for multiplicity of bloodstream expression sites: use of alternative sites brings into play different host-range phenotypes encoded in the different alleles of some expression site associated genes[28]; and the pool of silent sites is important for recoding much of the implicit *VSG* information as a prelude to a VSG switch,[29] as discussed further below. *MVSGs* are in RNA polymerase I monocistrons,[30,31] an unusual arrangement in kinetoplastids, where apparently all other protein-coding genes are within polycistronic transcription units. The *MVSG* expression sites appear to have been derived by degeneration of bloodstream expression sites,[32,33] consistent with the theory that the need for the *MVSG* subsystem arose as a consequence of immunity against the bloodstream subsystem.

Decoding the Implicit to State Explicitly

With these components ensuring exclusive expression in the bloodstream, the task for the trypanosome is to access the silent archive with a degree of selectivity. Our knowledge of how this is achieved is fairly scant. Gene conversion lodges an extra copy of a silent *VSG* in the expression site, eliminating the *VSG* already there. The silent copy is retained unmodified in its locus, to be inherited and used in future infections, while the temporary expressed copy will be eliminated at a subsequent VSG switch. The duplication (conversion) process normally involves the cassette, stretching from the 70-bp repeats to either within the region of homology extending from the C-terminal coding region to immediately beyond the open reading frame or, in the case of minichromosomal genes, to the end of the chromosome, possibly including also the telomere repeats.[34]

The gene conversions may not require much more than the parasite's standard homologous recombination pathways. For trypanosomes that switch at background rates (laboratory-adapted lines), standard RAD51 homologous recombination is used,[35] but the same remains to be tested thoroughly in rapidly switching trypanosomes. It has been hypothesized that high-rate switching is achieved through use of a dedicated endonuclease, which cuts the expression site,[36,37] initiating the routine creation of a gap by the homologous recombination system, followed by repair synthesis from a silent VSG. There is no evidence yet for an endonuclease, and other mechanisms are possible.[6]

Underlying the ordered expression of *VSGs* is the relative frequency with which each is activated. Those with the highest activation probability will be copied throughout infection, but once antibodies have arisen, any individual unfortunate enough to activate such a gene will be eliminated immediately, leaving other switchers to survive. This effect is cumulative as infection proceeds so that eventually even the most inefficiently activated gene has its turn. One main determinant of activation probability is the frequency with which different silent locus types engage in the switch process. The most frequently activated genes are those duplicated from subtelomeres,[24,38] while later in infection, array genes become duplicated.[24,39] Only rarely is a transcriptional switch observed. The high frequency associated with subtelomeric donor genes possibly arises from a tendency of subtelomeres to associate with each other,[40] creating the conditions for gene conversion from a donor into an expression site subtelomere. The probability for activation of intact array genes in general is lower, presumably depending instead on the standard homology search machinery of homologous recombination. Within the subtelomeric gene and the array gene groups, there is further imposition of order. It is unknown how this occurs for the subtelomeric genes; perhaps different degrees of homology between their flanks and what is occupying the expression site at that time influence recombination frequency. For the array genes, there appears to be a remarkable tool for spreading timing of activation—incomplete *VSG* genes (figure 5.3). Of the first 874 *VSGs* in the array archive fully annotated, only 35 (4%) are intact and predicted with certainty to encode fully functional VSGs. Some 79 (9%) are intact genes, but apparently encode "atypical" (many probably defective) VSGs, while 760 (87%) are pseudogenes.

Pseudogenes: An Implicit Resource

What is at first sight a disastrous situation for the trypanosome may in fact be a device for generating even greater diversity. After the phase where subtelomeric gene duplications are routinely detected, trypanosomes express mosaic *VSGs* assembled typically from three or possibly even more incomplete genes (figure 5.4).[41–43] With the recent discovery that most silent *VSGs* are flawed, what has been considered to be a rare, unremarkable event in antigenic variation might in fact be one of its fundamental features—if it can be shown that these sequences are actually used to any great extent. There are a number of important ramifications, impacting on the effective extent of antigenic variation, the timing of activation, the ordered activation of the archive, the existence of so many expression sites, and perhaps even transmission strategy.

Formation of functional mosaic contingency genes is not restricted to trypanosomes. In fact, it is quite common (figure 5.4). The bacterium *Neisseria gonorrhoea* (which causes gonorrhea) displays antigenic variation of its pilin expressed from the *pilE* locus.[44,45] Archival information is encoded by, typically, nearly 20 silent *pilS* genes at other loci,[46,47] which donate patches to novel mosaic alleles in the *pilE*. Two or even three silent genes can contribute to individual mosaics.[47] *Borrelia burgdorferi* (Lyme disease) typically has 15 silent *vls* partial genes, which donate patches to the functional, expressed copy, so leading to variation in the bacterium's surface coat.[48,49] Although *B. hermsii* tends to use complete gene replacement in its *vmp*-encoded surface coat antigenic variation system, some of its archive genes are incomplete and contribute to mosaic gene formation in the

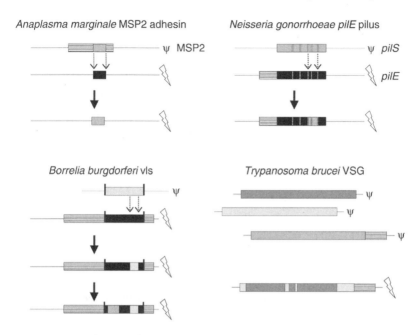

Figure 5.4. Mosaic gene formation in antigenic variation. The flow of information from pseudogenes to transcriptionally active genes. (Pseudo)genes are shown as boxes and flanks as horizontal lines. Sequence encoding variable regions of proteins are shown as gray or, in the initially expressed gene, as black. Conserved regions are horizontally hatched. Pseudogenes, usually truncated but in some cases frameshifted, have the symbol Ψ, while active genes have the lightning symbol. Dashed arrows depict what are thought to be gene conversion events and large arrows show switch events (the steps in the trypanosome switch cannot be elucidated). In the case of *B. burgdorferi*, direct repeats flanking the variable region are shown as vertical lines. Specific functions of proteins other than being involved in antigenic variation are stated, except where none has been demonstrated. (Data or schemes on which these diagrams are based are from references 42, 44, 49, and 67.) In all these systems, a case has been made that gene conversion mediated by homologous recombination, rather than mediated by specific "hotspot" sequences, copies information from the silent to the active gene, preserving the silent copy intact for use in future infections. However, there is little direct evidence to support these hypotheses, and in some cases alternatives have been proposed.

expression site.[49,50] The MSP2 adhesin on *Anaplasma marginale* (severe bovine anaplasmosis) undergoes antigenic variation, from an archive of 5 to 20 silent, incomplete genes, which donate partial information to the gene in the active expression site.[51,52]

It is not just bacteria where mosaic formation has been detected. The fungal pathogen *Pneumocystis carinii* (pneumocystis pneumonia) has a family of MSP (major surface protein)-related MSR genes that are incomplete, raising the possibility that they contribute to mosaic gene formation.[53,54] For protozoan parasites, evidence is emerging that *Babesia bovis* (redwater fever) *ves* genes undergo partial conversions associated with switching.[55,56] Some parasites require intact contingency genes because they switch amongst them transcriptionally rather than via gene conversions.[25] Nevertheless, genome sequencing and further experimentation, such as analyzing gene sequences expressed in antigen variants, might reveal short mosaic patches in these and other parasites.

Theoretically, mosaicism can yield enormous numbers of variants because there is a combinatorial use of the information. For example, it has been suggested that, based on the observation in one strain of four segmental conversions per gene and nine silent alleles, *Anaplasma* potentially could generate $10^{3.8}$ variants.[52] A similar calculation for *Neisseria* suggests more than 10^7 possible variants.[46] If the same calculation is applied to trypanosomes, based on just three events per expression site allele and the known near half of the archive, then a staggering total of more than 10^8 is predicted—which could be an order of magnitude greater over the whole genome. However, this calculation is not very meaningful as it is based, wrongly, on modular recombination between different sequences (and basis on each variable amino acid gives an impossibly astronomical prediction)[48] and assumes that any of the pseudogenes is equally capable of interaction with all other array sequences, which is not the case. Additionally, there must be significant constraints on what combinations actually produce functional proteins or genuinely distinct epitopes. Despite these caveats, the possibilities for variation probably far exceed what is required.

Why do pseudogenes exist and why are there so many? There are a number of possible reasons. First, they possibly add significantly to the scale of antigenic variation through combinatorial expansion of the archive while economizing on genome size. Second, they can help impose ordered expression, by being expressed later in infection due to a requirement in many cases for more than one conversion event. Third, their influence on order can be much more significant. Whereas the cassette system used by intact silent genes relies on gene flanks, mosaic formation has to involve sequence homologies in the variable domain-encoding region. Related genes can therefore become involved in a pathway of sequential conversions, although with the trade-off that many attempted switches will fail to provide complete epitope replacement. Fourth, a further benefit, which is of importance for transmission in the field, is that different pathways could be produced in different hosts. Upon invasion of a previously infected host, it can be imagined that the trypanosome population will encounter existing antibodies against the variants already expressed in that host, bringing about selection for an alternative series of VSGs that have arisen as a consequence of a stochastically different initial set of recombinations between pseudo*VSG*s. This can be tested experimentally by combined experimental and bioinformatics analysis. At first glance, a recently revealed trypanosome minor recombination pathway that operates via microhomology[57] appears to be perfect for creation of mosaic VSGs; indeed, *VSG* recombination patches possibly as short as 14 bp have been observed in mosaic genes[43] and similar, but distinct combinations can occur in independent infections.[43]

Generation of mosaic genes may require the existence of many expression sites. Thus, any trypanosome activating a *VSG* that is defective, or that has replaced the important epitopes only partially, will be eliminated by antibodies. At the same time, assembly of an intact coding sequence from pieces must involve trial and error, with many abortive products. The more silent templates involved in the assembly process, the longer it will take for the independent, chance recombinations to occur. For success, it would be better to undertake this tinkering in silent expression sites. A key difference from doing so in the active site is that coding error in the active site will lead rapidly to death, whereas the silent expression site fostering the assembly will be propagated at division, allowing more time for further conversion events. Occasional, subsequent activations of that site, or coding sequence conversions to an active site, could perform a functional trial of the mosaic gene.

The Transmission Stage

The implicitness in the *MVSG* subset is less complicated than that of the bloodstream *VSG*s, and it is expressed simply by transcriptional activation. Each *MVSG* has its own promoter, and there is a random activation from the pool as the metacyclic stage differentiates.[58] The *MVSG* archive, although somewhat fixed, is not impervious to change, and there is a slow turnover of the gene set. It is thought that there is occasional gene conversion by *VSG*s from other subtelomeres, at a rate up to 0.03 of the archive per fly per transmission, preventing the onset of herd immunity in the field.[59]

Shadow Boxing: Immune Response versus Antigenic Variation

It is ironic that the contenders, the acquired immune response and antigenic variation, bring shared logic to the conflict, which is a typical evolutionary arms race. Both use implicit information, in the form of incomplete coding sequence, as a means of condensing enormous information potential within the confines of a genome. Both use DNA rearrangements to generate diversity, with general recombination machinery playing a major role; whether the trypanosome has a specific initiating enzyme, equivalent to the RAGs of Ig splicing, is unknown. As described in chapter 9, immunoglobulins (Ig) achieve diversity in part by combinatorial (combinations of spliced gene segments), combinational (heterodimerization), and junctional (segment joining) assembly. The VSG also achieves diversity through combinatorial and, possibly, junctional assembly, although as a homodimer it does not use combinational diversity. One difference is that somatic point hypermutation is important for Ig diversity (chapter 10), but that cannot serve trypanosomes, where complete epitope change is necessary.[60] Both systems face the problem of imprecision in their recombination, leading to frameshifted failures and to cell wastage. There is a hierarchical effect in both systems, arising from different recombination signal efficiencies for Ig genes (chapter 9) and different homology levels between *VSG* genes. As the heavyweight in this encounter, due to its having to deal with much more than the trypanosome, the immune response utilizes more mechanisms and generates much more diversity, but it is intriguing to see the similarities. Depending on the particular host-infection combination, either contender is capable of winning.

Other Eukaryotic Antigenic Variation Systems

A question that arises from time to time is whether or not antigenic variation, and contingency systems in general, are programmed. This depends very much on how the term "program" is defined: whether it means a list of items or a definite plan of events. As the term is ambiguous, it is probably better not applied in the context of antigenic variation. Although some contingency systems are termed as programmed because the silent components are present in the genome and just need to be read out,[5] clearly the execution of these processes does not resemble classical, specific programs of gene expression in which a specific signal triggers a predictable response, such as in animal development.[1] Instead, the sequence of events has a strong stochastic element, which is a fundamental feature. As the most important point is to convey clearly the features of importance, "templated" might be more appropriate as a description of those systems that contain the silent information in archival form.

For other parasites, it is known that antigenic variation employs implicit coding, but there are some main differences from the trypanosome system. This is best illustrated in the malaria parasite, *Plasmodium falciparum*.[61] Like many other parasites, it lives inside host cells and achieves some interactions with the environment by synthesizing proteins it inserts in the host cell plasma membrane. A number of these proteins undergo antigenic variation. The best understood, PfEMP1, has a fairly conserved role in binding to various host cell surface molecules, resulting in an intramolecular conflict between conservation of binding capacity and diversification of epitopes. This is resolved, at least in part, through different regions of the molecule being assigned these different roles, similar in principle to what happens in a variety of other systems, including gp120 of HIV-1, on which antigenically variable loops shield invariant host cell receptor binding regions,[62] and the Igs, as discussed in chapters 9 and 10. Intracellular life changes the dynamic of the interaction between the parasite and the immune response, as parasites leave cells and reinvade others routinely, via different developmental stages. Thus, there are breaks in the continuity of interaction between infected cells and the immune response, perhaps placing a lower demand on the parasite's system. Also, chronicity of infection with intracellular parasites can be due to the parasite persisting "hidden" in host cells, again removing the need, such as in trypanosomes, for continual production of new variants. Certainly, *P. falciparum* has a PfEMP1-encoding gene (*var*) archive considerably smaller than the *VSG* one in trypanosomes, and all 60 or so genes are intact. Contingency systems classically are seen as using mutagenesis of one type or another for stochastic activation, but *var* genes achieve expression through transcriptional control, which is possible because they are intact.[63] *var* genes share with *VSGs* the feature of allelic exclusion, and appear to achieve exclusive transcription through use of a specific nuclear location and chromatin remodeling.[64,65] Taking a general view, it might be concluded that stochastic, differential transcriptional activation can be added to the types of trigger important in contingency systems.

Roles of Subtelomeres in Antigenic Variation

Most, but not all, of the silent archives associated with antigenic variation are subtelomeric, including those bacterial cases that use linear plasmids. One main explanation proposed for this localization is that differential activation of archive genes exploits the unusual stochastic silencing control, known as the telomere position effect, that sometimes pertains just upstream of telomeres. The second main explanation is that the tendency of subtelomeres to engage in ectopic recombination, whereby subtelomeres of different chromosomes recombine with each other, is exploited to maximize variability in the archive gene family. There is evidence that both pertain.[4,64,65]

Functionally Important Features of Antigenic Variation

One of the main weapons for the pathogen is that it is a proliferating population. Preemptive onset of diversity in the population occurs in all the systems, and the presence of many variants simultaneously in the host decreases the chance of antibodies terminating the infection, whether as infection proceeds or on transmission to a new, partially immune, host. Proliferation is essential to repopulate the host following each antibody-mediated decrease in the parasite population. It may also help activate the information retrieval mechanism, for example through the coordination of DNA recombination and repair pathways with the cell cycle.

Spontaneity of the switching process, whether recombinational or transcriptional, ensures that variation is preemptive of the selecting agent. The use of recombinational or transcriptional activation systems may be determined by the nature of the archive: where pseudogenes are involved, recombination is essential, otherwise transcriptional activation appears to be able to provide what is required, as long as it is stochastic.

For many, but not all, antigenic variation systems, an information archive is core: it not only provides considerable scope for variation, but also in some cases helps determine an ordered readout that probably optimizes use of the archive. Why do not all systems use an archive? Genome size may be key. Archival systems can achieve full epitope replacement in one or just a few steps where mosaic gene information is involved. Having a genome large enough to accommodate a large archive is easy for eukaryotes, and it is apparently straightforward to increase the genome size as necessary, as seen with the trypanosome. For bacteria, whose genomes normally are circular, this can be more of a problem. Some, such as *Borrelia* species, possibly have solved the problem by creating linear plasmids that appear to hold the implicit and expressed antigenic variation genes.[49] Pseudogene information, besides helping to impose some order on expression, as in the trypanosome, may act to economize on the use of the genome. The use of archives to economize in this way, rather than to play a role in ordered expression, is more likely in bacteria, which tend to activate pseudogenes by duplicating from common, short direct flanking repeats, rather than via the coding sequence itself.

For viruses, the small genome size precludes archives that are large enough to be effective—while there are many variants in a viral population, there is only one structural antigenic variation gene in each viron's genome and it is active.

Even with very high point-mutation rates and other rapid means of diversifying genetically, such as in influenza virus, full epitope replacement cannot be achieved during the single infection. Antigenic variation therefore has a different basic role in viruses. Epitopes are altered, but with the outcome of enhancing transmission in a partially immune host population. The very high mutation rate is indiscriminate, so there is enormous attrition of the population, but the huge rates of proliferation compensate. Viruses do not have the luxury of archives of silent information in their antigenic variation systems, but higher organisms do, enabling them to tackle infection head-on. Genome size is thus a major determinant of how the strategy of antigenic variation is used to the benefit of the pathogen. In the very highly evolved trypanosome system, it appears that the viral strategy of displaying diversity to drive through the host population has been rediscovered, in the *MVSG* subsystem and perhaps also in the potentially enormous mosaic gene system. Whether this applies for bacteria is unknown.

Conclusions

The survival and success of parasites and other pathogens is linked closely with the great extent to which they have encoded phenotypic potential in their nuclear genome. Besides the simple ability to switch antigens preemptively, greater levels of complexity can be achieved due to selection imposed in classical arms races and their capacity to develop and use implicit information. Parasitism also involves the surrogate use of host genes, such as for the provision of metabolic intermediates, which is well illustrated by the reduced complexity of some pathogen genomes. The powerful counterselections between pathogens and hosts will continue to drive toward economy of use of information, with implicit encoding providing a significant resource capable of delivering instant solutions to pressing problems.

Acknowledgments   I thank the Wellcome Trust for funding, including a Principal Research Fellowship. I am grateful to Mark Carrington for providing the structure image for figure 5.1. I also thank colleagues for discussions, and hope that the omission of many references will be excused as being due to space limitation.

6

The Role of Repeat Sequences in Bacterial Genetic Adaptation to Stress

Eduardo P. C. Rocha

Overview

Because stress is a major component of the life history of living species, many mechanisms have been created to deal with it. Even though stress is often unpredictable, it may be overcome by mechanisms allowing the generation of genetic variability upon which natural selection can act. As a result, recent work has been aimed at identifying the elements and the mechanisms allowing the generation of such genetic variability. Among these, repeated sequences have a prominent role. Bacteria can evolve very quickly through intrachromosomal recombination between repeats, which increases significantly in times of stress. In this chapter, the instabilities produced by repeats in genes of stress response and what results from it are described. While some of these instabilities lead to deletion of genetic material, the loss of this information is not irreversible as mechanisms have evolved that facilitate its recovery. The resulting genetic variability allows the fast adaptation of bacterial populations to stresses, such as host defenses, or to toxic agents, such as antibiotics.

Introduction

We describe selection as acting on the polymorphisms existent in natural populations to select the fittest elements, yet fitness is intrinsic to a specific set of environmental, physiological, and ecological parameters.[1] As a consequence, a mutation selected by the fitness it contributes in one environment may be counterselected because it results in decreasing fitness under other circumstances.[2] Many examples of such antagonistic pleiotropy in bacteria involve a conflict between adaptation to fast growth and survival in stressed or starved populations.[3] For example, the lag phase, usually necessary to re-adapt bacteria to new growth conditions, tends to shorten when bacteria are systematically grown in rich medium in semibatch cultures.[4] In this case, the lag phase disappears because the advantages it confers, adaptation to new conditions, have no fitness contribution in the experimental setup, whereas faster start of exponential growth is highly adaptive. Similarly, cells evolving under a defined set of nutrients tend to lose the capacity to use other nutrients that they do not encounter under these conditions.[5–7]

Stress often is faced in nature, whether from lack of nutrients, arrival of competitors or predators, or the presence of toxic substances. The survival of stressed organisms can be facilitated by the induction of a stress response that is specific to the individual stressor (or the class of stress encountered), or by a more general stress response. Stress can be frequent or rare. If a certain stress is faced much too infrequently, the functions specifically targeting such response will lack a selection pressure for maintenance over the long term. Hence, they are likely to be lost by drift before the recurrence of the stress. As such, dedicated stress responses are stably kept only if the stresses they tackle are sufficiently frequent.

Some stresses can be counteracted by the induction of a well-defined set of functions. This may be called deterministic stress response. For example, heat causes lethality mostly by decreasing the stability of the tertiary structures of macromolecules. Because of this, the heat-shock response is partly composed of a significant increase in chaperones that help to refold unfolded proteins.[8] Stalled replication forks occur in 18% of *Escherichia coli* replicating cells.[9] This stress can be managed by a set of responses involving the recombination and replication machinery.[10] In these two cases, the stress response includes induction of the expression of a set of genes coding for macromolecules involved in targeting the specific stress in the least disruptive possible way.

When stress is dealt with in a predictable way with a well-defined biochemical system, for example chaperones for heat shock, we can consider it a deterministic response to stress. It must be noted that there is some abuse in this nomenclature as molecular and evolutionary events are intrinsically subject to some unpredictability, in other words they are stochastic.[11,12] However, not all stresses can be tackled in such a deterministic way. A typical example is the response to stress resulting from an arms race between a pathogen and its host, such as between a pathogenic bacterium and a human. Because both contenders are capable of generating variability (see discussions in chapters 3, 5, 9, and 10), which may overcome the other's response, the reaction to such stresses must include a stochastic element allowing the generation of a range of responses, instead of a well-defined one. In such cases, adaptation depends on the development of strategies aimed at generating appropriate variability required for natural selection.[13]

The development of strictly deterministic responses is also much less efficient if that stress occurs rarely or is intrinsically stochastic. Under these circumstances, a mechanism that generates genetic diversity may increase the chances that at least a subset of the population acquires an adaptive mutation and prospers. The ways in which this can be done is the subject of many of the contributions to this book. This chapter focuses on the role of repeats capable of generating variability by intrachromosomal recombination processes in the stress-response genes of bacteria.

Sources of Variability

Genetic diversification in bacteria has three main sources: point mutation (including small insertions and deletions), horizontal transfer, and intrachromosomal recombination. Single point mutations may have profound effects on the phenotype, for example when they inactivate genes, but in most cases they underlie a slow evolutionary process of adaptation by specialization or optimization of a pre-existing gene sequence.[14,15] Horizontal transfer of genetic information between different cells may result in a very fast evolutionary process if the adaptive genes are available in the environment and can be transferred in a functional form to the genome.[16-18] This is particularly efficient when catalyzed by elements aiming at promoting the acquisition of such genes, such as integrons,[19] and has occurred very frequently in evolution (see chapter 7). Intrachromosomal shuffling of genetic information may create chimerical genes, modify regulatory sequences, or generate large duplications or deletions of genetic material.[20] These events take place through different recombination processes, which may or may not involve a specific enzymatic machinery. Intrachromosomal recombination only depends on sequences present in the genome and produces changes at a local level, leaving the remaining genome nearly unaffected (apart rearrangement effects).

Intrachromosomal recombination events between closely spaced repeated elements do not require RecA, the key enzyme of homologous recombination. These recombination processes may occur either by misalignment of the DNA template and daughter strands during replication or single-strand annealing, and historically they have been included under the term "illegitimate recombination."[20,21] In the literature, one also finds the term illegitimate recombination linked with cut and join reactions at nonhomologous sequences mediated by recombinases. Here it shall be used in the narrower sense of RecA-independent recombination taking place between small closely spaced repeats. This type of recombination may create long tandem repeats from smaller ones, thus providing more targets for intrachromosomal homologous recombination in the genome (i.e., large repeats).[22,23] The frequency of homologous recombination, while dependent upon sequence homology and the number of sequence repeats, is largely independent of the distance between the repeats. Hence it can operate between very distant repeats producing large chromosomal rearrangements.[24,25] Inverted repeats in particular may accelerate genome shuffling by producing chromosomal inversions.[26]

Both types of recombination are antagonized by the mismatch repair system (MRS) because it recognizes short mismatches in the heteroduplex DNA recombination intermediates. This is particularly important for homologous recombination and slippage of repeats composed of tandems of small motifs (=3 nt). Hence recombination is free to produce more sequence variability in bacteria lacking MRS.[27-30] Because horizontal gene transfer often relies on homologous recombination, mutators that lack MRS show higher frequency of gene transfer.[31] *MRS*⁻ mutators also have higher rates of point mutations, because they are less capable of repairing DNA mismatches.[32,33] Interestingly, as described below, slippage between closely spaced repeats frequently inactivates MRS genes.[32,34-36]

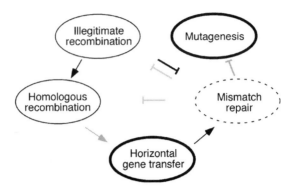

Figure 6.1.   Interaction between different elements involved in the generation of genetic variation. Arrows indicate a positive action, broken arrows indicate antagonistic action, black arrows indicate an interaction involving genetic change (e.g., recovery of MRS genes by horizontal transfer), and gray arrows indicate a physiological response (e.g., MRS antagonism of recombination between divergent sequences). The dashed circles indicate elements that decrease during starvation or stress, and bold circles indicate elements increasing in importance at the onset of such conditions.

The fundamental role in bacterial adaptation of the generation of genetic variation through recombination between repeats is best understood in the context of response to stress or starvation (figure 6.1). Under these conditions, one often observes an acceleration of the generation of genotypic diversity by one or more of the available mechanisms.[37] The frequency of horizontal transfer increases because stress induces lytic cycles of prophage,[38] transposition of insertion sequences,[39] and natural transformation in many bacteria.[40] Transfer of conjugative plasmids increases as *E. coli* cells leave the exponential growth phase.[41] The frequency of replication-independent point mutations also increases as a result of increased mutagenesis, if the bacterium is under stress from toxic agents, and down-regulation, or inactivation, of MRSs at the onset of stationary phase.[42–44] As described below, such inactivation is the result of small deletions, probably through recombination between close repeats. Such mutators have higher mutation rates, more intrachromosomal recombination, and more horizontal gene transfer. Thus, the generation of MRS deficiency puts together an evolutionary fast lane for mutator populations, which contributes diverse types of recombination mechanisms. As can be easily seen from the relations between the elements in figure 6.1, stress leads to a major shift in the equilibrium between these elements, resulting in increased generation of genotypic variability.

Multiple Origins and Roles of Repeats in Genomes

Repeats can recombine in different ways (figure 6.2, table 6.1). Homologous recombination depends on the RecA protein and the existence of two copies of a DNA repeat with a core of identity exceeding 20–30 nt.[45] Such repeats have a

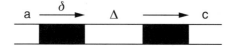

Figure 6.2.   Repeats and spacer. Two repeats of length δ are separated by a spacer of length Δ. The types of recombination events that can take place between them depend on these two parameters (see table 6.1).

probability smaller than 1‰ of occurrence in genomes of the size of bacteria's if their presence were due solely to random assortment of the genetic text.[46] In fact, they are found frequently, usually as part of duplicated genes, such as insertion sequences,[47] rDNA operons,[48] or included in intergenic regulatory sequences.[49] The density of long repeats is very different between genomes and is correlated with lifestyle: more repeats are present in genomes requiring constant generation of genotypic variability or alterations in gene dosage.[23,50]

The frequency of RecA independent recombination between repeats decreases exponentially with the distance between the copies of the repeat, both in *E. coli* and *Bacillus subtilis*,[51,52] and increases with the length of the repeated elements.[53,54] It is also affected by the nature of the repeat sequence. Experimental studies on different bacteria and phage have shown a significant number of recombination events for 8 nt repeats at a distance of up to 987 bp,[55] for 18 nt repeats at a distance of 2313 bp,[52] and for 24 nt repeats at a distance of 1741 bp.[56] Although repeats separated by more than 2 kb usually are refractory to recombination by slippage or single-stranded annealing, an exception has been reported where two copies of a 100 nt repeat recombined at distances up to 7000 bp.[51] This type of recombination results in local genetic diversity and typically local rearrangements. Nevertheless, as indicated above, this can result in the creation of larger repeats, which then can undergo homologous recombination and thus interact across large genomic distances.[22]

Table 6.1.   Repeats can generate genetic diversity by recombination mechanisms, which depend on the nature of repeats, simple repeats of self-replicating elements (such as transposons), and on a set of parameters such as the distance between the repeats and the repeat lengths. Different mechanisms result in different types of chromosomal change.

Element	Δ	δ	Mechanism	Outcome
Close repeat	<~1–2 kb	>~8 nt	Strand misalignment	Conversion, duplication, deletion
Simple sequence repeat	0	>~6 nt	Strand misalignment	Duplication, deletion
Tandem repeat	0	>~6 nt	Strand misalignment	Duplication, deletion
Long repeat	any	>25 nt	Homologous recombination	Conversion, duplication, deletion inversion
Insertion sequences	—	—	Transposition	Rearrangements, gene inactivation

Small repeats, which can increase local variation, usually are divided in different classes (table 6.1). Among these, simple sequence repeats (SSRs), tandem repeats (TRs), and closely spaced repeats (CRs) play important roles in generating diversity at the loci that contain them. SSRs and TRs are tandem repeats of small and large motifs (< 6 nt and = 6 nt, respectively), and making this sharp distinction between them is a slowly fading tradition. When such repeats are located in regulatory sequences, their slippage is associated with the variation of gene expression in bacterial pathogens. In this case, phenotypic variation is obtained by varying the gene or genes being expressed at a given moment (see also chapter 3). [57,58] These elements also may be found in genes, where they produce sequence length variation if the motifs are multiples of 3 nt or frameshifts if they are not. In one study, both SSRs and TRs were found to be, respectively, 50% and 41% more abundant in the genomes of pathogens than in the genomes of nonpathogens.[50] Closely spaced repeats are small (> 8 nt) and spaced by less than 1–2 kb. Contrary to simple changes in length with SSRs or TRs, recombination between two copies of a CR can lead to the deletion or duplication of the sequence between the repeats. In this case, deletion cannot be reverted by intrachromosomal recombination (unless the entire element is repeated in the genome).

CRs exhibit lower recombination frequency than SSRs or TRs because of the larger distance between the repeats. However, they can be far more abundant (for example, in *E. coli* K12, 383 CRs have been identified but only eight SSRs)[50] and so in aggregate they may account for a larger share of the overall slippage potential of genomes. Other types of small repeats have been identified. Cryptic simple repeats[59] are regions of DNA with a very biased composition toward some nucleotides, resembling the vanishing SSRs.[50] Spaced interspersed repeats (SPIDRs, also called CRISPs) are multiple-copy CRs (of ~30 nt), separated by spacers (of ~36 nt) comprised of sequences that are not repeated.[60] SPIDRs exist in many genomes of prokaryotes, and even the small genome of *Mycoplasma gallisepticum* (< 1 Mb) contains one such elements, which occupies nearly 5 kb. In general, there are large numbers of repeats capable of engaging into illegitimate recombination in bacterial genomes, and the precise numbers vary significantly between closely related genomes. For example, the pathogenic *E. coli* O157:H7 contains 45% more CRs and more than twice the number of TRs than the commensal *E. coli* K12.[50]

Variability through Recombination between Closely Spaced Repeats in Stress-response Genes

Small size repeats are overrepresented in stress-response related genes in the genome of *E. coli*, which facilitates the generation of genetic variability. We looked for close repeats longer than 8 nt and separated by less than one kilobase in the genome of *E. coli* K12 and found that the average gene has 7.9 such repeats, which is significantly more than the expected value, given sequence composition (6.5, $P < 0.001$). A particularly high overrepresentation was found among genes implicated in recombination, repair, and regulation.[34] Interestingly, horizontally transferred genes, as defined by codon usage analyses,[16] show lower frequency of such repeats, indicating that high repeatability is a feature of evolutionarily conserved genes in *E. coli*. We then put together a set of 196 genes related to different stress responses, such as the ones implicated in heat shock, MRS, or response to starvation (the complete list is available at http://wwwabi.snv.jussieu.fr/~erocha/).[61] This set contains an average of 10.5 repeats per gene, much more than the average gene ($P < 0.001$), indicating that stress-response genes contain higher than average potential for variation.

The functional class showing the most significant overrepresentation of repeats is the one related to recombination, where eight out of 19 genes showed statistically significant overrepresentation of close repeats.[34] Figuring among these genes are key elements of all known homologous recombination pathways in *E. coli*: *recBCD*, *recFOR*, and *recE* (which is a phage gene). We also find elements involved in resolving Holliday structures and recovery of stalled replication forks, such as *recG*, and exonucleases such as *sbcC*. As a longer gene will have more repeats simply by chance, the measure of significance of the number of repeats in individual genes depends on gene size. Yet the simple number of repeats in these genes is impressive, with many genes showing over 20 repeats. Among other genes involved in stress response that show overrepresentation of repeats are three genes involved in mismatch repair (*mutHSL*, see below). Surprisingly, repeats also are overrepresented in many genes involved in DNA synthesis, notably the replicative polymerase (*dnaE*) and the error-prone polymerases (*dinB*, *polA*, and *polB*) (see chapter 2). Finally, several regulatory genes also over-represent close repeats, notably *rpoS*, the sigma factor involved in the activation of genes in stationary phase. Mutations in many of these genes are known to cause mutator phenotypes, some when partially or totally deleted, others when overexpressed, and others in both cases.[62,63] Table 6.2 and figure 6.3 show a small sample of genes containing a significant number of close repeats ($P < 0.05$).

Recombination between close nontandem repeats may result in deletion of genetic material or in the amplification of genetic regions. Some of the genes highlighted by this analysis were previously shown to engage in such events (see below). For many other genes, the actual extent of genetic instability caused by the repeats, and its effect on protein function, is an area for further research. One should note that in-silico predictions of regions of genetic instability, through the identification of repeats, are seriously hampered by the current knowledge of the factors constraining and stimulating recombination. Several examples have emphasized that accurate predictions of recombination frequency will have to take into account the nature not only of the repeats but also the sequence context in which they reside. For example, in *E. coli* polymerase I mutants, CTGG and GTGG sequences are found to flank sites of frequent spontaneous deletion (see also chapter 11).[64] In addition, deletion between short repeats is facilitated by the presence of palindromic sequences that, by pairing, physically approximate the repeats.[65] The stability of repeats depends on both the composition of repeats and the direction of the replication. For example, in *E. coli* a tandem repeat of CTGCAG is less unstable than a tandem repeat of CGGCCG.[66] Finally, it has been observed that transcription levels also influence recombination rates. Active transcription increases the deletion rate of CTG·CAG tract repeats in *E. coli* plasmids

Table 6.2   A subset of genes implicated in stress responses that significantly over-represent close repeats (CRs) ($P < 0.05$). The mutator column indicates if the gene was found to lead to mutator phenotypes when deleted (D)[62] and/or over-expressed (O).[63] The statistical analysis account for gene length, but the table shows absolute counts. *, essential genes (only small deletions are viable); +, mutator effect not demonstrated but likely.

Gene	CR	Function	Mutator
appY	10	Regulator for growth-dependent gene expression	O
dinB	10	Error-prone polymerase	O
dnaE/N*	26/12	Main DNA polymerase (PolC)	D
gyrA/B*	23/25	DNA gyrase	D
katG	23	Detoxification catalase	—
lon	25	Heat-shock protein	—
mutH/L/S	5/25/22	Mismatch repair system	D (all)/ O (*mutS*)
polA*	30	Error-prone polymerase	D
polB	25	Error-prone polymerase	D
priA	19	Primosomal protein	D+
recB/C/D	32/26/21	Homologous recombination (RecBCD pathway)	D
recF	15	Homologous recombination (RecFOR pathway)	D
recG	22	Resolution of Holliday junctions	D/O
recQ	19	DNA helicase	D+/O
rpoB/C*	47/62	RNA polymerase	D
rpoS	8	Starvation response regulator	O
sbcC	43	Degradation of DNA (exonuclease)	—
sodB	10	Detoxification: superoxide dismutase	D
uvrA/B	18/19	Nucleotide excision repair	D+

(see also chapter 11).[67] Many of these dependencies are not sufficiently well characterized to estimate the extent to which repeats identified in silico only by parameters of length and distance actually do lead to genetic instabilities. It also is unclear if (and which) smaller repeats generate genetic instabilities. In spite of these challenges, many repeats identified in silico have later been confirmed experimentally to produce genetic instability.[68–71]

The *muts* Model Locus

When genetic material between closely spaced repeats is deleted, information formerly encoded between the repeated sequences is permanently deleted from the genome (figure 6.4). As an example, consider the generation of MRS deficient mutator strains, in which we showed previously that close repeats are overrepresented. Natural isolates of pathogenic and commensal *E. coli* show high frequencies of mutators,[32,33] which result from their ability to explore the fitness landscape more rapidly.[72] In other words, a mutator strain has a higher probability of generating a mutation that is adaptive because it generates more mutations, and when it does the mutator phenotype will increase in frequency by hitchhiking with it.

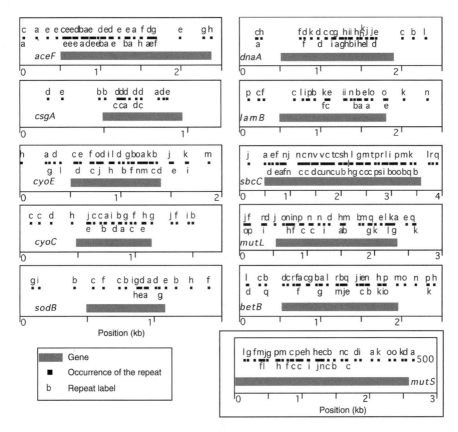

Figure 6.3.   Distribution of close direct repeats in the 10 stress response genes of E. coli *K12 most overrepresenting repeats + muts.* Black boxes represent repeats and gray boxes represent genes. Different occurrences of the same repeat are marked with the same letter. (Reprinted with permission from Oxford University Press.)[34]

However, because a mutator allele also generates a load of deleterious mutations, an adapted mutator strain will have a higher selective death rate than an adapted strain that has lost the mutator phenotype. Therefore, once adaptation to the new environment has occurred, selection for a reduction in the mutation rate will be renewed.[73]

Frequent deletions in *mutS* and *mutL* have been found in clinical and commensal isolates of *E. coli, Salmonella,*[32,33] *Pseudomonas,*[35] *Neisseria,*[74] and *Streptococcus.*[75] In enterobacteria, *mutS* is close to *rpoS*, and this region has been extensively investigated for rearrangements, recombination, and deletions.[36] As would be expected if sequences were prone to gain and loss due to the high frequency of repeats in this region, the regions between *mutS* and *rpoS* vary very significantly in length and gene content between genomes. For example, the two genes are contiguous in *Vibrio cholerae* and *Yersinia enterolitica,* yet separated by more than 12 kb in some *Salmonella* groups.

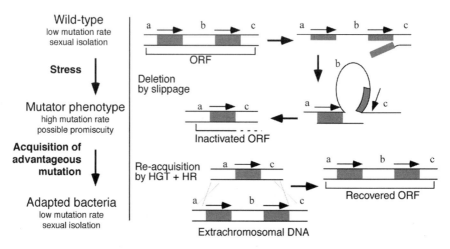

Figure 6.4.   Model system for the transient genotypic inactivation of mutator genes. Mutator genes are constantly being generated in bacterial populations by slippage between repeats, which tends to increase under stress or starvation. Mutators can explore a more substantial range of its fitness landscape, increasing the likelihood of finding an adaptive mutation. However, once this is found, mutators are at a disadvantage relative to nonmutators. Because mutators increase rates of recombination and stress increases horizontal transfer, the wild-type mutator gene may be easily recovered.

Recovering Genes

Loss of MRS Favors Horizontal Transfer

Generally, whenever a gene deletion has the effect of increasing horizontal gene transfer, it can revert by homologous recombination with an intact copy from another individual. This is because the implicated genes are nearly ubiquitous in the bacterial world and because, as indicated above, horizontal transfer is more frequent in MRS defective strains.[76] As a result, the loss of a MRS gene leads to an increase in the probability of a recovery of the same gene by recombination with exogenous DNA. The sequence of genes coding for *mutS* and *mutL* in *E. coli* show evidence for this scenario of frequent gene loss and reacquisition by horizontal transfer.[77,78] Once *mutS* function is reestablished, sexual isolation rises and the overall mutation rate drops.

Uptake Signal Sequences

"Marking" their DNA with uptake signal sequences (USSs) is another strategy utilized by some bacteria to allow these genes to be easily recovered by horizontal transfer.[79] Certain naturally transformable bacteria specifically uptake DNA that contains 9–10 nt long USSs, surrounded by a degenerated sequence motif.[80] There is strong evidence indicating that these sequences facilitate the binding and transport of the exogenous DNA into the cell. USSs are statistically far more abundant in genomes than would be expected.[81] For example, there are nearly 1500 of such 9 nt signals in the *Haemophilus influenzae* genome when only eight were expected. Because the genome has so many USSs, and these are specific to narrow phylogenetic groups, such as species or genera, this allows bacteria to access to incoming DNA of closely related species at a higher frequency than that of unrelated organisms. An interesting prediction for this system is that if a gene has a higher than average number of USSs, it should also have a higher probability of recombining with homologous exogenous DNA.[79] This does not result from a positive association between USSs and recombination, but simply from their role in facilitating DNA uptake. Many stress-related genes show higher than average frequency of USS signals in the genomes of all bacteria for which the USS is known: *Neisseria*, *Haemophilus* and *Pasteurella*.[79] Notably, USSs, which are overrepresented in the genome as a whole, are even further overrepresented in above-mentioned genes *mutS* and *mutL* in the three genomes. This strongly suggests selection for signals allowing the re-acquisition of these genes after loss (figure 6.4).

Repeat-mediated Gene Loss Is Not Reversed in the Absence of Horizontal Transfer

In contrast, if genomes are under sexual isolation, deletions will result in permanent gene inactivation. In this case, the larger the number of repeats in a gene, the more likely its deletion from the genome. An example of this is provided by *Buchnera*, which are sexually isolated endosymbiotic bacteria proliferating in the very protected intracellular environment of aphids.[82] The comparison of two completely sequenced genomes of these bacteria confirmed that genes with a higher frequency of closely spaced repeats were indeed more likely to be partially or totally deleted.[50] As long as these deletions are not strongly counter-selected, or hitchhike with a mutation providing higher fitness, they will become fixed in the population and result in ongoing genome degradation. Other genomes under the process of net gene loss, such as the ones of *Rickettsia* or *Mycobacterium leprae*, also show prevalence for small deletions that lead to gene inactivation.[83] This may explain why these genomes have often lost nonessential stress-response genes: there is lower selection pressure for their maintenance, and their repetitiveness, which may well have been positively selected in an ancestor, now render them particularly prone to deletion.

Phenotypic Variation through Duplications and Deletions

One often associates small intragenic deletions and duplications with loss of function. However, many small deletions and duplications, while affecting gene activity, do not in fact abolish it. This is particularly likely if these events take place near the carboxyl terminus, where frameshifts have a minimal impact on the protein sequence, or when deletions or duplications in the coding region involve a segment whose length is a multiple of three nucleotides, thus not producing frameshifts. For example, there are multiple different *vlp* genes in *Mycoplasma hyorhinis* that engage in antigenic variation strategies by slippage of in-frame tandem repeats.[84] These repeats correspond to multiple codons and their slippage does not lead to frameshifts, but rather to different efficiency to escape growth-inhibiting host antibodies.

Several cases have been described of partially deleted or amplified stress-response genes, whose proteins remain partially functional. For example, starved *E. coli* cells were found to adapt through a small 46 bp duplication in the *rpoS* gene, which results in a "reduced function" phenotype (although the physiological basis for the fitness gain that underlies selection for this duplication is not yet clear).[85] In natural isolates of *Pseudomonas putida*, a third of the *mutS* gene is missing, which gives a mutator phenotype intermediate between *mutS*+ and *mutS*−.[86] In fact, experimental deletion analysis of the *mutS* gene of *E. coli* indicates that several large regions of the protein can be deleted, affecting to varying degrees the DNA binding and dimerization properties of the protein.[87] Therefore, different tuning of mutator phenotypes may be provided by deletions or amplifications produced by recombination between repeats. The artificial deletion data of *mutS* suggests that systematic search for strains with intermediate functional levels may lead to more examples of partially deleted genes. One such survey of *mutS* genes among *Acinetobacter* species revealed a large number of small insertions and deletions, with corresponding differences in the activity of the MRS.[88]

As mentioned previously, mutator genes can be mutagenic both when they are (partially) inactivated and when they are overexpressed. Overexpression can arise due to genetic changes in regulatory regions[63] or from transient tandem gene amplification.

Several mechanisms have been proposed to explain such tandem amplification of one or a group of genes. All these models start with a first step of recombination between repeats flanking the gene.[89] Once the gene is duplicated, its number of copies can quickly increase, and decrease, by homologous recombination. Other models propose that misalignment during DNA replication results in recombination intermediates that can directly replicate multiple copies of the gene.[90] Because large tandem duplications are unstable and prone to deletion, selection for gene dosage usually is necessary to maintain multiple copies of the gene in a bacterial cell.

Increasing the level of gene activity through duplication appears to be particularly effective in dealing with certain stresses. For example, large tandem duplications, up to 40 copies, of β-lactamase genes were observed in *E. coli* cells grown with ampicillin.[91] Amplification was suggested to have occurred through slippage between 12 bp flanking repeats sequence, which disappears when the selective pressure is removed.[92]

Is There Selection for Repeats in Stress-response Genes?

The neo-Darwinian paradigm usually sustains that mutations are random, in other words they are blind to any possible future advantage.[93] In contrast, this chapter described mechanisms by which global or local evolvability may increase as the result of mechanisms generating genetic variability operating in stress-response genes. This is mostly obtained through deletion or amplification of these genes, or parts of these genes, mediated by recombination between repeats. As such, if increased mutation rates are adaptive when bacteria face stress, one may expect the observed overabundance of these repeats in stress-response genes to be under positive selection acting on the beneficial adaptive mutation that may result from hypermutation or hyperrecombination phenotypes. An alternative explanation is that repeats that mediate recombination events are present in stress-response genes by chance or because of protein sequence or structural requirements.

Chance is unlikely to justify the overrepresentation of repeats in these genes, for, as described above, stress-response genes have more repeats than expected given their length and sequence composition. They also have more repeats than the average gene in the genome. Protein structural requirements do not appear to be responsible for the overrepresentation of repeats for two reasons. First, even if a repeat of the same amino acids was required by the protein structure, due to the degeneracy of the genetic code, the same amino acids could be encoded without the need for a repeat.[94] Second, one would also expect that if structural or functional elements were responsible for repeats, then these should cluster within known functional modules or between repeated structural motifs. This was not observed in the *mutS* gene in *E. coli*.[34] If protein function drives gene repetitiveness, then repeats should not only be abundant, but the specific repeat sequence should be more conserved than the rest of the gene, which was also not observed. It is only the large number of repeats—and not the location or sequence in the gene —that is conserved.[34] This is a strong argument in favor of selection for the presence of repeats (see also chapters 3 and 4). Naturally, the observation that, in competent bacteria, these genes also contain larger than average abundances of USSs[79] reinforces the hypothesis that repeats in these genes are under selection because of their ability to generate sequence diversity, upon which selection can act.

Conclusions

The role of repeats in the generation of antigenic variation in pathogens (see chapters 3 and 5) is well known. It is the role of this chapter to draw attention to other purposes for which repeat-generated genotypic variability has been selected. For example, in *Mycoplasma pneumoniae* many parts of the P1 operon, which codes for the essential major adhesin, are repeated in the genome,[95] generating diverse P1 sequences.[96] Although early reports indicated that these elements were linked to antigen variation, the analysis of genome data at the light of experimental analysis of patients' antibodies suggests that sequence variation is instead allowing the variation of niche tropism. This is because the repeats are present in regions that can alter the adhesive properties of the protein and not in the regions targeted by the host antibodies.[97]

There are many similar examples of the recruitment of repeats to generate adaptive genetic variability (see for example chapters 3 and 4). This chapter has focused on stress-response genes. It is interesting to notice that among these genes, the ones involving genome maintenance functions, such as recombination and mismatch repair, show a higher overrepresentation of repeats than other genes, for example the ones involved in cold shock or heat shock. This suggests that repeat-mediated stochastic responses are more relevant for the former. As discussed in the beginning of this chapter, heat and cold shock, on the one hand, can be tackled with precise responses aimed at fundamentally invariant problems (e.g., protein folding). On the other hand, genome maintenance genes are the major regulators of the rate at which the genome changes. As a result, they can influence the rate by which other genes vary. For example, MRS mutators not only have higher mutation rates but, by decreasing the recombination barriers, they also effectively increase variability at regions throughout the genome that contain repeats. This can result in variability in a wide range of genes, from other stress-response genes to contingency loci involved in pathogenicity.

A major question still remaining to be conclusively answered concerns the role of what Werner Arber has termed "second-order selection" (see discussion of second-order selection also in chapters 3 and 4) in the variability of the stress-response genes. Population genetics data and theory suggest that mutator phenotypes in stressed populations result from their more rapid exploration of the fitness landscape. If this is so, it is likely that repeats generating variability in these genes are also under positive selection. Evolutionary processes do not involve foresight. Hence, these repeats are overrepresented in genes because they were useful in past stress-response events. This means that one can identify genetic elements for which variability is important through the analysis of genome sequences. In this sense, the potential of variability is imprinted in genomes through the selection for repeat sequences. Their role and effects in the genome evolution are now open to elucidation.

Acknowledgments   I am most indebted to the many people with whom I have discussed the issues contained in this review, notably François Taddei, Guillaume Achaz, Francisco Dionísio, Alain Blanchard and Ivan Matic. Previous versions of this review also benefited from the comments and criticisms of Francisco Dionísio, Tone Tønjum, Christopher Bayliss, and Lynn Caporale.

7

The Role of Mobile DNA in the Evolution of Prokaryotic Genomes

Garry Myers, Ian Paulsen, and Claire Fraser

Overview

Prokaryotic organisms are notoriously changeable—they adapt to new environmental niches quickly, adroitly exploiting opportunities as they arise. The acquisition and dissemination of antibiotic resistance by numerous bacteria is the classic example of this adaptability. Such phenotypic fluidity derives from the plastic and dynamic nature of the prokaryotic genome. Prokaryotes have evolved a repertoire of genomic shuffling processes, including mechanisms of gene rearrangements, acquisition, and deletion within and between organisms that can rapidly and dramatically alter the nature and behavior of the organism. Mobile DNA exemplifies the well-established concept that gene products encoded by an organism can directly effect genetic change in that organism. The information that encodes a subset of these reshuffling events subsequently survives by improving fitness, and thus become fixed in descendants, enabling future generations to capture additional information through this advantageous mechanism, opening a doorway to novel pathways of evolution. The "modern synthesis" in evolutionary theory describes selection as acting on randomly generated variation. However, as genome-shuffling events provide a rapid route to the acquisition of important traits, such as resistance to antibiotics, selection has favored the evolution of many types of genomic shuffling events for which outcomes that are favorable to the organism become more probable than random.[1]

Introduction

The importance of mobile DNA in prokaryotic biology has long been appreciated. Indeed, the original experiments by Avery et al.[2] demonstrating DNA to be the material responsible for heredity were based upon the transformation of avirulent pneumococci to virulent forms—the elegance of these groundbreaking experiments was firmly based on the capability and propensity of prokaryotic organisms to exchange and make good use of foreign DNA. More recently, lateral gene transfer has been recognized as a major contributor to the evolution of bacteria. However, despite this long-standing appreciation, the sheer scope of the effect of mobile DNA on prokaryotic genomes has only come into focus since the advent of whole-genome sequencing. It has become apparent that mobile DNA mediates a significant proportion of genomic shuffling processes, forming a potentially significant framework for the evolution of the prokaryotic genome. For example, the genome of the *Escherichia coli* K12 laboratory strain MG1255 was published in 1997.[3] Subsequent sequencing efforts of pathogenic *E. coli* began with the expectation that additional genome sequencing would reveal only minor differences between strains. In fact, analysis of the enterohemorrhagic *E. coli* (EHEC) and uropathogenic *E. coli* (UPEC) genome sequences demonstrated the opposite: each strain is dramatically different from each other, to the extent that only approximately 39% of proteins are common between EHEC, UPEC, and the benign K12 strain.[4] This was a startling result. In fact, existing genomes of *E. coli* strains exhibit a remarkable mosaic structure that has resulted from numerous, independent insertions and/or deletions, mediated by extensive lateral gene transfer. Welch et al. concluded that the EHEC and UPEC genomes were as different from each other as each pathogen was different from the benign K12 strain[4]; for three organisms that are ostensibly the same species, the extent of these differences is a remarkable observation.

This theme of extensive modification of prokaryotic genomes by mobile DNA is being reiterated as more genome sequences are completed, particularly with the growing availability of multiple closely related genomes. In this chapter, we explore the variety of prokaryotic mobile DNA elements and processes that are largely responsible for producing such genomic heterogeneity, in the context of whole genome sequencing and analysis.

Chromosomal Rearrangements

Selection

Selection is widely held to play a central role in determining the outcome of changes to prokaryotic genome structure. Some changes will be "neutral," with no discernable effect on organism fitness. Changes that increase or decrease fitness may be detected at greater or lesser frequency than the changes actually occur. In particular, nonsynonymous mutations in genes that encode proteins essential for cellular function may be lethal and not be seen in progeny. Genetic drift (the increase or decrease of the frequency of a given mutation in a population due to chance alone) is considered to be a major factor determining genome change in small populations by causing the fixation of somewhat deleterious mutations following a population bottleneck. Analysis of prokaryotic genome sequences from the insect endosymbionts *Buchnera* and *Wigglesworthia glossinidia* may indicate that at least some bacteria also are directly affected by random genetic drift, contributing to their extreme genome reduction (but see also Chapter 6 for an alternative explanation). This may be explained by the absence of selection on genes not required for survival in the intracellular niche, and also related to their localization, a reduced opportunity for participating in lateral transfer.

The role of selection in the loss of larger genomic regions also has been established in the evolution of pathogenic *Shigella* from *E. coli*, through "pathoadaptive mutations."[5] The product of lysine decarboxylase (encoded by *cadA*), cadaverine, blocks the functionality of *Shigella* plasmid-encoded enterotoxins (acquired by lateral gene transfer) and is considered an antivirulence factor. Several *cadA*-inactivating alterations including insertion sequences (ISs), disruption by phage insertion and other gene rearrangements have been recognized in the various *Shigella* species.[5] The convergent evolution of such pathoadaptive mutations implies that the environmental niche (in this case host tissues) allowed for selection of clones with increased fitness and virulence through inactivation of *cadA* functionality and hence enterotoxin activity. (See chapters 3 and 4 for an expanded discussion of selection.)

Several biochemical systems act to minimize change to genomic content, for example DNA repair systems.[6] The proficiency and specificity of distinct prokaryote DNA repair systems is likely to result in different patterns of change on the genomic scale. Indeed, mutations within certain genes can in turn influence overall mutation frequency. For example, there is positive selection for defects in the DNA mismatch repair systems of some pathogenic bacteria as part of the process of adapting to new environments (see chapter 6).[7] Natural isolates with subpopulations possessing increased mutation rates are commonly observed (e.g., *Pseudomonas aeruginosa* in cystic fibrosis lung infections).[8] These subpopulations have a distinctly higher frequency of mutation due to error-prone repair of DNA mismatches. For pathogenic *E. coli*, this may also provide a selective advantage for high-frequency mutational change that allows a rapid response to a new ecological niche in a host.[9] After a period of higher mutation rate has enabled the organism to adapt, positive selection for avoidance of further deleterious mutations will lead to reversion to mismatch-repair proficiency. As described in chapter 6, an important mechanism by which reversion may occur is through recombination with another, intact, gene, suggesting one reason for the observed degree of chromosomal mosaicism in *E. coli* and perhaps explaining why phylogenetic trees of repair genes from Gram-negative bacteria are inconsistent with trees built with typical housekeeping genes.

Frameworks for Efficient Exploration: Homologous Recombination and Slippage

The central thesis of this chapter is that the plasticity observed in many prokaryotic genomes is an outcome of a variety of processes, particularly mobile DNA. However, it should be kept in mind that analyses of primary data (i.e., the genome sequence) can give insight into only some of these processes because whole genome sequencing takes only a snapshot of a genome at a moment in time—some events do not leave footprints in a particular genome sequence. Moreover, a single genome sequence obtained in a laboratory should not be treated as an endpoint in an organism's evolution, nor is it necessarily representative.

Insertion and deletion of genomic regions is a principal source of genome plasticity, with repeat sequences being an important focus for these events. Prokaryotic repeats can be separated into two major types: microsatellites and longer repeats.

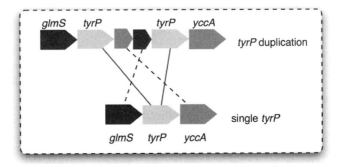

Figure 7.1.   Duplication of the Chlamydophila pneumoniae tyrP *locus corresponds to disease type.* Multiple *tyrP* genotypes are associated with respiratory infections; the single-copy phenotype is associated with cardiovascular conditions. Remnants of *glmS* and *yccA*, and distribution of single-nucleotide polymorphisms in *tyrP*, suggest the single-copy genotype is ancestral. *glmS*, glucosamine-fructose-6-phosphate aminotransferase; *tyrP*, tyrosine/ tryptophan-specific transport protein; *yccA*, integral membrane protein (possible permease). (Data used with permission from R. Belland.)

Microsatellites (also discussed in chapters 3 and 4) consist of small oligonu-cleotides, ranging from mono- to pentanucleotide in size, repeated many times head-to-tail. Longer repeats include transposable elements, large tandem repeats, and spaced repeats. Several proposed mechanisms exist for the expansion of tandem repeats, including slipped stand mispairing (page 44) unequal crossover via homol-ogous recombination, circle excision with reinsertion, and rolling circle.[10] In addi-tion, nucleotide composition bias will affect the rate of tandem repeat formation.[11]

Prokaryotic large tandem repeats may be generated by microsatellites acting as primers, a mechanism identified as central to the production of long repeats in eukaryotes.[11,12] Large repeats often play a role in antigenic variation and, in extreme cases, may represent a major fraction of some genomes. For example, in the reduced genome of *Mycoplasma genitalium*, repeats of a three-gene operon encoding the adhesin MgPa represents more than 4% of the genome.[13] The sequence similarity of these repeats ranges from 78% to 90%, allowing homologous recombination between representatives. Another example of the biological importance of large tandem repeats is the *tyrP* gene of *Chlamydophila pneumoniae* (figure 7.1),[14] which encodes an important aromatic amino-acid transport protein. Genome sequencing of strains isolated from around the world revealed that each genome has one to three tandem repeats of this gene, with an association of two or three repeats to respiratory isolates, and one repeat unit in cardiovascular disease isolates.

Microsatellites often are found in prokaryotes—indeed, polynucleotide repeats are the most common microsatellites—and typically are among the most variable features of bacterial genomes. Microsatellites can influence gene function. For example, in phase variation in *Haemophilus influenzae*, expression of the pilus struc-ture is controlled by bidirectional promoters separated by a poly-AT microsatellite.[15] When the microsatellite has nine units, the appropriate spacing between the promot-ers is produced, allowing efficient expression of the pilus genes. When there are any other numbers of units, the spacing is incorrect and the pilus structure is not produced (see chapter 3 for discussion of microsatellites in prokaryotes, and chapter 4 for effects in eukaryotes of variation in the number of tandem repeats).

Inversions around Chromosome Replication and Termination Origins

A striking feature of genome structure that has become more evident with genome sequencing is the presence of symmetrical recombination around the replication origin and replication terminus regions, the so-called "X-plot" (figure 7.2). Originally noted in the Enterobacteriaceae, this phenomenon is widely observed in bacteria.[16,17] One explanation for this phenomenon is that double-strand breaks, occurring through errors in processing of the bidirectional replication fork, promote recombination between DNA strands at sites approximately equidistant from the origin.[18] The frequency of these inversions appears to be greater at the terminus than at the origin, possibly because of a higher likelihood of DNA breaks at the end of replication.[16,19] A consistent observation is that many genes with globally conserved functions are located near the origin, with a higher concentration of hypothetical genes and genes possibly acquired by lateral transfer near the terminus.[19-22] Possible reasons for this observation include a higher mutation rate, lower expression level, or a higher frequency of insertion of exogenous DNA at the terminus. Moreover, there may be selective pressure on "housekeeping" genes to move away from the replication terminus to the extent that it is a region with a higher rate of mutation. An imbalance of housekeeping genes at the origin and nonessential genes at the terminus may serve to reinforce this bias. The excess of conserved genes at the replication origin also may contribute to selection against replication-directed symmetrical inversion, which may account for the lower number of these events observed at the origin than at the terminus. This underscores the complexity of factors that influence the dynamics of prokaryotic genome structure.

Transposable Elements

DNA recombination can result in translocation of a DNA sequence (an insertion sequence or transposon) from a donor site to a nonhomologous target site (see also chapter 8).[23] Recombination can also promote other types of DNA shuffling events, such as deletion, inversion, or gene fusion. As noted earlier, the presence of multiple copies of an insertion sequence can provide a target for homologous recombination, potentially leading to deletions, inversions, fusions, and duplications. Transposition may modify gene expression through deletion or insertion into genes or by supplying an active promoter that triggers the expression of downstream genes.[23] (See also chapter 8 for a discussion of the potential for effects on gene expression resulting from insertion of transposable elements.)

Transposition can occur at defined sites and cause a specific biological outcome; for example, rather than purely random "cutting and pasting," site-specific recombinases catalyze DNA exchange at specific target sites. This lends support to

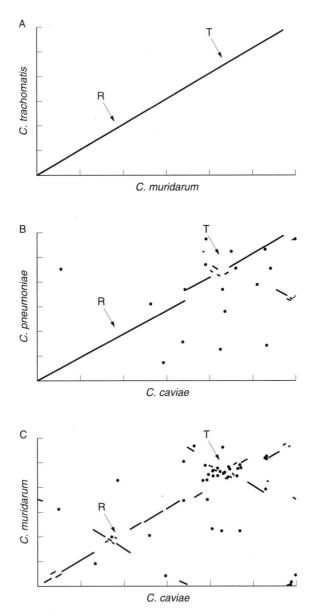

Figure 7.2. Comparison of the Chlamydiaceae showing breaking of chromosomal symmetry around the replication origin and terminus regions. (A) *Chlamydia trachomatis* versus *C. muridarum*. (B) *C. pneumoniae* versus *C. caviae*. (C) *C. muridarum* versus *C. caviae*. Axes are marked with 200 bk graduations. R and T indicate the origins of replication and termination, respectively (Reprinted with permission from reference 19).

the concept that prokaryotes have developed mechanisms that enable exploration of the infinitely variable possible routes through evolutionary pathways that are more efficient than random.[24]

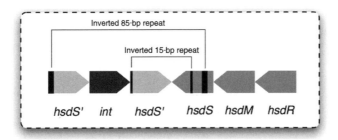

Figure 7.3. Organization of the polymorphic type I restriction enzyme (hsdS)
operon of S. pneumoniae *TIGR4.* Genes marked *hsdS'* are partial *hsdS* genes (pseudogenes). *int* is an integrase gene and *hsdS, hsdM,* and *hsdR* are the specificity, modification,
and restriction subunits respectively. The inverted 85 bp repeat and inverted 15 bp repeats
are marked. Clones were found that were fusions of the *hsdS* and the *hsdS'* pseudogenes,
with the boundary being either the 85 bp or 15 bp repeat. This process may allow the generation of variant genes that enable different target specificities for the restriction enzyme,
a possible adaptation to protect against introduction of undesirable foreign DNA, or perhaps
a way for bacterial populations to sample different foreign DNA.

In many instances, such recombination inverts the DNA between two or more
target sites found close together in the genome. One example is the modulation in
the specificity subunit of a type I restriction system in *Streptococcus pneumoniae.*[25]
A site-specific recombinase, situated next to the *hsdS, hsdM,* and *hsdR* genes,
which encode the specificity, methylation, and restriction subunits of the tripartite
restriction enzyme, can cause inversion at two sites within the *hsdS* subunit and
nearby *hsdS* pseudogenes (figure 7.3). This structure was discovered by examination of the variant sequence clones found in random shotgun sequencing; these
variants have four distinct nucleotide sequences in the region that encodes the
specificity of the enzyme. It is plausible that these variant genes enable different
target specificities for the restriction enzyme, which may be an adaptation to protect against introduction of undesirable foreign DNA. Alternatively, this may have
evolved as a population-scale mechanism that enables different members of the
bacterial population to "sample" different pieces of a foreign genome through variation of target specificities of restriction endonucleases. Other well-characterized
examples of site-specific recombinase activity influencing a specific biological
outcome include the *Salmonella hin* flagellar phase variation system, the
Moraxella piv pilus variation system,[26] and the multiple outer surface protein variations in *Bacteroides.*[27,28]

Other less studied transposable elements or other potential agents of genome
shuffling include introns, retrons, and inteins. Type I and type II introns that can be
spliced from transcribed RNA molecules, commonly found in eukaryotes, have
also been found in prokaryotes, albeit far less frequently.[29] In contrast to introns,
inteins, which are genetic elements present in protein-coding sequences, are
spliced from translated proteins using a self-catalyzed protein splicing reaction (see
http://bioinformatics.weizmann.ac.il/~pietro/inteins/ and http://www.neb.com/neb/
inteins.html). Inteins are found broadly across eukaryotes, bacteria, and archaea.[30]

Some inteins also encode an endonuclease component that is not involved in protein splicing but is thought to mediate transfer of the intein by cleaving intein-less alleles at specific locations, encouraging repair with the intein to eliminate the cleavage site (this process is also termed "homing"). Inteins are generally considered "selfish" genetic elements as there is no evidence for any direct benefit for the host; thus it is considered that the processes of selection will constantly act against them.[30] One possible reason for their persistence in the face of selective forces is that inteins have a tendency to be encoded within highly conserved regions of proteins (such as active sites)—removal of an intein in such regions would require specific mechanisms that would not introduce deletions or frameshifts. This predominance of inteins in such regions probably reflects the impact of selective forces. The nature of their location within proteins suggests inteins have, or had, the potential to benefit their host by affecting the behavior of the host protein. However, such benefit has yet to be identified. In this context, it is hypothesized that inteins are relics from early organisms and participated in domain shuffling at the protein level, and perhaps also at the DNA level, by providing regions of conservation for recombination.[30] Such combinatorial *trans*-splicing would allow smaller genomes to address a larger number of sequences and increase the rate of selection of useful combinations.[30]

Retrons are small genetic elements that can replicate through an encoded reverse transcriptase.[31] Many prokaryotic genomes also contain complex repeat structures (i.e., BOX and RUP elements in *S. pneumoniae*,[25] Correia elements in *Neisseria*,[32-34] and REP elements in *E. coli*).[35] These repeats, often numerous, are often dispersed to several genomic locations in different strains, suggesting movement within the genome.

The BOX and RUP (repeat unit of Pneumococcus) elements of *S. pneumoniae* are found in intergenic regions and make up more than 3% of the genome.[25,36] While the function of these elements is not known, there has been some discussion of a possible role in the regulation of neighboring genes.[36] For example, BOX elements have been suggested to form stable DNA secondary structures that may form the binding site for a protein that subsequently modulates the expression of downstream genes.[37] Consistent with this suggestion is the observation that the inactivation of the BOX element upstream of the *comA* gene produces an *S. pneumoniae* mutant incapable of competence.[37] RUP elements are thought to be active insertion elements that do not encode their own transposase but rather have been transactivated by the action of an IS element[38]; moreover, RUP elements are often associated with IS elements, leading to the suggestion that RUP elements contribute to genome flexibility through sequence rearrangements.[38] Despite this, the actual function of these elements remains obscure while the mechanisms for their maintenance and expansion are considerably less well understood than for IS elements.

Insertion Sequences, Composite Transposons, and Conjugative Transposons

ISs are transposable elements that are frequently encountered within bacterial genome sequences.[23] The basic IS element is recombinatorially active and consists of a site-specific recombinase (transposase) gene and flanking DNA sequence, usually bound by inverted repeats that mark the sites for recombination.[39,40] Often, there are short direct repeats flanking the insertion site that are generated through replication (duplication) of short single-stranded DNA ends created during transposition.

IS elements may mobilize as individual units, or as complex composite transposons containing multiple units of the same or different IS elements.[39,40] In some cases, mobile insertion cassettes consisting of a DNA sequence flanked by inverted repeats may be transposed in trans. Several bacterial genomes contain multiple identical IS sequences suggestive of recent expansion through nonconservative transposition.[41,42] These elements then become targets for homologous recombination, leading to deletions, inversions, fusions, or duplications.

Transposition mechanisms and specificity of target DNA sites vary enormously across different family of elements. For example, currently the only feature discerned for insertion sites of IS1 is a preference for an AT bias at the insertion site, rather than a specific nucleotide sequence.[43] (Chapter 1 discusses the effect of biased sequences on the physical chemical properties of DNA.) In contrast, IS7 and IS30 insert at a particular sequence.[44,45] Such target specificity may be an adaptation to long-term survival within specific hosts by avoiding insertion within regions vital for host survival. For example, there are 84 IS elements identified within the *S. pneumoniae* genome, however only two are inserted within genes of known function.[25] However, as noted earlier, any sequenced genome is a single instance of many; apparent avoidance also may derive from insertion of IS elements into genes that subsequently have been eliminated.

While the action of IS elements and composite transposons leads to reorganization within a genome, conjugative transposons (which combine the properties of transposons, bacteriophage, and conjugative plasmids) can directly mediate the lateral transfer of genes between prokaryotic organisms.[23] Conjugative transposons excise from and integrate into DNA in the same manner as transposons. Like plasmids, conjugative transposons form a covalently closed circular intermediate transferred by conjugation, but unlike plasmids, they do not replicate when they are not integrated into a chromosome or plasmid. Conjugative transposons utilize similar excision and integration mechanisms as phage but do not form viral particles. However, a key property of conjugative transposons is that they can mobilize the transfer of other DNA, such as plasmids inserted into the prokaryotic chromosome.[23]

Gene Cassettes and Integrons

Gene cassettes and integrons often are termed "natural cloning and expression systems;" where the gene cassette represents the insert and the integron is the expression vector (figure 7.4).[23] While many of the genes found in cassettes are antibiotic-resistance genes, other genes also are mobilized by this mechanism. Lateral transfer of such cassette-encoded genes occurs when an integron bearing one or more of these cassettes is incorporated into a broad-host range plasmid, which is subsequently transferred into another host. It has become apparent that the integron/gene-cassette system represents a major mechanism by which prokaryotes capture and express new genes and sets of genes.

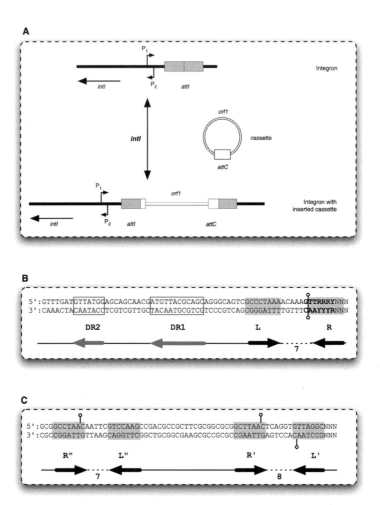

Figure 7.4. *The integron/gene-cassette site-specific recombination system.* (A) Schematic overview of recombination. Cassettes are integrated into the *attI* recombination site (*attI* is represented by a filled box; *attC* by an open box). The *intI* gene, encoding a tyrosine recombinase, is transcribed from the promoter P1 while promoter P2 directs transcription of integrated gene cassettes. *Orf1* represents a phenotypic marker (e.g., an antibiotic resistance gene). The integron/gene-cassette mechanism differs from conjugative transposons (which encode the mobilizing recombinase along with the mobilized genes) in that gene cassettes are non-autonomous genetic elements that rely upon the nonmobilizable recombinase gene within the integron to move around. (B) Organization and sequence of a typical *attI* site (type I integron). The sequence in bold is the conserved *intI* recognition sequence; the vertical line indicates the site of most recombination events. The boxed regions (direct repeats; DR1 and DR2) indicate accessory DNA binding sites for the recombinase. Regions indicated by L and R (shaded boxes) represent recombinase binding sites within the core recombination site. (C) Organization and sequence of a typical *attC* site. The vertical line indicates the site of most crossover events. The shaded regions indicate conserved *attC* regions and represent the inner portion on the *intI* binding site. (Adapted with permission from Recchia, G. D. and Sherratt, D. J., in *Mobile DNA II* [eds. Craig, N. L. et al.] pp.162–176, ASM Press, Washington, D.C., 2002.)

An integron is a genetic unit that includes elements of a site-specific recombination system that is able to capture and mobilize genes (the gene cassettes).[46] The recombination system includes an integrase gene (*intI*) and an adjacent *attI* site; integrons also possess a promoter for the expression of the cassette genes. The gene cassettes usually include a promoterless gene and an integrase-specific recombination site (*attC*; also termed a 59-base element) (figure 7.4).[46,47] Unlike transposons, gene cassettes are not flanked by inverted repeats or duplications of the target sequences.[48] Cassettes may exist either free in a circular form or may be integrated at the *attI* site of an integron. A key distinction from the site-specific recombination system of bacteriophage is that the integrase is encoded by the recipient (integron) and not the mobile element (gene cassette). Moreover, many distinct cassettes are recognized by the integrase encoded within the integron, multiple integrations may occur, and many cassettes encoding a variety of properties may be contained within an integron.[48]

Whole-genome sequencing has identified integrons and associated gene cassettes in the chromosomes of a variety of prokaryotes, including *Pseudomonas, Vibrio, Xanthomonas*, and *Shewanella* spp.[49] *IntI* homologues have been identified in *Treponema denticola, Geobacter sulpherreducens, Acidithiobacillus ferrooxidans*, and *Nitrosomonas europaea*, suggesting integron activity is present in these genomes as well.[49] The diversity of the integron/gene-cassette system is determined by the size and scope of the cassette pool and the rate at which integrons can sample this pool. Recent PCR-based studies of the environmental gene-cassette pool have demonstrated that a substantial and largely unrecognized "metagenome," including both protein-coding and nonprotein-coding sequences, exists within a pool of prokaryotic organisms in the environment.[49] Essentially, integrons are thought to enable the means for bacteria to perform "combinatorial genetics" upon this metagenomic pool of gene cassettes.[49] By integrating into other genetic elements, such as plasmids and transposons, integrons and gene cassettes are able to move within and between prokaryotic cells. Although the extent of influence that the integron/gene-cassette system exerts on the prokaryotic genome is not yet determined, the abundance of gene cassettes suggests the impact is substantial.

Lateral Gene Transfer

Lateral gene transfer (horizontal gene transfer) is recognized as a major factor in the evolution of prokaryotes. Three basic mechanisms for the acquisition of exogenous DNA by lateral gene transfer are: bacteriophage-mediated transfer (transduction), conjugation, and transformation/plasmid-mediated nonconjugative transfer (figure 7.5). DNA transformation, described by Oswald Avery and collaborators,[2] is the process by which bacteria acquire naked DNA from the environment. Certain bacterial species have "competence" genes that facilitate binding and uptake of DNA, often bearing specific sequence signatures. The best described competence systems are found in *Bacillus, Streptococcus, Neisseria*, and *Haemophilus*. Following uptake by these species, exogenous DNA may be rescued by recombination-mediated integration into the genome. Regulation of competence then is a primary mechanism

DNA uptake

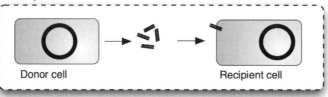

Transduction

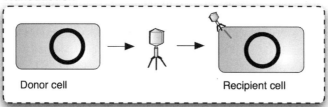

Conjugation - plasmid transfer

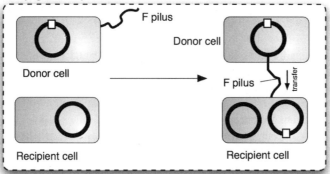

Retrotransfer

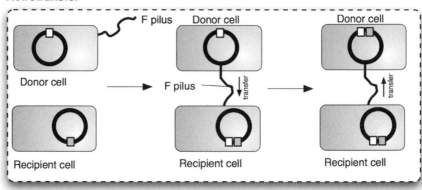

Figure 7.5. Schematic representation of selected mechanisms for lateral gene transfer in prokaryotes.

for bacteria to control the rate of exploration of a potential evolutionary pathway. For example, *Bacillus subtilis* regulates competence through a quorum sensing mechanism that is switched on when a high density of *B. subtilis* cells are present.[50] This maximizes the probability that gene exchange will occur primarily with other *B. subtilis* cells. Another mechanism of competence regulation is exhibited by *Neisseria* spp, where only DNA sequences bearing a specific sequence signature will be taken up. This sequence is overrepresented in the *Neisseria* genome, in effect carrying information that marks a piece of DNA as likely originating from *Neisseria*, thus enabling *Neisseria* cells to acquire DNA from other *Neisseria* cells preferentially (additional discussion of uptake signal sequences can be found in chapter 6).[51] Even without specific DNA-uptake genes, some bacteria can still be competent for acquisition of environmental DNA under certain circumstances. For example, the discovery by Mandel and Higa that *E. coli* was transformable in the presence of calcium chloride was a key innovation in the development of molecular cloning techniques.[52]

Plasmids and Conjugation

Plasmids and their transmission between bacterial cells represent a major source of lateral gene transfer and genomic variation. Plasmid transfer may be mediated by the plasmid itself or by other, conjugative, plasmids present in the same cell. Additionally, plasmids may be mobilized by co-integration with a conjugative plasmid or conjugative transposon. Plasmids that are transmissible in this manner are well known for their role in the spread of antibiotic resistance genes throughout bacterial populations, particularly for those plasmids with a broad host range. The pathogenic members of the Enterobacteriaceae, including pathogenic *E. coli, Yersinia pestis*, and *Y. enterocolitica*, often possess plasmids bearing specific virulence traits.[23]

The mechanism of conjugative transfer of plasmids in the Enterobacteriaceae has been well studied in the laboratory through the F plasmid (figure 7.5).[53] Conjugative transfer of the F plasmid is initiated by first producing a single-stranded copy through rolling circle replication, where the one strand of the double-stranded DNA molecule is nicked and the unnicked strand is used as a template. Single-stranded DNA then is transferred from the donor cell through a plasmid-encoded type IV secretion apparatus, which forms a bridge into the plasmid-free recipient. PilE, also encoded by the plasmid, serves to stabilize interactions between cells to increase DNA transfer efficiency. Once transferred to the recipient, the linear single-stranded DNA circularizes and replicates to reconstitute the original double-stranded molecule. Much of the chromosomal mosaicism that has been observed amongst enterobacterial genomes can be directly attributable to the action of conjugative plasmids.[23,53]

Conjugative DNA transfer also may take place between phylogenetically separated organisms, including organisms as distant as *E. coli* and yeast[54]—intriguingly, this points to at least partial conservation of the components required for transfer and thus may also be a selective advantage, enabling "access" to a broader genetic pool.

Despite this apparent functional conservation, most conjugative plasmids are unable to be maintained stably in a broad range of hosts; while in effect this can be a "suicide" DNA transfer, where the transferred plasmid is lost from the recipient shortly after conjugation, short segments of DNA may be integrated into the new genome by the process of transposition or recombination, thus stably transferring information borne by the plasmid even if the plasmid itself does not "survive." A process of retrotransfer (also termed "shuttle transfer" or plasmid-mediated "gene capture," or even described as a kind of bacterial "hermaphroditism") also has been observed, in which the newly arrived plasmid undergoes conjugative transfer back to the original donor cell.[55] DNA sequences transferred from the recipient chromosome to the plasmid can end up in the original donor cell. In this manner, plasmid-containing bacteria can use retrotransfer to "pickpocket" genes from their neighbors. Although discovered under laboratory conditions, this process has been demonstrated to occur naturally in the environment.[56]

The Ti plasmids of *Agrobacterium tumefaciens* are another well-characterized example of conjugative plasmids. While Ti plasmids transfer a self-mobilizable DNA segment to plant genomes using genetic machinery similar to the F plasmids,[57,58] many self-mobilizable plasmids found outside the Enterobacteriaceae contain few, if any, F-like conjugation determinants. The enterococcal pheromone-induced plasmids encode a set of gene products that enable the host cell to sense the presence of exogenous chromosomally encoded pheromone secreted by plasmidless cells.[59-61] In response to pheromone detection, conjugation functions are activated that transfer the plasmid to the plasmidless strain. Other functions on the plasmid prevent self-induction in response to the cells' own pheromone. This form of mobilization is a complex bacterial behavior controlled by sophisticated cell–cell signaling. There are numerous families of these plasmids, each responding to different pheromones, enabling rapid dissemination of bacterial traits to other enterococci.[59-61] The recent publication of the *Enterococcus faecalis* genome demonstrated that such pheromone-induced plasmids and conjugative transposons were key elements in the medically significant acquisition and possible dissemination of resistance to vancomycin,[42] an antibiotic of last resort for treatment of bacterial infection with multiple resistances.

Plasmids also may change a cell's genomic content by directly integrating into the chromosome. Such integration events have been found in numerous genome sequences, including *Enterococcus faecalis*, *Shigella flexneri*, *Y. pestis*, and *E. coli*.[4,42,62-64] These integration events may directly affect the virulence phenotype of the organism. In the case of *S. flexneri* and *E. coli*, insertion of the 220-kb plasmid, encoding virulence factors, occurs preferentially within the *metB* gene, creating methionine auxotrophy.[23] The act of insertion alters the expression of the plasmid-encoded virulence factors, rendering the strains noninvasive. Exact excision can occur, leading to restoration of virulence, but inexact excision may also occur, leaving plasmid remnants within the genome. This latter process can also lead to the formation of pathogenicity islands (see below).

In addition to changing the genomic content of a prokaryotic cell by integration and excision events, plasmids will also exchange sequences via homologous recombination, or indeed actually fuse. IS elements borne on plasmids often are the targets for such recombination events. Sequence analysis of enterobacterial virulence plasmids demonstrates the mosaic structure of these plasmids.[65,66] These plasmids also contain many complete and incomplete IS elements and bacteriophage genes, indicating that there has been substantial exchange of genetic material between different replicons during their evolution. This points to an ongoing and dynamic process that is likely to be a reflection of the plasticity of the larger chromosome and again may indicate that bacterial populations "explore" a set of evolutionary pathways less randomly than prescribed by currently accepted evolutionary theory.

Bacteriophage

Bacteriophage provide another significant means of lateral gene transfer of foreign DNA into bacterial genomes (figure 7.5). DNA contained within phage is resistant to loss of biological activity for many years, even in an external environment hostile to bacterial cells. As phage often are able to integrate into a chromosome via lysogeny, rather than destroying their hosts after infection, their environmental stability increases the significance of phage as an agent of bacterial evolution.[23] Upon integration, the prophage represses transcription of its genes and enters a dormant phase. Prophage may be revived following derepression, often linked to a host cell perturbation, when they excise from the genome and enter the lytic phase, resulting in host cell lysis and release of viral progeny.

Phage generally exhibit a much narrower host range than conjugative plasmids, likely due to the need of the phage to attach to a specific receptor on the outside of the bacterium. Nevertheless, the contribution of bacteriophage to lateral gene transfer has been substantial. Lateral gene transfer of virulence by the toxin-converting bacteriophage of Shiga toxin producing *E. coli* (STEC) is a well-known example.[67] The family of Stx-phage bears the Shiga toxin genes as a fundamental part of their genome and is able to distribute the *stx* genes throughout the spectrum of enteric *E. coli* strains, converting commensal *E. coli* into a pathogenic form.

Other notable examples of bacteriophage-mediated lateral gene transfer include the *Salmonella* phage SopE[68] and the serotype-converting phage of *S. flexneri*.[69,70] SopE is a broad host range phage of Gram-negative bacteria and encodes SopEΦ, a type III secretion effector protein translocated by the type III secretion system (a system that itself has spread widely between phylogenetically disparate organisms by lateral gene transfer mechanisms) contained within the *Salmonella* pathogenicity island SPI-1. Thus, if SopEΦ transfers the *sopE* gene into other bacteria with a type III secretion system, it will supply potentially novel effector proteins that can interact with host cells.[68] Similarly, the serotype-converting phage of *S. flexneri* contain genes for glycosyl transferases, which modify the O-antigen structure. Consequently, these serotype-converting phage can create antigenic variation in *Shigella*, contributing to survival and virulence of the bacteria.[69,70]

Remnants of phage and phage-like genes are often observed in association with virulence genes within genome and plasmid sequences.[23] It seems likely that these were formerly intact phage bearing these virulence genes; subsequent deletion or mutation of the phage genes has led to the fixation of the virulence genes, and hence their associated phenotype, within the bacterial genome or plasmid.

Lateral Gene Transfer without Mobile Elements

When a gene is associated with a mobile element, there is little doubt as to its identity as a laterally transferred gene. Controversially, lateral gene transfer often is inferred in the absence of obvious mobile elements (i.e., the gene is a remnant of ancient lateral gene transfer but evidence of a transfer mechanism has been lost through recombination mechanisms)[71] such that the true extent of lateral gene transfer is a topic of heated debate.[72] Using approaches such as incongruities in gene phylogeny or nucleotide composition differences from the rest of the genome, up to 25% of genes in certain microbial genomes are estimated to have been acquired by lateral gene transfer.[73] Lateral gene transfer was even claimed to be responsible for numerous apparent bacterial genes within the human genome[74]— this claim has since been refuted.[75–77] In the absence of association with mobile elements, true lateral gene transfer needs to be distinguished from other contextual phenomena that cause genes to look anomalous, such as unusual rates of evolution, strong selection, or gene loss.[78]

It is likely that the role and extent of lateral gene transfer in the evolution of bacterial genomes is not only part of a greater and more complex process,[71] but is also a key source and driver of adaptive variation that in turn may modulate how bacterial populations can "explore" new evolutionary directions.

Genomic Islands

Comparative genomic sequencing analyses led to the discovery of large islands of inserted genes that appear to insert and move as a unit. Originally called "pathogenicity islands" (PAI), as the first examples contained genes essential for virulence in pathogenic *E. coli* strains,[79] they are now more correctly termed "genomic islands," as the genes contained within these regions are responsible for a much wider range of phenotypes. Genomic islands often have a different G+C content relative to the host genome. They frequently are flanked by direct repeats and are associated with tRNA genes and IS elements at their boundaries. Genomic islands often feature mobility genes, such as IS elements, integrases, transposases, or plasmid replication origins, reflecting a complex history of insertion, deletion, and excision of phage, transposons, conjugative plasmids, and conjugative transposons.

Losing (or acquiring) a genomic island is a major step in the evolution of a bacterial pathogen. For example, obtaining an enterobacterial PAI can, at a stroke, transform an avirulent enterobacterial strain into a pathogen.[23] Many genomic islands bear complex gene clusters that enable the bacteria to interact with the host cell via adhesion and invasion or to survive an otherwise hostile environment, such as iron limitation or the intracellular milieu.

Nevertheless, despite the widespread nature of genomic islands, particularly within species and genera, it is not clear what mechanisms govern the acquisition of a genomic island. Many genomic islands possess phage integrase genes, suggesting a phage-like integration/transfer mechanism. The two pathogenic strains of *V. cholerae*, O1 and O139, have arisen through the acquisition of phage-borne factors, the toxin-coregulated pilus (TCP) and the cholera toxin (CT), that influence *Vibrio* fitness.[80] The evolution of these pathogenic strains (notably O139) is a rapidly ongoing process, involving rearrangements of the so-called CTX-Φ element encoding the latter factor. The genetic element encoding TCP (also termed VPI, for "*V. cholerae* pathogenicity island") has been described as the genome of an active filamentous phage (VPI-Φ),[81] however, the nature of this island as a phage element is no longer widely accepted.[82] Alternatively, many genomic islands possess IS elements, indicating a transposition-like mechanism of transfer of at least portions of the island.[23] Similarly, many islands are obviously remnants of integrated plasmids, where presumed loss of the associated mobility genes has served to fix the island within the chromosome.

Genomic islands can influence genome structure not only by their presence or absence but also through variation within themselves. For example, the aerobactin-encoding island of *Shigella*, SHI-2, is found in two different *S. flexneri* strains that have been sequenced.[63,64] While most of the island is identical between the *S. flexneri* strains, the region 3′ to the aerobactin operon is significantly different. Even larger variation of this region has been observed in other *Shigella* species.[83] Thus, this genomic island appears to be a composite island, which may have evolved a "core" region with a divergently evolved variable region that suits the environment specific to each strain and/or species. Indeed for many bacteria, this combination of a conserved core region and divergently evolved variable regions may be regarded as an analogy for the genome itself.

Conclusions

Whole genome sequencing has borne many fruits since its inception over 10 years ago, including insights into metabolism and physiology, virulence and disease pathogenesis, as well as providing a framework to study the nonculturable microbial world. A key outcome of this leap has been the recognition of the extent and diversity of lateral gene transfer through the action of mobile elements and its subsequent effect on prokaryotic evolution. However, for all their proven utility, single-genome sequences are essentially "frozen snapshots in time." Current work in many sequencing centers is focused on genome sequences of multiple strains of a bacterial species. This finer level of comparison will reveal the extent of the fluid nature of bacterial genomes, further demonstrating how information that enables DNA to be mobile is critical to the evolution of prokaryotic chromosomes.

8

Eukaryotic Transposable Elements: Teaching Old Genomes New Tricks

Susan R. Wessler

Without transposable elements we would not be here and the living world would probably look very different from the one we know.[1]

Overview

When transposable elements were discovered in maize by Barbara McClintock over 50 years ago they were regarded as a curiosity—now they are known to be to the most abundant component of probably all eukaryotic genomes. As such, they make up the vast majority of the output of genome sequencing projects. The availability of so much new information has fueled a revolution in their analysis and studies of their interaction with the host. In addition to discovering transposable elements, McClintock also uncovered three ways that the elements can alter genetic information: by restructuring the genome through element-mediated chromosomal rearrangements; by inserting into and around genes and, in the process, generating new alleles; and by imposing their epigenetic marks on flanking chromosomal DNA. In the context of this book, what is implicit about transposable elements is that their presence and extraordinary abundance in genomes promotes a myriad of genome-altering events. By presenting recent case studies that illustrate each of the three modes of action, this chapter brings the reader up to date on the molecular consequences of transposable element activity on host gene expression and genome evolution.

Introduction

Transposable elements (TEs) are fragments of DNA that can insert into new chromosomal locations, and often make duplicate copies of themselves in the process. With the advent of large-scale DNA sequencing, it has become apparent that, far from being a rare component of some genomes, TEs are the single largest component of the genetic material of most eukaryotes. They account for at least

45% of the human genome and 50–90% of some plant genomes (reviewed in references 2 to 4).

TEs were discovered in maize by Barbara McClintock more than a half century ago as the genetic agents that are responsible for the sectors of pigmentation on otherwise colorless mutant kernels.[5] Each sector of colored tissue arises from the mitotic products of a single cell where a TE, which had inserted into and inactivated a gene whose expression is necessary for kernel pigmentation, has excised. Subsequent analysis of mutant alleles from *Drosophila melanogaster*, yeast (*Saccharomyces cerevisiae*), *Caenorhabditis elegans*, and other model eukaryotic organisms furnished the raw material from which molecular biologists could isolate active TEs (TEs have been extensively reviewed in reference 6). Active elements like these, however, constitute only a tiny fraction of the TE complement of the genomes of these model organisms and of most other eukaryotes. Instead, the genomes of higher eukaryotes are filled with thousands, even millions, of seemingly inactive TEs. However, as will be discussed, both active and inactive TEs can impact the evolution of genome structure and the regulation of gene expression.

TE Classes and Mechanisms of Transposition

Eukaryotic TEs are divided into two classes, according to whether their transposition intermediate is RNA (class 1) or DNA (class 2) (figure 8.1). For all class 1 elements, the element-encoded transcript (mRNA) forms the transposition intermediate. In contrast, with class 2 elements, the element itself moves from one site to another in the genome. Each group of TEs contains autonomous and nonautonomous elements. Autonomous elements have open reading frames (ORFs) that encode the products required for transposition. In contrast, nonautonomous elements do not encode transposition proteins but are able to transpose because they retain the *cis*-sequences necessary for transposition. Integration of almost all TEs results in the duplication of a short genomic sequence (called a target site duplication, or TSD) at the site of insertion.

Eukaryotic DNA (class 2) transposons usually have a simple structure with a short terminal inverted repeat (TIR) (around 10–40 bp, but can be up to about 200 bp) and a single gene encoding the transposase. Transposase binds in a sequence-specific manner to the ends of its encoding element (called an autonomous element) and to the ends of nonautonomous family members. Once bound, transposase initiates a cut-and-paste reaction whereby the element is excised from the donor site (generating an "empty site") and inserted into a new site in the genome. There are several possible fates for the empty donor site that can lead to different outcomes for the host. Repair of the double-strand break at the empty site can be precise (leaving no trace of the element or TSD) or imprecise (leaving a so-called "transposon footprint" of a few to several base pairs or deleting adjacent host DNA). Increase in element copy number occurs when the transposon sequence is restored to the empty donor site templated by the DNA sequence of the sister chromatid. This mechanism also can replace autonomous elements at the empty site with nonautonomous elements

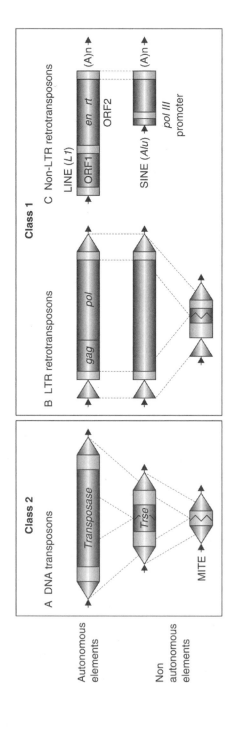

Figure 8.1. Structural features and classification of eukaryotic transposable elements. Elements are divided into two classes, according to whether their transposition intermediate is RNA (class1) (B and C) or DNA (class 2) (A). Class 1 elements are further divided into two groups on the basis of transposition mechanism and structure: (B) LTR retrotransposons, and (C) non-LTR retrotransposons. Each class contains autonomous and nonautonomous elements. Autonomous elements encode proteins required for transposition (gag, capsid-like protein; pol, reverse transcriptase; ORF1, a gag-like protein; en, endonuclease; rt, reverse transcriptase). Nonautonomous elements do not encode these proteins but retain the cis-sequences necessary for transposition. Target site duplications are the black arrowheads flanking each element, and the inverted repeats at the termini of class 2 elements (A) and the direct repeats at the ends of LTR retrotransposons (B) are represented by large gray arrowheads.

if, for example, a deletion occurs during template-mediated repair. The elements studied by McClintock, including the *Ac/Ds* and *Spm/dSpm* families, are DNA transposons capable of insertion and excision.

Class 1 retroelements can be divided into two groups on the basis of transposition mechanism and structure. LTR retrotransposons have long terminal repeats (LTRs) in direct orientation that can range in size from around 100 bp to several kb. Autonomous elements contain at least two genes, called *gag* and *pol*. The *gag* gene encodes a capsid-like protein and the *pol* gene encodes a polyprotein that is responsible for protease, reverse transcriptase (RT), RNase H, and integrase activities. LTR retrotransposons resemble retroviruses in both their structure and mechanism of transposition (called retrotransposition). An element-encoded transcript that initiates from a promoter in the 5′ LTR and terminates in the 3′ LTR is transported to the cytoplasm. There it serves as both mRNA and template for double-strand cDNA that is transported into the nucleus where it can then integrate into the genome, leading to massive increases in copy number (thousands, even hundreds of thousands). The host can mitigate this increase in genome size by mediating homologous recombination between the identical or near-identical LTRs of full-length elements, generating a much shorter solo LTR (where "solo" LTR refers to only one copy of the sequence that normally is repeated at the termini of the element). The efficiency of solo LTR formation depends on several factors, including the efficiency of host recombination mechanisms and the length of the LTR. LTR retrotransposons compose the largest fraction of most plant genomes, where they appear to be the major determinant of the tremendous variation in genome size.

Non-LTR retrotransposons are divided into the autonomous long interspersed elements (LINEs) and the nonautonomous short interspersed elements (SINEs). LINEs encode two ORFs, which are transcribed as a bicistronic mRNA composed of ORF1 (an RNA binding protein) and ORF2 (endonuclease and RT activities). Both LINEs and SINEs terminate by a simple sequence repeat, usually poly(A). LINE transcripts initiate at a promoter within the 5′ end of the element and terminate at or often downstream of the simple repeat sequence. SINEs are characterized by an internal RNA *pol III* promoter (in contrast, all protein coding genes, including those in LINEs, have *pol II* promoters) near the 5′ end. SINEs are a heterogeneous group of elements that range in length from 90 to 300 bp and are derived either from a variety of tRNA genes or from 7SL RNA. There is increasing evidence that SINEs rely on LINEs for the machinery necessary for their amplification.

LINEs amplify by an interesting mechanism (called target primed reverse transcription, TPRT) that appears to have played a critical role in the evolution of eukaryotes, especially primates.[7] As with LTR retrotransposons, an element-encoded transcript is transported to the cytoplasm where it serves as mRNA. However, unlike LTR element encoded transcripts, which serve as template for reverse transcription in the cytoplasm, non-LTR transcripts re-enter the nucleus where chromosomal DNA that has been nicked by the element-encoded endonuclease primes reverse transcription of the transcript into DNA (figure 8.2). SINEs have been spectacularly successful at utilizing LINE machinery to propagate; that is, a SINE transcript instead of a LINE transcript is inserted at the chromosomal nick. In addition, processed pseudogenes are thought to arise through TPRT utilizing

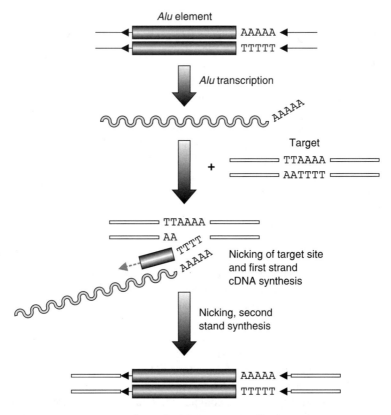

Figure 8.2. Transposition (also called "retroposition") of non-LTR retrotransposons by target primed reverse transcription. An *Alu* element is shown; the same mechanism can also explain the increase in copy number of LINEs and other SINEs. An *Alu* RNA transcript (wavy line) anneals to a nicked site in the genome (the target) and the 3' OH of the T residue at the nick is used to prime first strand synthesis of a cDNA copy of the *Alu* transcript by reverse transcriptase (gray rectangle and dotted arrow; probably encoded by an *L1* LINE). Presumably, nicking of the other DNA strand must precede second strand synthesis. Black arrowheads represent the target site duplication, both at the original site (thick lines) and at the target site (thin open boxes). (Adapted from Batzer M. A. and Deininger P. L. *Alu* repeats and human genomic diversity. *Nat. Gen. Rev.*, 2002; 3: 370–379, box 1.)

LINE machinery to generate and insert cDNA copies of cellular mRNAs into the genome. A plausible model has also been proposed for the origin of introns through TPRT. As will be discussed below, non-LTR retrotransposons are the dominant element type in mammalian genomes, where they appear to account for most of the species-specific differences.

Changing Views of the Impact of TEs on Evolution

Why do TEs predominate in most genomes and how much have they influenced the evolution of life? Ever since the discovery of TEs in maize, speculation has centered on their possible role in genome evolution. McClintock called TEs "controlling elements" because her observations of mutant phenotypes led her to propose that TEs normally controlled maize development.[8] This idea was rejected and she later proposed that TEs were part of a global stress response that could potentially restructure genomes and promote survival ("genome shock").[9] In time, TEs were recognized as ancient components of all genomes (both prokaryotic and eukaryotic). Their ubiquity and mutagenic potential led some—especially neo-Darwinian selectionists—to propose that they originated and thrived because they were important tools of evolution and were essential (integral) genome components (discussed in reference 10).

The field was transformed in 1980 with the publication of two influential papers heralding the view that TEs were selfish or junk DNA and that their evolutionary success could be explained solely by their ability to replicate themselves.[11,12] As stated by Orgel and Crick:[12]

> When a given DNA or class of DNA of unproven phenotypic function can be shown to have encoded a strategy (such as transposition) which ensures its genomic survival then no other explanation of its existence is necessary. The search for other explanations may prove if not intellectually sterile, ultimately futile.

These views had a chilling influence on the field of transposon biology, leading many investigators to change their research focus from the impact of TEs on their host to the characterization of TEs and transposition mechanisms.

From Genetics to Genomics

The selfish DNA theory held until it became clear that TEs usually made up the largest fraction of the genomes of multicellular eukaryotes. Instead of there being one or two TEs near a gene, some human genes were found to contain up to 100! This revelation led to an entirely new set of questions about how organisms and their TEs coexist. What emerged was a new synthesis of prior ideas and current data, whereby TEs and their hosts are seen as being in an arms race—with the TEs trying to increase their copy number and the host attempting to protect its genetic information from mutation. This arms race leads to the development of genetic novelty that can be co-opted by the host. This view has been nicely summarized by Labrador and Corces:[1]

> As replicative sequences, TEs are kept in check by their environment—which is the genome. Natural selection is thus responsible for the existing diversity of TEs and for the many different ways they employ to interact with their hosts. In the same manner TEs benefit from their hosts and evolve, improving their replicative efficiency. And, because evolution is an opportunistic process, the host benefits also from the genetic variability offered by TEs. As a result of the selective process, TEs have become a natural component of modern genomes, and their endurance is due not only to their ability to replicate themselves inside the cell but also to the fact that the eukaryotic genome found in these elements an excellent tool that is constantly used to generate evolutionary novelties and to maintain its own integrity.

The remainder of this chapter will discuss features of TEs that have been co-opted by their hosts by first defining the major mechanisms of genomic restructuring mediated by TEs and then providing examples, or case studies, that highlight different mechanisms.

Three Major Ways TEs Restructure Genomes

McClintock's characterization of the genetic behavior of TEs revealed three distinct ways that TEs could restructure the host genome.

TE-mediated Chromosome Breakage and Rejoining

TEs were first observed in maize as specific sites of chromosome breakage (called *dissociation* or *Ds*) that could initiate the breakage–fusion–bridge cycle.[13] The name of this cycle derives from three events: DNA breakage at the chromatid stage; fusion of the broken ends to produce a dicentric chromosome; and formation of a bridge when the two centromeres of the dicentric chromosome are pulled to opposite poles during mitosis. We now know that this phenomenon is due to the ability of DNA transposons to mediate nonhomologous recombination events when transposase-generated single- and double-strand breaks are repaired (reviewed in reference 14). In addition, there is growing recognition that homologous recombination between the thousands, tens of thousands, or even hundreds of thousands of related TEs dispersed throughout eukaryotic genomes has had a major impact on genome structure and gene content. (See the case studies sections below.)

TEs as Insertional Mutagens

The ability of TEs to knock out or alter gene function via insertion has been recognized since McClintock's analysis of spotted corn kernels over 50 years ago. However, while TEs as insertional mutagens were originally thought to be rare, genome sequencing projects have revealed that the vast majority of normal plant and mammalian genes harbor several TE insertions. This is largely because the majority of eukaryotic transposons are small nonautonomous elements whose insertion into genes can alter rather than knock out gene function. Who would have guessed that over 200,000 of the 1 million *Alu*s (SINEs) in the human genome are in (human) genes? As will be discussed below, TEs are prominent components of a large fraction of the regulatory regions of the genes of higher eukaryotes. (See the case studies sections below.)

TEs and Epigenetic Regulation

McClintock was also the first to note that the activity of some TEs cycled between active and inactive states, a phenomenon she later called "change in phase."[15] Change in phase occurred during a single plant generation or from one generation to the next. In addition, many investigators determined that endogenous inactive

TEs could be reactivated by a variety of stresses (e.g., following the breakage–fusion–bridge cycle, chemical mutagenesis, and radiation). These observations led eventually to the recognition that TEs are the targets of inactivation by the host via epigenetic mechanisms that, as will be discussed in a later section, interfere with TE activity by preventing the production or accumulation of TE-encoded RNA. TEs that are inactivated by epigenetic mechanisms are said to be "silenced." As illustrated below, a host's attempts to protect its genetic information from insertional mutagenesis by silencing its TEs may result in epigenetic alterations that affect the activity of the genes it is trying to protect (see the case studies sections below).

Huge Variations in TE Content May Impact Evolution

An unexpected finding from the analysis of genome sequences is that TE content varies from species to species in two important ways: by the classes of TEs present and their fractional representation in the genome, and by the level of TE activity. The yeast *Saccharomyces cerevisiae* has only LTR retrotransposons (called *Ty* elements), and the vast majority are solo LTRs, generated by the yeast's very efficient homologous recombination machinery. In mammalian genomes, class 1 non-LTR retrotransposons predominate, with class 2 DNA transposons making up less than 5% of the TE fraction (reviewed in reference 16). A remarkable 25–30% of the human genome is derived from just two families of non-LTR elements: *L1* (*LINE-1*), with more than 500,000 copies (~17%); and the much smaller *Alu* (a SINE), with approximately 1–1.4 million copies (~10%). The genomes of flowering plants, including both monocots (e.g., grasses such as rice and maize) and dicots (e.g., *Arabidopsis* and tomato), have a rich collection of both class 1 and class 2 elements, with LTR retrotransposons comprising the largest fraction of most characterized genomes (reviewed in reference 17). *C. elegans* and *Drosophila melanogaster* also have both class 1 and class 2 elements, but class 2 elements predominate in the former and class 1 in the latter.

Genome-wide activity of TEs also varies from species to species. Given the rich genetic analysis of TE-mediated mutations in maize, *Drosophila* and *C. elegans*, it is not surprising that these genomes were found to contain many young, active TE families. Flowering plants, especially members of the grass clade (e.g., rice, maize, barley, and wheat), are in an epoch of TE-mediated genome diversification, with the participation of many families of active class 1 and class 2 elements. In contrast, while extant mammalian genomes have many fewer active TE lineages, TE activity varies dramatically between species. For example, although *L1* elements make up approximately the same fraction of the human and mouse genomes (~20%, ~500,000 copies), only about 80 to 100 *L1*s are active in humans while approximately 3000 are active in the mouse. The functional consequence of this difference is reflected in the fact that new mutations due to insertion of retroelements are rare in humans (about 1 in 500 human mutations) but represent approximately 10% of all mutations in the mouse. Analysis of the age of *Alu* elements in the human genome (as estimated from the amount of sequence identity) indicates that the insertion rate may have been 100-fold higher (one new insertion per primate birth) earlier in primate evolution.

One of the objectives in presenting the following case studies is to illustrate how variations in TE content from species to species may influence both the quality and quantity of genomic variation and, in this way, impact the trajectory of genome evolution. To this end, the case studies are organized into one of the three categories of TE-mediated genomic change already discussed: recombination, insertion mutagenesis, and epigenetic.

Case Studies of TE-mediated Recombination

TE-mediated Inversions in Drosophila buzzatii

The *Drosophila* genus is known for a remarkable level of chromosomal inversions in natural populations, both within and between species. The best-characterized example at the molecular level is a large inversion containing about one third of the euchromatin found on chromosome 2 of *D. buzzatii*, a species found throughout South America.[18] Two chromosomal forms, the ancestral, *2 standard (2st)*, and the derived, *2j*, are both found in populations at high frequency, possibly due to fitness advantages offered by each form; the *2j* derivative conditions a larger body while the *2st* form develops more quickly (figure 8.3). Thus, this is an example of a balanced polymorphism. The origin of these adaptive changes has been investigated by determining first how the inversion arose and then how it impacts the host phenotype. With regard to origin, the *2j* chromosomal inversion was caused by ectopic (homologous) recombination between two *Galileo* elements found in opposite orientation on *2st*. Perhaps significantly, *Galileo* is a member of the *Foldback* group, which is known to promote chromosomal rearrangements during transposition. The *2j* chromosome is itself highly polymorphic due to mutations that occurred subsequent to the original inversion. Many of these secondary mutations are insertions of TEs. The maintenance of these variants of *2j* is probably due to reductions in recombination associated with chromosomal inversions.

Having defined the chromosomal lesions, the next step was to understand the connection between the inversion and the host phenotype. Two models have been proposed to explain why some inversions are maintained at high frequency in natural populations. The "coadaptation" hypothesis posits that inversions maintain favorable allele combinations in heterozygotes because, as mentioned above, of a reduction in recombination in the inverted region. In contrast, the "position effect" hypothesis proposes changes in the expression or function of genes located near or within inversion breakpoints. In support of the position effect model, Puig et al.[19] identified an ORF (CG13617) whose 3' end is located just 12 bp from the inversion breakpoint and whose expression is reduced in embryos homozygous for the *2j* chromosome. Surprisingly, the silencing of CG13617 appears to be caused, like the original inversion, by a TE. Specifically, a transcript originating in another *Foldback*-like TE (called *Kepler*) that inserted into the *Galileo* element (after the original inversion) is

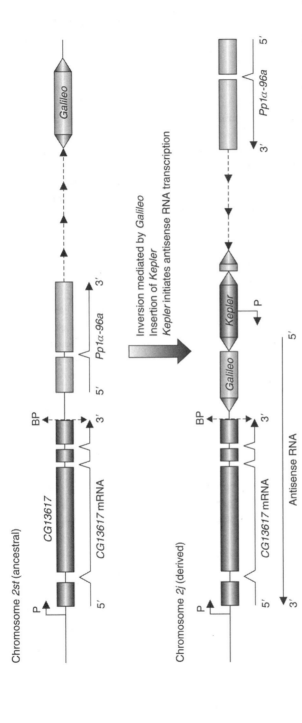

Figure 8.3. Transposable elements mediate both a chromosomal inversion and altered gene expression in natural populations of Drosophila buzzatii. Part of ancestral chromosome 2 *standard* (*2st*) is shown at the top and derived chromosome *2j* at the bottom. The *2j* inversion was caused by ectopic recombination between two *Galileo* transposable elements found in opposite orientation on *2st* (only one element and one breakpoint [BP] are shown). The breakpoint occurred between the 3′ end of a gene of unknown function called *CG13617* and the 5′ end of a gene called *Pp1α-96a*. Subsequent insertion of another element, *Kepler*, into the *Galileo* element near one breakpoint introduced a promoter that initiated an antisense transcript believed to be responsible for a reduction in *CG13617* mRNA in embryos of *D. buzzatii* with the *2j* chromosome.

147

transcribed across the breakpoint into CG13617, generating an antisense mRNA that appears responsible for gene silencing. The authors suggest the involvement of TEs at three stages—the inversion origin, the subsequent insertions into the inverted DNA, and the position effect—may contribute to the evolutionary success of inversions.

Ty-mediated Adaptive Rearrangements in Yeast

In multicellular eukaryotes such as *Drosophila*, it is much more challenging for the investigator to go beyond analyzing extant strains like those described in the previous section. It would be ideal if one could grow populations of eukaryotes under stress conditions, select survivors that adapt best, and determine whether and how the genome has been rearranged. Fortunately, such experiments have been done in *Saccharomyces cerevisiae* (yeast), where isogenic strains were grown in a chemostat under conditions of glucose-limitation for 100 to 500 generations and the survivors (so-called "evolved" clones) analyzed using DNA microarrays for changes in gene copy number and expression.[20]

Of eight evolved clones examined, the vast majority of the chromosomal rearrangements (deletions and translocations) had a TE-related sequence at the breakpoints. The genome of *S. cerevisiae* contains only one elements class, LTR retrotransposons (called *Ty* elements), and the vast majority of the approximately 330 copies in the genome are present as solo LTRs, not full-length elements. As with the *D. buzzatii* inversion, most of the yeast rearrangements appear to be due to ectopic recombination between *Ty* sequences. While the study does not offer direct evidence that the rearrangements promote fitness, the fact that multiple strains share the same breakpoint despite their independent origins points to the rearrangement as the basis for adaptation. For example, three strains shared a breakpoint at a *Ty* sequence near *CIT1,* which encodes citrate synthase, an important enzyme in the TCA cycle. The authors speculate that the rearrangement may activate *Ty* sequences that lead to the derepression of *CIT1* in the presence of glucose. As a key regulator of the TCA cycle, *CIT1* activation may promote the derepression of other genes in the TCA cycle and result in the adaptive phenotype. The authors further speculate that the yeast genome may be populated with *Ty* element sequences that have become fixed because they offer selective advantages, in this case the promotion of adaptive chromosomal rearrangements in response to glucose limitation.

Alu-mediated Chromosomal Rearrangements in Humans

The increasing availability of genomic sequence, especially from primates and other mammals, provides an unparalleled opportunity to unravel the contributions of TEs to genome variation, hereditary disease, and speciation in complex organisms. Humans offer several advantages in this regard. There is not only a complete genome sequence but also increasing amounts of sequence from multiple individuals, including healthy people and those with genetic disorders. Almost 50% of the genome is derived from TEs, mostly *L1* and *Alu* elements. Finally, an extraordinary array of community resources (e.g., expression and metabolic profiles) is available to elucidate the functional significance of TE-mediated genomic alterations (both within human populations and with close relatives, such as the chimpanzee).

As mentioned in an earlier section, although human genomic DNA is largely derived from TEs, fewer than 100 *L1* and *Alu* elements may be currently active (capable of transposition) (reviewed in reference 16). With so little activity, it was reasonable to assume, as many did, that the impact of TEs on genome evolution might be inconsequential. Recent studies make it clear that this assessment was dead wrong. As it turns out, TEs do not have to be active to be responsible for genomic rearrangements. What makes the human genome particularly susceptible to TE-mediated rearrangements is that with only two element families reaching huge copy numbers, the genome is packed with homologous sequences that are potential sites of unequal recombination and other mechanisms that promote rearrangements. Moreover, the enrichment of *Alu* in GC rich DNA (gene-rich regions) means that rearrangements are particularly likely to affect genes, with consequences for both disease and evolution.

Low Copy Repeats in the Human Genome

One of the most surprising findings from the sequence of the human genome is the discovery that as much as approximately 5–6% is comprised of low copy repeats (LCRs) (also called "segmental duplications").[21] LCRs are 10–250 kb in length and have greater than 95% sequence identity with each other, suggesting that they evolved over the past 35 million years. Most LCRs are dispersed, not tandemly arranged.

Evidence that LCRs are especially dynamic regions of the human is provided by their frequent association with hereditary diseases.[22] It appears that once generated, the LCR is more likely to undergo additional rearrangements due, in part, to nonallelic homologous recombination. Compared with other sequenced animal genomes, the human genome is enriched for longer LCRs (>10 kb) that preferentially contain genic sequences. Diseases caused by alterations in LCR sequences include Williams–Beuren, Prader–Willi, Angelman, and cat-eye syndromes.

What is the mechanism(s) underlying LCR formation? A role for *Alu* has long been suspected and has been documented for several deletions and rearrangements associated with genetic diseases (reviewed in reference 23). The abundance of *Alu*s, especially in genic regions, makes it a prime candidate for mediating LCR formation via transposition and for subsequent LCR instability via unequal homologous recombination between *Alu* elements. In addition, the apparent formation of most LCRs since 35 million years ago (mya) coincides with the timing of bursts in *Alu* activity, which resulted in the enormous number of insertions of two *Alu* subfamilies, *AluS* (25–45 mya) and *AluY* (35 mya to present).

Data for a role of *Alu* in the origin of LCRs was provided by Bailey et al.,[24] who performed a comparative analysis of thousands of LCRs extracted from the human genome sequence. They found that *Alu* sequences appeared more frequently in LCRs than would be expected by chance and, more importantly, members of the young *Alu* subfamilies (*AluS* and *AluY*) were much more frequently associated with the LCR junctions than expected. They concluded:

We propose that the primate-specific burst of *Alu* retroposon activity (which occurred 35–40 mya) sensitized the ancestral human genome for *Alu-Alu*-mediated recombination events, which, in turn, initiated the expansion of gene-rich segmental duplications and their subsequent role in nonallelic homologous recombination.

The significance of gene-rich segmental duplications in the evolution of primates and of rodents is suggested by recent comparative analysis of gene content in humans, rats, and mice. While all three organisms have the same set of approximately 30,000 protein-coding genes, major differences in gene content arise from species-specific expansion and divergence of gene family members, especially those of possible adaptive significance, such as genes involved in olfaction and pathogen defense.[25]

Case Studies of TE Insertions: Diversifying Genes and Gene Expression

Prior to the advent of genome sequencing, there were many reports of alterations in gene structure or function due to particular TE insertions. Several instances were reported, for example, of TEs in promoters influencing transcription initiation, or TEs in introns influencing pre-mRNA splicing patterns. Reports like these were dismissed by some as being anecdotal and rare and, as such, providing only marginal support for a significant role for TEs in evolution. The reader is referred to some reviews where these prior studies have been summarized and discussed.[26–28] The following section summarizes a few recent studies where large quantities of genomic sequence data have been used to argue that TEs may be influencing the expression or altering the function of hundreds, perhaps thousands, of host genes.

TE Insertions into Regulatory Regions of Human Genes: Guilt by Association

Arguments have been made for the importance of TEs in human evolution based solely on the number of insertions in regulatory regions, many having occurred after divergence from our last common ancestor. The sheer number of insertions is staggering: over 20% of human genes have TEs (mostly *Alu*s) in their 5' and 3' noncoding sequences, and approximately 25% of the entries in the human promoter database contain a TE-derived sequence.[29] At this time, very little is known about the actual role, if any, of TE sequences in the regulation of individual human genes. (See chapter 4 for evidence that some TEs may promote the origination of simple sequence repeats, which in turn may facilitate a mode of mutation that can quantitatively and reversibly adjust gene activity.) Furthermore, because most of the insertions occurred over 5 mya, it is likely that their impact on gene expression will never be known due to the accumulation of other mutations since the insertion event. The following case study describes the very recent insertion of hundreds of TEs into rice genes, thus permitting an examination of the impact of insertion on gene and genome evolution.

Miniature Inverted-repeat Transposable Elements Can Rapidly Diversify Rice Genes

Miniature inverted-repeat transposable elements (MITEs) are a special class of nonautonomous DNA elements that are found in genomes at very high copy number, where they are preferentially in gene rich regions (reviewed in reference 30). What appears to make them "special" is how they originate and amplify. The majority of characterized nonautonomous class 2 (DNA) elements are more than 1 kb in length and can amplify to moderate copy number (usually fewer than 50 copies in a genome) after they arise by deletion from an autonomous (transposase-encoding) element. In contrast, MITEs are short (usually less than 500 bp) and appear to amplify from one or a few elements to over 1000 elements in a very short period of time (perhaps only a few hundred years). While class 1 *L1* and *Alu* elements are the most common TEs in the introns and regulatory regions of human genes, class 2 MITEs are the most common TEs in plant genes (they also are abundant in certain animal genomes, including insects and fish). Unlike *Alus*, where all one million plus elements have a common origin, there are many distinct MITE families of independent origin in a single genome, each with hundreds or thousands of related elements.

Plant genes have, on average, very short introns (about 200 bp, although there are plant introns longer than 3 kb) compared with their mammalian counterparts (about 2.5 kb on average). There is some evidence that plants cannot efficiently splice long introns; a requirement for short introns may be one reason why short elements such as MITEs (and, less frequently, SINEs) predominate in plant genes. That is, there is a good chance that a MITE insertion into a plant gene will not disrupt gene expression. This apparently is the case as there are hundreds of normal genes already in databases with MITE sequences in their introns, 5' and 3' untranslated regions, and in their promoters. There also are numerous examples, especially in the grasses (e.g., rice and maize), of alleles that differ, in part, due to the presence or absence of MITEs. Unfortunately, in most of these cases the MITE insertion occurred a long time ago (perhaps over a million years) and it is virtually impossible to distinguish the impact of insertion on gene expression (if any) from the effect of other sequence changes that have accumulated in each allele.

To assess the impact of MITE insertions on genome evolution, it first was necessary to identify MITEs that are still transposing. Such a family recently was found in rice.[31–33] The rice genome is being sequenced because rice is the most important source of calories for humans and, fortuitously, because rice has the smallest genome among the cereals (~430 Mb, maize ~2500 Mb, barley ~5000 Mb). Whole genome draft sequences are available for *japonica* and *indica*, two of the three subspecies of rice that have been independently domesticated from wild relatives.[34,35] A 429 bp MITE called *mPing* is active in both *japonica* and *indica* rice. The difference in the estimated copy number of *mPing* elements in a *japonica* (Nipponbare) and an *indica* (93-11) genome (70 versus 14) suggested the recent amplification of this MITE family, perhaps during domestication. Furthermore, analysis of several *japonica* cultivars found that the temperate *japonicas* contained the highest number of *mPing* elements (over 1000 elements in a few cultivars!) whereas the tropical *japonicas* contain the least (many have only a single element).

This dramatic difference in *mPing* copy number between the two subgroups of *japonica* is significant because the temperate and tropical cultivars are thought to have diverged from a common ancestor since domestication (5000 to 7000 years ago). The two varietal groups are adapted to radically different temperature and water regimes: the tropical cultivars flourish in tropical and subtropical environments whereas the temperate cultivars represent an evolutionary extreme, having been selected for productivity in cool, temperate zones with very short growing seasons. Thus, in a situation reminiscent of McClintock's genome shock theory,[9] stress activiation of *mPing elements* during the domestication of temperate *japonicas*, followed by their preferential insertion into genic regions, might have diversified these cultivars and hastened their domestication by creating new allelic combinations that might be favored by human selection.

As with most of the case studies in this chapter, the impact of the bursts of *mPing* insertions on genome evolution is unclear at this time. What is clear is that the thousands of new insertions, presumably into gene rich regions of the genome, will be the focus of detailed analyses to determine which, if any, contributed to adaptation and/or domestication.

L1 *Insertions May Down-regulate Human Gene Transcription*

This review provides case studies that support the view that genetic novelty is an important outcome of the competition between TEs and their hosts. Given the recent availability of the human genome sequence, many findings regarding the impact of TEs are, like the examples above, preliminary in nature but still intriguing because of the potential to impact a very large number of genes. The possible impact of *L1* elements on human gene transcription provides a fitting example. Recall that *L1* elements are autonomous retrotransposons that encode two ORFs. The rarity of *L1*-encoded RNA and protein in vivo has been a longstanding puzzle given that 20% of the genome is derived from *L1* sequences. Clearly, the host tightly regulates *L1* expression, but how? Results from two studies indicate that *L1* RNA accumulation is inhibited in two ways: by premature polyadenylation within *L1* sequences, or by a block in the movement of RNA polymerase while *L1* sequences are being transcribed.[36,37]

L1 sequences are found in around 79% of human genes; the vast majority is in introns. Prior to these studies it was reasonable to assume that most *L1* sequences were spliced with the surrounding intron from human gene transcripts and, as such, were phenotypically neutral. In light of the new findings, Han et al.[36] suggest an alternative model whereby *L1* sequences regulate gene expression on a global scale by serving as a "molecular rheostat" of transcription levels. According to this model, RNA polymerase may slow down or even terminate while transcribing *L1* sequences in the introns of certain genes due to features of the sequence that have yet to be determined. In support of this hypothesis, their computational analysis of genomic sequence and gene expression data revealed that genes with more *L1* insertions accumulate, on average, less mRNA than genes with fewer *L1* insertions.

Alu *Exonization: Diversifying Human Genes for Good and Bad*

While humans, like other mammals, are estimated to have only around 30,000 protein-coding genes, the diversity of the proteome is much greater than this number because a large fraction of human genes (~40–60%) produce more than a single type of mRNA through alternative splicing (reviewed in reference 38 and chapter 15) (see also discussion of RNA editing in chapter 14). Alternative splicing in mammals is attracting increasing attention because species-specific splicing has been found to generate transcripts that encode distinct proteins from what appears to be the same gene. Furthermore, species-specific alternative splicing often is due to the presence of an *Alu* insertion in a human gene versus, for example, its mouse homolog.

First, some definitions are necessary. Exons that occur in all the transcripts from a single gene are called "constitutive" exons, while those that are not in all transcripts are called "alternative" exons. Among alternative exons are "major form" exons, which occur in a majority of the transcripts from a gene, and "minor form" exons, whose occurrence is less frequent and often rare. Comparative analysis of gene expression in mice and men show conservation of 67–98% of constitutive and major form exons but only 15–28% of minor form exons. One reason for this difference is that the hundreds of thousands of *Alu* elements that are in human but not mouse introns (mice introns have their own families of TEs that are not in human introns) have a significant impact on human splicing patterns.

Recent whole-genome analysis of human transcripts has revealed that some *Alu* elements in internal introns of protein-coding genes have become exons; a process that has been called "*Alu* exonization" (figure 8.4). Specifically, over 5% of alternatively spliced exons in humans are derived from *Alu* sequences.[39,40] Virtually all of the *Alu* exons are minor form exons that are transcribed from alternatively spliced genes. In this way, *Alu* sequences added to protein coding regions generate new protein isoforms that can be tested by evolution while the normal protein product is still being synthesized.

This sounded like an ideal situation to evolve new proteins until it was noticed that a few human diseases are caused by *Alu* exons that have become constitutive, that is, the *Alu*-containing transcript is the only transcript produced. In one case a patient with ornithine amino transferase deficiency had sustained a single base pair mutation that activated a cryptic 5′ splice site in an *Alu* element located in an intron of the ornithine aminotransferase gene. The mutant gene produced a constitutive *Alu* exon that encoded a truncated protein due to an in-frame stop codon in the *Alu* sequence. Another human disease, Alport syndrome, was found to result from a single base pair mutation that activated a cryptic 3′ splice site in an intron of a gene (*COL4a3*), which encodes collagen type IV, *a*3. This mutation transformed an *Alu* element that was never exonized into a constitutive *Alu* exon. It has been estimated that a staggering number of *Alu* elements, about 80,000, located in the introns of protein-coding genes may be, like these examples, a single base change away from becoming constitutive exons. While only a subset of these mutations might produce a disease phenotype, given the number of opportunities, it is likely that

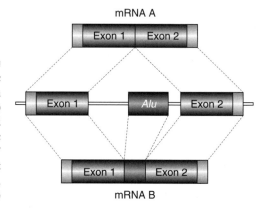

Figure 8.4.   Exonization of Alu *sequences.* A gene with an *Alu* element inserted into an intron can produce a major transcript (mRNA A) and a minor alternatively spliced transcript (mRNA B) where *Alu* sequences have been "exonized." (Adapted from Makalowski W. Not junk after all (Perspectives) *Science*, 2003; 300: 1246–47.)

Alu exonization will ultimately be shown to be not only a source of diversity in human populations but also a significant causative agent of human genetic diseases.

TE-mediated Exon Shuffling

Unlike any other genomic component, TEs routinely move from one genomic location to another. Transposition may provide opportunities to rearrange host sequences though the acquisition of host DNA from one locus and its transposition, along with the TE, to another site in the genome. As such, TE activity could provide a mechanism for exon-shuffling, a 25-year-old hypothesis positing that new genes are assembled by stringing together fragments of existing genes.[41] The acquisition of host genes during the transposition of retroviruses is an important part of viral evolution. While LTR retrotransposons, which are structurally and mechanistically related to retroviruses, also have been reported to acquire fragments of host genes, these examples appear to reflect extremely rare events. In contrast, the two case studies below describe mechanisms of transposition where the acquisition of host sequence is a frequent outcome and, as such, may provide long sought after mechanisms for exon shuffling.

L1-mediated 3′ Transduction

As illustrated in figure 8.2, human LINE *L1* transposes through an RNA intermediate by a mechanism called target primed reverse transcription (TPRT). Recall that during TPRT, element-encoded transcripts are inserted into nicked sites in the chromosome, where they are copied into double-stranded DNAs (see figure 8.2). Interestingly, *L1* elements contain weak polyadenylation signals near their 3′ end. During transcription of active elements, this signal is often bypassed in favor of stronger transcription stop sequences in flanking host sequences (figure 8.5). If these readthrough transcripts are reverse transcribed and reinserted back into the genome, a copy of the 3′ host DNA also will be inserted into the new chromosomal

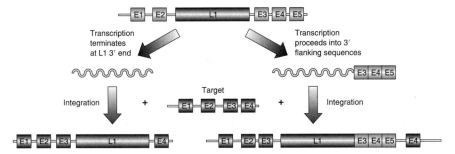

Figure 8.5.   Transduction of host sequences by the human L1 *retrotransposon.*
An L1 element is shown in the second intron of a hypothetical gene whose exons are light
gray boxes (E1–E5). Integration of a new element copy via target primed reverse transcription (see figure 8.2) into a target gene (dark exons E1-E4) without 3′ transduction of host
sequences (on the left) or with 3′ transduction of downstream spliced exons (on the right) is
shown. In this example, E3–E5 are inserted into the coding region of a hypothetical target
gene. (Adapted from Boeke J. D. and Pickeral O. K. Retroshuffling the genomic deck. *Nature*,
1999; 398: 108–109, figure 2.)

locus in a process that has been called L1-mediated 3′ transduction.[42] This mechanism can potentially move any non-L1 sequence that happens to be next to an
active *L1* element, including parts of regulatory regions, exons, or introns, to
another chromosomal locus, where it could alter the regulation or function of a
gene at or near the insertion site.

The impact of this mechanism on human genome evolution has begun to be
addressed by determining the frequency of 3′ transduction events in the complete
human genome sequence. Initial studies focused on so-called "young" *L1*s, which
represent the most recent insertion events. The rationale for this strategy is that the
majority of 3′-transduced DNA is expected to be of no use to the host and will
rapidly accumulate mutations that, over time, will make them unrecognizable.
Analysis of young *L1*s established that the frequency of 3′ transduction was a
remarkably high 15–20% of all insertions, which led the authors to estimate that
about 1% of human genomic DNA may have arisen in this way.[43,44]

Pack-MULEs: Exon Shuffling
Mediated by a DNA Transposon

DNA transposons are known to be important vectors in the transfer of genes
between bacterial cells. As described in chapter 7, TE- and bacteriophage-mediated
horizontal gene transfer now is recognized to be a significant factor in bacterial
evolution. However, until recently there have been only a few tantalizing reports of
eukaryotic DNA transposons containing fragments of host genes. A few of these
reports involve *Mutator* elements, a family of DNA transposons that were first isolated from maize mutant alleles. One nonautonomous *Mutator* element, called

Mu1, was subsequently found to harbor a small fragment of a maize gene of unknown function.[45] That *Mu1* elements bearing this gene fragment have attained copy numbers of around 50 indicate that the captured DNA does not (at least in this case) interfere with transposition. The significance of this observation will soon become apparent.

The availability of complete genome sequences from two other plants, *Arabidopsis* and rice, led to the identification of *Mutator*-like transposable elements (known as MULEs) in their genomes and a handful of MULEs were reported to contain gene fragments. The name Pack-MULEs was given to these chimeric elements. MULEs are now recognized as a component of many eukaryotic genomes and are especially prevalent in higher plants. The potential impact of gene capture by MULEs on genome evolution was assessed by using a computational screen to identify Pack-MULEs in the rice genome.[46] Over 3000 Pack-MULEs containing fragments from over 1000 cellular genes were identified. About 5% of the Pack-MULEs are represented in cDNA collections, providing evidence that they are expressed. Comparison of the cellular genes and their Pack-MULE-borne counterparts indicates that fragments of genomic DNA have been captured, rearranged, and amplified over millions of years. About one fifth of the identified Pack-MULEs contain fragments acquired from multiple genomic loci, thus demonstrating their potential to promote exons shuffling through the duplication, rearrangement, and fusion of diverse genomic sequences (figure 8.6).

Case Studies: Epigenetic Consequences of TE Insertions

As discussed above, McClintock discovered that maize TEs could be reversibly inactivated (epigenetically silenced). Inactivation could be for only a few cell divisions or it could be heritable for several generations.[15] Molecular analyses of members from three different maize TE families (*Ac*, *Spm*, and *Mutator*) later established that active elements differed from inactive elements in that the latter were not transcribed and their ends (containing the transposase gene promoter and the element TIRs) were hypermethylated at the C^5 position of cytosine residues (reviewed in reference 47). Interestingly, like inactive TEs, DNA methylation also can be epigenetically inherited because methylation patterns are copied during replication.

It took over a decade for the mechanistic connections between DNA methylation, element transcription, and the influence of TE activity on flanking gene expression to finally emerge. The early molecular results, along with other related phenomena, sat on the shelf for almost a decade as scientists unraveled the mechanisms underlying epigenetic regulation. This is now a very hot area of research, with hundreds, perhaps thousands, of publications each year. Rather than summarize the current state of knowledge, three case studies are presented in which a combined genetic and genomic approach has begun to open the black box of the epigenetic control of TEs by their host and the consequences for host gene regulation.

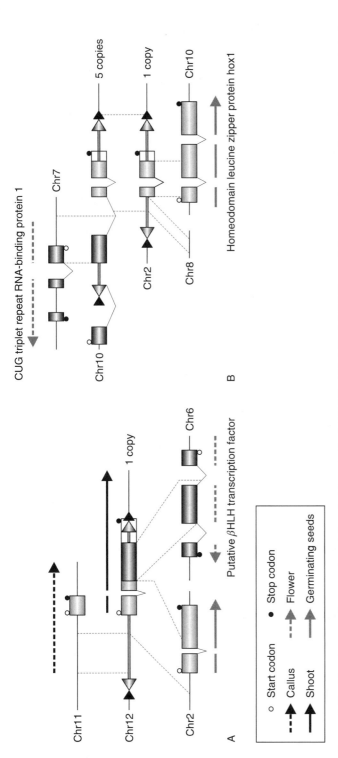

Figure 8.6. Structure and genomic origin of chimeric Pack-MULEs. (A) A Pack-MULE containing gene fragments from three genomic loci including one intron. (B) Possible step-wise formation of a chimeric Pack-MULE: a Pack-MULE on chromosome 10 with sequences acquired from three loci (on chromosomes 7, 8, and 10) and an apparent intermediate element (on chromosome 2) with the gene fragments from chromosomes 8 and 10. Pack-MULE terminal inverted repeats are shown as gray arrowheads and target site duplications are shown as black arrowheads. Homologous regions are associated with dashed lines. Long dotted arrows indicate sequences matching cDNAs from the designated tissues. Light gray boxes represent exons (or part of an exon) where the origin of the sequence is not clear. The gene name is given for putative genes and hypothetical proteins; all other genes encode unknown proteins. (Adapted from reference 46, figure 3.)

157

RNAi Regulates Transposition in the C. elegans Germline

The *C. elegans*[48] genome harbors several families of class 2 TEs belonging to the *Tc1/mariner* superfamily. Distinct element families in *C. elegans* are designated Tc1, Tc2, Tc3, and so on, and Tc1, which has around 31 copies, is the most abundant family in the genome. In most isolates of *C. elegans*, transposition of Tc1 and the other Tc families is repressed in the germline but not in the soma. However, in the Bergerac strain, transposition of Tc elements occurs both in the soma and the germline and Tc1 insertions are the major cause of spontaneous mutation. That transposition could occur in the germline led the laboratory of Ron Plasterk to undertake both forward and reverse genetic screens to identify the gene(s) responsible for repressing germline transposition. While many loci identified by these screens are of unknown function, and others are only indirectly involved in repression, the screen identified a group of gene products that are now known to be part of the epigenetic regulatory process called RNA interference (RNAi).

In eukaryotes from yeast to plants to animals, RNAi has been shown to be a mechanism of posttranscriptional gene silencing (PTGS) that is triggered by double-strand RNA (dsRNA), which can be a by-product of genome perturbation. For example, insertion of a TE or foreign DNA (viruses or transgenes) next to a host gene may promote transcription into the gene on the antisense strand. The dsRNA formed when sense and antisense transcripts anneal is a substrate for an RNase III-like enzyme (called dicer or DCR-1) that cleaves dsRNAs into short interfering RNAs (siRNAs) of about 21–24 nt. Interaction of siRNAs with the RNA-induced silencing complex (RISC) leads to the degradation of RNAs that are specified by the siRNAs; in this example, the transcript of the host gene next to the TE insertion site. Recall from a previous section that antisense transcription promoted by a TE at an inversion breakpoint in *D. buzzatii* was implicated in the downregulation of a host gene adjacent to the breakpoint. We will return to these siRNAs in the next case study (additional discussion of siRNAs can be found in chapters 13 and 14).

The repression of Tc1 transposition in the germline implies that this process is very efficient and that dsRNA derived from Tc elements must be readily available to maintain repression. But what is this dsRNA and how does it originate? Unlike most class 2 elements, the Tc1 transposase gene does not appear to have its own promoter. Instead, transposase expression may rely on transcripts initiated in flanking host sequences that read through the entire element. The dsRNA trigger for RNAi is thought to arise from intramolecular pairing between their TIR sequences. Interestingly, while the internal sequences of other Tc families in *C. elegans* are distinct, their TIRs are very similar. Thus, siRNAs produced by the synthesis and processing of readthrough transcripts from a single Tc element may be sufficient to silence all Tc elements in the genome.

TEs and Heterochromatin
Formation in Arabidopsis thaliana

Recall that inactive maize TEs are hypermethylated as are the TEs of other organisms, including other plants and mammals. However, not all organisms methylate their DNA; among these exceptions are the model organisms *C. elegans, Drosophila*, and yeast. If *C. elegans* can repress its TEs without methylating them, why do most other organisms methylate their TEs? While there is still no definitive answer to this question, the search for an answer in plants has helped reveal connections between methylation, histone modification, and RNAi.

Because silencing of both TE and transgenes in plants had been associated with DNA methylation, methylation became the focus of many genetic screens in the model plant *Arabidopsis thaliana*. In one screen, mutagenized plants assayed for reduced methylation of repeat sequences (each plant was tested by Southern blot!) led to the identification of a gene, called *decrease in DNA methylation 1 (DDM1)*.[49] In addition to reduced DNA methylation, the mutant *DDM1* background gave rise to both developmental and unstable mutations that were not linked to the *ddm1* locus. One mutant, called *crab*, was found to be due to the insertion of a transposable element that was normally transcriptionally silent and transpositionally inactive in *A. thaliana*. While this study demonstrated that an *A. thaliana* transposon that must have been epigenetically silenced was released in the *DDM1* mutant background, the association between DNA methylation and TE inactivity remained unclear when it was determined that *DDM1* encoded a putative chromatin remodeling protein, not, as might be expected from its mutant phenotype, a methyltransferase. Recall that higher organisms have transcriptionally active chromatin and transcripitonally inactive or silent chromatin. The latter can often be observed cytologically as regions of heterochromatin, located around centromeres, telomeres, and, in some organisms, interstitially (where it is called "knobs"). While active and inactive chromatin often are defined biochemically by measuring acetylation and deacetylation of histones, respectively, actively transcribed chromatin can be identified by the methylation of lysine 4 in the amino terminus of histone 3 (H3mK4) while inactive chromatin is marked by methylation at lysine 9 (H3mK9). Recent studies indicate that mutations in *DDM1* decrease the methylation of H3mK9 in all loci tested.

To review up to this point, the research had shown that the TEs of *C. elegans* are not methylated but are rendered inactive through the production of siRNAs that targets TE transcripts for degradation. In *A. thaliana* (and in other plants and mammals), TEs are methylated and their chromatin contains H3mK9. In *A. thaliana* with mutant *DDM1*, TEs are both activated (H3mK9 decreases) and demethyated (H3mK4 increases). These results suggested that the product of the *DDM1* gene and a DNA methyltransferase are in a complex that can recognize TEs and silence them by methylating their DNA and associated histones. However one big question remained; what does this complex recognize when it distinguishes between TEs and genes?

This question was addressed by comparing a duplicated region in the *A. thaliana* genome where one duplicate is euchromatic and contains 33 genes and the other is heterochromatic (a chromosomal knob) and contains 8 of these 33 genes and also 73 TEs that inserted after duplication. In an experimental tour de force, the lab of Rob Martienssen compared the two regions by microarray analysis (in 1 kb segments) with respect to transcription, and histone and DNA methylation in both wild-type and mutant *DDM1* backgrounds.[50] They concluded that under the control of DDM1, TEs and repeats were responsible for heterochromatin formation because, as they put it: "In a *ddm1* background, TEs adopted gene-like chromatin properties and the majority were expressed." Therefore, DDM1 distinguishes TEs and related repeats from host genes. Given that many TEs, like host genes, encode proteins and both have similar GC content, the basis for discrimination was suspected to be sequence-based, which brings us back full circle to RNAi. In support of a role for an RNAi based mechanism for TE recognition, they detected siRNAs from the TEs that were inactivated in wild-type and became activated in *DDM1* mutants. Thus, as in *C. elegans*, *A. thaliana* TEs appear to be transcribed to generate siRNAs, which guides DNA and histone methylation a process that involves DDM1 and a DNA methyltransferase, leading to heterochromatization and inactivation of the TEs.

Both in this instance and in others, TEs seem to be responsible in large part for the regions of heterochromatin in the genomes of most eukaryotes. For this reason, it has been suggested that TEs may be essential to centromere and telomere function as they too are sites of heterochromatin. This has been discussed in recent reviews[51–53] and will not be reviewed here. In keeping with the theme of this chapter, the last case study demonstrates that a mechanism to maintain centromeres and silence TEs can turn on the host to inadvertently silence its own genes.

LTR Retrotransposons Can Silence Adjacent Host Genes: from Polyploidy to Cancer

Organisms use a variety of interrelated epigenetic mechanisms, such as RNAi and DNA methylation, to inactivate the TEs in their genome. This strategy is absolutely essential in the very large plant and animal genomes where the majority of TEs are class 1 retroelements (LTR and non-LTR retrotransposons, figure 8.1) that require transcription for mobility. However, rendering TEs immobile is only one reason to repress their transcription. Equally important is to prevent so-called "readout" transcription into host genes. As discussed earlier in this chapter, LTR retrotransposons are the most abundant component of many plant genomes, accounting for an astounding 70% of the maize genome, for example. In addition, about 8% of the human genome is made up of a group of elements called human endogenous retroviruses (HERVs), which are derived from ancient infections of exogenous retroviruses. Like LTR retrotransposons, HERVs are flanked by LTRs (see figure 8.1 for the structure of LTR retrotransposons). Recall that a promoter in the 5'LTR initiates transcription into the element, producing an RNA that is reverse transcribed into a cDNA that can integrate elsewhere in the genome. However, because the two LTRs are identical at the time of insertion, the same promoter in the 3'LTR (or in a solo LTR) may initiate readout transcripts into flanking host sequences. The danger to the host is that their genes can be silenced if they are adjacent to LTRs that promote antisense readout transcripts (like the one promoted by the Kepler element into a flanking *D. buzzatii* gene in figure 8.3).

For this reason, the host must be able to protect its genes from the hundreds of thousands or millions of LTR-borne promoters that can be scattered throughout its genome. Fortunately, the host defends itself against both mobility and readout transcription by methylating the 5' and 3' LTRs and rendering their promoters inactive. The retrotransposons in the genomes of plants and animals are said to be hypermethylated relative to the genes. However, more and more chinks in the host armor are being revealed as new studies report examples of what McClintock called "genome shock,"[9] where host DNA becomes hypomethylated and LTRs are transcriptionally reactivated.

Reactivation of LTRs can be potentially good or bad for the host. One of the major shocks to the integrity of a genome occurs during the formation of polyploids when the genome doubles. While polyploids were once thought to be rare, whole-genome sequencing has revealed evidence for one or multiple polyploidization events in the history of all plants and many animals (reviewed in reference 54). To understand the earliest events in polyploid formation, researchers have turned to domesticated plants such as wheat and cotton where synthetic polyploids can be created in the field and the genome-wide impact on TEs can be analyzed in the laboratory. In one study, genome-wide analysis of newly formed wheat polyploids revealed that widespread changes in DNA methylation in the LTRs of the Wis 2-1A retrotransposon activated readout transcription into adjacent wheat genes and, in some cases, led to gene silencing.[55] Because newly formed polyploids have a duplicate set of genes, gene silencing is probably not a serious threat and may in fact facilitate the successful merger of two genomes into one.

In contrast, the reactivation of HERV elements in humans may exacerbate an already bad situation. Hypomethylation of DNA is a consequence of many cancers and other human diseases (reviewed in reference 56). Recent studies have demonstrated that the methylation status of several HERV LTRs is altered and that some become transcriptionally reactivated in some human tumors. While the consequences of readout transcription on the cancerous state is currently under investigation, the wealth of human genomic resources promises to make this a lively area of future research.

Conclusions

The thread that connects most of the case studies presented in this chapter is that TE-mediated genomic diversification is a by-product of the arms race between TEs and their hosts and that the novelty generated by this arms race has facilitated the evolutionary success of the host. When seen in this light, the term "coevolution" may be a more accurate term than "arms race" in describing the interaction between TEs and their hosts.

Scientists rely on negative controls to validate their experimental results. To bolster the argument that TEs are a significant component of the evolutionary success of eukaryotes, a comparison could be made between eukaryotes with and without TEs. Unfortunately, such a comparison is not possible at this time because all characterized sexually reproducing eukaryotes have TEs. But nature may have provided the next best thing, a eukaryote that has evolved a defense mechanism that appears to be completely successful in preventing the amplification of TEs. The eukaryote is the model organism *Neurospora crassa* and the mechanism of defense is called repeat-induced point mutation (RIP). RIP first detects duplications and then mutates them by changing up to 30% of their G·C base pairs into A·T pairs (reviewed in reference 57). In fact, not a single intact TE was detected in the draft sequence.[58] But, the apparent consequence is that *N. crassa* has few highly similar duplicate genes. Of around 10,000 predicted protein-coding genes, there are only six pairs (12 genes) with greater than 80% nucleotide identity. Thus the route to evolution of new function through duplication and gradual divergence of genes is blocked by RIP. In addition, other TE-mediated mechanisms illustrated in the case studies, such as allele diversification via TE insertion and TE-mediated exon-shuffling, presumably cannot occur either. However, *N. crassa* does exist and has managed to survive and flourish for a very long time; clearly it has evolved mechanisms to evolve without TEs. What those mechanisms are will surely be the subject of future studies.

In addition, further studies will almost certainly reveal new ways that TEs diversify genomes. The availability of increasing amounts of eukaryotic genome sequence has permitted our first glimpse of the relationship between host genes and TEs. What is most striking about these initial results is that each genome has a different story to tell. As discussed above, there is tremendous variation in the TE content of the characterized genomes with respect to overall TE composition and level of activity. This variation reflects the distinct evolutionary trajectory experienced by each species. In this regard, we have only begun to understand how the coevolution of host genes and TEs impacts the mode and tempo of evolution. This situation is reminiscent of the statement at the end of each episode of *Naked City*, a TV show that I watched as a child: "There are eight million stories in the Naked City. This has been one of them."

9

Immunoglobulin Recombination Signal Sequences: Somatic and Evolutionary Functions

Ellen Hsu

Overview

Vertebrates defend themselves against the onslaught of a vast and evolving array of pathogens by drawing from a vast antibody repertoire that is encoded by somatically generated DNA sequences. This diversity can be achieved because it is implied, rather than explicity encoded, in the germline genome. This chapter describes our current understanding of the rules by which this diversity of immunoglobulin combining sites is created in B lymphocytes.

Introduction

The adaptive immune systems of vertebrates are characterized by a primary response mounted specifically against the invading pathogen and, upon reinfection by the same pathogen, a secondary response that is more rapid and of greater magnitude. The specificity and memory are based on the possession of lymphocytes: circulating cells with surface receptors expressing a vast, anticipatory repertoire of ligand combining sites directed against potential pathogens. Antibody-producing B lymphocytes express immunoglobulin (Ig) on their cell surfaces while effector and cytotoxic killer T lymphocytes express T-cell receptor (TCR). For humans, estimates of the number of different receptors range from about 10^8 to 10^{14}, which means that most lymphocytes have a unique receptor. Since the human genome carries, at the current estimate, 30,000 protein-encoding genes, it cannot encode so many antigen-binding receptors in the germline explicitly. In fact, the receptors are somatically generated from component parts by recombination in differentiating lymphocytes during the lifetime of the individual.

The Ig and TCR genes are not functional until the variable regions that form the antigen-combining site are assembled. This entails joining together the various gene segments that may be distantly located in a cut-and-paste process that involves double-strand DNA breakage and removal of the intervening DNA. The endonuclease action is site-specific, targeted by the recombination signal (RS) sequences flanking every gene segment that can be rearranged. One may thus consider the germline Ig or TCR gene segments (which number between a few dozen to a few hundred, depending upon the animal species) and their associated RS sequences, prime examples of how information can be encoded "implicitly" in the genome, the central theme of this volume. This chapter focuses on the RS elements, which contain information that directs and influences targeting, orientation, and ordering of the rearranging gene segments, generating the vast repertoire of Ig and TCR antigen receptors. Although rearrangement occurs somatically and normally only in lymphocytes, preassembled Ig genes have been discovered in germline DNA of the earliest vertebrates, sharks and skates; we shall include in our considerations on RS function an evolutionary role for the RS in targeting changes in germ cells and reshaping the genome.

In-depth reviews on the Ig/TCR rearrangement process and the structure and activity of the recombinase are contained in references 1 to 5. Although the rearrangement mechanism is the same in T and B cells, this chapter centers on the Ig genes in B cells in order to dovetail with chapter 10 on somatic hypermutation.

Assembling Genes in the Immune System

The basic Ig unit is composed of two heavy (H) and two light (L) chains; the first 110–120 amino acids at the N-terminal end of both polypeptides constitute the variable (V) region, which is the ligand-binding portion. The rest of the polypeptide, the constant (C) region, contains the effector functions of an antibody, such as complement-binding and Fc-receptor binding. The V regions of L and H chains are assembled by gene rearrangement; a V_L gene segment joins to a J_L gene segment to generate the V region of the L chain, and a V_H gene segment rearranges to the D and the J gene segments to form the V region of the H chain; the process is referred to as V(D)J recombination.[6] Figure 9.1 shows the relationships between the germline genes, the rearranged genes, the transcribed mRNA, and the H-chain polypeptide.

At the germline human IgH locus (figure 9.1), the various gene segments, V, D, and J, are present in multiple tandem copies.[7] This arrangement is similar in all tetrapods for IgH and, without the D gene segments, for IgL. The main organizational difference among the vertebrate classes is the number of H chain C region genes present.[8] As described in chapter 11, a B cell retains its V_H rearrangement but can change the class or isotype of the C region by a DNA recombination event called class switch recombination. Whereas the multiple Ig gene segments were generated in tandem fashion in mammals, reptiles, and amphibians, gene duplication occurred for entire Ig loci, $V–D–J–C_H$ or $V–J–C_L$ en bloc in cartilaginous fishes. In sharks there are 15 to 100 IgH loci, each one consisting of one V_H, two D, and one J_H gene segments, and exons encoding one C_H isotype. Elasmobranchs are representatives of the earliest vertebrates and have an alternative form of Ig organization that perhaps reflects a primordial Ig locus.

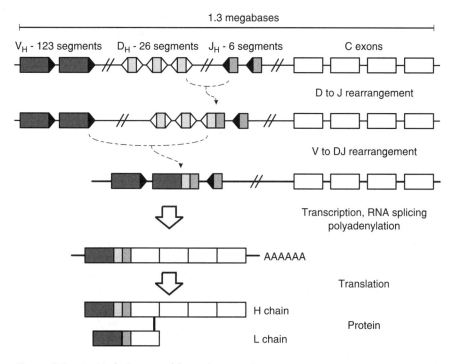

Figure 9.1.   Ig H-chain assembly and expression. The human germline H-chain locus spans about 1.3 Mb, from the 5′-most V_H gene segment to the last C region exon.[7] There are multiple C region genes (not shown) as well as 123 V_H gene segments, 26 D genes, and six J_H genes. The size of the locus and the number of genes depends upon the species. A rearrangement involving any D to the 5′-most J_H (line 1 to 2) and the 3′-most V_H gene segment to that DJ (line 2 to 3) would entail deletion of about 74 kb of intervening DNA. The precursor RNA is spliced to generate the mRNA (line 4), which is translated (line 5). A half-molecule consisting of one H and one L chain is shown. The black triangles represent 23-RS and the white triangles 12-RS (as described in detail in the section on rearrangement and in figure 9.2).

Diversity in the tetrapod primary antibody repertoire is generated several ways. There is the "combinatorial" diversity formed by different V+(D+)J; H and L chain pairing also provides "combinational" diversity. Most importantly, while the site of the double-strand DNA break is at the RS and precise, the process of rejoining the broken ends involves nucleotide deletions as well as insertions at the joints of $V_H D$, $D J_H$, and $V_L J_L$, generating "junctional" diversity. With imprecise joining sites, the chances are that in two out of three rearrangements the V to DJ or J joining will not result in an open reading frame. The failure to form an in-frame rearrangement at the IgH or IgL locus on one chromosome leads to continued recombination process on the allele; because B cells which have not successfully rearranged H and L chains do not survive, one consequence of imprecise joining is cell wastage. However, in most antibodies and all TCR, the V/(D)/J junction is important in ligand binding, so that the benefit derived is valuable diversity created by the wide spectrum of junctions, differing in sequence and in length.

This chapter describes the Ig gene assembly process, including what is known about the nature of the information the RS contain, what types of sequence diversity are generated, and the functional consequences of that diversity when viewed at the level of the protein. Although antibody specificities are generated in somatic tissues, we will speculate on recombination of Ig gene segments in shark germ cells that might lead to novel functions.

Rearrangement

The process of rearrangement involves DNA double-strand breaks followed by repair and rejoining. The endonuclease RAG (recombination-activating gene), consisting of two polypeptides RAG1 and RAG2, is necessary for DNA cleavage and is the only lymphocyte-specific component of the process.[9-11] The other cofactors are ubiquitous nuclear proteins or components of DNA repair pathways. Only the RS that flank V, D, and J are needed for recognition and rearrangement by RAG.

The RS consist of a well-conserved palindromic heptamer sequence (CACAGTG) and a relatively conserved nonamer sequence (ACAAAAACC). The heptamer (7-mer) and nonamer (9-mer) are separated by a spacer of either 12 or 23 nucleotides, with the 7-mer always flanking the coding region. Of the two gene segments selected for a rearrangement, one must carry an RS with a 12 bp spacer between the 7-mer and 9-mer (12-RS) and the other must be a 23-RS; this requirement is called the "12/23 rule."[12,13] The mechanistic basis by which it operates is not known. It also is not clear how the RAG brings together a 12/23 pair from the distantly located gene segments available, but once the RAG/12-RS/23-RS complex is formed, it is stable and does not dissociate readily. The double-strand DNA cleavage occurs only within this synaptic paired complex.

RAG is expressed in precursor T and B lymphocytes, but introducing it into any somatic cell does not initiate recombination at the endogenous loci. For TCR or Ig loci to rearrange, they must be activated in the course of a developmentally regulated program to become "accessible" to RAG. Accessibility consists of as yet undefined chromatin structural modifications that render the RS available to binding by RAG and a chromatin-associated factor HMG (high-mobility group protein, HMG1 or HMG2) (for a review, see reference 14). The role of HMG nuclear proteins is not entirely clear, but they do not have base sequence specificity and their binding is thought to distort the DNA favorably for interaction between RAG and the 23-RS.[15]

The double-stranded break involves two steps,[11,16,17] occurring at the 7-mer/coding end boundary of both RS and results in two double-stranded DNA breaks and the excision of the intervening DNA between them (see figure 9.2). The coding ends initially form a hairpin intermediate. The hairpinned coding ends are retained in the postcleavage complex, which is thought to serve as a scaffold in the recruitment of factors that process the DNA ends before joining.[21,22,23] The asymmetric opening of the hairpin creates an overhang, which can be retained (P region) or trimmed back. Additional sequence can be inserted by terminal deoxynucleotidyl transferase (TdT), a DNA polymerase that is unusual in that it does not require a

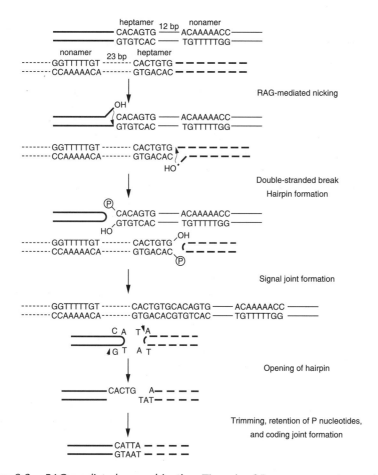

Figure 9.2.   RAG-mediated recombination. The pair of Ig gene segments containing a 12-RS and 23-RS are bound by RAG, which is in contact with both the 7-mer and 9-mer of the 12-RS and the 23-RS. With the formation of the synaptic paired complex, the first nick is introduced by RAG. The nicking occurs 5′ of the 7-mer end of the RS on the top strand of each of the two Ig gene segments, producing a 3′-hydroxy on the (coding) end of the Ig gene segment and a 5′ phosphoryl on the RS (signal) end. The result is a duplex nicked at either RS. The second step involves intramolecular transesterfication reactions where the 3′-OH attack the opposing phosphodiester bonds, causing the coding ends to become a covalently closed hairpins and freeing the blunt signal ends.[16] Cofactors that process the hairpins are present or are recruited. These include Ku70, Ku80, DNA-dependent protein kinase (DNA-PK), and the endonuclease Artemis. The heterodimeric Ku binds DNA ends, providing an end-alignment function, and Ku80 recruits and activates DNA-PK, which in turn phosphorylates and activates the nuclease activity of Artemis (reviewed in reference 5). Artemis opens the hairpin coding ends by nicking one strand 3′ of the tip of the hairpin.[18,19] The resultant single-stranded overhang consists of a portion of the coding end and its complementary sequence. Sometimes the overhang could be included as part of the final joined product and is observed as inverted repeat sequence (P region).[20] The DNA ends often are trimmed, probably by Artemis; TdT may insert nontemplated nucleotides (not illustrated). The joining together of the coding ends requires additional cofactors, among which include DNA ligase IV and XRCC4.

167

template (step not shown in figure 9.2). These random nucleotides, called N region,[24] can be one to 20 nucleotides long, are GC-rich, and are de novo sequences generated in the somatic cell. However, TdT expression is dependent upon the animal species and sometimes the stage of ontogeny.

At some point the coding ends become dissociated from RAG, although RAG retains the signal ends until they join to each other head-to-head and form a circular DNA molecule that will be lost with time (figure 9.2). The processed flaps on either coding end are joined to each other by components of the nonhomologous end joining (NHEJ) repair pathway to form the final VJ or VDJ rearrangement product. Impairment of these proteins in genetic deficiencies in humans, or from gene disruption experiments in mice, perturb the rearrangement process.[25]

Combinatorial Considerations

Combinations of Gene Segments

Although nicks can be introduced in 12-RS pairs or 23-RS pairs, complete cleavage and hairpin end formation occurs only with the 12/23 pair.[26] It is the second hydrolysis step completing the double-strand break that is dependent upon the 12/23 pairing.[27–29] Thus, rearrangement would be completed only when two different kinds of gene segments have been brought together, with the germline RS configuration directing the particular gene combinations to be joined. As shown in figure 9.1, the human H-chain Ig V and J genes all are flanked by 23-RS, so that direct recombination between them, excluding the D genes, does not occur. The D genes have 12-RS on both flanks, obliging the final combination to be V–D–J.

The 12/23-RS do not have to face each other, as depicted in figure 9.1. If they are present in the same orientation, recombination can occur by inversion (figure 9.3). Thus, in theory the human D genes can also recombine by inversion, which would double their diversification potential, but the existence of such sequences in the human antibody repertoire has been disputed.[30] However, D inversion can be found in other Ig systems. In sharks, some of the V_H loci carry V and J genes with 23-RS and 2 D gene segments, where D1 carries a 12-RS on the 5′ flank, and 23-RS on the 3′ flank and D2 has 12-RS on both flanks.[31] If guided by the 12/23 rule, then the following combinations should be possible: VD1D2J, VD2J, as well as VD2invJ, where the D2 is inverted because the D2 5′RS was involved in the joining to J and D2 3′RS joining to D1. Such examples were found in the nurse shark by our group, where one particular unique locus has been characterized. The contribution of P region, N region, and varying D gene reading frames is shown in figure 9.4, illustrating the extraordinary junctional diversity achieved, even with just four germline gene segments.

Why are inverted D sequences found in sharks but not in humans? The greatly reduced combinatorial possibilities in the shark, due to the mini IgH loci, may have been compensated by the D configuration that allows for D to D joins, D2-only joins, and inverted D2. This compensation also suggests that diversity of junctional sequences is a crucial component for a diverse antibody repertoire. In contrast,

Deletional V(D)J recombination   Inversional V(D)J recombination

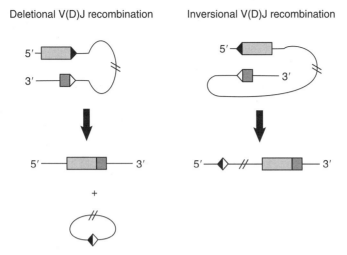

Figure 9.3.   Deletional or inversional recombination determined by the RS configuration. Rearrangement of the gene segments involves deletion of the intervening DNA (left) when the 12-RS (black triangle) and 23-RS (white triangle) are in convergent orientation, resulting in a joined VJ and circularized DNA containing the RS. An inversion occurs when the RS are in the same orientation (right).

there are plenty of D genes in the human IgH (figure 9.1); nonetheless, why do some theoretically possible RS combinations not rearrange, or occur below the level of detection? This is the central question taken up in the next sections: not all RS or RS combinations are equal.

Specificity of RAG Cleavage

A survey of the many different RS from Ig and TCR gene segments characterized in humans, mouse, and other vertebrates showed that the first three nucleotides of the 7-mer (CAC/GTG) were almost perfectly conserved,[32] in accordance with functional studies demonstrating that substitutions at these crucial positions affected recombination function.[33] X-ray diffraction studies on CA tracts, like this part of the 7-mer, show a shifted, distorted structure[34]; local unpairing at CAC/GTG may favor cleavage as well as enable the formation of a hairpin.[18,35] The 7-mers and 9-mers reveal similar characteristics regardless of whether they were from a 12 bp or 23 bp RS; this suggests that the 12-RS and 23-RS are functionally equivalent. In experiments testing the roles of the 7-mer versus the 9-mer, it was found that the efficiency and preciseness of the nick made by RAG is guided cooperatively by the two in concert. In the presence of the 7-mer alone, some nicks are introduced in the correct place, but when the 9-mer is alone, RAG induces nicking in the region of the absent 7-mer.[35]

The spacer sequences align the 9-mer and 7-mer on the same rotational phase of the DNA helix. Whereas the 12 bp spacer separates them by almost one turn of the helix, the 22 bp spacer separates them by two. This observation suggested that the

A

VH (23-RSS)　　　　　　(12-RSS) **D1** (23-RSS)　　　(12-RSS) **D2** (12-RSS)　　　　　　　(23-RSS) **JH**
-TGTGCAAGAGAC　　　ATACTACAGTGGT　　　　　ACATACTGGGATAG　　　　　　ACTATTTGATTACTGGGACAAGGG-
　C　A　R　D　　　　　I　L　Q　W　S　G　　　　T　Y　W　D　　　　　　Y　F　D　Y　W　G　Q　G
　　　　　　　　　　　　　　Y　Y　S　G　　　　H　T　G　I
　　　　　　　　　　　　　　　T　T　V　G　　　　　I　L　G　*
　　　　　　　　　　　　　　　　　　　　　　　CTATCCCAGTATGT (inv)
　　　　　　　　　　　　　　　　　　　　　　　L　S　Q　Y　V
　　　　　　　　　　　　　　　　　　　　　　　　I　P　V　C

cDNA
A5　　..............　　　catTACAGTGG　　　　　　ctCTGGGggg　　　　　..............　　　D1 + D2
CH7　..........GTC　　　aggTACTACAcgggg　　　　aATACTGGGgcc　　　----------　　　P region, D1 + D2
T16　.........--　　　　　　　　　　　　　　　　　　ccGATtcgat　　　GT...........　　　D2 alone, P region
E9　　.........-----　agCAGT　　　　　atatcttTATCCCAacaagagg　　　-----...G.G........　D1 + D2 inverted

Reading frames of D genes

B

Clone　　CDR3

A5　　DHYSGSGGDYFDY　　　D1 frame 2; D2 frame 2
CH7　　DVRYYTGNTGADY　　　D1 frame 2; D2 frame 2
T16　　ADSMYFDY　　　　　　D2 frame 1
E9　　QYISIPTRGLGY　　　　D1 frame 1; D2 inverted frame 3

Figure 9.4. Gene combinations determined by 12/23 RS pairing. (A) Top, the genomic arrangement of shark IgH locus V18 containing V_H–D1–D2–J_H. The flanks of the shark V_H gene segment, J_H gene segment, and its two D genes are shown, with the predicted amino acid sequences. Each D can be read in three frames (shown in single letter amino acid code), and for D2 there are an additional three if involved in inversions; asterisk indicates stop. Bottom, RT-PCR with primers specific for V18 sequences provided rearranged sequences showing junctions containing D1 and D2 (underlined sequences in A5, CH7, E9), or D2 alone (T16). P region (italicized, V_H flank in CH7 and J_H flank in T16), N region (lower case junctional sequence). An inverted D2 is underlined in E9. Dots indicate identity to germline sequence, dashes gaps. (B) Predicted amino acid sequences of the antigen-binding complementarity-determining region, CDR3. CDR3 begins at the third amino acid after the invariant C in the V and ends before the invariant W in the J. The actual contribution of the D gene is underlined; reading frames are indicated. It can be seen that although rearrangement events forming A5 and CH7 both involved D1 and D2 in reading frame 2, the resulting CDR3 are very different, due to trimming and N region addition.

170

spatial relationships between the 7-mer and 9-mer must be important.[13] When the spacer length was experimentally lengthened by more than 1 bp, cleavage by RAG was reduced or abolished; when the spacer length was adjusted to create another full turn (total 34 bp), cleavage levels improved, suggesting that the 7-mer and 9-mer must be on the same helical plane to interact, synergistically, with RAG.[35,36]

Thus spacer length is invariant in nature, a limitation expected from functional studies. However, Ramsden and coworkers[32] also reported that the spacer sequences themselves unexpectedly contain some conservation. (The importance of both the length and nature of "spacer" sequences also is discussed in chapter 1.) Subsequent studies supported the importance of spacer sequence, as differences can result in several fold differences in recombination efficiency.[37,38] An RS with consensus 7-mer/9-mer but a "poor" spacer bound half as much RAG, a modest difference which may not be sufficient to explain its ninefold drop in recombination efficiency.[39] This suggests that the recombination efficacy of the RS involves more than just RAG-binding. The experiments described below underscore this idea.

Some 12/23 Pairs Are Inhibited

The TCR β locus is organized as a series of Vβ gene segments flanked by 23-RS, one D gene segment flanked on the 5′ side by a 12-RS and on the 3′ side by 23-RS, and a series of Jβ segments with 12-RS. In this configuration, Vβ should be able to rearrange to Jβ as well as D, in accordance to the 12/23 rule. However, only VDJ and never VJ combinations are generated. Even when the D gene was experimentally excised, Vβ nonetheless did not rearrange to Jβ. When the 5′12-RS of D was substituted by the 5′12-RS of the Jβ, then no rearrangements took place between Vβ and the mutant D. When the 5′12-RS of one of the Jβ genes was replaced by the 5′12-RS of the D, then rearrangements between Vβ and that mutant Jβ were found. This meant that the 12/23 combination between Vβ and D was targeted but the one between Vβ and Jβ was inhibited.[40]

Exchange of the various elements (7-mer, spacer, 9-mer) between the permissive and nonpermissive RS showed that all three components had roles in variously inhibiting or permitting rearrangement between the Vβ and Jβ.[41] Thus, the patterns of recombination are due to the nature of the RS as a whole. Olaru and coworkers[42] obtained similar recombination results, but in addition tested RS binding affinity to purified RAG proteins. They found that binding by the Jβ 12-RS was actually more competitive than the D 12-RS—the opposite of what would be expected. They concluded that selective recombination involves selective formation of a synaptic paired complex, which includes but is not solely based on affinity for RAG.

If synapsis is not directly related to inherent RS RAG affinity, it may be that additional factors enhance the RS-RAG interaction. It has been reported that both Ku70/Ku80 and DNA-PK, thought to be recruited at the postcleavage stage to participate in repair functions, also are involved in enforcing the 12/23 rule.[19] This suggests that they are actually present before cleavage, the event that requires the presence of 12-RS/23-RS to proceed. These or other as yet unidentified proteins may interact with the 12/23 pair and, depending upon the particular RS sequences, may improve or inhibit the RAG interaction.

RS and the Antibody Repertoire

The relationship between efficient recombination and the nature of the RS may be perhaps be summed up as "good" RS readily form the synaptic paired as well as post-cleavage complexes whereas "poor" or cryptic RS are less inclined to form a stable complex, resulting in the untimely release of RAG and disruption of the process. Thus the RS not only serve to guide the recombinase to the site but also modulate the frequency with which the gene segment is recombined. A spectrum of efficiency in the RS thus exists at the Ig loci, and this results in rearrangements to certain gene segments more often than to others. Because the gene segment usage is nonrandom, the primary antibody repertoire of the animal will be biased (biasing of the initial repertoire by differing efficiency of recombination of different gene fragments also occurs for trypanosome surface antigens, as is discussed in chapter 5). Certain V_H genes[38] as well as V_L genes [43] are expressed in biased frequencies, which in some instances was experimentally demonstrated to be due to RS efficiency.

A survey of V_H RS among mammals showed that the RS did not evolve merely as the extension of the adjacent V_H coding region; it was observed that in the course of evolution the RS can be "swapped" between different families of V_H gene segments by unequal crossing over or other mechanisms. Hassanin and coworkers[44] suggested that, since the more efficient RS type was represented exclusively among mammalian species that carried few germline genes, the germline V_H with potential ligand-binding properties best suited to the animal's survival acquired the most efficient RS through this selection. In the same vein, Olaru[42] has argued that the poor RS sequence combinations were selected for in the TCR β system during evolution to exclude direct V to J recombination and insure the participation of the D gene.

So far, we have mostly been considering the mechanics of rearrangement and how gene usage may be modulated, to bring about the correct recombination product in some instances and in others perhaps to bias for certain gene segments promoting the animal's immune repertoire. However, combinatorial diversity by itself would be the simple multiplicative result of the number of gene segments in the animal's germline. What raises antigen receptor diversification to the truly significant levels is the generation of the highly varied junctions.

(It is useful to note here that the somatic component has been estimated to contribute to 90% of the antibody repertoire.[45] If the number of functional H and L chain gene segments were considered in humans [65 V_H × 26 D × 6 J_H for H chain, 40 Vκ × 5 Jκ for kappa L chain, 30 Vλ × 4 Jλ for lambda L chain], the combinatorial and combinational V_H/V_L gene pairs would be 3.2 × 10^6. Junctional diversity enhances this by about 3 × 10^7, bringing the potential total diversity to 10^{14}.)[46]

Impact of Junctional Diversity

Rearrangement Creates Loop Length Diversity

The V and C regions (figure 9.5) exist as autonomous folding units called domains. The V domain and any C domain both share a tertiary structure called the "immunoglobulin fold",[48] which consists of two β-pleated sheets packed face-to-face, usually

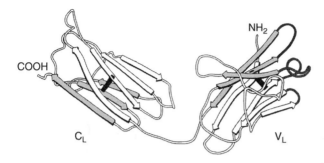

Figure 9.5.   Ribbon drawing of L chain polypeptide. V_L (right) and C_L (left) domains are shown in three dimensions, with β-strands illustrated as gray strips that are connected by loops and bends; the amino and carboxyl termini are labeled. The three CDR loops that form part of the antigen-combining site in the V region are shaded black showing their location at the N-terminal tip. (Adapted with permission from reference 47.)

bridged by a disulfide bond. Each sheet is formed by seven to nine anti-parallel β-strands connected by loops and bends of varying lengths and conformations (figure 9.5). The junctions of both the H and L chains, generated by the recombination process, are part of loops that are situated at the N-terminal tip of the antibody. These loops are called complementarity-determining regions (CDR). Altogether, six loops—two (CDR3) with sequence contributed from the H (V/D, D/J) and L (V/J) chain junctions and four integral to the germline V gene segments—form a continuous surface that constitutes the antigen-combining site. In H chains, HCDR3 is the most variable portion, in length and in sequence, of the entire polypeptide. Human HCDR3 lengths range from two to 26 amino acids.

Because they are at the surface, the loops can tolerate changes in both amino acid composition and overall size. Amino acid sequence variation can change the antibody specificity, but length variations would have an even greater influence on the conformation of the CDR3 loop. Combining sites containing a two-amino-acid HCDR3 versus a 26-amino-acid HCDR3 have radically different topologies. Thus the length changes that result from the rearrangement process are a unique type of diversification that cannot be compared to other somatic mechanisms (hypermutation or gene conversion) in terms of the impact on tertiary structure. Rearrangement introduces diversity, and at one site, directed by the RS. This site localization restricts the wide sequence variation to a region accommodated as part of a loop and that does not interfere with the main-chain folding of the V domain (see figure 9.5). Thus the site-specific RAG-mediated recombination ensured a unique type of diversification (sequence plus loop length variation) that occurred on a regular basis, and in one place that was tolerated in the protein structure.

The CDR3 examples in figure 9.4 give an idea of the extensive junctional diversity that can be achieved even with only four gene segments. Due to trimming and N region addition, two clones (A5 and CH7) consisting of the same combination of gene segments in the same reading frame can have completely different CDR3 sequences. CDR3 sequence and length diversity determine the topology of most ligand-binding sites; in fact, in TCR it is mainly CDR3 that is involved in antigen contact. Thus junctional diversity was likely a significant selection factor in the acquisition of V(D)J recombination early in vertebrate evolution.

Origin of Rearranging Genes

RAG1/2, TCR, and Ig are found only in jawed vertebrates, suggesting that recombining genes may have originated about 450 million years ago. The current theory is that RAG1/2 and the RS they recognize were part of a DNA transposon (see chapter 8 that had entered the vertebrate lineage through horizontal transfer.[49] This hypothesis explains the "sudden" appearance of RAG in evolution, as well as certain properties of RAG and Ig/TCR gene organization. When the V and J gene segments were first isolated and characterized, the inverted repeat sequences of the RS seemed reminiscent of recognition motifs at the termini of mobile elements.[12] With the identification of RAG1 and RAG2 as the recombinase, their unusual genomic organization—tightly linked and each one a continuous reading frame in a single exon—suggested a transposon origin.[10] Moreover, the RAG-mediated DNA strand breakage involves a single-step transesterfication that is chemically similar to the reactions described during retroviral integration and bacteriophage transposition (see chapter 7).[17] The strongest evidence supporting this theory is that RAG can in fact be induced to effect transposition of DNA with RS ends.[50-52]

A model of the evolution of Ig/TCR rearranging genes is outlined in figure 9.6.[2,12,50,53] A DNA transposon (a mobile element with RS flanks and probably encoding the RAG genes) entered the primitive jawed-vertebrate system. The RAG protein was expressed in the germ cells, and through its transpositional abilities RAG integrated this mobile element by chance into an ancestral V domain, splitting it in two segments, V and J. It is likely that the V gene already had an immune function requiring sequence diversification, perhaps by hypermutation,[54] since this mechanism precedes H and L chain divergence[49] and is as ancient as rearrangement. We suggest that rearrangement enhanced diversification in a unique way and was selected for. Subsequently, the V gene segments duplicated in tandem, carrying one terminus of the mobile element; likewise, the J gene segment also duplicated along with the other end, generating the organization that is today present in tetrapods (figure 9.6).

RS Directing Genome Changes?

In the course of evolution, RAG expression in most animals became restricted to precursor lymphocytes, where its transpositional capabilities were suppressed, either in the course of adaptation to the sole use of its V(D)J recombinational function, or because factors in the host environment regulate this unwanted activity. Initially, RAG would have been expressed in germ cells; it may have continued to be expressed in gonads after the V gene disruption. There is evidence for long-standing RAG-mediated activity on the fish germline, and in fact mature transcripts for both RAG1 and RAG2 have been found in abundance in zebrafish oocytes (Changchien and Hsu, unpublished results).

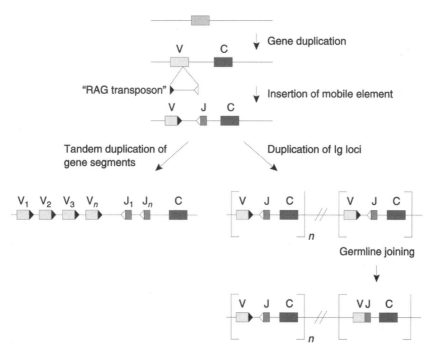

Figure 9.6.   The origin of antigen-receptor genes. The Ig superfamily is ancient; a primordial Ig domain gave rise to V and C domains. The RAG transposon entered the lineage of the primitive vertebrate, and inserted the mobile element with RS (indicated as triangles) into a V domain. With the establishment of V(D)J recombination, two alternative antigen receptor gene organizations developed. In general, tandem duplication of the gene segments occurred at the one IgH or IgL locus in tetrapods (left). In elasmobranch IgH and IgL, duplication of the gene segments along with the C exons resulted in multiple loci (right). Rearrangement of V gene segments at some loci to VJ (IgL) and to VD, VDD, or VDDJ (IgH) occurred in the germ cells; many of the fully assembled V regions appear to be functional.

In sharks, among the 100 to 200 IgH and IgL loci, some contain fully ($V_H DJ_H$ and, as shown in figure 9.6, $V_L J_L$) or partly assembled V_H region gene segments (like VD alone).[31] A comparison of unrearranged L-chain flanks with the joints of germline-rearranged VJ showed that the joining event appeared to be site-directed by the RS. Moreover, some junctions appear to carry P region sequence, suggesting that recombination may have occurred through a hairpin intermediate.[55] In other words, the germline-joined VJ junctions looked very much like the products of lymphocyte recombination; a phylogenetic tree showed that five of the six germline-joined VJ isolated from one shark arose from independent, unrelated events that occurred millions of years apart. The rearrangement of gene segments in elasmobranch germ cells apparently occurred repeatedly and is possibly ongoing.

Other re-configurations of Ig and RS have been examined, and it has been speculated that rearrangements in germ cells could lead to changes in antigen receptor gene organization and perhaps the generation of new sequences with flanking

RS—like D genes.[56] It is possible therefore that RAG not only initiated rearranging genes but also participated in reconfiguration of their gene segments in early vertebrates, for example the H-chain fused VD genes in chickens that are used as templates for gene conversion. Today, RAG activity in germ cells is probably limited to fishes; this is where germline V(D)J joining events are repeatedly discovered.

What is the biological significance of a germline-joined Ig? A few years ago we pointed out that since a preassembled Ig gene is de facto no longer tied to the rearrangement process for its expression, it may evolve a different function.[55] Such an example may be provided by an H-chain-like Ig gene recently characterized by Flajnik and coworkers.[57] IgM1gj is a single copy gene with a germline-rearranged VDJ. It carries three C region exons that are descended from exons 1, 3, and 4 of IgM H chain. It is transcribed and translated, prominently expressed in neonatal sharks; the polypeptide interacts with a L chain. However, it appears not to have a transmembrane exon and is only secreted. This means that this Ig-like molecule cannot be a cell surface receptor that is capable of transmembrane signaling. With this crucial change, the former IgM has evolved away from traditional Ig function, although it interacts with L chain and thus still forms an Ig-like combining site.

We suggest that this is one example of how the information implicit in RS may not only have a role in generating antigen receptors for the adaptive immune system but perhaps also an evolutionary one as well in directing RAG-mediated changes in the genome.

Summary

When the RAG transposon split an ancient "variable" region gene, this was the first step in generating rearranging genes and the adaptive immune system in early vertebrates. The RS are motifs recognized by RAG that serve to promote somatic recombination and create antigen receptors in lymphocytes. We suggest that during the evolution of vertebrates the RS also had a role in directing RAG-mediated changes in the germline.

Acknowledgments   Dedicated to Cecile and Abba Lichtenstein. I thank David Feliciano for designing the figures and Susanna Lewis for her critical evaluation and good company. I also wish to thank Lynn Caporale for her interest and her many helpful comments on this manuscript. This work was funded in part by grants from the National Science Foundation (MCB 0080098) and the National Institutes of Health (R01 GM068095).

10

Somatic Evolution of Antibody Genes

Rupert Beale and Dagmar Iber

Overview

Proteins evolve by the iterative alternation of mutation and selection over evolutionary time. Yet in a matter of days, the vertebrate immune system generates novel antibodies with high affinity for cognate antigens. This is accomplished by somatic mutation of the diverse pool of antibody genes generated from gene fragments as described in chapter 9, and clonal selection of high-affinity antibody producing B cells—a process of somatic rather than the familiar germline evolution. A spontaneously occurring process in DNA—deamination of deoxycytidine to deoxyuridine (C→U)—is catalyzed to produce mutations in antibody genes. A competitive environment is generated within the organism to select for mutations conferring high affinity—and to guard against self-reactivity. Thus the capability to produce high-affinity antibodies is an implicit property of the vertebrate genome.

The Biological Problem

Ehrlich's famous dictum, *corpora non agunt nisi fixata* ("a body will not work unless it is bound"),[1] poses a problem to the immune system: how can the germline genome encode the potential to bind to a vast array of distinct, rapidly evolving, infectious agents? Whereas the "innate" immune system attacks the problem by using receptors which bind to molecules substantially conserved amongst many different microorganisms,[2] the "adaptive" immune system attempts the more ambitious enterprise of evolving specific antibodies to recognize essentially any antigen that may be associated with a pathogen.

Antibodies consist of an amino-terminal antigen binding region called the V (variable) region and a carboxy-terminal C (constant) region, which mediates the effector properties of the antibody by directing other components of the immune system to eliminate the pathogen. The initial diversity generated by VDJ recombination is vast (see chapter 9) but not so vast that high-affinity antibodies to every possible pathogen are created. Fortunately, vertebrates have evolved a mechanism capable of generating high-affinity antibodies from low-affinity antibodies. This process, known as affinity maturation, is achieved by means of multiple cycles of somatic hypermutation and clonal selection. Somatic hypermutation is characterized by point mutations in the V region at all four bases at a rate approaching 10^{-3} per nucleotide, many orders of magnitude above background. Most mutations do not improve affinity, but some give rise to increased affinity variants (and also to variants that are self-reactive). A clonal competition process therefore selects those B-cell clones that are capable of producing higher-affinity antibodies and also guards against the possibility of selecting high-affinity self-reactive clones, as these might pre-dispose to autoimmune diseases such as rheumatoid arthritis.

The mechanism of somatic mutation and the means by which selection is achieved have provided immunologists with a formidable series of interrelated problems in molecular, cellular, and mathematical biology. Recent data regarding how the balance of activity of multiple DNA repair proteins affects the spectrum of mutations observed in the immunoglobulin variable region has shed considerable light on the mutation mechanism. This chapter therefore presents a fair amount of detail regarding the steps from initial mutagenic insult to the observed mutation. After discussing the mutation mechanism we then summarize what is known about targeting of this mutation, and finally address the problem of how high-affinity B-cell clones are selected.

Somatic Hypermutation

Deamination as a Mutagenic Process

The mechanism of somatic hypermutation of the immunoglobulin genes has been a conundrum for many years. This important problem has been partially solved in an elegant manner, and we thus present the key experiments below in approximately chronological order. Whereas many puzzles still remain, it is now generally accepted that somatic hypermutation is initiated by the enzymatically catalyzed deamination of cytosine in DNA. Before describing the experimental basis for the conclusion that catalyzed deamination of cytosine initiates hypermutation, it is worth considering how DNA cytosine deamination might lead to mutation. When C is deaminated, it becomes U. This process occurs spontaneously, and usually triggers a nonmutagenic base excision repair response. Initially in this response, uracil is recognized by a DNA repair enzyme, uracil DNA glycosylase (UDG), which excises the uracil to leave an abasic site, and ordinarily this is subsequently repaired correctly (see chapter 2). However, if replication takes place either before uracil excision or over the abasic site, a mutation can occur. When uracil is present in DNA it is read as thymidine—hence a C→T transition mutation will result; and as incorporation of any base is possible at an abasic site a C→T, C→A, or C→G mutation can occur. Thus a C:G pair can be changed first to a U:G mismatch, then an abasic site:G mismatch and either subsequently fixed as T:A, A:T or G:C mutations or repaired to the original C:G. Clearly, deamination of C could be responsible for mutations at C and/or G, depending upon whether the deaminated C is on the coding or template strand. This idea provided the inspiration for the experiments that demonstrate that somatic hypermutation is initiated by catalyzed DNA cytosine deamination. Of course, the above scheme can only account for mutations at C/G—the mechanism by which mutations at A/T occur is currently unknown. Mutations at C/G are called phase 1 mutations, and mutations at A/T are known as phase 2 mutations, for reasons discussed below.

Mutation, Recombination, or Conversion

Somatic mutation is the exclusive means by which mice and humans further diversify their antigen-binding repertoire after VDJ recombination. However, this does not represent the full extent of somatic modifications to antibody genes. Two other B-cell specific processes are mechanistically related to somatic hypermutation: antibody gene conversion and class-switch recombination. These apparently dissimilar processes arise as a result of differential repair of a common initiating lesion.

Antibody Gene Conversion

In contrast to mice and humans, some species (such as the chicken) utilize gene conversion templated on a number of pseudogenes downstream of the rearranged V gene to introduce additional diversity to the rearranged variable region (see also chapter 5 for a discussion of the use of pseudogenes to generate diversity). This appears especially important in the chicken since VDJ rearrangement generates little diversity in this species. Gene conversion is accompanied by somatic mutation and the relationship between the two has long been suspected to be intimate.

Class-switch Recombination

Class-switch recombination, which involves the deletion of some C regions, brings the V region next to constant regions with more appropriate properties for combating the infection at hand (see chapter 11; reviewed in reference 3). Thus the processes of antibody gene conversion and class-switch recombination give rise to very different final outcomes (see figure 10.1). However, both are initiated in the same way as somatic hypermutation, by the deamination of cytosine, and mutations occur during both processes. They are poorly understood at a mechanistic level after the initial events, and will be discussed here only in as much as they relate to somatic hypermutation.

The Importance of Editing

An important breakthrough occurred when a single gene called activation-induced deaminase (AID) with homology to an RNA editing enzyme called APOBEC1 was discovered to be necessary for both hypermutation and class switch recombination in both mice and humans.[4] Furthermore, the generation of V-region diversity by gene conversion in the chicken cell line DT40 also was shown to depend on AID.[5,6] As APOBEC1 was known to edit RNA, initial speculation centered on the possibility that distinct RNA targets would be edited and altered to generate the complex machinery hypothesized to be responsible for the apparently very different processes of point mutation and gene conversion. However, an alternative explanation was that AID's deaminating activity was focused directly on DNA encoding the V region, rather than on RNA.

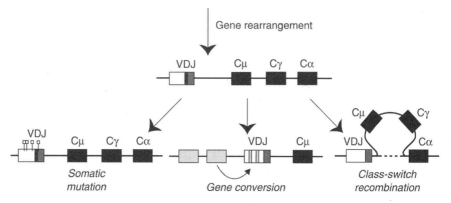

Figure 10.1.   Somatic mutation, gene conversion, and hypermutaton occur after VDJ rearrangement. After antigen encounter in human and mouse, the VDJ rearranged genes in those B cells which express an antigen specific low affinity IgM antibody are subjected to a second wave of diversification, this time by hypermutation. In chickens and rabbits, V region diversification is achieved by using neighboring genes as templates. Short gene conversion tracts are observed. In CSR, the constant (C) region for IgM can be looped out and a different C region can be brought into proximity with the V region, and thus new antibodies with the same antigen specificity but different effector properties can be formed. (VDJ rearrangement is discussed in chapter 9, and class switch recombination in chapter 11.)

DNA Suffers a C Change

AID is now known to be part of a DNA deaminase family of enzymes that act on retroviral DNA (see chapter 14). However, the prototypical family member APOBEC1 has an established physiological role as an RNA deaminase. The first strong evidence for AID acting directly on DNA, rather than on special RNAs expressed in lymphocytes about to undergo hypermutation, came when expression of AID in *Escherichia coli* gave rise to a mutator phenotype.[7] The mutator phenotype resulted from an increase in G/C to A/T point mutations (transition mutations at C/G). This is exactly as one would expect if the increase in mutation were due to deamination of C to U, as during replication A is inserted into the complementary strand opposite any U in DNA that is not repaired prior to replication. In *E. coli*, as in higher organisms, multiple mechanisms exist to repair G:U mismatches that arise from spontaneous deamination of C to U. In the base-excision repair pathway, uracil DNA glycosylase (UDG) recognizes U in DNA and excises it, leaving an abasic site. The abasic site triggers a repair response in which an AP-endonuclease nicks the DNA strand at the abasic site, an exonuclease chews back a patch of nucleotides thereby excising the abasic site, and a polymerase inserts the correct base using information retained in the undamaged complementary DNA strand (see figure 10.2). The mutator effect of AID was greatly enhanced in *E.coli* that, due to UDG deficiency, lack the ability to remove U prior to replication. This strongly suggested that AID is able to cause mutations by directly deaminating C in DNA. Subsequently, this was shown biochemically when human AID was shown to deaminate single-stranded DNA.[8]

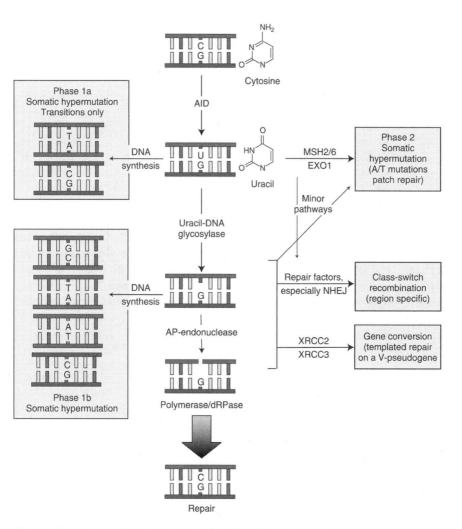

Figure 10.2. DNA deamination model of antibody diversification. AID deaminates C→U in DNA. Phase 1a mutation arises directly from deamination by reading of U as T by a replicative polymerase and results in C→T and G→A transition mutations only. Phase 1b arises as a result of replication over an abasic site, and thus can cause a C→N or G→N mutation. Phase II is dependent on MSH2/6, exonuclease I and polymerase Eta. It may arise from patch repair directed by MSH2/6 recognition of a G:U mismatch or a G:abasic site mismatch. It might also arise from AP-endonuclease dependent base excision repair. Gene conversion is predominantly triggered at a stage downstream of UDG, and is dependent on factors involved in homologous recombination such as XRCC2. CSR is partially dependent on MSH2/6 and exonuclease I, but shows a more pronounced dependence on UDG. It depends on nonhomologous end joining rather than homologous recombination. Finally, correct repair of the lesion may be achieved by base excision repair machinery. (AP endonuclease, apyrimidinic endonuclease; dRPase, 5′-deoxyribophosphodiesterase.)

Though provocative, the demonstration that AID had DNA cytidine deaminase activity did not prove that this was its physiological mechanism of action in lymphocytes. Independent evidence that AID deaminates V-region DNA came from testing the predictions of the model by examining of the effect of inhibiting, or genetically ablating, UDG on the observed pattern of V-region mutations. Whilst C→T (or G→A) transition mutations result from replication over unrepaired U (phase 1a mutations, see figure 10.2), as described above mutations from C to G or A (or from G to C or T) could be generated from the same initial lesion if U was removed by UDG, thus generating an abasic site, followed by the action of a polymerase capable of adding any nucleotide opposite an abasic site (phase 1b mutations, see figure 10.2).

As noted in chapter 2, the nucleotide insertion preference of the polymerases that replicate across the abasic site would influence the observed mutation spectrum. However, the spectrum of mutations observed will also be heavily influenced by the balance of different DNA repair pathways acting on a G:U mismatch. For example, if UDG activity were decreased, fewer abasic sites will be generated while more U remains in place, so mutation will be biased toward transition mutations (C→T or G→A) that result from insertion of an A opposite an unrepaired U on either the coding or complementary strand. A strong prediction would therefore be that if UDG were inhibited or knocked out, there would be a dramatic increase in the ratio of transitions to transversions at G/C (see figure 10.2).

Of Mice and Men—And Chickens

As described above, chickens normally undergo gene conversion as well as point mutation in their V regions. The chicken cell line DT40 provided the first confirmation for the model presented in figure 10.2. This remarkable line exhibits constitutive AID-dependent antibody gene conversion, and its hyperrecombinogenic phenotype enables gene disruption by homologous recombination, facilitating experimentation. Both the gene conversion and hyperrecombinogenic phenotype are dependent on DNA repair enzymes which mediate homologous recombination, such as the Rad51 paralogues.[9] When DT40 cells are genetically deficient in either of two of these Rad51 paralogues (XRCC2 and XRCC3) the balance between point mutation and gene conversion is altered, greatly increasing the rate of point mutation at C/G.[10] However, unlike the human and mouse, in which mutations at C/G are predominantly transitions, as would be expected if much of the U was left unrepaired, mutations in the XRCC2 deficient DT40 are predominantly G/C→C/G transversions. This suggests that, in wild-type DT40, most Us are processed into abasic sites that then trigger gene conversion. In the XRCC2 deficient lines used in this experiment, translesion synthesis by a polymerase capable of adding nucleotides opposite the abasic site (as discussed in chapter 2) results in a high proportion of G/C→C/G transversion mutations. If UDG were inhibited in the XRCC2 deficient DT40, one would expect a shift toward G/C→A/T transitions because the U that resulted from the deamination of C would remain in place and thus direct insertion of A on the new strand. In order to test this, the inhibitor of uracil DNA glycosylase (UGI) (found in bacteriophage which utilize U rather than T in their genomes) was expressed in XRCC2-negative DT40 and the pattern of antibody hypermutation was examined. Indeed, G/C→A/T transitions now predominated,[11] exactly as one would expect according to the model in figure 10.2.

This model was further tested by targeted disruption of the UNG gene coding for UDG in mice. An examination of the sequence of the Immunoglobulin V regions of these mice lent support to the model.[12] As expected, if U is left unrepaired prior to DNA replication due to the absence of UDG, 95% of their mutations at G/C are transitions to A/T, as opposed to only about 50% in litter mate controls. Intriguingly, pointing to its dependence upon the same initiating event, these mice were also seriously defective in class switch recombination. Latterly, a group of patients have been identified with mutations in the gene for UDG who have a phenotype similar to that of the mice.[13]

Hypermutation Hotspots

It had long been observed that antibody hypermutation especially targets C and G in particular nucleotide contexts. These sequences correspond to the preferred DNA targets of AID in biochemical assays. Biochemical[14,15] and bacterial genetic[16] studies have shown that the local sequence specificity of AID and the local sequence context of mutations at G/C in the V region are congruent. These studies now identify WRC (A or T, purine, C) to be the preferred target of AID. WRC motifs are not just concentrated in the V regions, but are enriched in the parts of the V regions that encode antigen binding regions of the antibody. Thus the residues most critical to the generation of a broad range of combining site diversity are especially likely to mutate. Intriguingly, the hottest hotspots are invariably WGCW, a sequence formed from overlapping WRC motifs on opposite strands, suggesting AID may simultaneously engage both DNA strands.[16]

Phase 2—A/T Mutations

Whilst AID's ability to deaminate cytosine in DNA can, as described above, readily account for mutations at C/G pairs, A/T pairs also undergo hypermutation. The initial hypothesis put forward to explain somatic hypermutation was erratic replication by especially error-prone polymerases misincorporating opposite all four base pairs (Brenner and Milstein 1966).[17]

However, it now appears that mutation takes place in two discrete phases. Most of what we know about somatic hypermutation concerns phase 1—mutations at C/G as discussed above—since these are directly dependent on the deamination of C in a C/G pair. The mechanism by which AID triggers hypermutation at A/T pairs remains a mystery. Deficiency, in mice and humans, of components of mismatch repair recognition and the exonuclease and repair proteins that follow it, do give some clues as to how it might proceed. For example, mice deficient in either of the two components of the eukaryotic single-base mismatch recognition heterodimer homologous to mutS (i.e., MSH2 or MSH6) have C/G biased hypermutation—mutation appears to proceed normally at C/G but not at A/T (in addition, they have defective class-switch recombination, reviewed by Martin and Scharff).[18] This was perhaps the first evidence in favor of two distinct phases of mutation, with all mutation initiated by AID and a second phase that involves mismatch repair.[19] In addition, mice deficient in exonuclease 1 exhibit a very similar phenotype to the mismatch repair-deficient mice. Exonuclease 1 is thought to excise a short patch of DNA adjacent to the mismatch recognized by the MSH2/6 heterodimer, which is replaced by the action of polymerases based on information in the complementary strand. Patients who are deficient in polymerase eta also have normal C/G mutations (and normal class-switch recombination), but reduced mutations at A/T. Thus genetic deficiencies in the mismatch repair pathway or DNA polymerase eta[18,20,21] all reduce phase 2 mutation but leave phase 1 intact. Furthermore, mice and humans deficient in UDG have normal mutations at A/T but all the C/G mutations are transitions (i.e., phase 1a, replication over U). The phenotype of mice deficient in both UDG and MSH2 is very instructive.[22] Whereas Msh2–/– mice show reduced mutations at A/T base pairs, these doubly deficient mice show a complete absence of mutation at A/T, and all the mutations at C/G are transitions—only phase 1a mutations (i.e., insertion of A opposite U) occur in these mice. These findings suggest that the G:U mismatch arising from a deaminated G:C pair is either recognized by UDG and the uracil is excised leading to a phase 1 mutation as discussed, or it is recognized as a mismatch by MSH2/6 and this leads to mutations at A/T mutations (i.e., phase 2 mutations).

The A/T specific nature of phase 2 mutations has prompted an intriguing suggestion that they arise as a result of misincorporation of U opposite A during MSH2/6-triggered repair in the region of a G–U mismatch.[22,23] According to this attractive but at present entirely speculative scheme, phase 1 mutation is triggered by deamination of C to form U, and phase 2 mutation is triggered by misincorporation of U. Both phases of mutation would then be dependent on excision of U and synthesis over the resultant abasic site to generate a full spectrum of mutations.

Targeting of Mutations

The very high levels of mutation seen in the V region could presumably not be tolerated genome-wide in lymphocytes. Early observations suggested that *cis*-acting elements were sufficient to direct hypermutation to any DNA sequence placed in the genomic context of the recombined V region.[24] On the other hand, some AID induced mutations are found outside the antibody locus, albeit at a much lower level (less than one tenth the level of somatic hypermutation in the V region).

BCL-6 was the first such gene described as undergoing mutation *in vivo*, and it seems possible that many actively transcribed genes in mutating B cells may accumulate mutations with a WRC preference, suggesting AID dependence. Hypermutation is transcription linked, with the probability of acquiring a mutation increasing rapidly shortly after the promoter and then tailing off with further distance.[25] Biochemically AID acts on single-stranded DNA but not on double-stranded DNA. Since C and G seem equally likely to mutate, then both the transcribed and the nontranscribed strand must become accessible in the B cell. It has been suggested that supercoiling may possibly have the effect of promoting DNA deamination on both DNA strands.[26] There is some indirect evidence that increasing the single-strandedness of the DNA in the V region may be partly responsible for targeting of mutations. A single-stranded DNA binding protein, replication protein A (RPA), has been identified as a binding partner of AID which stimulates the DNA deamination activity of AID on transcribed substrates in vitro.[27] It seems possible that a combination of transcription, supercoiling, and association with RPA could be sufficient to target mutations to the V region, but it is more likely that major components of the targeting mechanism are as yet undiscovered. There is no evidence at the time of writing for a V-region specific factor which recruits AID. Thus the mechanism by which targeting of mutations to the V region is achieved is currently unknown, and remains an important focus for research.

Selection

Targeting of mutations to the V region increases the fraction of effective mutations, that is of mutations that alter binding to the antigen at hand. However, in the immune system, much as in evolutionary biology, the great majority of all effective mutations are unfavorable. Such mutations may not only lead to loss of antigen specificity, but worse still may engender significant affinity to an autoantigen. Ongoing selection is therefore essential, both to eliminate self-reactivity and to select pathogen-specific antibodies that have acquired affinity-enhancing mutations.[28] While in the case of adaptive genome evolution selective forces derive from challenges from the external environment across generations, this cannot apply to a somatic evolution process. Rather, selective forces must originate from within the organism. The immune system thus had to evolve not only an elaborate mutation mechanism to generate variation in antibody genes but also an anatomic framework and a signaling infrastructure to enable selection of high-affinity variants.

The Basic Principles of Selection in the Immune System

The selective forces applied to the B cell are similar to those experienced at the whole organism level. For natural selection to drive genome evolution, entities must be able to produce similar offspring with variations in individual characteristics that affect the fitness of the offspring—that is its reproductive success. Likewise in the immune system, selection cannot act directly on antibodies (which cannot reproduce) but must act on B cells, which express the rearranged and possibly mutated antibody gene as B-cell receptor (BCR). B cells increase their proliferative rate upon stimulation by antigen via their BCRs, but low-affinity as well as self-reactive B cells are ultimately doomed to death by apoptosis. While apoptosis of autoreactive clones is crucial to guard against autoimmunity, apoptosis of low-affinity B cells increases the stringency of selection. The reproductive success of B-cell clones is thus represented by their ability both to proliferate and avoid apoptotic death.

In order to survive and proliferate, B cells need to pass a two-step selection process, during which they first acquire antigen and subsequently present fragments of this antigen to T helper cells.[29] The first step is dependent on BCR recognition of antigen, and thus serves to demonstrate binding capabilities of the BCR.

The second step of recruiting T cell help requires B cells to internalize BCR-bound antigen and target the antigen to specialized compartments (the MIIC) where the antigen is proteolytically processed into shortened peptides.

These peptides are subsequently loaded onto MHC class II molecules and transported to the B cell surface. Antigen specific T helper cells can recognize the peptide–MHC molecules (pMHC) and rescue the pMHC-presenting B cells from apoptosis. While antigen presentation by "professional" cells (cells that also provide co-stimulation) is likely to reduce the risk for B cells to acquire self-reactivity, it seems to be the requirement for B cell–T cell interaction that mainly helps guard against autoimmunity. This is because self-reactive T cells are eliminated during their development in the thymus[30] and should thus not be available in the periphery.

A Place to Meet

Antibody affinity maturation requires multiple interactions between antigen, B cells, and T cells. Usually antigen and antigen-specific B and T lymphocytes are all scarce at the onset of an infection and thus it is insufficient to rely on chance encounters. The immune system solves this problem of first contact by concentrating antigen in local lymph nodes through which lymphocytes pass.

After a proliferative burst in primary foci, a small number of B cells and T cells enter primary follicles and seed germinal centers. These germinal centers are microenvironments for antibody affinity maturation and greatly facilitate B cell selection and survival by increasing the number of B cell–antigen and B cell–T cell encounters. It should be noted that affinity maturation can, however, also be observed in the absence of germinal center reactions, though not with the same efficiency.[31]

The germinal center reaction is complex, as shown by figure 10.3, and fully reviewed by MacLennan.[29] Briefly, once the primary follicles are filled with proliferating B cells, the B cell blasts differentiate into so-called centroblasts.[32] Centroblasts are the cells that are thought to express AID and thus to hypermutate their antibody genes. Centroblasts differentiate into centrocytes, increase the expression of surface immunoglobulin, and are selected upon their BCR's affinity for antigen. Centrocytes are prone to apoptosis and will die rapidly unless they are rescued by follicular dendritic cells (FDCs), which retain the antigen in form of immune complexes (composed of antigen, antibodies, and complement)[29] and which provide temporary survival signals to antigen-specific centrocytes.[33,34] While germinal centers formed in the absence of FDCs fail to support affinity maturation and rapidly regress,[35] FDCs are not sufficient for long-term B-cell selection and survival. T cells need to finally rescue and select antigen presenting B cells.[32,36] Nonselected centrocytes undergo apoptosis and are quickly engulfed by macrophages; selected centrocytes may leave the germinal center reaction to become memory or plasma cells.[37] Back-differentiation of selected centrocytes into centroblasts enables a B-cell clone to go through multiple rounds of mutation and selection.[28,38]

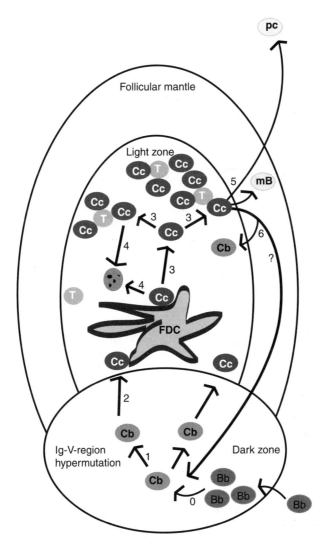

Figure 10.3.   A simple model of the GC reaction. A small number of B blasts (Bb)
initially seed the germinal centre reaction and exponentially expands before differentiating
into centroblasts (Cb). Centroblasts (Cb) continue to strongly proliferate (1), start hypermu-
tating their antibody genes and eventually differentiate (2) into centrocytes (Cc). Centrocytes
re-express IgM and are selected according to their affinity for antigen. Selection involves
antigen acquisition from follicular dendritic cells (FDCs) and subsequent presentation to T
cells (3). Failure to either acquire antigen or to find T-cell help results in apoptosis of B cells
(4). Selected B cells, on the other hand, either leave the germinal centre reaction to become
memory cells or plasma cells (5), or back-differentiate into centroblasts, thus enabling several
rounds of mutation and selection (6). It is currently unclear whether these back-differentiated
centroblasts migrate back to the initial pool of centroblasts which in some (but not all)
germinal centre reactions segregate into a so-called dark zone, while centrocytes and FDCs
are found in a so-called light zone.

The Mechanism of B-cell Selection

For affinity to increase during the course of the germinal center reaction, the affinity threshold for selection needs to increase (i.e., a higher affinity must be required for survival as selection proceeds over time). The extent of increase required for survival appears to be determined by clonal competition within the B-cell population. Accordingly, low-affinity B-cell clones can persist when competition with high-affinity B-cell clones is reduced. Experimentally, this is achieved in germinal centers of transgenic mice that bear a prerecombined antigen-specific (low-affinity) V_HDJ_H transgene, which is also not well-targeted by somatic hypermutation.[39,40]

Principally, B-cell clones might compete for antigen or for T-cell help—or both. Older models have particularly stressed the role of B-cell competition for scarce antigen[41] and suggest that antigen removal during the immune response as well as emerging serum antibodies would lead to a gradual increase of the affinity threshold.[42,43] However, affinity maturation is hardly compromised by alterations in the initial amount of antigen deposited on FDCs[44,45] or by lack of secreted antibodies.[46] Competition for antigen thus appears inadequate to explain B-cell selection for all the conditions under which it occurs.

In contrast, recent experiments suggest that B cells can discriminate between different affinities of binding on an individual basis and signal accordingly.[47]

Given that T-cell responsiveness to antigen-challenged B cells correlates with binding affinity,[48] it is conceivable that the BCR binding affinity is translated into a cellular signal that affects the ability of B cells to recruit T-cell help. One way that the signaling strength of B cells may affect T-cell responsiveness is an enhanced presentation of pMHC,[49,50] which may be enabled by an improved targeting of antigen to the processing compartments. Targeting to and accession of the processing compartments requires signaling through Ig-α and Syk (as well as downstream effectors).[51] The degree of Ig-α and Syk phosphorylation is affinity dependent[47] thus suggesting a link between BCR affinity and the level of pMHC presentation on the B-cell surface.

The Basis for the Efficiency of the Somatic Evolution of Antibodies

Starting with weak binders, the immune system is capable of generating high-affinity antibodies with remarkably few mutation and selection steps. Key to this efficiency are both the evolved antibody structure, which permits large (targeted) mutation rates, and the stringency of the selection mechanism.

Antibodies Evolved to Evolve

Increased mutation rates are in general harmful as most mutations are detrimental (as discussed in chapter 4). In the case of antibodies, mutations are targeted to the V region, and in particular to the complementarity-determining regions (CDRs). The CDRs form loops that mediate contact with the antigen (chapter 9). Targeting of mutations to the CDRs increases the fraction of mutations which alter affinity and limits the number of mutations that may disrupt other essential structural elements of the antibody—a cause for a large proportion of detrimental mutations in other proteins.[52] The affinity of antibody binding is increased both by direct changes in the binding site as well as by mutations that lie more distant and alter the conformation of the binding site. While some single mutations can increase an antibody's affinity by tenfold, affinity maturation is the result of several additive changes; high-affinity antibodies harbor on average six to eight mutations.[53] Considering that in protein engineering approximately six to eight appropriately targeted mutations are required to alter binding specificity,[54] the optimized character of antibody affinity maturation becomes clear.

Efficient Selection

Another evolved feature of the immune system is its effective selection strategy. There are several key features. First of all, and most importantly, the multiple rounds of mutation and selection enable a stepwise affinity maturation process, which strongly facilitates the generation of high-affinity antibodies.[38] Analysis of fitness landscapes in evolutionary biology suggests that a potential danger of such stepwise evolution process is the trapping in local optima.[55] However, the parallel but spatially separate evolution of many germinal center reactions in the follicle (each seeded by few B blasts) avoids trapping all lineages of B cells in what may prove to be an unfavourable local optimum that lacks paths to high-affinity antibodies.

But B cells must not only be selected for their ability to produce high-affinity antibodies. There also must be a mechanism that guards against emerging autoreactivity. This challenge is met by the second key feature of the selection strategy, that B-cell selection depends not only on the acquisition of antigen but also on T-cell help. The requirement for T-cell help may also enable the large robustness of affinity maturation to variations in the initial amount of available antigen since competition for interaction with T cells (which make up only 5–10% of all germinal center cells) may drive affinity maturation over a wide range of antigen concentrations.

The third feature is that the short lifespan of centrocytes, and the ability of T cells to confer both death and survival signals increases the efficiency of B-cell selection. In mice in which apoptosis is impaired either by decreased ability to generate an apoptotic signal (mice that lack Fas) or by increased suppression of apoptosis (mice that overexpress Bcl-x), low-affinity B cells persist in the germinal centre reaction and the selection of high-affinity B cells is impaired.[56,57]

Conclusions

The vertebrate immune system must adapt to pathogens that are both rapidly reproducing and evolving. Adaptation on the time scale of host generations is not sufficient. What has evolved, on the time scale of host generations, is an immune system that is organized to generate the diversity required to adapt to pathogens rapidly within the lifetime of an individual. Rapid adaptation can be enabled by a situation-dependent increase in the local mutation rate, which generates functional variation.

As a result of targeted mutation and an efficient somatic selection mechanism, antibodies somatically evolve that will bind with high affinity to any attacking pathogen, including one never before encountered in the evolutionary history of the host.

Whilst an organism inherits a single genome, it can generate B cells with varied genomes. This is achieved by VDJ recombination as described in chapter 9 and a subsequent targeted mutation mechanism that exploits fundamental properties of DNA mutability and repair, and which acts on an easy-evolvable antibody structure. The selection mechanism is fine-tuned to avoid autoreactivity and promote rapid generation of high-affinity antibodies. Although our current understanding of these processes appears to be sufficient to generate and select high-affinity antibodies to a desired specificity in vitro, this requires a much longer time span and far more cumbersome selection methods than employed by the immune system.[58] The elegance of the biological process thus belies its bewildering complexity.

11

Regulated and Unregulated Recombination of G-rich Genomic Regions

Nancy Maizels

Overview

One of the surprises emerging from genomic sequencing has been the unevenness of the genomic landscape. Even simple features, like base composition, strand bias, and identity and density of repetitive elements, are nonuniformly distributed among chromosomes and along specific chromosomes. The unevenness of the genome can be seen either as reflecting the dictates of biological function, or as a curious characteristic without biological rationale. The more we learn about mechanisms responsible for genomic maintenance and evolution, the less accidental genomic sequence appears to be, and even simple sequence motifs become informative about function.

One of the simplest of sequence motifs, the G-run, confers on G-rich regions the ability to form an alternative structure, G4 DNA. G4 DNA can promote interactions between regions which are G-rich but otherwise unrelated in sequence, enabling G-rich genomic regions to participate in pathways that contribute to genomic variability. This is both useful and perilous. Immunoglobulin class switch recombination provides an example of a regulated, targeted, region-specific recombination event that depends on G-rich sequences. In contrast, unregulated or unchecked formation of G4 DNA can result in replication errors and translocations, and may be a potent force for genomic variability both within an individual and on evolutionary time scales. This chapter summarizes what we know about G4 DNA: what it is, how it forms, how it is eliminated, and how it contributes to biological functions.

G4 DNA: An Alternative Structure Formed by G-rich Motifs

The *B*-form DNA duplex is an elegant structure for storage of genetic information, but DNA has considerable potential to form structures that depart—sometimes

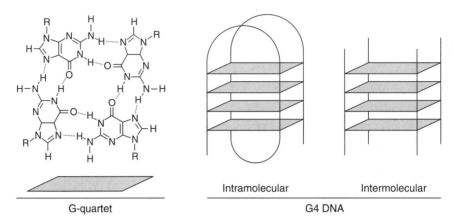

Intramolecular   Intermolecular

G-quartet   G4 DNA

Figure 11.1.   G4 DNA, a four-stranded structure stabilized by G-quartets. G-quartets (right) are planar arrays of four guanines. Each guanine forms four hydrogen bonds, two with each neighbor. The bond at the N7 position is diagnostic of a G-quartet. In G4 DNA, G-quartets stabilize interactions between four DNA strands. G4 DNA can form as a result of intramolecular or intermolecular interactions, as shown. The shaded parallelograms represent G-quartets. G4 DNA forms most readily in DNAs that contain runs of three or more consecutive guanines, although G-quartets can be formed within sequences with shorter G-runs (2 nt).

dramatically—from this canonical double helix. Potential for structure formation depends upon DNA sequence, as reviewed in chapter 1. Among the simplest sequence motifs prone to form stable, noncanonical structures are guanine runs (G-runs), which spontaneously assemble into four-stranded structures stabilized by interactions among guanines (see figure 11.1).[1] These structures are variously referred to as G-quadruplex, G-tetraplex, or G4 DNA. For simplicity, the term G4 DNA will be used here to describe the variety of four-stranded structures stabilized by G-quartets. G4 DNA forms spontaneously in G-rich oligonucleotides in vitro, and it also forms intracellularly.

G-quartets and the Structural Potential of G-rich Nucleic Acids

The basic unit of G4 DNA is the G-quartet, a planar array of four guanines stabilized by hydrogen bonding.[2] The chemical hallmark of a G-quartet is the hydrogen bond between the ring N7 of one guanine and the exocyclic amine of a neighboring guanine. In contrast, guanine N7 is not hydrogen-bonded in single-stranded or duplex nucleic acids. G-quartets can stabilize structures in DNA, and also in RNA. DNA structures stabilized by G-quartets are four-stranded. The interacting guanines may all be on a single nucleic acid molecule (intramolecular G4 DNA; figure 11.1), or they may be on two, three or four DNA molecules (intermolecular G4 DNA; figure 11.1).

G4 DNA is Structurally Distinct from Duplex DNA

In G4 DNA, four phosphodiester backbones encircle a core of very large diameter, defined by the G-quartets.[3,4] The strands may be antiparallel, as in duplex DNA, or parallel.[5] In parallel G4 DNA, the phosphodiester backbones are separated by grooves of equal size, which are even narrower than the minor groove of B-form DNA. Cations, especially K^+, fit snugly in the hole formed by the guanine bases at the center of a G-quartet, and this contributes dramatically to stability. The K^+ concentration in a mammalian cell is 120 mM, so the intracellular ionic environment will maintain G4 DNA structures. G4 DNA is very stable, deriving stability from hydrogen bonding within quartets and stacking between guanines. G4 DNA resists thermal denaturation, but can be denatured by treatment with alkali or by removal of K^+. Because G4 DNA is so different in structure from duplex DNA, it is resistant to attack by most nucleases and helicases. Moreover, G4 DNA (or G4 RNA) is not included in most standard algorithms designed to reveal structural potential of nucleic acids, so this structure can easily be ignored or overlooked.

G-rich Oligonucleotides Spontaneously Form G4 DNA in Vitro

G-rich oligonucleotides spontaneously form G4 DNA in solution. This occurs readily in physiological salt conditions at room temperature—so readily that even laboratory stocks of synthetic oligonucleotides partition into an equilibrium mix of G4 and single-stranded DNA. In fact, G4 DNA was "discovered" when synthetic oligonucleotides bearing sequences from the immunoglobulin switch regions were found to migrate anomalously upon gel electrophoresis.[1] Oligonucleotides bearing sequences of other naturally occurring G-rich repeats also form G4 DNA, for example the vertebrate telomeric repeat, $(TTAGGG)_n$.[6] G4 DNA formation approaches 100% following overnight incubation in a concentrated solution (1 mg/ml) at 60 °C, in high salt. G4 DNA is identified by footprinting with dimethylsulfate (DMS). DMS attacks the N7 of guanine, which is accessible in single-stranded and duplex DNA, but hydrogen-bonded with the exocyclic amino group of a neighboring guanine in a G-quartet (figure 11.1). Resistance to DMS cleavage is diagnostic for G-quartets and G4 DNA.

Dynamic Formation of G4 DNA and Genomic Variability

In principle, any DNA strand that contains at least four runs of guanines has the potential to form intramolecular G4 DNA. The length of the guanine runs and the distance between them will determine the efficiency of G4 DNA formation and the stability of G4 DNA once formed. The ability of the telomeric repeat $(TTAGGG)_4$ to form stable G4 DNA shows that runs consisting of as few as three guanines can support G4 DNA formation, within the appropriate sequence context. Formation of G4 DNA does require that guanines be free to interact with other guanines to form G-quartets. This cannot occur in duplex DNA, so under most circumstances DNA

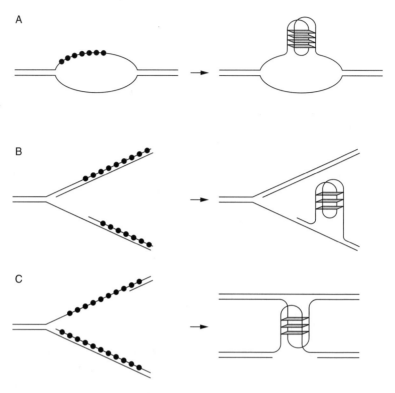

Figure 11.2.   Consequences of G4 DNA formation during transcription and repli-cation. (A) Formation of G4 DNA induced by transient denaturation accompanying tran-scription. Left: a transcription bubble formed within a DNA duplex; filled circles indicate G-runs. Right: G4 DNA stabilized by intramolecular G-quartets (parallelograms). (B) Intra-molecular formation of G4 DNA in a nascent G-rich lagging strand. If unrepaired, this struc-ture could lead to expansion of a G-rich region. Symbols as in (A). (C) Intermolecular formation of G4 DNA. This could result in sister chromatids that fail to separate at metaphase.

within the genome is protected from G4 DNA formation. G-rich regions are released from the prison of the Watson–Crick duplex during transient denaturation that accompanies transcription and replication. This means that G4 DNA formation is a dynamic process and subject to transcriptional and cell-cycle controls.

G4 DNA Induced by Transcription

Transient DNA denaturation that accompanies transcription can induce G4 DNA formation (figure 11.2a). Moreover, because transcription is tightly regulated, G4 DNA formation can also be regulated. As discussed below, the immunoglobulin switch (S) regions provide an example of G-rich regions that do not encode protein but which must be transcribed for regulated DNA recombination to occur. Transcription of these and other G-rich regions has been shown to induce forma-tion of G4 DNA in vitro and in vivo.[7]

G4 DNA Induced by Replication

DNA undergoes transient denaturation at the replication fork. The single-stranded region is protected by single-strand binding proteins, but protection is imperfect. Replication thus has the potential to induce formation of both intra- and intermolecular G4 DNA.

Expansion and Contraction of G-rich Regions

If a G-rich region forms G4 DNA upon transient denaturation during normal DNA replication, factors associated with the replication apparatus must unwind the structure for faithful replication to proceed. If G4 DNA is not unwound, it may cause genomic expansion or contraction by a mechanism similar to that which causes instability of simple sequence repeats, including monomers, dimers, and triplet repeats associated with human disease (see chapter 4).

Intramolecular formation of G4 DNA during replication can have two different results. An expansion of the G-rich region will occur if G4 DNA forms on the nascent lagging strand (figure 11.2b); or a contraction if G4 DNA forms on the template strand (not shown). As discussed below, there are numerous examples of unstable G-rich repeats.

Synapsis of G-rich Regions and Failed Chromosome Segregation

Intermolecular G4 DNA formation at a replication fork can create a structure that is effectively synapsed (figure 11.2c). If not resolved, this would leave sister chromatids joined at metaphase and result in chromosome nondisjunction. Among the factors that ensure chromosomal disjunction are helicases of the RecQ family, which as described below are critical for maintenance of G-rich genomic regions. Structures similar to that shown in figure 11.2c, called "quadriradials," are evident in human cells lacking BLM helicase, a RecQ family member. In these cells, the level of sister chromatid exchange is also greatly elevated,[8] which may result from attempts to repair replication intermediates like those shown. In yeast, mitotic chromosome segregation is similarly impaired by mutations of the related helicase Sgs1.[9]

G-rich Genomic Regions

Certain chromosomal regions are G-rich and contain sequences with considerable potential to form G4 DNA. G4 DNA can contribute to genomic variation, as described above; and it can also provide a structural target for specific factors that regulate gene expression or promote recombination. Distinctively G-rich regions with specialized functions in vertebrate cells include: the immunoglobulin heavy chain switch regions; the telomeres; the genes that encode ribosomal RNA (rDNA); and simple sequence repeats and minisatellites. G-rich regions are also found within single-copy genes.

A Immunoglobulin switch (S) regions

Murine

Sμ　GCTGAGCT<u>GGGG</u>TGAGCTGA

Sγ3　<u>GGGG</u>ACCA<u>GG</u>CT<u>GGG</u>CAGCTCT<u>GGGGG</u>AGCT<u>GGGG</u>TA<u>GG</u>TT<u>GGG</u>AGTGT

Sγ1　ACCCA<u>GG</u>CAGAGCAGCTCCA<u>GGGG</u>AGCCA<u>GG</u>ACA<u>GG</u>T<u>GG</u>AAGTGT<u>GG</u>TG

Sα　<u>GGG</u>ATGAGCTGAGCTA<u>GG</u>CT<u>GG</u>AATA<u>GG</u>CT<u>GGG</u>CT<u>GGG</u>CT<u>GG</u>TGTGAGCT<u>GGG</u>TT

Human

Sμ　TATACT<u>GGG</u>CTTAGCT<u>GGGG</u>CT<u>GGG</u>C

Sγ4　CA<u>GGGG</u>CA<u>GG</u>T<u>GGGGGG</u>CA<u>GG</u>A<u>GG</u>A<u>GG</u>AGCA<u>GGGGG</u>AGCTCTT<u>GG</u>AGCTCA<u>GGGG</u>

Sε　T<u>GGG</u>ATTGAGCTGA<u>GG</u>A<u>GGG</u>CTA<u>GG</u>CT<u>GGGGG</u>AGAGACTGACGACGG<u>AC</u>A<u>GGG</u>TTA

B Telomeric repeats

Vertebrates	TTA<u>GGG</u>
C. elegans	TTA<u>GGC</u>
T. thermophila	TTTT<u>GGGG</u>
S. pombe	TTACA<u>GG</u>
S. cerevisiae	T<u>G</u>$_{1\text{-}3}$

C Human simple sequence repeats and hypervariable minisatellites

Fragile X	C<u>GG</u>
D1S7 (MS1)	A<u>GGG</u>T<u>GG</u>AG
D4S43	<u>GGGG</u>A<u>GGGGG</u>AAGA
Insulin-linked repeat	ACA<u>GGGG</u>TGT<u>GGGG</u>
D1Z2	CCT<u>GGGGG</u>NGNGTGCTGTTCCA<u>GG</u>CTGTCAGA<u>GG</u>CTC

Figure 11.3.　G-rich chromosomal regions. This figure shows examples of G-rich chromosomal regions, with G-runs underlined. (A) Representative sequences from murine and human immunoglobulin switch (S) regions. (B) Telomeric repeats. (C) Simple sequence repeat and hypervariable minisatellite sequences from the human genome.

Immunoglobulin Heavy Chain Switch Regions

The immunoglobulin heavy chain switch (S) regions comprise a G-rich chromosomal domain with specialized function in the mammalian immune response. As discussed in greater detail below, the S regions participate in immunoglobulin class switch recombination, a regulated, region-specific recombination process in which an expressed variable region is joined to a new constant region, deleting many kilobases of DNA. S regions consist of degenerate G-rich repeats, which are from 2 to 10 kb in length and conform to a loose consensus. Examples of representative sequences from murine and human S regions are shown in figure 11.3a.

Telomeres

In almost all eukaryotes, the telomeric repeat contains runs of Gs (figure 11.3b), and oligonucleotides bearing telomeric repeats readily form G4 DNA in vitro.[10,11] As G4 DNA is distinct in structure from duplex DNA, regulated formation of G4 DNA at the telomeres could produce a structure that is recognized specifically by telomere binding proteins. Telomeres terminate with G-rich single-stranded tails, which are the primers for telomere extension by telomerase. G4 DNA formation within the telomeric tails could protect them from degradation by exonucleases, or prevent addition of new telomeric sequence by telomerase.

Ribosomal DNA

Eukaryotic ribosomal RNA (rRNA) is transcribed from a family of repeated DNA sequences, the ribosomal DNA (rDNA). rDNA repeats are G-rich on the nontemplate strand, not only within the regions that encode ribosomal RNA but also in the spacer regions. Transcription of the rDNA occurs in a specialized compartment within the nucleus, called the nucleolus. The rDNA can be very actively transcribed, with polymerases separated by as little as 100 bp.[12] In principle, passage of each polymerase could be accompanied by denaturation and reannealing of the duplex, but this would be likely to diminish the overall efficiency of transcription. Alternatively, as suggested by the discovery that the abundant nucleolar protein nucleolin binds tightly to G4 DNA, regulated formation of G4 DNA by the G-rich nontemplate strand could leave the template strand exposed for rapid transcription.[13]

G-rich Simple Sequence Repeats and Hypervariable Minisatellites

As discussed in chapter 4, the human genome is replete with simple sequence repeats, microsatellites and minisatellites which exhibit pronounced length instability. Some unstable repeats are G-rich and readily form G4 DNA. Examples include the CGG triplet repeat in the fragile X gene, *FMR1*, as well as longer, unstable G-rich minisatellites (figure 11.3c). One of these, the MS1 minisatellite (locus D1S7), is one of the most variable of human minisatellites. Allele lengths of MS1 can vary from 60 to more than 1000; allele-length heterozygosity exceeds 99%,[14] and the germline mutation rate of this minisatellite is at least 5.2% per gamete.[15] Because the consensus repeats in these minisatellites all contain at least one run of guanines (underlined) these sequences would be predicted to form G4 DNA, and structure formation will be enhanced by longer guanine runs, like those found in the minisatellite repeats. G4 DNA formation in vitro has in fact been directly confirmed for CGG repeats[16] and for two G-rich "variable number tandem repeats" (VNTRs), D4S43, and the insulin-linked hypervariable repeat.[17]

Expansion of G-rich repeats in single copy genes can lead to neurological disease. *FMR1*, the gene that encodes the fragile X mental retardation protein (FMRP), provides one example.[18] The 5′ untranslated region of FMR1 contains CGG repeats that undergo expansion, leading to disease. Higher repeat lengths are associated with diminished protein expression, resulting in a range of symptoms including cognitive deficiency, perseverative thinking, anxiety, ataxia, tremor, and traits such as prominent ears and flexible finger joints. Interestingly, the *FMR1* gene itself encodes a protein that regulates translation in neuronal cells by binding to a subset of RNAs which contain G4 structures.[19] The possibility that there may be positive selection for the variation generated by expansions and contractions of triplet repeats such as CGG, and of other repeat sequences, is discussed in chapter 4.

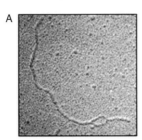 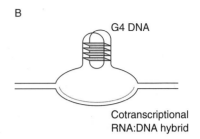

Figure 11.4. G-loop formed within a G-rich template. (A) Electron micrograph of a G-loop formed following intracellular transcription of a plasmid carrying a G-rich telomeric repeat. Plasmid DNA was purified, linearized by restriction digestion, and visualized by Kleinschmidt spreading. (Image reprinted from reference 7 with permission of Cold Spring Harbor Press.) (B) G-loop, showing G4 DNA formed in the G-rich strand and the cotranscriptional RNA/DNA hybrid formed on the C-rich strand.

The selective advantage of certain mutable repeat sequences in bacteria is described in chapters 3, 6, and 7.

Intracellular Formation of G4 DNA

Despite the readiness with which G-rich sequences form G4 DNA in vitro, the possible biological significance of G4 DNA was unclear because G4 DNA had not directly been identified in living cells. Intracellular formation of G4 DNA since has been established in experiments that used electron microscopy to visualize structures formed in individual molecules.[7] G-rich templates were created by cloning short (0.4–1.1 kb) DNA fragments derived from the mammalian telomeric repeat and the immunoglobulin class switch regions, as well as synthetic repeats, just downstream of an inducible promoter in a plasmid gene. Upon transcription, either in vitro or in vivo, characteristic looped structures formed (figure 11. 4a). These loops contain G4 DNA on the G-rich nontemplate strand, and an RNA/DNA hybrid on the C-rich template strand (figure 11.4b). The G-rich strand of the loops was shown to contain G4 DNA by two criteria: probing with a recombinant derivative of nucleolin that has great binding specificity for G4 DNA[13]; and cleavage by the G4 DNA-specific nuclease, GQN1.[20] The C-rich strand was shown to carry a stable RNA/DNA hybrid, as predicted, by experiments which used either gel electrophoresis[21] or atomic force microscopy.[22]

The loops formed upon transcription of G-rich regions have been called "G-loops" to emphasize that they contain G4 DNA and that their unusual structure depends upon G-richness of the transcribed region.[7] Formation of G-loops and G4 DNA is dependent not only on transcription but also transcriptional orientation, and occurs only if the C-rich strand is the template strand for transcription. Indeed, this is the physiological orientation of transcription in both the G-rich rDNA and immunoglobulin S regions. Formation of G-loops is very efficient. Following intracellular induction of transcription of G-rich templates carried on a plasmids,

15% of isolated plasmid molecules visualized in the EM contained G-loops and G4 DNA. The readiness with which G4 DNA forms provides support to the notion that biology takes advantage of regulated G4 DNA formation.

Cotranscriptional RNA/DNA Hybrids May Contribute to Genomic Instability

Two distinct structures form in transcribed G-rich templates: G4 DNA, and a stable RNA/DNA hybrid.[7] The RNA/DNA hybrid forms cotranscriptionally, and its formation reflects template sequence composition. Hybrid formation requires a C-rich template strand, which in turn reflects the unusual stability of rG/dC base pairs compared with the 15 other naturally occurring ribo/deoxy base-pair combinations.[23] Persistence of an RNA/DNA hybrid will prevent renaturation of the template and nontemplate strand. This could in turn contribute to regulated formation of G4 DNA.

Cotranscriptional hybrid formation may be an underappreciated source of genomic instability in somatic cells, which employ several mechanisms to prevent their formation or persistence. Both single- and multisubunit RNA polymerases appear designed to prevent hybrid formation, as in both classes of enzymes the nascent RNA is forced to exit through a topological tunnel which separates the RNA transcript from the DNA template strand.[24,25] Cells contain a variety of enzymes that can specifically digest the RNA strand of an RNA:DNA hybrid, leaving the DNA strand intact. These include RNase H, FEN-1 (essential for DNA replication), and EXO-1 (involved in mismatch repair).[26-28] In addition, eukaryotic cells contain a conserved polypeptide complex, THO/TREX, which is associated with the transcription apparatus and functions to prevent formation of stable cotranscriptional hybrids.[29,30] Cotranscriptional hybrid formation may have particular impact on genes of unusual sequence composition, as GC-rich transcripts are especially dependent upon THO/TREX. In yeast, deficiencies in subunits of the THO/TREX complex result in increased genomic instability, suggesting that the RNA/DNA hybrid itself or structures formed within the DNA strand may be targets for recombination and repair pathways.

Programmed Alterations in Genomic Structure: G-rich Regions in Immunoglobulin Class Switch Recombination

The most dramatic example of regulated function of G-rich regions in genomic rearrangement is in mammalian immunoglobulin heavy chain class switch recombination. Switch recombination is one of the three distinctive processes that irreversibly modify genomic structure and sequence in B cells. Early in B cell development, site-specific cleavage and rejoining of V(D)J segments, mediated by RAG1 and RAG2, produces the functional antigen receptor (chapter 9). Later, upon B cell activation, the immunoglobulin loci embark upon two additional distinct and irreversible genetic alterations. In somatic hypermutation (chapter 10), targeted mutagenesis alters the sequences of the expressed heavy and light chain variable regions. Hypermutation is coupled with selection for B cells expressing high-affinity antigen receptors, which

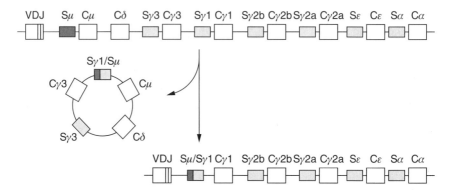

Figure 11.5.   Immunoglobulin heavy chain class switch recombination. The murine heavy chain locus is shown before and after recombination from µ to γ1; this switches the isotype of the expressed immunoglobulin molecule from IgM to IgG1. Immunoglobulin class switch recombination is a regulated DNA deletion event that occurs in activated B cells. Switch recombination joins a rearranged and expressed variable (VDJ) region to a downstream constant (C) region. Recombination junctions occur in G-rich switch (S) regions. Circles containing deleted sequences can be recovered from B cells that have completed switch recombination.

enhances the efficiency and specificity of the immune response. In class switch recombination, regulated DNA deletion replaces one heavy chain constant region with another, optimizing clearance of antigen from the body.

As diagrammed in figure 11.5, switch recombination joins a rearranged and expressed variable (VDJ) region to a new downstream constant (C) region, deleting the DNA between as an excised switch circle.[31] The V region of an immunoglobulin molecule interacts with antigen, while the C region determines how antigen is removed from the body. Because switch recombination changes the C region without affecting the V region, the result is to alter how an immunoglobulin molecule removes antigen without altering its specificity for antigen.

A Cytidine Deaminase Is Critical to Switch Recombination

Expression of a single polypeptide, activation induced deaminase (AID), induces both switch recombination and somatic hypermutation.[32,33] In activated B cells, AID deaminates C→U at transcribed immunoglobulin loci; this lesion is then attacked by uracil DNA glycosylase (UNG), which creates an abasic site (chapter 10).[34,35] AID is the only B-cell-specific factor required for either switch recombination or somatic hypermutation,[36,37] and downstream events depend upon ubiquitous factors key to other pathways of DNA repair and recombination.

G-rich Switch Regions in Class Switch Recombination

Switch recombination is induced and targeted by transcription of G-rich regions, called S regions, located just upstream of those C regions that participate in switch recombination: Cµ, Cγ, Cε, and Cα (figure 11.5). S regions are 2–10 kb in length, and consist of highly degenerate repeats. The repeats differ in length and sequence from one S region to another (figure 11.3a). The junctions produced during switch recombination lie within S regions, and are heterogeneous both in sequence and in the sites of breakpoints within both the donor and the acceptor S region.[38] However, the S regions and switch junctions are located within introns, so the imprecision of switch recombination does not affect the translation frame, length, or structure of the immunoglobulin heavy chain polypeptide encoded by the VDJ and C region exons.

The use of G-rich regions to promote recombination is the rule, but there are exceptions. In frogs, S regions are not G-rich, but contain many reiterations of the sequence motif AGCT.[39] These regions support active recombination, not only in frogs but also in mammals, apparently because the AGCT motif is a preferential target for deamination by AID.[40] In mammals, there is no G-rich S region upstream of Cμ, and alternative RNA splicing—rather than DNA recombination—regulates Cμ expression. (Alternative splicing is discussed in chapter 15.)

Switch Transcription and Targeting of Switch Recombination

Switch recombination requires that both S regions that will carry out recombination be transcribed simultaneously.[41] A dedicated promoter/enhancer upstream of each S region allows that region to be transcribed in response to specific cellular/extracellular signals. Transcription of a specific S region is activated by a signaling cascade, which is triggered by binding of T cell-derived cytokines or lymphokines to cognate receptors on the B cell surface. This signal is relayed to the nucleus by a pathway that culminates in binding of transcription factors to specific S region promoters; for example, in B cells switching from Cμ to Cγ1 (figure 11.5), transcription would be activated at the Sμ and Sγ1 promoters. The effect of transcription is in *cis*: if an S region promoter is deleted, that S region becomes inactive for recombination; and in heterozygotes carrying a promoter deletion on only one allele, recombination is affected only on that chromosome.[31]

As described above, mammalian S regions are G-rich on the nontemplate (top) DNA strand. Efficient switch recombination requires that transcription occur in this physiological orientation,[42] which is the orientation that supports formation of G-loops and G4 DNA. Deamination by AID requires a single-stranded region of DNA sequence. The presence of G4 DNA might enhance deamination by AID, by prolonging denaturation of the top and bottom strands, and/or by transiently impeding replication and thereby increasing exposure to AID, which binds to a component of the replication apparatus.[43]

G4 DNA as a Structural Target for Switch Recombination

Switch recombination is unusual among regulated recombination processes because it is region-specific, and not sequence-specific. In addition, switch recombination requires that the two S regions targeted for recombination be actively transcribed. These observations have given rise to models postulating that regulated formation of nucleic acid structures at the transcribed S regions produces targets for recombination factors. As described above, G4 DNA forms within transcriptionally activated mammalian S regions. G4 DNA has been shown to be a structural target for two factors shown by genetic analysis to be important to switch recombination, BLM helicase (see below) and the mismatch repair factor MutSα. MutSα (a heterodimer of Msh2/Msh6) functions in the mismatch repair pathway to recognize base mismatches and small DNA loops,[44] and it also functions in switch recombination. Ablation of either subunit decreases levels of switch recombination and diminishes heterogeneity of switch junctions.[45-47] MutSα associates with transcribed S regions activated for switch recombination in activated B cells. Recombinant MutSα binds with high affinity to G4 DNA formed upon transcription of the S regions, and can be visualized by electron microscopic imaging bound to G-loops formed upon transcription of S regions in vitro.[48] Strikingly, MutSα can promote interactions between G-loops in two transcriptionally active S regions, creating a structure very like that observed upon synapsis of recombining DNA. This suggests that G4 DNA formed in the activated S regions in switching B cells provides a structural target for MutS, which recognizes this target to promote DNA synapsis and recombination.

Potential for Formation of G4 DNA at Single Copy Genes: c-MYC

The mammalian genome contains many regions that are G-rich and have the potential for G4 DNA formation. A key current challenge is to determine whether and how this affects genomic stability and gene function. Considerable interest has been drawn to the c-MYC gene, which encodes a transcription factor that is a key regulator of cell proliferation. Deregulated c-MYC expression contributes to tumorigenesis in many different tissues, and it can arise through a variety of mechanisms, including translocation, mutation, and amplification. The c-MYC gene contains strikingly G-rich regions, raising the possibility that G4 DNA formation within c-MYC is a key determinant of genetic stability. In B cell lymphomas, c-MYC frequently translocates to the immunoglobulin S regions, suggesting that c-MYC could be a target for enzymes that promote switch recombination. This possibility is supported by recent evidence that G-loops and G4 DNA form upon transcription of c-MYC, and map to exon 1 and intron 1, the zone that translocates in B cell lymphomas.[49] It will be of interest to learn whether potential for G4 DNA formation characterizes other proto-oncogenes that undergo frequent translocation.

G4 DNA within the c-MYC Promoter: A Therapeutic Target?

Oligonucleotides bearing a G-rich element from the c-MYC promoter form G4 DNA in vitro, and structure formation appears to impair transcription, suggesting a possible avenue for therapeutic intervention: small molecules that stabilize G4 DNA could repress c-MYC transcription, and prove useful as antitumor drugs.[50-52] The success of this or other strategies employing G4 DNA-binding ligands for therapeutic purposes will almost certainly require some mechanism for targeting the ligands to specific genomic regions.

c-MYC *Translocates in Switching B Cells*

The c-*MYC* gene is a frequent target of translocation in B cell lymphoma, a tumor of activated B cells.[53] The most common translocation, t(8;14), moves c-*MYC* to the immunoglobulin heavy chain locus and thereby deregulates expression of this oncogene. Like the heavy chain switch regions, c-*MYC* is actively transcribed in activated B cells; and the c-*MYC* gene contains numerous G-runs within the first exon and intron, where translocation junctions map. G4 DNA forms within this region of c-*MYC* upon transcription, and this may render c-*MYC* a target for translocation by the switch recombination apparatus in activated B cells.[49]

Regulated Formation of G4 DNA at Telomeres

Telomeres are structures at the very ends of chromosomes that are critical for genomic stability.[54-56] Telomeres are synthesized by telomerase, a specialized reverse transcriptase, which extends telomeres using the 3′ end as a primer. Among their many functions, telomeres provide buffers against sequence loss during replication, protect the ends of chromosomes from digestion by exonucleases, affect cellular lifespan, and distinguish normal DNA ends from DNA breaks.

Telomeric repeats in essentially all eukaryotes are G-rich and have considerable potential to form G4 DNA (see figure 11.3b). This has led to considerable speculation about how telomere biology could take advantage of regulated G4 DNA formation.[10,57] The presence of G4 DNA could: prevent degradation of the telomere tails by enzymes that attack single-stranded DNA; inhibit telomere replication by telomerase by making the 3′ end inaccessible; provide binding sites for specific factors that specifically recognize G4 DNA; or stabilize t-loops,[58] lariat-shaped structures formed when telomeric tails interact with regions within the telomeric repeats. Intracellular formation of G4 DNA has been shown to occur in telomeric repeats carried in plasmid substrates in which transient denaturation is induced by transcription.[7] This may provide a model for structure formation at telomeric tails, or within telomeric repeats during their replication.

Recombination-dependent Telomere Maintenance

Cell proliferation requires that telomeric repeats be replenished. As most cells in an adult lack telomerase activity, telomere maintenance is a limiting step in immortalization and tumorigenesis. In most tumors, telomerase activity is reactivated for maintenance of telomeric repeats. However, in a minority of tumors (5–10%) there is no measurable telomerase activity. In these tumors, telomeres are maintained by an alternative pathway, called the ALT pathway, which depends upon recombination to replenish telomeric sequence.[59,60] Genetic components of the ALT pathway have been identified in yeast, where two distinct arms of the ALT pathway have been identified. One arm is dependent upon homologous recombination and *RAD51* gene function; the other arm functions in the absence of *RAD51* but requires *SGS1*, which encodes a helicase of the RecQ family (discussed below).[61-63] The genetic requirements for telomerase-independent telomere maintenance in mammalian cells have not been defined. This pathway is of particular interest because its components of this pathway may also participate in other recombination pathways targeted to G-rich regions, such as switch recombination (see above).

Helicases That Target G4 DNA and Potential Instability of G-rich Regions

The potential for G4 DNA formation to contribute to genomic instability is most dramatically illustrated by the impact of mutations in helicases essential for maintenance of G-rich chromosomal regions. G4 DNA differs from canonical *B*-form duplex DNA in structure, and it therefore is not a substrate for enzymes that attack duplex DNA, including potent helicases such as RecBC.[64] However, distinct helicases are identified with function at G-rich genomic regions, including members of the RecQ family, and helicases related to RTEL/DOG-1.

RecQ Family Helicases at the rDNA, Telomeres, and S Regions

RecQ family helicases are members of a highly-conserved enzyme family homologous to *Escherichia coli* RecQ.[65] They share a central helicase domain and two additional regions of homology, the RQC and HRDC domains. There is only one RecQ family member in *E. coli* (RecQ), one in *Saccharomyces cerevisiae* (Sgs1), one in *Schizosaccharomyces pombe* (Rqh1), two in chickens (WRN and BLM), and five in humans. Three of the human RecQ family members are deficient in diseases characterized by profound genomic instability: BLM in Bloom syndrome, WRN in Werner syndrom, and RTS (or RECQ4) in Rothmund–Thompson syndrome. Malignancy is the most striking shared disease outcome of human deficiencies in RecQ family helicases, reflecting the critical role of these enzymes in maintenance of genomic stability.

G4 DNA is actively and preferentially unwound by RecQ helicases, including human BLM and WRN, yeast Sgs1, and *E. coli* RecQ.[64,66-69] RecQ family helicases associate with the replication apparatus,[70] which may reflect a critical role for G4 DNA unwinding in replication. If G4 DNA forms upon transient denaturation of a G-rich region, it must be resolved for replication to proceed (see figure 11.2). Failure to unwind G4 DNA could result in instability of the G-rich region, due to either expansion or contraction; or to elimination of a structured region by repair factors.

RecQ helicases act on a variety of structured substrates, including Holliday junctions, bubbles, and flaps.[71,72] However, the importance of G4 DNA unwinding activity to RecQ helicase function in vivo is highlighted by the roles for these helicases in maintenance and expression of G-rich regions. The eukaryotic RecQ helicases Sgs1, BLM, and WRN all localize to the nucleolus, where the G-rich rDNA is transcribed. In *S. cerevisiae*, Sgs1-deficiency results in reduced proliferative capacity, popouts of rDNA circles, nucleolar fragmentation, and impaired rDNA transcription and replication.[73–76] In human cells, WRN interacts with pol I to accelerate rDNA transcription.[77]

As discussed above, switch recombination requires transcription of both S regions targeted for recombination, which may cause formation of G4 DNA. Bloom syndrome, which results from BLM helicase deficiency, is characterized by growth retardation, immunodeficiency, impaired fertility, and predisposition to a wide variety of malignancies.[78] The immunodeficiency associated with Bloom syndrome results from impaired class switch recombination: following immunization, appearance of IgG antibodies is delayed, and the levels in serum are below normal. G4 DNA formed upon S region transcription could, if not resolved, halt replication (figure 11.2a) and cause cells to arrest at a DNA replication checkpoint. This would diminish the number of B cells that effectively completed switch recombination. Nonetheless, the cells that did complete switch recombination would proliferate, producing IgG antibodies, albeit on a delayed timetable and at reduced levels, as observed in the immunodeficiency associated with Bloom syndrome.

RecQ family members also have telomeric functions.[69] In yeast lacking telomerase, telomeres can be maintained by recombination-dependent pathways, and *SGS1* is essential to one pathway of recombination-dependent telomere maintenance.[61,62,79] In human cells, WRN associates with the key telomere binding protein TRF1[80]; and expression of a dominant negative WRN mutant causes telomere loss even in the presence of active telomerase.[81] The details of WRN telomeric function have yet to be worked out, but they are particularly intriguing because, as discussed above, telomere length affects cellular lifespan, and premature aging is one of the most salient features of Werner syndrome, which results from the absence of WRN helicase.

RTEL and DOG-1: Related Factors but Distinct Functions at G-rich Regions.

Recently, the potential for G4 DNA formation to occur within telomeric repeats has been highlighted by identification of an essential gene which regulates telomere length in mice.[82] Mutation of the murine *Rtel* gene results in embryonic lethality. At the cellular level, similar to telomerase knockout mice, *Rtel–/–* cells display profound chromosomal instability, evident as end-to-end chromosomal fusions, broken chromosomes, chromosome fragments, and telomere-less chromosomes.[83] However, *Rtel* does not appear to be required for stabilization of nontelomeric G-rich repeats.[82]

Rtel is homologous to the gene *dog-1*, identified in the nematode. While *dog-1* does not appear to have telomeric function,[84] it is necessary for maintenance of polyguanine tracts.[84] Ablation of *dog-1* expression results in genome-wide destabilization of polyguanine tracts more than 18 nt in length. The proteins encoded by *Rtel* and *dog-1* contain a conserved helicase motif as well as a domain for interaction with proliferating cell nuclear antigen (PCNA), part of the replication apparatus. This suggests that RTEL and its homologs may function by unwinding G4 DNA that could otherwise impair replication,[82,84] as has been proposed for BLM helicase[64] and other RecQ family members. However, the biochemical activities of RTEL, DOG-1, and related factors have yet to be defined.

Homology-independent Synapsis and Genomic Evolution

The discussion above has focused largely on a few classes of G-rich chromosomal regions which have been functionally well-characterized. Nonetheless, there are many G-rich regions scattered throughout the genome, and their potential for G4 DNA formation may have important implications for biological function. Intermolecular G4 DNA can synapse two genomic regions that are G-rich but are otherwise unrelated in sequence. Synapsis stabilized by G4 DNA contrasts with synapsis stabilized by complementary A–T and G–C base pairing, which requires extensive sequence-dependent homology.

Meiosis

Intermolecular G4 DNA formation may stabilize pairing of homologous chromosomes in meiosis.[1] This proposal, made when G4 DNA was first discovered, has received tantalizing experimental support. RecQ helicases, which have essential functions in G-rich chromosomal regions (see above), also are critical for meiosis. Humans with Bloom syndrome have impaired fertility; and flies lacking the Drosophila homolog of human BLM are sterile, due to homologous chromosomes becoming interlinked and failing to segregate correctly.[85] Mutation of *SGS1*, which encodes the yeast RecQ helicase, similarly impairs chromosome segregation in meiotic and mitotic cells.[9] In addition, in yeast, a meiosis-specific protein, HOP-1, is critical to meiotic pairing and synaptonemal complex formation.[86] HOP1 also binds to G4 DNA with high affinity,[87] and can promote synapsis of double-stranded DNA via G-quartet formation.[88]

G-rich Regions, an Evolutionary Resource

As discussed above, sophisticated mechanisms for transcription and replication have evolved to minimize the effects of DNA structure on replication fidelity in living cells. Nonetheless, stable interaction between unrelated G-rich regions occurs readily under physiological conditions in vitro,[89] and even a very low level of G4 DNA-mediated synapsis in a living cell could alter genomic structure by promoting recombination events between unrelated sequences. Because formation of G4 DNA is a powerful mechanism for promoting interactions between nonhomologous genomic regions, it may well be useful on the evolutionary time scale. Exactly this sort of mechanism could promote the shuffling of exons so critical to the complex architecture of vertebrate genes.[90]

Conclusions

In one view, the propensity of G-rich regions to form G-quartets and G4 DNA can be seen as a flaw in the nearly impeccable chemistry of base pairing. In another view, G-rich regions can be seen as an evolutionary resource because they are a potential source of genomic variation. In this latter view, selective pressures might tend to exclude G-rich sequences from certain classes of regions while favoring them in others. G-rich motifs, like the uneven genomic landscape itself, can then be seen not only as tokens of the genome's past, but also as potential players in its future.

Acknowledgments   I am grateful to my colleagues and members of my laboratory, especially Michelle Duquette, Michael Huber, and Erik Larson, for thoughtful discussions and critical insights; and to the U.S. National Institutes of Health for supporting our research on class switch recombination and dynamic formation of G4 DNA (R01 GM39799 and R01 GM65988).

12

The Role of the Genome in the Initiation of Meiotic Recombination

Rhona H. Borts and David T. Kirkpatrick

Overview

Initiation of recombination is an essential and highly regulated process. Recombination during meiosis is important for enabling exchange of information between the paternal and maternal contributions to a genome. In addition, in most organisms, the absence of meiotic crossovers causes homologous chromosomes to segregate aberrantly. Recombination is not randomly distributed along or between chromosomes. Rather, the genomes of many species contain regions that have an elevated level of recombination ("hotspots") or a decreased level ("coldspots"). Factors that determine the level of recombination at a specific site are the primary sequence, the configuration of the chromatin, and the nature and function of the DNA sequence. Also, prior recombination initiations repress additional events in their vicinity. The probability of recombination within a region of the genome is very context dependent. It can be modulated by factors that include the nutritional state of the cell, the base composition of the region, and the level of primary sequence variation present between the maternally and paternally inherited chromosomes, as well as the action of DNA repair proteins on sequence mismatches occurring in regions of recombination between maternal and paternal chromosomes. Thus, initiation of meiotic recombination is tightly controlled. Due to the importance of the DNA sequences themselves and their interaction with meiotic proteins, there are regions where the probability of meiotic recombination can be considered to be encoded implicitly in the DNA sequences as a result of interactions with the meiotic protein complement.

Meiotic Recombination: An Overview

Meiosis Ensures Continuity of Chromosome Number

Meiosis is an ordered process common to all sexually reproducing organisms. The genomes of diploid organisms have two copies of each chromosome, termed

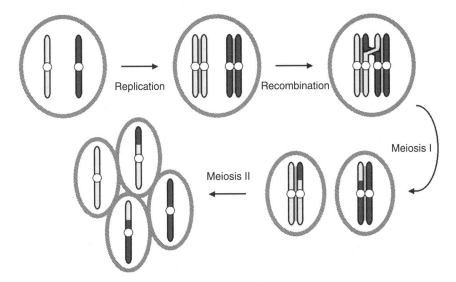

Figure 12.1. The stages of meiosis. A diploid nucleus is shown, although for clarity only one pair of homologous chromosomes (light gray and dark gray) is indicated. During premeiotic replication, the two homologs are duplicated, forming sister chromatids that are held together. Early in meiosis, recombination between the homologs is initiated, leading to a crossover product that helps to properly align the homologs. In meiosis I, the crossover is resolved and the homologues segregate away from each other. In meiosis II, sister chromatids segregate, leading to the formation of four haploid gametes.

homologs, one inherited from each parent. In order to prevent a doubling in chromosome number every generation, the number of chromosomes must be halved when forming gametes. Haploid gametes from each parent subsequently fuse to form a genetically unique diploid organism. This process is a complex interplay between events at the DNA level and the proteins facilitating the appropriate separation of chromosomes. Failure to undergo chromosome segregation properly leads to sterility, spontaneous loss of a fetus, or genetic disorders (e.g., Down syndrome) associated with an imbalanced chromosome complement.

Meiosis consists of two divisions, preceded by DNA replication (figure 12.1). The products of this replication are two identical copies ("sister chromatids") of each chromosome that, in contrast to their segregation during mitosis, are held together throughout the first meiotic division. Following replication, the homologous chromosomes pair, recombine, and segregate away from each other. Meiosis I is the "reductional" division: the information content is halved. At each position in the genome, each gamete will receive information from only one parent.

Due to recombination, the homologs that assort in meiosis I are not identical to the chromosomes inherited originally from each parent. Instead, each chromosome is a composite of genetic information obtained from both parents. The second division, meiosis II, is "equational:" the sister chromatids separate to each gamete, resulting in four cells, each with a haploid genome.

Recombination Provides the Mechanical Link for Proper Chromosome Segregation

Recombination during meiosis is initiated by a double-strand break (figure 12.2).[1] Following cleavage, 3′ single-stranded regions are formed by either a 5′ to 3′ exonuclease or a coupled helicase–endonuclease activity[2-4]. The single-stranded end invades the homolog in the region of sequence identity, displacing the native strand to form a region of biparental 'heteroduplex' DNA. The invading 3′ end serves as the primer for DNA repair synthesis; DNA replication will further displace the native strand. The invading strand and the displaced strand form a crossed intermediate known as a Holliday junction, after Robin Holliday, who first described the structure.[5] Capture of the second free end will form another Holliday junction. This crossover is the site of physical association between the two homologs and acts to tie the two chromosomes together.

The reciprocal exchange of DNA at the site of Holliday junctions leads to the formation of a proteinaceous structure that is visible first as a recombination nodule and then as a chiasma (plural: chiasmata). Chiasmata allow the spindle attachments to orient properly and to pull the homologs apart during chromosome segregation. Failure to crossover leads to nondisjunction, and aneuploid gametes when both homologs are pulled to the same pole. The number and distribution of crossovers is highly regulated to ensure that every chromosome gets at least one crossover. Following chiasmata formation and homolog alignment, the Holliday junctions must be resolved prior to homolog separation during meiosis I, to prevent chromosome damage. Enzymes responsible for resolving mitotic Holliday junctions have been clearly identified in bacteria, but their identification and characterization in eukaryotes has only just begun.[6] Cleavage of the crossovers in opposite planes results in two recombinant chromosomes, in which the flanking markers are no longer in their parental arrangement (figure 12.2).

The Crossover/Noncrossover Decision

Meiosis I requires the presence of at least one chiasma per chromosome for proper homolog alignment, placing a constraint on the factors that control the initiation of recombination. However, many double-strand breaks (DSBs) do not form a crossover. The time at which it is decided whether a DSB will result in a crossover or noncrossover product occurs very early in the recombination process (figure 12.2).[7,8] Until very recently, it was thought that both crossovers and noncrossovers arose late, when double Holliday junctions were resolved; cleavage of both crossovers in the same plane would result in noncrossovers, while cleavage in opposing planes would give rise to recombinant crossover products. It is now thought that there are two parallel pathways, with one leading to crossovers and the other to noncrossovers (figure 12.2). Both classes are initiated by a DSB and require strand invasion by one of the 3′ ends to form heteroduplex DNA. Subsequent to strand invasion the pathways differ. Crossovers result when the second end is captured by the homolog. Noncrossovers are thought to form through a mechanism called "synthesis dependent strand annealing" (SDSA), in which the newly replicated DNA unwinds and pairs with the resected DNA on the other side of the DSB.

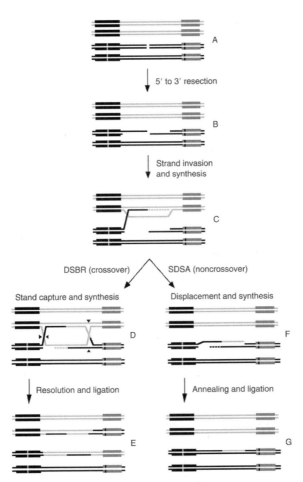

Figure 12.2. **Models for meiotic recombination pathways.** The intergenic region
between two genes is depicted. The first gene is shown as a solid black box and the adjacent gene is dark gray. The homologous chromosomes are indicated by light gray or black
lines. A heterozygous point mutation in each of the genes on the black homolog is shown
by a vertical line. (A) A double-strand break (DSB) occurs on one copy of the black
homolog. (B) The 5′ ends of this break are degraded, leaving long 3′ ends. (C) The gray
homolog is invaded by the 3′ single-stranded tail, displacing the gray strand of the same
polarity and forming a single Holliday junction. The dashed lines indicate repair synthesis
initiated at the end of the invading strand. Following the initial Holliday junction formation,
two different outcomes are possible. If the recombination event is designated to generate a
crossover product, (D), the second 3′ single-stranded tail is captured and two Holliday junctions are formed. DNA replication, as indicated by dashed lines, can then reconstitute the
lost DNA sequences. Cleavage of the Holliday junctions at the locations indicated by the
arrowheads will create a recombinant, crossover, product (E) following resolution and ligation. However, if the recombination event is designated to generate a noncrossover product,
(F), the invading strand is displaced, and the extended end of the molecule can bridge the
DSB to allow the reconstitution of the broken homolog. Gene conversion is still possible upon
re-annealing (G) if the newly synthesized region now contains sequence differences.

211

The approximately 100 crossovers per yeast meiosis are not randomly distributed among and within the chromosomes, implying that the regulation of DSBs and crossover formation is monitored across large regions of the genome. Within chromosomes, crossovers are relatively evenly distributed as a result of a process known as "interference." Interference can be observed experimentally, when the observed number of double crossovers in a defined interval is less than expected from the observed number of single crossovers, indicating that something is preventing the formation of a second crossover when the interval already contains one.

Two models have been proposed to account for interference: a "counting" model[9] and a model based on chromosome architecture.[10] The counting model, based on mathematical modeling, hypothesizes a mechanism in which every crossover is separated from the next by a fixed number of noncrossovers.[9] The chromosome architecture model proposes that chromosomes are under stress due to internal mechanical forces generated by interactions between chromatin elements and the chromosomal architecture. Tension is relieved locally by crossover formation, with the subsequent lack of tension preventing the formation of additional crossovers nearby. This model can also guarantee that each chromosome receives at least one crossover by modulating the stress level or the level at which a response occurs.[10]

In yeast and other organisms, including plants and humans, it is also thought that there are two types of crossovers, those that display interference and those that do not (reviewed in references 9 and 10). Noninterfering crossovers arise from cleavage of the strand invasion structure (figure 12.2c) by an endonuclease called Mus81/Mms4p.[6] This type of crossover is required to successfully fit the counting model of interference to experimental data from *Saccharomyces cerevisiae*.[9]

Mismatch Formation and Repair Leads to Non-Mendelian Segregation Events

Heteroduplex DNA is formed when single-stranded DNA from one homolog is complexed with single-stranded DNA from the other homolog. If the paternally and maternally derived strands contain sequence differences, mismatches will occur in the heteroduplex region (figure 12.3). These can be as small as a base–base mismatch, or very large, as might result from the insertion or deletion of a transposable element. Yeast cells possess DNA repair factors that recognize and repair these small mismatches[11] or large loops,[12] as reviewed previously.[13,14] At the site of a mismatch, Mendelian (2:2) segregation will be restored (a "restoration" event) or a non-Mendelian ratio will be observed (a "gene conversion" event), depending on the direction in which the mismatch is repaired. In yeast, if the mismatch is not repaired, the spore inheriting the unrepaired DNA will segregate the two alleles in the first round of DNA synthesis and mitosis following spore germination. The colony formed from this spore will have sectors expressing the phenotype of each allele. This is termed a "post-meiotic segregation," or PMS, event. Therefore, the amount of recombination at a particular site in the genome can be monitored by determining the total amount of non-Mendelian segregation (gene conversions plus PMS events) at that locus, with gene conversions representing mismatches in which the sequence on the invading strand is replaced (figure 12.3) and PMS events representing a failure to recognize or repair the mismatch.

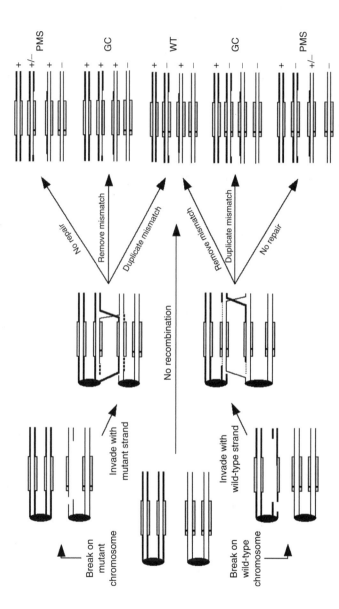

Figure 12.3. Mismatch repair during meiosis leads to gene conversion and PMS events. Patterns of aberrant segregation associated with meiotic recombination at *HIS4*. Chromosomes are shown as double-stranded DNA molecules. A gene is shown as a gray rectangle, with the internal vertical black line indicating a mutant sequence that will form a mismatch when incorporated into heteroduplex DNA (as shown). Dashed lines represent regions of DNA replication. The initiating events (DSB formation and strand resection) can occur on either the wild-type or the mutant homolog. The invading strand contains the sequence difference that will generate the mismatch. The segregation pattern of the resulting spore colonies when transferred to medium without histidine is shown on the right: (+), wild-type; (−), mutant; (+/−), sectored colony containing both wild-type and mutant cells. The type of aberrant segregation is indicated next to the segregation pattern: GC, gene conversion; WT, wild-type; PMS, postmeiotic segregation.

While one might expect a 1:1 ratio of repair of a heteroduplex mismatch—replacement of the original sequence with the second sequence (a gene conversion event), or replacement with the original sequence (a restoration event, figure 12.3), this ratio is not observed for mismatches that occur very near to the site of DSB formation. Instead, heterozygous mismatches near the DSB preferentially undergo conversion repair,[15] but as the distance between the initiating site and the mismatch increases, the frequency of restoration repair becomes predominant.[16,17] A second gradient exists in that sequences further from the initiating DSB are less likely to fall within the region of single-stranded DNA, and thus less likely to be included in heteroduplex DNA following strand invasion.[3,14] The combination of these two gradients gives rise to a phenomenon known as a "gradient of polarity." The steepness of the gradients vary from site to site, for reasons that are not fully understood.[3,15] Thus, sequences that are close to sites of DSBs on the homologous chromosome are more far more likely to be donors in gene conversion events than are other sequences in the genome, which potentially impacts the inheritance of hotspot sequences in subsequent generations.

Primary DNA Sequence Influences Meiotic Recombination

The Frequency of Recombination Initiation Varies from Locus to Locus

To assess the frequency of recombination in an interval, two types of outcomes can be measured. Non-Mendelian segregation of a heterozygous marker, as indicated by gene conversion or PMS, is the result of recombination at a specific position in the genome and represents a nearby initiation. A nonparental association of linked heterozygous alleles ("crossover recombination") could have been initiated anywhere between the markers, and its frequency is a composite of a number of factors including the number and intensity of initiations over the entire interval and the frequency with which an initiation is associated with a crossover.

Initial work on the initiation of meiotic recombination was entirely genetic, utilizing loci that provided a sufficient number of meiotic events for genetic (and later physical) analysis. In yeast, these loci included *ARG4*, *CYS3*, *HIS2*, *HIS4*, and an artificial *HIS4LEU2* construct. All of these turned out to be "hotspots" in that they had higher than average levels of recombination. In addition to traditional genetic means of monitoring recombination frequency, the initiating DSB can be monitored by physical assays, such as Southern analysis[18-21] or whole-genome microarray analysis,[22] to determine the positions and frequency of the DSB (figure 12.4). At each locus studied in detail, the strand cleavages do not always occur at the same position, but rather are distributed over a small interval of 100 to 500 base pairs.[19,23-25]

The product of the *SPO11* gene, a member of the type II class of topoisomerases, was identified as the endonuclease responsible for meiosis-specific DSBs,[26] but the factors determining the exact site of cleavage have not been determined. In *S. cerevisiae*, an open chromatin configuration (as evidenced by increased accessibility to micrococcal nuclease or DNase I) is found in regions

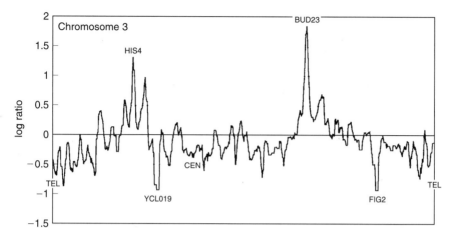

Figure 12.4.   Distribution map of DSB formation on chromosome 3. The graph depicts the data from a microarray analysis of DSB formation, as indicated by Spo11p binding,[22] across chromosome 3. The log ratio of the signal from meiotic DNA compared with total genomic DNA is plotted versus position along the chromosome. A positive value indicates an increase, while a negative value indicates a reduction. The locations of the centromere and telomeres are shown, as are the genes containing the two strongest hotspots and coldspots. (The figure was kindly provided by Piotr Mieczkowski and Thomas D. Petes.)

exhibiting a high level of DSB formation.[27,28] In certain regions, the amount of open chromatin was found to increase specifically at meiosis.[29] However, regions with the highest degree of open chromatin do not correspond exactly with the position of the DSB by Southern analysis, indicating that, while the open chromatin appears to correlate with DSB formation, it is not the indicator of the site of DSB formation. In sharp contrast with results from bacteria, where specific DNA sequences such as chi stimulate recombination directly (reviewed in reference 30), sequencing of DSB regions did not reveal any sequence motifs common to all hotspots. A consensus sequence of approximately 50 bp (the CoHR motif) was proposed to correlate with DSB formation,[30] but recent experiments at *HIS2*, in which the consensus sequence was deleted, showed no effect on DSB formation upon loss of the CoHR sequence.[31] Given the lack of a specific target sequence, and the distribution of DSBs across an approximately 100 bp region at those hotspots where the precise locations of DSB formation is known,[19,23-25] the location specificity for DSB formation is likely to arise from a confluence of inputs. In keeping with this, mutation of a number of residues in Spo11p alter the distribution of DSB sites within the *HIS4/LEU2* hotspot, implying that Spo11p itself plays a role in selection of DSB sites but is not the primary determinant.[32]

Hotspots for Meiotic Recombination

Physical and genetic analysis of multiple hotspots in the yeast genome has led to the classification of hotspots into three types: alpha, beta, and gamma.[22,33,34] Alpha hotspots occur in intergenic regions and require the presence of transcription factors for activity. Beta hotspots are marked by regions of open chromatin and have often been associated with the insertion of foreign DNA sequences into the genome. Gamma hotspots are those associated with a high level of G+C DNA content and are the least characterized of the three types. The hotspot types are not necessarily mutually exclusive. For example, an alpha hotspot can also be considered to be a gamma hotspot depending on the local DNA composition.

Alpha Hotspots

Alpha hotspots are defined as those that are dependent on the presence of transcription factors. For example, at the *HIS4* locus, where alpha hotspots were first described, three transcription factors bind in the promoter region, and are required for both transcription and meiotic recombination: *GCN4*[35], *BAS1/BAS2,* and *RAP1*.[36–38] While transcription factors are required, transcription of *HIS4* does not appear to be required for recombination, as deletion of the TATA box has a minimal effect on recombination.[38] During meiotic recombination initiation, the transcription factors at alpha hotspots are thought to open up the chromatin and recruit recombination initiation or chromatin-modifying factors to the site.

Since they result from the binding of transcription factors, alpha hotspots would be expected to occur at many sites in a genome. Microarray and Southern analysis of recombination initiation sites indicates that the majority of sites in yeast are found in intergenic regions, where transcription factors usually bind.[21,22] However, as most intergenic sites are not meiotic recombination hotspots, it is probable that not all transcription factors function in initiating meiotic recombination. In *Schizosaccharomyces pombe*, the Mts1/Mts2 transcription factor binds to the well-characterized *ade6-M26* hotspot and is necessary for hotspot activity.[39]

Beta Hotspots

Beta hotspots represent regions of the genome that exhibit a high level of open chromatin and meiotic recombination that is thought to arise from DNA structure, rather than dependence upon transcription factors.[40] For example, an insertion of 12 copies of CCGNN into the *HIS4* promoter,[40] and four copies of the CGGATCCG repeats associated with the *HIS4/LEU2* insertion hotspot,[19] create strong meiotic recombination hotspots. Both of these hotspots are artificial constructs and do not appear to bind any known transcription factors. Plasmid and bacterial sequences introduced into the genome also stimulate recombination,[41,42] as do human repetitive minisatellite sequences.[43,44] Insertions may create hotspots by opening a region of chromatin, allowing the recruitment of recombination initiating factors such as Spo11p, without the intervention of a transcription factor.

Beta hotspots may be characterized by the presence of repetitive DNA sequences. In addition to the examples above, other repetitive DNA sequences have been associated with recombination events. Repetitive sequences have been the source of rearrangements throughout eukaryotic evolution, although their contribution to speciation is less clear.[45] The recent availability of complete human, mouse, and rat genomes, and large data sets for pig, cattle, cat, and dog, has allowed a much higher resolution examination of the breakpoints between syntenic blocks (regions of homologous sequences that are conserved intact between species).[46,47] These blocks were originally thought to become unordered through a random chromosomal breakage model. However, the breakpoints between blocks reside in or adjacent to repetitive DNA sequences that appear to have been used multiple times as the various species diverged. These same sequences are sites of genomic rearrangement leading to human genetic disease,[48] suggesting that they are indeed "hotspots" for recombination. Translocations that have been observed experimentally in a number of model organisms often map to repetitive sequences found on virtually all of the chromosomes.[49-51] Interestingly, one of these, the yeast Ty element, is a recombination "coldspot" (see below). Experiments in these organisms have begun to elucidate the underlying mechanisms responsible for the observed connections between beta hotspots, repetitive DNA, syntenic blocks, and genetic diseases.

Gamma Hotspots

In the meiotic DSB microarray analysis by Gerton et al.,[22] a significant number of DSB hotspots occur near to or at peaks of high G+C content; these hotspots have been called "gamma hotspots."[34] For example, eight regions with a significantly elevated G+C content were found on chromosome 3; seven of the eight hotspots that mapped to chromosome 3 were within 2.5 kb of these G+C regions. While the mechanism by which gamma hotspots initiate recombination is not known, Petes[34] has suggested that the high G+C content at gamma hotspots may be sufficient to transiently pause replication forks during premeiotic DNA synthesis. The stalling of the forks may lead to local chromatin modifications that subsequently act in recombination initiation. Alternatively, yeast cells may contain proteins that specifically recognize genomic regions with an elevated G+C content and facilitate initiation of recombination, perhaps through chromatin modifications. Regions that are G-rich also have the potential to form DNA structures during transient denaturation that accompanies transcription or replication, and these structures may provide targets for recombination enzymes or facilitate chromosome pairing (see chapter 11).

Coldspots for Meiotic Recombination

While hotspots represent regions of the genome with an elevated level of recombination, coldspots are regions with a depressed level of recombination. In the past, they did not receive as much attention as hotspots, for in order to identify a coldspot it is necessary to be able to detect a locus at which there are significantly fewer non-Mendelian segregations than the average (which in yeast is estimated to be approximately 3–4%), and thus a very large data set is required for genetic analysis. However, regions that exhibit significantly fewer DSBs than the genome on average (figure 12.4) now can be identified by microarray analysis.[22] Microarray analysis of the distribution of meiotic DSB sites demonstrated that almost half of these coldspots cluster in a few locations in the genome—the telomeres and centromeres.

DSB formation at these locations was significantly reduced; 10 out of 40 coldspots were associated with telomeres, while eight out of 40 were associated with centromeres.[22] Individual centromeres[21,22,52] and telomeres in a *rad50S* strain[53] previously have been shown to be depressed for meiotic recombination. The mechanisms underlying this suppression have not yet been determined but may be linked to a change in the timing of replication in the region.[53]

In *S. cerevisiae*, the Ty retrotransposon has been associated with regions that exhibit a reduction in recombination.[54] Introduction of a Ty element approximately 300 bp from the *HIS4* alpha hotspot reduced the frequency of meiotic recombination. After insertion of Ty, chromatin regions that were normally in an open configuration were much less susceptible to DNase I treatment in strains with the Ty insertion, indicating altered chromatin configuration at *HIS4*.[55] Another naturally occurring coldspot has been identified and characterized in *S. pombe*. The region between the mating type loci *mat2* and *mat3* exhibits a simultaneous suppression of recombination and transcription,[56] and mutations that increase transcription in the interval also increase the frequency of crossovers.[57,58] Thus, these coldspots may result from formation of an altered chromatin state.

Finally, an artificial coldspot was formed by the insertion of multiple copies of a CCGNN pentameric sequence.[40] As described above, 12 copies of this sequence form a beta hotspot at *HIS4* locus when they replace the wild-type promoter. However, introduction of four copies of the beta hotspot (48 copies of CCGNN) significantly reduced the level of meiotic recombination. DNase I hypersensitivity experiments indicated that both the 12- and 48-copy insertions generated regions of open chromatin. The insertion of the $(CCGNN)_{48}$ tract suppressed the activity of an introduced meiotic recombination hotspot. It is possible that the larger $(CCGNN)_{48}$ is acting as multiple sites for recombination initiation, preventing the formation of a complete initiation complex at any of the sites. In support of this model, competition between adjacent hotspots has been observed at other loci, including *HIS4*[59] and *ARG4*.[60]

Alteration of the Hotspot Sequence
Affects Recombination Initiation

The recombination-initiating ability of hotspots can be altered by mutation. Deletions of regions known to be important for meiotic DSB formation can either eliminate the DSB or alter the positioning of the lesion, as demonstrated at *ARG4*.[61] Small changes also have significant impact on alpha hotspots; point mutations in the *HIS4* binding site for the *RAP1* transcription factor that prevent its binding significantly decreases meiotic recombination.[36] Introduction of three binding sites in place of the normal promoter increases hotspot activity, indicating that Rap1p can substitute for all of the other *HIS4* transcription factors, if bound in sufficient concentration.[37] Deletion analysis of *RAP1* indicated that the DNA-binding and transcription activation domains of the protein were required for hotspot activity.[33] In *S. pombe*, single nucleotide changes led to the creation of the *ade6-M26* hotspot when an Mts1/2 transcription factor binding site was formed.[39] Nucleotide polymorphisms have also been demonstrated to moderate the activity of a human recombination hotspot.[62] Taken together, these data indicate that alpha hotspots are susceptible to mutations that alter their activity; selective pressure may act via this mechanism.

The Influence of Genome Wide Sequence Divergence on Recombination

Gene conversions result from the repair of mismatches in heteroduplex DNA. However, if the number of sequence differences within a region exceeds a certain threshold, recombination does not occur. Regions with a high level of sequence divergence ("homeology") arise when related species are mated and made to undergo meiotic recombination. These hybrids are sterile,[63] but this sterility can be partially alleviated by deletion of the mismatch repair system. The mechanism for this "antirecombination" effect may be the unwinding of heteroduplex DNA containing "too many" mismatches. The mismatch repair system is thought to assess the extent of homology between the interacting sequences, rejecting sequences of insufficient homology, thus preventing crossing over and elevating chromosome missegregation. Thus, the mismatch repair system is a component of the barrier to interspecific hybridization.

The degree of divergence at which this phenomenon occurs varies from species to species. In yeast this level can be as low as 0.16%,[63,64] while in humans much greater divergence is tolerated.[62] The variation in threshold level from species to species may be an evolutionary response to the degree of diversity within and amongst populations. Antirecombination at the level of sequence divergence seen in yeast would have dire consequences for humanity.

To aid in the study of the effects of sequence divergence, single or chimeric chromosomes from other yeast species can be introduced into S. cerevisiae, creating "partial hybrids." These experiments have demonstrated that the higher the degree and extent of homeology, the greater the effect on crossing over and chromosome segregation.[65] Analysis of the surviving spores[66] in such experiments suggests that the degree of sequence divergence differentially influences the success of strand invasion (figure 12.2c) and strand capture (figure 12.2d). This hypothesis is supported by the demonstration that strand invasion and capture involve different subsets of recombination proteins.[67] The consequences of this differential response range from complete sterility due to chromosome missegregation or partial sterility due to chromosome loss. These activities of the mismatch repair system have been hypothesized to play a central role in the mechanisms by which species arise[68] and are maintained as separable groups, as described above and reviewed in reference 14.[14]

Global Influences on Recombination

Chromatin State Is Important for the Initiation of Recombination

The importance of chromatin and chromatin-modifying proteins in recombination initiation and meiosis has been appreciated for years. As described above, early work in yeast indicated that DSB formation was linked to an open chromatin configuration.

Also, mutation or deletion of genes known to have chromatin-modifying activity often lead to sterility due to failure to complete meiosis successfully. The inability to complete meiosis has limited the experiments that could be conducted successfully, but recent technical advances have allowed researchers to examine the meiotic role of some chromatin-modifying proteins. The *SET1* H3 methyltransferase,[69] which methylates lysine 4 of histone H3 has been implicated in DSB formation, as DSBs at two loci tested in *set1Δ* were reduced. Similarly, the *RAD6-BRE1* complex, which adds a ubiquitin molecule to a C-terminal lysine of histone H2B, also influences DSB formation.[70] Indeed, it has long been known that *RAD6* mutants display locus-specific defects in rates of gene conversion.[71] Deletion of *RAD6* or *BRE1*, as well as mutations in H2B that prevent ubiquitylation, reduce DSBs in a locus specific manner, especially at strong hotspots. The pattern of DSBs remains unchanged, indicating that the histone modification is not involved in designating the site of the hotspot but rather the site's susceptibility to cleavage. The chromatin modifications made by Set1p and Rad6-Bre1p are associated with opening up of chromatin, which is consistent with protein loss leading to a reduction in the frequency of DSBs. It is likely these and other chromatin modifications yet to be identified impact the function of all three types of hotspots. The specific modifications that modulate coldspots remain to be identified, but chromatin structure appears to be important for recombination suppression at coldspots.

Premeiotic DNA Replication and DSB Formation Are Linked

Recombination initiation also is affected by alterations in premeiotic DNA replication. Mutations that disrupt replication initiation sites on the left arm of chromosome 3 delay replication and DSB formation on that arm, but not elsewhere in the genome.[72] This delay in DSB formation is accompanied by a delay in the appearance of the meiosis-specific increase in open chromatin, as assayed by sensitivity to micrococcal nuclease. These results indicate that premeiotic replication may be required for meiosis-specific alterations in chromatin and subsequent DSB formation, and that the influence of replication on DSB formation is region-specific. As telomeres and centromeres have long been known to replicate later than most genomic loci, the observed link in these experiments between replication delay and DSB formation may be pointing to a more general level of regulation affecting large areas of each chromosome. It is yet to be experimentally determined whether all coldspots occur in late replicating regions of the genome.

Metabolic State

In addition to the chromosomal and genomic inputs into the initiation of meiotic recombination discussed above, the environment and the metabolic state of the cell have a direct impact on the decision to initiate the meiotic pathway and the level of recombination.

Meiosis is induced in *S. cerevisiae* by the local environment. Starvation for fermentable carbon sources and nitrogen triggers a signal transduction cascade that activates meiosis-specific genes required for recombination and sporulation.[73] The metabolic state of the cell also influences recombination levels. In yeast, transcription of *GCN4* is induced by starvation for amino acids, purines, or glucose. The *GCN4* protein is a leucine zipper transcriptional activator controlling the transcription of approximately 500 genes.[74] As mentioned above, Gcn4p is required for hotspot activity at *HIS4*. Surprisingly, meiotic recombination frequencies at *HIS4* are affected by deletion of genes involved in leucine or lysine biosynthesis. The ability to synthesize lysine decreases gene conversion while the ability to synthesize leucine increases gene conversion. Both of these effects are via an increase in Gcn4p. In the first instance, the inability to make lysine is perceived as a starvation signal and induces Gcn4p, leading to activation of recombination at *His4* and presumably other loci containing Gcn4p binding sites. The ability to synthesize leucine creates an imbalance in the isoleucine/valine amino acid pools, which is also known to induce Gcn4p.[35] Directly increasing or decreasing the amount of Gcn4p by genetic alterations increases or decreases the level of gene conversion at *HIS4*. The *GCN4* transcription factor presumably acts on alpha hotspots, influencing their meiotic activity in response to cellular conditions. In support of this model, DSB-associated genes, as determined globally by Gerton et al.,[22] are enriched twofold specifically for the presence of binding sites for *GCN4* protein.[35,74] They are also enriched for genes involved in metabolism, suggesting that hotspot activity may be influenced by the intracellular environment.

Gene conversion at *HIS4* is also increased by the absence of the adenine biosynthetic pathway.[35] Part of this increase is due to the effect of starvation for adenine on induction of Gcn4p. The other part is dependent on the transcription factor Bas1/Bas2p. As discussed above, Bas1/Bas2p is also required for maximal recombination at *His4*.[37] It has been shown that adenine itself interferes with the formation of the Bas1/Bas2p complex,[75] therefore its presence will suppress recombination at *HIS4*. Thus, the frequency of meiotic recombination at alpha hotspots is directly connected to the cellular environment through their dependence upon transcription factors and the metabolic states that regulate them.[35,76]

Interactions between the Various Levels of Control: Questions and Puzzles

The Hotspot Paradox

As described above, hotspots are regions of the genome that stimulate meiotic recombination, and many different factors influence the relative "strength" of each hotspot throughout the genome. Alleles of a hotspot region that differ by only a few nucleotides can have widely divergent influences on recombination frequency. For example, alterations in the *HIS4* promoter region can change the level of meiotic DSB formation and aberrant segregation.[33,36,37,59] Thus, maternal and paternal alleles have the potential to differ widely in their ability to stimulate recombination. In yeast and other model organisms, it is possible to ensure that the two alleles are identical, but this is rarely the case in nonmodel organisms such as humans, where it has been demonstrated that specific nucleotide changes can affect initiation.[62]

Variation in the level of recombination initiation between paternal and maternal hotspots leads to a paradox for hotspot propagation. If one allele is stronger than another, the stronger allele will initiate recombination more frequently. The DSB-repair model[1] for recombination postulates that the initiating hotspot acts as a recipient of information (figure 12.3). Thus, the DNA sequence at the stronger initiating hotspot will be replaced with the sequence from the weaker hotspot. If this cycle of selection against strong hotspots is repeated without interruption, the hotspot will eventually be lost entirely. Simply stated, any imbalance in the ability of a particular DNA sequence to stimulate meiotic recombination should result in the replacement of the stronger sequence with the weaker sequence.

Recent analyses of recombination frequencies in orthologous regions of human and chimpanzee genomes[77,78] give support to the existence of the hotspot paradox. Statistical evaluation of large linkage disequilibrium (LD) data sets have demonstrated that meiotic recombination hotspot locations do not occur at the same position in humans and chimpanzees, even though the total amount of recombination is similar, and the two genomes differ by only 1.5%. Taken together, these data indicate that sites of recombination initiation evolve more rapidly than the genome as a whole, as predicted by the hotspot paradox.

Mechanisms for Hotspot Maintenance

A number of factors can influence the retention of hotspot sequences in the face of the hotspot paradox. These include the presence of counterselection for the retention of hotspot sequences, the frequency in the population of alleles at a particular locus that exhibit a range of hotspot activity, and the frequency at which new alleles arise that have an elevated or reduced level of recombination-stimulating activity relative to that for the surrounding DNA sequences.

The alpha hotspots described above are the least likely to succumb to the hotspot paradox because they are dependent on DNA-binding transcription factors for recombination initiation. As these proteins are also required for transcription initiation, alterations in the DNA sequence that greatly reduce or enhance their ability to bind will be deleterious to the cell, and will be selected against. Thus, the tendency to lose the hotspot sequence via meiotic recombination will be balanced by the vegetative drive to maintain the sequences. An additional factor may be the sequence-nonspecific nature of the DSB; because the breaks occur over a region, rather than at a specific nucleotide sequence, sequences necessary for the hotspot activity may not be involved in gene conversion events if the initiating break is relatively distant from those sites.

In contrast, beta hotspots do not have counterselection to maintain their presence in the genome. Beta hotspots are often the result of introduction of DNA sequences into a particular region, leading to a disruption of the local DNA and chromatin. Thus, they are likely to be present only on one of the two homologs, and will be lost if they initiate recombination. In the case of beta hotspots, the hotspot paradox may be beneficial to the genome as it would act to eliminate the foreign DNA insertions.

Gamma hotspots, unlike beta hotspots, are likely to escape the hotspot paradox. Gamma hotspots are the result of a confluence of elements, most notably an elevated G+C content in the region, that combine to allow elevated levels of meiotic recombination initiation at that locus. Because of the large region that contributes to the hotspot, gene conversion events at the site of DSB formation are not likely to decrease the ability of the hotspot to initiate recombination as the gene conversion events will not include enough of the involved sequences. In fact, analysis of all of the available gene conversion events and their outcome indicates the existence of a slight bias toward gene conversion events that result in the replacement of T or A with G or C bases.[79] This bias may be an evolutionary response to the biased mutation of G/C bases to T/A bases. The existence of a gene conversion bias toward G+C would favor the retention of gamma hotspots. Finally, when the *ARG4* gene and its associated promoter region were embedded within a G+C-rich region of the genome, deletion of the sequences known to be sites of high levels of meiosis-specific DSB formation did not lead to the loss of recombination initiation,[60] but instead the DSB moved to the adjacent sequences. Similarly, a plasmid sequence acting as a beta hotspot exhibited a different level of recombination at each locus into which it was inserted.[42] Recombination frequency was significantly correlated with the G + C content of the surrounding 30 kb of DNA into which the plasmid was inserted.[80] This is as might be expected, if the controlling elements were not absolutely dependent on the inserted sequences, but instead were determined by the genomic context of the entire locus.

Conclusions

The generation of crossovers during meiosis is absolutely essential for the proper segregation of chromosomes in the subsequent meiotic divisions; each homolog pair must have at least one chiasma. Crossovers and chiasmata form as the direct consequence of the DSBs that initiate recombination, and multiple breaks occur on each chromosome to ensure that at least one crossover forms. Thus, the formation of DSBs is vital, both for successful completion of meiosis to form the gametes required for the next generation and to ensure that there is genetic diversity upon which selection can act. However, while DSBs often occur in specific regions (hotspots) of the genome, their exact location is not dependent strictly on the DNA sequence at that position, but rather is influenced by the context in which the sequence is located. The location of DSBs is the result of a complex interplay of many factors: the local DNA sequence, its ability to bind transcription factors, the degree to which the chromatin in the region is in an open configuration, the G+C content of the region, the presence or absence of other DSBs in the local region, and environmental factors, including the metabolic state of the cell and the temperature the cell is experiencing during meiosis. Thus, while in theory any locus in the genome can undergo recombination, some loci are very likely to rearrange, while others rarely do. Additionally, gene conversion events are directly coupled to repair of the DSB, and thus regions of the genome containing meiotic recombination hotspots are likely to be altered in their primary sequence at high frequency, as long as the recombining homologs are not too divergent. The interplay of all of these elements gives rise to a genome that has the potential to be very dynamic from generation to generation, but which exhibits numerous levels of control that act to regulate the frequency and location of alterations, and upon which selection can act.

13

Nuclear Duality and the Genesis of Unusual Genomes in Ciliated Protozoa

Carolyn L. Jahn

Overview

Ciliated protozoa have two kinds of nuclei, micronuclei and macronuclei, which are functionally distinct. The genome of the macronucleus is derived from the micronuclear genome by a highly controlled developmental process of DNA rearrangements. In contrast to developmentally regulated genome rearrangements studied in other organisms, the ciliate rearrangements are global and occur throughout the genome. The diversity of rearrangement mechanisms apparent in different ciliates indicates that nuclear duality provides an unusual "playing field" for genome evolution. The characterization of genome remodeling in ciliates offers a completely different view of how genomes function and challenges our understanding of what a genome is. The biology of ciliates has spurred remarkable discoveries in the area of RNA splicing (the self-splicing rRNA intron of *Tetrahymena*), telomere biology (the first telomeres sequenced and the identification of telomerase in *Tetrahymena*), and the "histone code" (the functions of histone acetylation and methylation in transcription). The seemingly "bizarre" nature of their genome processing holds equal promise.

Introduction

Reorganizing the genome is a fundamental life process in ciliated protozoa. These organisms have evolved a highly sophisticated system of maintaining, within each unicellular individual, two different genomes within two kinds of nuclei. Named for their sizes, macronuclei and micronuclei are functionally distinct: a true division of labor has evolved. Macronuclei are the site of most transcriptional activity. The macronucleus is frequently referred to as a "somatic" nucleus because it does not participate in sexual processes yet drives the clonal expansion of cells. In contrast, the micronucleus serves as a transcriptionally silent "germline" that carries out meiosis leading to sexual exchange of haploid micronuclei between organisms. What makes ciliates so amazing is that the macronuclear genome is derived from the micronuclear genome in a highly orchestrated developmental process that is carried out during the sexual phase of the life cycle.

The "DNA processing" that generates the macronuclear genome is extraordinary in both quality and quantity. In all of the ciliates described in this chapter, the processing involves DNA deletion, chromosome fragmentation with de novo telomere addition and DNA amplification. In some ciliates, the processing is so extensive that every gene undergoes all three of these processing steps. The DNA processing is reproducible from individual to individual and generation to generation, thus it is a fundamental genetic mechanism in these organisms. Genome remodeling can differ dramatically between different ciliate species: some are completely dependent on processing to generate functional genes while others are not; and some species have taken DNA processing to another level, where gene segments are reordered in the deletion process ("unscrambling"). This chapter gives an overview of the different types of processing and summarizes research that addresses the following questions. First, are there specific DNA sequences that drive the DNA processing, and do different organisms utilize the same sequences? Second, what are the mechanisms involved and do they differ between organisms? Third, what aspects of sequence recognition are encoded versus non-encoded or determined by epigenetic mechanisms? Fourth, how is the processing controlled in a developmental stage and nucleus-specific way? And finally, how did such an elaborate system evolve?

The genome rearrangements in ciliates are frequently described in the context of "developmentally controlled gene rearrangements," such as mating type switching in yeast or immunoglobulin gene rearrangement in mammals. Although processing in ciliates is as controlled, precise, and reproducible as these other rearrangements, it differs substantially from these other cases in that it is global; it occurs throughout the genome and involves tens of thousands of different sequences that can be reliably rearranged in an identical manner each time a cell forms a new macronucleus. Because of this global aspect, ciliate genome remodeling may be more related to the processes of transcription and meiotic recombination, which are global, restricted to certain DNA sequences, and controlled in a highly developmental stage specific way.

Nuclear Duality

The unique structure of ciliate genomes and the DNA processing that generates one genome from the other have arisen as a result of nuclear duality. Hypothetically, all ciliates arose from a protozoan with multiple identical nuclei that initially became differentiated as to whether or not they divided mitotically.[1-3] This proposed ancestral form, with multiple identical nuclei, remains characteristic of only one ciliate class existing today. In most ciliates, nuclear duality involves differentiation between the two genomes at the DNA sequence level and a clear separation of nuclear functions as a result of differences in nuclear, chromosome, and chromatin

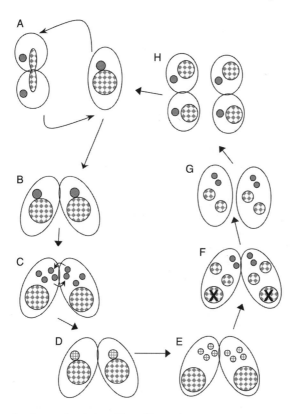

Figure 13.1.   A schematic of the ciliate life cycle. (Representative of major events but not specific to any particular organism.) Cells divide vegetatively by fission (A), with a macronucleus dividing as cell division occurs. During the sexual phase of the life cycle, two cells undergo pairing (conjugation), as in (B), followed by meiosis of the micronucleus. Haploid micronuclear meiotic products are exchanged between the cells (C), to form a new zygotic diploid micronucleus (D). This zygotic micronucleus divides (E) and gives rise to new macronuclei and micronuclei (F). At this time, the old macronucleus is degraded (F, G). After completion of macronuclear development, cell division occurs to reconstitute the vegetative state (H).

structures. The macronucleus is polyploid and divides by an "amitotic" mechanism, where the DNA is partitioned between the two daughter cells as cell fission occurs (see figure 13.1a). Micronuclei are diploid (in cases where it is known) and divide mitotically. As a consequence of nuclear duality, the cell cycle is complex in ciliates because it involves two different nuclear cycles: macronuclei and micronuclei replicate and divide on independent schedules that are individually coordinated with cell growth and division.[4]

As described in the introduction, nuclear duality results in a division of labor between micronuclei and macronuclei, where one is transcriptionally active and the other silent; one divides mitotically and the other is amitotic. These distinctions between the kinds of nuclei indicate that two major functions of the genome, transcription and the partitioning of genetic material in an ordered mitotic and meiotic segregation, occur in two independent organelles: the macronucleus and micronucleus, respectively. As described below, this is accompanied by differences in genome structure and the content of DNA sequences in the two kinds of nuclei. Thus many ciliate biologists believe that the sequence differences between macronuclei and micronuclei could inform us of the DNA sequence context of these divergent nuclear and genomic functions. Our current knowledge of eukaryotic chromatin and chromosome structure suggests that transcription and segregation compete with one another: the former requires decompaction to allow accessibility of transcriptional machinery, whereas the latter requires a high degree of compaction so that entire chromosomes can be moved around independently. As ciliates evolved from an organism that performed both mitosis and transcription within a single nucleus, we might expect that certain meiotic and mitotic functions have been lost or suppressed in the macronucleus, and likewise that transcriptional activity and the flow of genetic information from DNA to RNA to protein has been lost or inhibited in the micronucleus.

Many of the functional differences between these nuclei can be attributed to what is frequently referred to as "epigenetic" mechanisms: heritable changes arising from genomic functions separate from the DNA sequence. For example, many of the histone modifications that have been shown to play a role in transcriptional activation or silencing or in regulation of mitosis were first documented in *Tetrahymena*, where they showed macronuclear or micronuclear specific localization.[5,6] Nuclear transport functions also play a role in the functional distinction between a macronucleus and a micronucleus. Thus, this is clearly a case where two genomes are functioning within two different nuclear organelles. Although this chapter focuses on the genomes of ciliates, it is important to remember their subcellular context.

Genome Remodeling as a Developmental Process

The macronuclear genome is derived from the micronuclear genome. This is a developmental process that occurs during the sexual phase of the ciliate life cycle (refered to as "conjugation") (see figure 13.1b–h). Typically, when two cells are induced to mate (or "conjugate"), micronuclear meiosis is followed by sexual exchange of haploid micronuclei and fusion of a haploid nucleus from the partner cell with a resident haploid nucleus to generate a "zygotic" micronucleus. This zygotic nucleus divides twice and the four descendants differentiate into micronuclei and macronuclei. As these new macronuclei form, the old macronuclei that were present prior to conjugation are destroyed. Thus, the somatic nuclear lineage is reestablished from the germline lineage after sexual exchange and creation of a new zygotic nucleus. From a developmental biology perspective, these single-celled organisms carry out a nuclear "germline versus soma differentiation" that parallels the cellular differentiation in multicellular organisms.

Evolutionary Relationship between Different Lineages

Genome remodeling has been characterized in a variety of ciliate organisms and it is therefore important to point out their phylogeny. Comparisons between macronuclear and micronuclear genes and genomes have been carried out for at least eight different organisms belonging to four different genuses and representing two out of ten distinct classes of ciliates (phylum Ciliata, kingdom Protista). These include: *Oxytricha fallax*, *O. trifallax*, *O. nova*, *Stylonychia lemnae*, *Euplotes crassus*, *Tetrahymena thermophila*, *Paramecium tetraurelia*, *P. aurelia*, and *P. primaurelia*. Sequence analysis of ribosomal RNA genes indicates that these classes diverged early in the evolution of eukaryotes.[3,7] Thus, time of divergence of the class Oligohymenophora (such as *Paramecium* and *Tetrahymena*) from the Spirotrichea (such as *Oxytricha*, *Stylonychia*, and *Euplotes*) is equivalent to the time of divergence of the common ancestor of a plant (such as rice, *Oryza sativa*) and an animal (such as a frog, *Xenopus laevis*) (this comparison is taken from reference 7). Even within a class, the distance is great. For instance, *Paramecium* and *Tetrahymena* are as distant as any extremes within metazoans, such as *Caenorhabditis elegans* and humans.[8] It is notable that differences in codon usage appear to have arisen among ciliate species.[9] For example, within the spirotrichs, Euplotids use TAA and TAG as stop codons and TGA encodes cysteine, whereas *Stylonychia* and *Oxytricha* utilize TGA as a stop codon and TAA and TAG to encode glutamine. Thus, the evolutionary time separating these organisms has allowed shifts in coding. It seems highly likely that the processes of genome rearrangement have evolved independently over a huge expanse of time as well. This chapter highlights some of the similarities and differences in these processes in the organisms mentioned above, with an emphasis on exploring the diversity of ways these organisms manipulate their genomes. Because a lot is known about *Tetrahymena thermophila* and *Euplotes crassus* and these two organisms are very different, these two cases are emphasized. This is not an exhaustive review since several detailed reviews are available.[10–15]

Types of DNA Processing during Macronuclear Development

At the DNA sequence level, the differentiation of a micronucleus into a macronucleus involves extensive DNA elimination (chromatin diminution) as well as DNA amplification. Two types of DNA processing occur that result in DNA elimination: breakage with telomere addition (chromosome fragmentation), and precise breakage with rejoining (deletion). Thus, eliminated DNA either lies between two fragmentation sites or is internal to a macronuclear chromosome and referred to as "internal eliminated sequences," or IESs. The amount of processing and developmental timing of the processing differs dramatically in different ciliates (see figure 13.2). The difference between micronuclear and macronuclear genomes is greatest in spirotrichs, such as *Euplotes*, *Stylonychia*, and *Oxytricha*. These organisms eliminate 90–95% of the micronuclear genome during macronuclear development. In addition, fragmentation and telomere addition occurs at sites surrounding every gene to generate 10,000

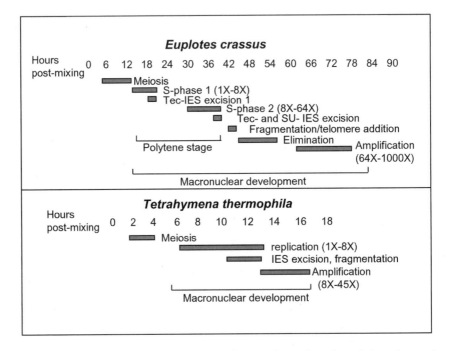

Figure 13.2.   The timing of developmental events in conjugation. Timing of events is referred to in "hours post-mixing" of two different mating types to initiate the sexual phase of the life cycle. The sexual phase includes meiosis followed by macronuclear development, as shown for *Euplotes crassus* and *Tetrahymena thermophila*. In *E. crassus*, during two S-phases in the developing macronucleus, polytene chromosomes are formed. Deletion (IES excision) events occur during the latter half of each S-phase, and chromosome fragmentation and most of the sequence elimination occurs at the end of the polytene replications. In *T. thermophila*, specific events in the DNA processing to form a new macronucleus are not separable because of the short timespan involved.

to 20,000 different "gene-sized" (500 bp to 20 kb) macronuclear chromosomes. Because the precursors to the macronuclear chromosomes make up such a small proportion of the micronuclear genome, they are frequently referred to as the "mac-destined sequences" (MDSs). In contrast to the spirotrichs, the *T. thermophila* micronuclear genome is only slightly more complex than the macronuclear genome, and approximately 15% of the sequences are eliminated. These sequences are typically referred to as "mic-limited." The fragmentation process occurs less frequently, resulting in macronuclear chromosomes containing numerous genes per chromosome (600 kb average). In addition to elimination of sequences, DNA amplification results in polyploidy in the macronucleus. In spirotrichs, each linear DNA molecule is present in thousands of copies per macronucleus (1000 average in *Euplotes*, 10,000 average in *Stylonychia*). In *Paramecium*, the ploidy is as high as the spirotrichs (1000C) whereas in *Tetrahymena* macronuclear ploidy is only 45C. The DNA replications that accomplish these amplifications differ in number and timing in different ciliates. In all of these organisms, the linear chromosomes bearing the ribosomal RNA genes are amplified to a higher degree than other chromosomes.

Chromosome Fragmentation and Telomere Addition

In general, genome remodeling results in smaller chromosomes and hence an increased number of chromosomes in the macronucleus. This ranges from five micronuclear chromosomes being fragmented to around 250 in *Tetrahymena* to 50 to 100 micronuclear chromosomes being fragmented to 20,000 unique minichromosomes in spirotrichs. Since each chromosome is linear and capped with a telomere at each end, ciliate macronuclei are "telomere rich." Much of what we know about the structure and function of telomeres originated from work with ciliates.[16–18] Nevertheless, we still have much to learn about how these organisms recognize where to cut their micronuclear chromosomes and how they add telomeres at these fragmentation sites.

Sequencing of the ends of macronuclear chromosomes revealed C_4A_2 repeats at *Tetrahymena* telomeres and C_4A_4 repeats at spirotrichs telomeres (see figure 13.3a). The telomeres in *Tetrahymena* are typically 50 to 70 repeats of C_4A_2 and they are not uniform in size. In contrast, the telomeres in spirotrichs consist of C_4A_4 repeats and have a very uniform structure: 20 or 28 base pairs of double-stranded repeats and a 3' single strand 16 or 14 base pair overhang of G_4T_4 repeats in Oxytrichids or Euplotids, respectively.

One of the surprises when micronuclear precursors to macronuclear chromosomes were first sequenced was that no telomeric repeats were present at the sites where chromosome fragmentation occurs. This became even more perplexing when Greider and Blackburn identified telomerase in *Tetrahymena* and showed that it added repeats to a preexisting end but did not have de novo synthesis activity with a nontelomeric sequence.[16] Subsequently telomerase extracts that support telomere addition to nontelomeric ends with defined structures, such as a single-stranded end, have been characterized.[17] In *E. crassus*, two different high molecular weight complexes containing telomerase have been defined: one from vegetative cells and one from cells with developing macronuclei. Only the latter is capable of de novo telomere addition. Interestingly, identification of TERT (i.e., the telomerase reverse transcriptase component) genes in *E. crassus* demonstrated that one of the genes is expressed only during macronuclear development.[19] Thus, this telomerase and its interacting proteins may be differentiated for de novo telomere addition.

In *Tetrahymena* and *Euplotes*, de novo addition occurs by specific breakage of chromosomes at a defined "chromosome breakage sequence" (CBS) (figure 13.3b). The CBS in *Tetrahymena* triggers fragmentation, resulting in loss of the CBS sequence in the process. The telomeric repeats are added at variable positions, which may indicate that nucleases are active at the ends prior to telomere addition. Mutational studies of the *Tetrahymena* CBS indicate that the 15 bp is highly stringent (i.e., very few substitutions are tolerated).[20] Any mutations that affect fragmentation also affect telomere addition. As these two processes were not distinguishable by mutation, this suggests that fragmentation and telomere addition are coupled and dependent on the CBS.

In contrast, the *Euplotes* CBS, with a well-conserved core TTGAA consensus, results in cutting at a specific distance and is usually retained in the macronuclear sequence (figure 13.3c). As a consequence, when the telomeres of macronuclear DNA molecules were sequenced the "TTGAA" sequence was identified as a conserved

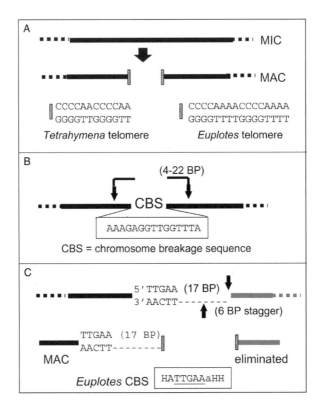

Figure 13.3.   The chromosome fragmentation and de novo telomere addition in Euplotes crassus *and* Tetrahymena thermophila *utilize consensus sequences known as "chromosome breakage sequences" (CBSs).* (A) The types of telomeric repeats specific for each organism. (B) The 15 bp *Tetrahymena* CBS. The arrowheads designate where cleavage occurs relative to the CBS. The DNA containing the CBS is eliminated in *Tetrahymena*. (C) The *Euplotes* CBS consensus and the staggered cleavage relative to one mac-destined sequence. Both the eliminated DNA and the mac-destined sequence gain telomeres. (H = A or C or T.)

sequence that occurred 17 bp internal to the C_4A_4 repeats. Sequence analysis of micronuclear precursors to macronuclear molecules, showed that those MDSs that lacked the TTGAA had the sequence in the flanking eliminated DNA at a spacing 11 bp from the site of telomere addition. This led to the hypothesis that the CBS could trigger a 6 bp staggered cut in the DNA. Characterization of ends produced as fragmentation intermediates has verified the existence of this 6 bp staggered cut.[21] In some cases, one CBS and fragmentation site produces the ends of two different macronuclear chromosomes, whereas in others eliminated DNA lies between two macronuclear ends. Surprisingly, telomeric repeats are also added to the eliminated sequences.[22] Thus, the telomeres do not designate what should be retained as a macronuclear sequence.

In contrast to *Tetrahymena* and *Euplotes*, there are no obvious consensus sequences associated with de novo telomere addition in *Paramecium*, *Stylonychia*, and

Oxytricha. Nevertheless, in the latter two spirotrichs, the site of telomere addition occurs in a nearly precise location ("micro heterogeneity," or within 10 bp). Biases in the nucleotide content (purine versus pyrimidine) have been noted in the sequences adjacent to the fragmentation site in the retained macronuclear chromosomes, but whether this influences the fragmentation process or perhaps reflects sequence biases important for transcriptional control, is unknown. In *Paramecium*, fragmentation and telomere addition occur within 0.2–2 kb regions and alternative telomere addition regions separated by up to 20 kb can be used in generating the same macronuclear chromosome. Several instances of telomere addition at sites where IES excision occurs have been demonstrated in *Paramecium*, which indicates that these two processes are at least temporally closely associated.

IES Deletion

Like the chromosome fragmentation events, the deletion events in ciliates are abundant and are highly reproducible from cell to cell and generation to generation. In many cases the deletions result in rejoining of coding sequences in a manner that resembles removal of introns by RNA splicing. Nevertheless, the mechanisms involved do not appear to be as conserved as that of RNA splicing and different ciliates have different "rules" in terms of the sequence requirements and the mechanisms of deletion. (RNA splicing is discussed in chapter 15.)

In spirotrichs, approximately 10% of the eliminated DNA exists internal to the sequences that are destined to become the macronuclear chromosomes and comprise the deleted sequences referred to as IESs (see figure 13.4a).[12] Most genes in spirotrichs contain one or more IES; thus, to maintain the reading frame, precise excision is required for gene activity. There are two major classes of IESs: transposons (or repeated "transposon-like" sequences) and small unique sequences, which are referred to as SU-IESs. In the spirotrichs both classes are abundant. For instance, in *E. crassus* there are two related families of transposons, Tec 1 and Tec 2, which each exist in 5000 to 7000 copies per genome. Approximately one third of these elements occur as IESs that must be precisely removed in order to reconstitute genes. In addition, there are an estimated 40,000 SU-IESs per genome. Similar or greater numbers of SU-IESs are found in all of the spirotrichs examined to date. In *O. trifallax*, an abundant transposon family, the telomere bearing elements (TBEs) (so-called because they have telomeric repeat sequences at the ends of the element) exist both within and outside genes. Both the TBEs in *O. fallax* and the Tec elements in *Euplotes* are sufficiently intact to identify open reading frames within the elements. For example, both of these element families possess a transposase ORF that is related to the Tc1-mariner family of elements that is widespread in eukaryotes and has prokaryotic counterparts. Although no sequence analysis has been carried out on repetitive elements in *O. nova*, a 25 kb element that is repeated around 1000 times also interrupts genes as IESs. Thus a notable characteristic of micronuclear genomes in spirotrichs is the disruption of genes by both large, highly repetitive, transposon-like elements and small unique sequence elements, both of which must be precisely removed to generate functional genes.

In *Tetrahymena*, most of the eliminated DNA is removed via deletion of IESs, with a minority of the DNA being lost at fragmentation sites. IESs range in size

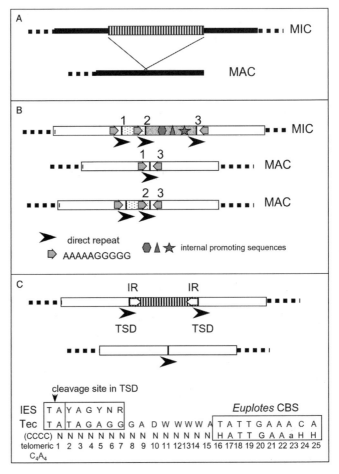

Figure 13.4.   IES deletion. (A) The deletion of internal eliminated sequences (IESs). (B) The *Tetrahymena thermophila* "M-region," which has been extensively characterized by mutagenesis and DNA transformation. Three types of sequences that participate in processing are designated. First are the direct repeats that are the sites of the recombination events: one copy of the direct repeat is retained in the macronuclear chromsome. Second are the flanking A_5G_5 sequences that are essential for deletion and which position cleavage at approximately 45 bp from the sequence. Third are sequences that occur within the IESs that promote deletions in a non sequence specific way. The M-region is typical of *Tetrahymena* IESs as it exhibits alternative deletions either between sites 1 and 3 or between sites 2 and 3. (C) The deletion of Tec element IESs in *Euplotes crassus*. The target site duplication (TSD) consists of a TA (shown as an arrowhead): one copy is retained in the macronuclear sequence. Tec elements possess 700 bp inverted repeats (IR). At the ends of their repeats is a consensus sequence that occurs in *Paramecium* and *E. crassus* SU-IESs, which is known as the "TAYAG" consensus (Y = C or T; R = A or G; H = A or C or T; D = A or G or T). The Tec element inverted repeat ends also contain a "TTGAA" consensus that matches the *Euplotes* CBS (see figure 13.3c). The spacing of this sequence is similar to the spacing in macronuclear chromosome ends relative to the telomeric repeats (C_4A_4).

234

from about 0.5 kb to over 25 kb. The largest of these IESs belong to a family of elements that are retrotransposon-like.[23] These elements are not bounded by the same sequence at all of their insertion sites, thus they may have inserted within pre-existing IESs. Another family in *Tetrahymena* is similar to TBEs in *O. fallax* and has telomeric repeat sequences at the ends of its inverted repeats. Although many of the IESs in *Tetrahymena* are repetitive (like the Tec transposons in *E. crassus*), many different repetitive sequence families (and types of transposons) are represented and, to date, no common features characteristic of all IESs have been defined. Transformation of *Tetrahymena* has allowed definition of the *cis*-acting sequence requirements for excision of IESs. In all of the cases examined, there is clearly a role for sequences that exist outside the IESs, in the flanking mac-destined DNA. For instance, in the well-characterized "M region" IES excision, repeated sequences of A_5G_5 (figure 13.4b) are necessary for excision.[15] Mutation of the A_5G_5 abolishes excision; when A_5G_5 is moved to a new position it generates a new deletion endpoint at similar spacing (~45 bp) from the element. Although flanking sequences have been demonstrated to play a role in the deletion of other IESs in *Tetrahymena*, none of these possess the A_5G_5 tracts and no consensus sequences in these flanking regions have been found. Thus, a variety of sequences trigger deletion. The mutational studies have shown that IES sequences themselves contribute in some less sequence-specific way to deletion because they cannot be substituted by "foreign" (non-*Tetrahymena* IESs) sequences.[10] As described below, the IESs are transcribed bidirectionally during meiosis and the resulting double-stranded RNAs are involved in the elimination of the IESs. Thus, the sequences within the IESs may play a role in this "RNAi"-based mechanism. The *Tetrahymena* IESs examined to date do not show precise excision: there are typically multiple deletion endpoints and alternative deletions. In concert with this imprecision, most IESs in *Tetrahymena* do not occur within coding regions, although one has been found within an intron. Many ciliate genes contain introns, so it is notable that these organisms readily distinguish introns from IESs, even though they resemble each other in many ways, for instance in AT content.

Although *Paramecium* is more closely related to *Tetrahymena* than the spirotrichs, it has SU-IESs that are very similar to those in the spirotrichs. The *Paramecium* genome is estimated to have around 60,000 SU-IESs that frequently interrupt genes and therefore must be precisely excised. As discussed below, consensus sequences in these elements suggest that they have arisen in the same way as the *Euplotes* IESs. The *Paramecium* IESs are somewhat smaller and even more AT-rich than the spirotrich SU-IESs.

Consensus Sequences and Their Role in IES Deletion

The precision of IES deletion, particularly in spirotrichs, where coding regions must be reconstituted, suggests that the process requires highly specific sequence recognition. Thus, it was expected that consensus sequences, comprising the sites recognized by the deletion machinery, would be apparent in the boundaries of the deletions. Surprisingly, the consensus sequences are minimal. Thus, once again, the ciliates break the mold of what molecular biologists expect in a genetic process.

Consensus sequences at the boundaries of IESs have been identified in *Paramecium* and *Euplotes* (figure 13.4c).[12] Initially, a TA target site duplication (TSD) was noted at the ends of all *E. crassus* IESs. Comparison of *Paramecium* IESs identified a "TAYAGYNR" consensus, which matches the highly conserved sequence at the ends of the *E. crassus* Tec elements (TATAGAGGG: note that the first TA is the TSD). Comparisons of all of the SU-IESs from *E. crassus* found evidence for conservation of the TAYAG consensus at their boundaries as well. This consensus matches the consensus inverted repeats ends of Tc1-mariner family elements, suggesting that SU-IESs may all be derived from this type of transposon family in *Paramecium* and *E. crassus*. Mutations in the TAYAG consensus in *Paramecium*, either an A to G transition in the TA or C to T transition at the "Y" position, prevent excision. Thus this consensus is critical to IES removal, not just a leftover relic from a transposon insertion.

As described above, the TTGAA consensus is the *E. crassus* chromosome breakage sequence involved in fragmentation (figure 13.3c). Despite clear evidence for a role as a CBS, a role outside of fragmentation and telomere addition is indicated by the presence of this sequence at identical spacing relative to the required cleavage site within the TA TSD in the Tec element inverted repeats (figure 13.4c). Furthermore, the sequence is found in many SU-IESs at the 17 bp spacing in at least one end of the IES. Since the chromosome fragmentation and telomere addition sites appear to require a different type of staggered cleavage than we would predict for IESs, this may be a sequence that facilitates cleavage at a site 17 bp away by two different types of cleavage machinery (one for IES excision and one for fragmentation). This could mean that a protein that recognizes this site interacts with multiple DNA processing enzymes. Interestingly, proteins with "AT-hook" domains interact with AT-rich sequences and preferentially contact a consensus TTGAA motif.[24,25] These proteins typically cause a bend or kink in DNA at these sites and have been characterized as facilitators of DNA recombination events. Unfortunately, the amino acid sequence of an AT hook domain does not allow searching for AT hook protein genes using "reverse translation" and gene hunting by PCR with degenerate oligonucleotide sequences. Nevertheless, this consensus could be recognized by proteins with similar DNA bending activities and thereby play a role in generating a "hypersensitive site" that is cleaved by the fragmentation and IES excision machinery.

Reaction Intermediates and Products Define Potential Mechanisms

Characterization of precursors and products is part of defining a reaction mechanism. Finding intermediates provides additional clues to the enzymatic steps involved. For IES deletion, defining both the structure of the IES after deletion and the structure of macronuclear genomic product can elucidate how the DNA is cut to release the IES. Identifying intermediates helps to define what type of DNA break is made to initiate IES removal and whether each end of the IES is cleaved by the same mechanism. Analyses of reaction intermediates and excised products has indicated that diverse mechanisms of IES excision exist in different ciliates.[12]

In *E. crassus*, analysis of the products of Tec element and SU-IES excision indicates that these sequences are excised by the same mechanism. Both Tec elements and SU-IESs from circular products that have an unusual structure with a region of heteroduplex DNA 10 bp in length between two copies of the TA-TSD. This heteroduplex presumably is generated by a staggered cut, which would result in precise reconstitution of the MDSs with a single copy of the TA-TSD. Sequences from each side of the IES in the MDSs pair with each other to form the heteroduplex region in the circles (see figure 13.5a). Analyses of circular products and the cleavage intermediates of TBE excision in *O. fallax* indicate that a staggered cut at one end is followed by reaction at the other end of the IES (a proposed transesterification) (figure 13.5b). The circular products of TBEs contain one copy of the TSD. The characterization of intermediates and products in *Tetrahymena* suggest a similar cleavage and transesterification, but the IES is released as a linear molecule (figure 13.5c). The cleavage preferentially occurs 3' to an adenine and has a 4 bp staggered end in *Tetrahymena*. In general, the conclusion from these studies has been that cleavage and religation are different in each organism and therefore are most likely carried out by different enzymes. However, there is a remote possibility that cleavage is carried out by a common enzyme that can recognize different "altered" DNA structures, for instance hairpins versus cruciforms, which might form different intermediates and cleavage products. It also is important to note that the excision mechanisms are acting within the context of each organism's array of DNA repair and recombination machinery and that multiple enzymes may be involved; thus some mechanistic features could be shared and not others, and the outcome may be determined by the balance between these activities.

Origins of IESs

Many lines of evidence point to the IESs being derived from transposable elements. First, any transposons that exist in micronuclear genomes are eliminated during macronuclear development. Several different types of transposons have been identified in ciliates and regardless of how they function as a transposon they are subject to elimination. To date, there is no evidence that the enzymes encoded by these elements are functioning in their elimination. Many of these elements have decayed to the point that their genes are probably nonfunctional, although at least some, when examined with respect to replacement versus synonymous codons, do show signs of having been selected for function. In addition to recognizable transposon families, many other IESs are repetitive sequences, especially in *Tetrahymena*, which we might assume are even more degenerate versions of transposons. Thus, in general, the SU-IESs are thought to be completely degenerated transposons that no longer bear familial sequence similarity. This seems particularly true of the relationship between transposons and SU-IESs elements in *E. crassus*. The Tec transposons are highly abundant "TA" transposons related to the Tc1-Mariner family (i.e., the TA target site duplication). As mentioned above, most of the SU-IESs also have a TA target site duplication at their ends and a TAYAG-like sequence at their ends. Because, both Tec elements and SU-IESs are excised to form circular molecules of the same structure, they most likely are eliminated by the same mechanism. However, whether this mechanism is dependent on this conserved structure is presently untested.

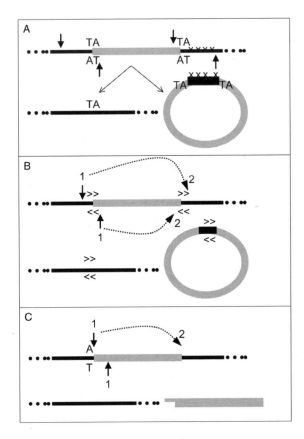

Figure 13.5.   Proposed mechanisms of excision for IESs. (A) *Euplotes crassus* Tec elements and SU-IESs. (B) *Oxytricha trifallax* TBE elements. (C) *Tetrahymena thermophila* IESs. In each case, arrowheads designate cleavage sites on each strand of DNA and the numbers indicate which cuts occur first or second. The dashed line arrows between the first and second cleavages in (B) and (C) indicate that the ends produced by the first cleavage are active at the second cleavage in a proposed transesterification. In (A), the staggered cut at each side of the Tec element or SU-IES results in a heteroduplex region in the circular excised product that is surrounded by the TA target site duplication. The heteroduplex region consists of sequences from each of the flanks. A single copy of the TA target site duplication remains in the chromosome after precise removal of the IES. In (B), the circular product contains precisely one copy of the target site duplication, with the other copy remains in the chromosome. In (C), the predominant product of excision is a linear form of the IES.

Although transposons have clearly played a role in generating IESs, debate exists surrounding their role in generating the mechanisms of DNA processing. Did IES excision evolve from a transposition mechanism when a ciliate host assimilated transposon genes, and did this system then evolve to recognize a wide range of sequences? Or, are we looking at "host" mechanisms of DNA recombination and repair that evolved in a unique direction in ciliates such that they are regulated developmentally and are highly efficient at removing IESs? In either case, once a mechanism exists that efficiently removes transposons, the micronuclear genome becomes a "safe haven" for transposons, where they are free to expand without being selected against due to detrimental effects on their host. As noted below, the spirotrich micronuclear genomes are very large, which suggests that their genomes have continuously expanded.

Extreme IESs and Unscrambling

A group of ciliates, including *Oxytricha* and *Stylonychia*, exhibit an extreme form of IES deletion that removes IES DNA while reordering coding sequences.[14,26,27] This process is referred to as "unscrambling" because the segments of DNA that encode a gene, the MDSs are scrambled relative to their order within the macronucleus where they comprise a functional gene.

The first scrambled gene identified was an actin gene from *O. nova* (see figure 13.6), which has nine MDS segments separated by IESs. Only one of the nine IESs is a typical SU-IES, where after excision its flanking MDSs are joined to each other. The other eight MDSs must be reordered in the process of IES excision. The boundaries of these eight scrambled MDSs contain 4–13 bp sequences that can be aligned with a matching sequence in another MDS to reorder the coding sequences such that the proper coding region is assembled and contains a single copy of each repeat (figure 13.6b). In this sense, these boundaries behave like the target site duplications in the typical spirotrich IES.

Out of 10 micronuclear genes that have been characterized in *O. nova*, three are scrambled. The same genes are scrambled in *Stylonychia*. Thus, it is expected that this may be a common feature of micronuclear genes in *Oxytricha* and *Stylonychia*. The most highly scrambled gene identified so far is the DNA polymerase alpha gene, which is split into 45 scrambled segments in *Oxytricha nova* and 48 segments in *Stylonychia lemnae*. In this micronuclear precursor, the MDSs are in a nonrandom order, with "odd-numbered" MDSs separated from "even numbered" MDSs. This has led to a model whereby pairing of an uninterrupted gene with an AT-rich DNA segment leads to insertion of multiple IESs and subsequent rearrangement of IESs all at once (see the review by Prescott for a model).[26] Thus, this suggests a very different origin than the hypothesized transposon origin of the "normal" SU-IESs. Comparison of actin and polymerase genes from multiple species of *Oxytricha* and *Stylonychia* has shown that a scrambled organization exists in other species, but the number of MDSs and IESs can differ and IESs and "guide" sequences accumulate sequence differences. Furthermore, the guide sequences can shift or "slide" 1 to 20 nucleotides. Thus, this arrangement is in flux and some lineages have a higher propensity to add IESs than others. The combination of IES addition by the proposed AT-rich DNA pairing and by transposon insertion can lead to different outcomes in each organism.

DNA Amplification

Although DNA amplification does not alter the DNA sequence of the genome, changes in copy number can affect the DNA processing events of macronuclear

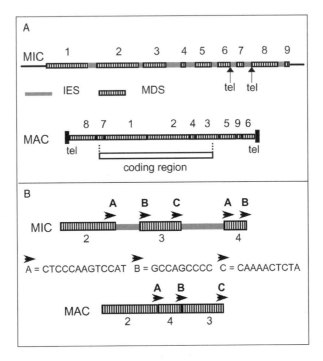

Figure 13.6. DNA unscrambling. (A) The arrangement of IESs and mac-destined sequences (MDSs) in the *Oxytricha nova* scrambled micronuclear precursor of the actin I gene, with the two sites of telomere addition designated by arrows (tel). The MDSs are numbered according to their order in the micronuclear genome. The unscrambling process reorders them in the macronuclear sequence. (B) The repeats at the ends of three of the MDS, shown as arrowheads labeled A, B, C, with the sequences that they represent shown below. In the reordering of the MDS, one copy of each of these repeats is retained in the coding region.

development. Amplification of genes is a mechanism common to many organisms, either as a normal part of their life cycle or as a response that provides a selective advantage under altered environmental conditions. In ciliates it most likely allows the macronucleus to transcriptionally provide for the cell growth of these large single cells. Because some amplification occurs prior to DNA processing in the spirotrichs, it may have been critical in evolving the precise DNA processing that reconstitutes coding regions of genes.

In spirotrichs, amplification occurs at multiple stages (see figure 13.2). At the start of macronuclear development, multiple rounds of DNA replication occur to form cytologically visible polytene chromosomes. The degree of polyteny differs in different spirotrichs, leading to DNA contents from 25C to 110C. In contrast, *T. thermophila* replicates the DNA once before DNA elimination (i.e., 4C). When DNA rearrangements in *E. crassus* were staged relative to the two S-phases of this polytene stage (see figure 13.2), it was found that deletion events occur during the latter half of each S-phase, while chromosome fragmentation and most of the sequence elimination occurs at the end of these two polytene replications.[12] The two polytene S-phases result in differential amplification of sequences. The MDSs, which are clustered in the genome, replicate in both phases and thus reach a maximum ploidy of 64X, whereas large regions of eliminated DNA that are devoid of MDSs replicate only during the second phase and thus increase eightfold or less prior to their elimination. Replication timing can play a role in defining transcriptionally active versus silenced chromatin domains: for instance, in mammalian cells the heterochromatic, inactive X chromosome replicates later than the transcriptionally active genes. In ciliates, the early replication of the MDSs versus late replication of much of the eliminated sequences might allow distinctions between euchromatin and heterochromatin and might be a determinant for recognition of retained (active) versus eliminated (silenced) sequences. After the DNA processing and elimination in spirotrichs, multiple rounds of replication increase the ploidy of the macronuclear linear DNA molecules to 1000–10,000X. In *T. thermophila*, a similar increase via replication at the end of macronuclear development occurs and increases the final copy number of macronuclear chromosomes to 45.

The presence and degree of polytenization in different organisms may play a role in the fidelity of DNA processing. Polytenization increases the number of DNA strands available for recombination and thus increases the probability of a "favorable" outcome. In an evolutionary sense, this allows a less-accurate system to evolve toward a more-accurate system, but it also allows for rapid evolution, as new combinations of coding sequences can arise and provide diversification of genes in the macronucleus. In the spirotrichs, where precise excision is required to reconstruct coding regions and the genes can exist in a scrambled state, the ability to generate multiple products from the polytenized DNA may be essential for ensuring that at least one functional copy of a gene is faithfully transmitted and amplified in the macronucleus. Since the macronucleus divides amitotically, the multiple alleles of genes assort randomly during cell division. With polytene chromosomes, the multiple rearranging strands of DNA can generate as many as 64 or 128 different "alleles" of a given gene, which are then further amplified at the end of macronuclear development. During vegetative growth of the cells, different progeny can receive different numbers of copies of each "allele" during the amitotic division. The cells carrying higher copy numbers of the correctly rearranged copies of genes could have a selective advantage.

Chromatin Structure and DNA Processing

Given the magnitude of DNA processing events in ciliates, it was expected that there would be DNA sequences that specified the rearrangements and that consensus sequences would be identified by DNA sequence analysis of multiple rearrangement sites. Although this is partially true, the absence of obvious consensus sequences in *Tetrahymena* and *Oxytricha* IESs and at *Paramecium* and *Oxytricha* chromosome fragmentation sites suggests that there is not a strict requirement for specific sequence recognition in DNA processing. The consensus sequences identified in IESs are fairly short and would be expected to be found frequently in the genome, not exclusively at sites of DNA processing. This suggests that chromatin structure and DNA structure might be determinants of sequence selectivity.

Tetrahymena presents the most complete story of a chromatin-mediated mechanism of rearrangement in ciliates. Multiple lines of investigation have come together to suggest that DNA elimination in *Tetrahymena* is guided by RNA interference (RNAi)[28] mediated heterochromatin formation.

First, three proteins (Pdd1, Pdd2, Pdd3) were identified that are made only during conjugation.[29-32] These proteins colocalize with sequences that are destined to be eliminated (i.e., mic-limited DNA) in condensed, heterochromatic structures in the developing macronucleus. The sequences of the corresponding genes defined two of these as chromodomain proteins. Chromodomain proteins identified in *Drosophila* and mammals are associated with heterochromatin and the chromodomain has been shown to interact with RNA. Knockouts of the genes for these *Tetrahymena* chromodomain proteins demonstrated that they were required for DNA deletion. Second, bidirectional transcription of IESs was observed during meiosis, indicating that double-stranded RNAs (dsRNAs) from the IESs could accumulate.[33] Third, characterization of a homolog of *Drosophila piwi* (or argonaute in *Arabidopsis*), which is a gene known to be involved in RNA interference and transcriptional silencing, demonstrated that a *Tetrahymena piwi* homolog (TWI1), is made only during conjugation and localizes similarly to the Pdd proteins.[34] Knockout of the *Twi1* gene blocks deletion and inhibits chromosome fragmentation. Many of the previously documented silencing processes in fungi, plants, and animals that involve *piwi* homologs also involve small dsRNAs produced by the RNaseIII-like enzyme called "dicer." Similar small dsRNAs that corresponded to mic-limited sequences were identified in *Tetrahymena* during macronuclear development. Accumulation of these small RNAs is dependent on TWI1 function. Finally, a histone modification that is associated with RNAi mediated heterochromatin formation in *S. pombe*, lysine 9 methylation of histone H4, was shown to colocalize with Pdd proteins and DNA undergoing elimination.[35,36] Mutation of this methylation site in histone H4 blocks both development of the macronucleus and IES elimination.

Based on the above findings and the known mechanisms of RNAi in other organisms (see figure 13.7), a model for DNA elimination in *Tetrahymena* was developed, called the "scan RNA model."[34] Double-stranded RNA, transcribed from mic-limited sequences during meiosis, is digested by a dicer-like enzyme and then associates with an RNA-induced silencing complex (RISC)-like complex, which moves to the old macronucleus. While in the old macronucleus, these dsRNAs are "scanned" such that sequences homologous to macronuclear sequences are eliminated and only dsRNAs corresponding to true mic-limited sequences remain. These scan RNA complexes then move to the developing macronucleus, where they assemble a heterochromatin structure that specifically targets their homologous mic-limited sequences for deletion.

Yao et al. carried out two tests of this model.[37] They injected dsRNAs, corresponding to non-*Tetrahymena* sequences that existed in the micronucleus but not the macronucleus (as NEO gene interruptions used create knockout mutations) and dsRNAs corresponding to known macronuclear genes, into conjugating cells at about the time that the "scan RNAs" move between the old and new macronucleus.

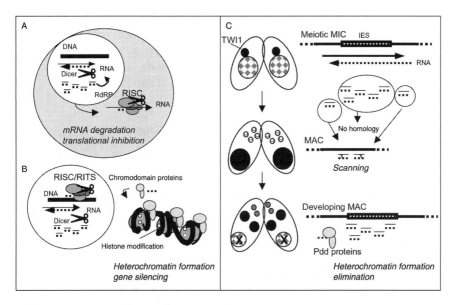

Figure 13.7. RNA interference (RNAi) results in gene silencing through multiple mechanisms, all of which are homology-dependent and triggered by double-stranded RNA (dsRNA).[28] Common to all of these mechanisms is the cleavage of dsRNA to small interfering dsRNAs (siRNAs) by an evolutionarily conserved RNaseIII-type enzyme named "dicer," shown in (A). These small dsRNAs can lead to gene silencing in the cytoplasm by inducing degradation of the homologous mRNA or by inducing translational suppression by interacting with the 3' untranslated region of a mRNA. This is mediated by a multiprotein RNA-induced silencing complex (RISC), shown in (A). The argonaute proteins (also known as *piwi* in *Drosophila*) are evolutionarily conserved components of RISC. In some organisms, additional dsRNAs are produced by an RNA-dependent RNA polymerase (RdRP), shown in (A), thus amplifying the signal. A nuclear form of gene silencing is mediated by a RISC-like argonaute protein containing complex called RITS (RNA-induced initiation of transcriptional gene silencing) which mediates chromatin remodeling and heterochromatin formation in the nucleus, shown in (B). Chromatin remodeling relies on a specific methylation pattern of histones and involves "chromodomain" or polycomb-like proteins that can interact with RNA to form a highly condensed transcriptionally inactive chromatin structure. In *Tetrahymena*, the identification of "chromodomain" proteins (Pdd proteins), an argonaute homolog (TWI1), and small dsRNAs produced from bidirectional transcripts of IES has led to a "scan RNA model,"[34] shown in (C). During meiosis in the micronucleus, the IESs are transcribed bidirectionally across the IES. The resulting dsRNAs are cleaved to small (26–31 bp) dsRNAs, which move to the "old" macronucleus. There, they are scanned against the existing macronuclear DNA sequences such that dsRNAs with homology to macronuclear DNA are degraded and those without homology (i.e., corresponding to the mic-limited, IES sequence) are retained. The dsRNAs (and TWI1 and Pdd proteins) move from the old macronucleus to the developing new macronucleus where they form heterochromatin structures at the homologous DNA sequences (the IES). The three stages of conjugation shown roughly correspond to figures 1b, 1c, and 1f.

243

In both cases they observed that the corresponding genes are deleted from the new macronucleus. In other words, dsRNA that enters the new macronucleus is "interpreted" by the new macronucleus as a sequence that was made in the micronucleus but was not protected by the old macronucleus and thus should be deleted. This experiment demonstrated that dsRNA is sufficient to direct DNA deletion.

"Scanning" and the Role of the Old Macronucleus in Macronuclear Development

The "scanning" concept in the *Tetrahymena* model is an attempt to explain several studies that demonstrate that the old macronucleus plays a role in determining which sequences are eliminated in the new macronucleus.[38–40] One of the first instances was a *Paramecium* mutant, d48, which lost a surface antigen gene in the macronucleus because of chromosome breakage and generation of a new end. This mutation subsequently caused the same aberrant fragmentation, even when only wild-type sequences existed in the micronucleus that gave rise to the new macronucleus. Thus the mutation behaved as a cytoplasmically inherited trait. The old macronucleus was implicated in this epigenetic inheritance when it was found that injection of wild-type sequences into the macronucleus resulted in a wild-type chromosome fragmentation (reversion of the mutation).

Whereas microinjection of a wild-type sequence repaired the DNA processing in the d48 mutant, another effect is observed when IESs are injected into macronuclei. In this case, the injected sequences inhibit removal of the IES during development in a sequence-specific and concentration-dependent manner, as if they were titrating their homologous sequences in some way. It is not yet clear whether an RNAi mechanism underlies some or all of these effects. Nevertheless, that proteins such as Pdds and TWI1 move from the old macronucleus to the developing macronucleus points to the communication between these nuclei. Since it is possible that DNA, RNA, and proteins could move between nuclei, there may be many ways that the old macronucleus influences the differentiation of a micronucleus into a new macronucleus.

Micronuclear Genome Organization

Taken together, the data that have been accumulated on genome structure and mechanisms of DNA processing in ciliates indicate tremendous diversity. In addition to major differences seen at the level of individual genes and IESs, there are huge organizational differences between micronuclear genomes and macronuclear genomes in different organisms. For instance, comparison of micronuclear genomes and the scale of genomic processing in *T. thermophila* and *E. crassus* (see figure 13.8) indicates that these represent two extremes of DNA processing. *T. thermophila* eliminates about as much DNA as *E. crassus* (or other spirotrichs) retains. Thus, in a simplistic sense, *T. thermophila* needs to select out and differentiate the

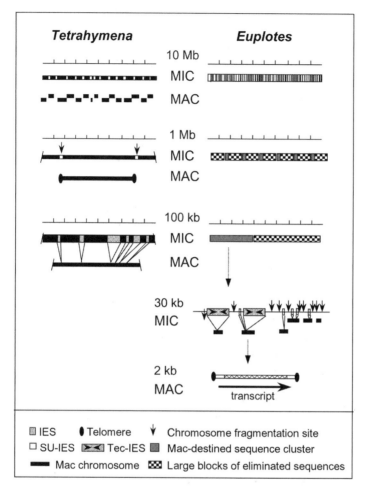

Figure 13.8. *A comparison of the scale of processing events in* Tetrahymena thermophila *and* Euplotes crassus. (Note that stepping down the page the scale expands tenfold.) Macronuclear chromosomes in *T. thermophila* average 0.6 Mb and thus are visible at the 10 Mb scale. In contrast, the average size for *E. crassus* is 2.2 kb, and thus the *Euplotes* macronuclear chromosomes are shown at the very bottom. There are approximately 200 chromosome fragmentation sites in *T. thermophila,* whereas there are approximately 20,000 in *E. crassus.* In *Tetrahymena,* approximately 15% of the micronuclear genome is eliminated either at breakage sites (short stretches of DNA) or as IESs (the majority of eliminated DNA). There are approximately 6000 IESs. In *Euplotes,* the majority of the eliminated DNA occurs as large blocks of DNA, which contain both unique and repetitive sequences, and occur as domains that are separate from regions containing macronuclear destined sequences (MDSs), which are clustered together in the genome. Approximately 35,000 IESs consisting of Tec elements (~15,000) and SU-IESs (~20,000) occur with in the MDSs within these clusters. Thus these regions are sites of intense DNA processing.

eliminated sequences from those that are retained. In *E. crassus*, however, the converse is required, and the MDSs must be distinguished from the rest of the genome.

An additional point that is obvious from figure 13.8 is that the processing events in *E. crassus* are highly clustered in the genome: MDSs are grouped together and large stretches of eliminated DNA exist in between clusters. This organization is typical of spirotrichs and most likely reflects differences in how the eliminated and retained sequences evolved and how they are recognized. The clustering of the MDSs may be an indication that "gene by gene" chromosome fragmentation arose early on in the spirotrich lineage such that once these regions were designated for processing, other regions of the genome could expand in different ways. The micronuclear genomes of spirotrichs are very large relative to other organisms at their level of evolutionary branching (i.e., ten times that of Oligohymenophera, such as *Tetrahymena* and *Paramecium*, and corresponding to three times that of *Drosophila*, and one-fifth that of mouse). Most of this extra DNA is within the long stretches that are devoid of MDSs. These regions contain both repetitive and unique sequence DNA and probably have been distinct from the "gene" regions (i.e., MDS clusters) for a very long time. Little is known about how these sequences are eliminated, except that they differentially replicate during the polytene stage and disappear at the time of chromosome fragmentation in *E. crassus*.

The Role of Transcriptional Silencing and Heterochromatin

Returning to the functional distinctions between macronuclei and micronuclei, it is important to note that the sexual phase of the life cycle is the one time when the "division of labor" is relaxed. Micronuclei become transcriptionally competent in terms of gaining transcriptional machinery prior to DNA rearrangements. Transcription and the initiation of transcriptional competence in the micronucleus play a significant role in DNA processing in *Tetrahymena*. Heterochromatin formation on IESs in the developing macronucleus relies upon aberrant bidirectional transcription of IESs in the micronucleus to identify them as excisable. This transcription occurs in the meiotic micronucleus prior to formation of the zygotic nucleus that is the precursor to the new macronucleus (see figure 13.1). In contrast to what is observed in *Tetrahymena*, the spirotrichs may utilize transcription to differentiate the MDSs from the rest of the genome. Since each macronuclear molecule corresponds to a transcription unit, the assembly of a transcriptionally competent chromatin structure could define the genomic regions where processing occurs. The IESs may be distinguishable as disruptions of the transcriptionally competent chromatin state and telomere addition may select boundaries of transcription.

The genome-processing mechanism in *Tetrahymena*, which involves RNAi marking IESs for deletion, may have evolved from the RNAi-mediated genome surveillance mechanisms for silencing transposons in organisms, such as fungi, plants, and animals,[41–43] which arose very early in eukaryotic evolution. It remains to be seen whether all ciliates have retained this pathway; in yeast, this pathway exists in *Schizosaccharomyces pombe* but has been lost in *Saccharomyces cerevisiae*.[44] Nevertheless, *S. cerevisiae* exhibits transcriptional silencing via heterochromatin formation.[45] There are multiple ways to form heterochromatin and different transposons can be silenced by different mechanisms.[46] Thus, different ciliates may use different ways to form heterochromatin to target elimination, and possibly within one species, different sequences may be eliminated using more than one type of heterochromatin structure.

Although heterochromatin formation seems to be a likely mechanism for differentiating eliminated and retained sequences, it is difficult to conceive of heterochromatin playing a role in excision of SU-IESs in *Paramecium* or spirotrichs since this process requires differentiation of chromatin by modifying histones in one or more nucleosomes. Many SU-IESs are smaller than a single nucleosome and so could not be differentiated even at this level. It seems plausible that these small sequences are recognizable because of their DNA structure, as mentioned above. The SU-IESs are AT-rich sequences (70–80% AT), which can make them bendable. In addition to the possibility that they could be recognized by proteins that specifically interact with AT-rich sequences, they may adopt a structure that disrupts or phases chromatin structure such that they are nucleosome free and "hypersensitive" to processing machinery.

Conclusions

The genome remodeling of ciliates would not exist without the evolution of nuclear duality in these organisms. Thus, it is important to keep in mind that these genomes are interpreted through the organelles in which they reside. Silencing of the micronucleus during vegetative growth of cells was a critical step in setting up the possibility of evolving a highly "rearranged" genome. The separation of functions and the derivation of a macronucleus from a micronucleus sets up a "symbiosis" of sorts for the two genomes and organelles. Just as the symbiotic origin of mitochondria has allowed highly unusual patterns of genome evolution in these organelles, the distinctions between nuclei in ciliates provide a very different evolutionary "playing field."

14

Editing Informational Content of Expressed DNA Sequences and Their Transcripts

Harold C. Smith

Overview

A preliminary annotation of eukaryotic genomes has suggested that there are far fewer genes encoding mRNA than predicted from the number of proteins expressed in cells (the proteome). In fact, the coding capacity of genomes is expanded through conditionally activated mechanisms. These mechanisms are regulated in species- and tissue-specific manners and include, for example, mutation and recombination of DNA, use of alternative promoters, alternative pre-mRNA splicing, RNA editing, alternative polyadenylation, and mRNA turnover. It is likely that a substantial fraction of the genome encodes processes that diversify expressed sequences. The increasing awareness and acceptance that a simple linear analysis of DNA sequences is not sufficient to annotate the genome's full coding capacity represents a significant change in the scope of hypotheses that will drive research in the twenty-first century.

This chapter discusses select aspects of RNA (and DNA editing) with a goal of providing the reader with a sense of the exciting new research frontiers that have opened due to developments in this area. RNA editing is defined as a co- or post-transcriptional process that changes the nucleotide sequence in RNA from that complementary to the DNA from which it was transcribed, through mechanisms that involve either base modification, substitution, deletion, or insertion.

Discovery of RNA Editing

Once the table of codons was described in the 1960s, researchers assumed that they could simply translate a DNA sequence into the sequence of amino acids in proteins. This view of the informational content of the genome was shaken up by the discovery of intervening sequences in the late 1970s (see chapter 15). However, once introns were incorporated into our thinking, exon splicing and the removal of intervening noncoding intronic sequences was considered by and large the major means of diversifying the proteome. In this mind-set, once splice sites were identified in a gene, all of the protein-coding information could be translated from the linear DNA sequence. Yet the mechanism of coding for several proteins or protein variants remained enigmatic until mRNA editing was discovered. Unlike numerous covalent modifications of the sugar or base moiety of nucleotides in mRNA, ribosomal RNA, and transfer RNA (known generally as RNA modification) that already were known at this time, RNA editing had the potential to directly change the sequence and/or half-life of the protein encoded by the mRNAs.[1,2]

The potentially broad significance of the discovery by Rob Benne and colleagues[3] of RNA editing in flagellated protozoa known as kinetoplastids (referred to as Trypanosomes) was not immediately appreciated, although this discovery demonstrated unprecedented posttranscriptional uridine nucleotide insertions in mitochondrial mRNA. These edited nucleotides were not explicitly encoded by the DNA sequence that was transcribed into mRNA, yet they were absolutely required to induce frame shifts that established the correct reading frame of several mRNAs. Within the next few years, Stuart and colleagues[4] made the startling and widely noted discoveries that in some cases 50% of the protein coding sequence in mRNAs from Trypanosome mitochondria were added through editing, and in fact some of the DNA-encoded uridines were deleted. Mitochondrial genomic DNA revealed few or no full-length sequences corresponding to the mature mitochondrial mRNA sequences that encoded several essential proteins in the respiratory chain of enzymes. Neither were these proteins encoded in nuclear genomic DNA. Instead, mature mRNAs encoding full-length and functional protein sequences were constructed from partial or rudiments of mRNA encoded in the mitochondrial genome (genomic partial genes known as crypto genes) that were expressed and subsequently processed by multiple U insertions and deletions. Furthermore, each mRNA contained numerous editing sites, and each site was specified by a unique *trans*-acting small RNA (referred to as guide RNA, gRNA) containing complementary sequence to the mRNA just 3′ of the editing site, a region in the mRNA known as the anchor sequence (the mechanism is summarized in figure 14.1). The mitochondrial genome of Trypanosomes consists of cantenated maxi- and minicircular DNAs[5]. Crypto genes (and most of the mitochondrial transcribed sequences) are encoded within the maxicircular genome whereas guide RNAs are encoded largely on the minicircles.

Fewer than ten years after Benne's seminal work, the broad scope of RNA editing was evident as it had been discovered to affect numerous RNAs in phylogenetically diverse organisms including: the mRNA encoding mammalian transmembrane glutamate-gated ion channels and apolipoprotein B lipid carrying protein, plant mRNAs from chloroplasts and mitochondria, RNA viral genomes and numerous classes of RNAs in slime molds, amoeba, and yeast (select mechanisms are summarized in table 14.1 and figure 14.1). These editing events were in many instances more subtle than the extensive insertions seen in kinetoplastid mitochondrial mRNAs, and in most cases involved modification or substitution of individual nucleotides, resulting in various nucleotide transitions and transversions.[1,2]

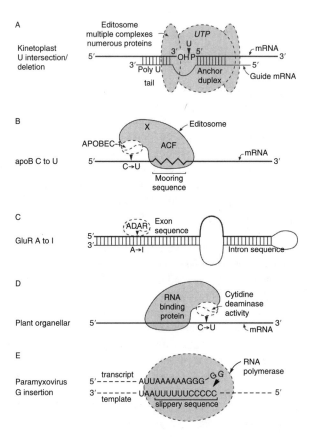

Figure 14.1.   Models of select mRNA editing mechanisms. Macromolecular complexes involved in RNA editing are shown for a few mechanisms: only the most general aspects of each editing mechanism are indicated. For most editing mechanisms, the protein composition of the editosomes has not been fully characterized. For each model, an example of an edited RNA substrate or the organelle in which a group of RNAs are edited is stated to the left. (A) The editing complex or editosome for U insertion or deletion consists of multiple subcomplexes, each containing several proteins (suggested as gray ovals) and involving distinct enzymatic activities for insertion and deletion editing. The anchor duplex determines the site of editing, and mismatches between the guide RNA and the substrate (looped out region) are thought to determine the actual nucleotide position of editing. For other editing mechanisms (B–D), the part of the editosome involved in editing site recognition and binding to the catalytic subunit are shown in gray and the catalytic subunit is indicated separately. For apoB mRNA editing (B), the RNA binding protein ACF (see figure 14.2) binds to the 11 nucleotide mooring sequence and binds and positions APOBEC-1 for editing the appropriate C. The deaminases for C to U and A to I editing function as dimers. ADARs bind to double-stranded RNA (C) and deaminate A to I within duplex regions (shown here as exon sequence in duplex with an adjacent and intron sequence). ADARs have autonomous double-stranded RNA binding activity. Therefore, unlike apoB or plant organellar C to U editing (C), A to I editing is believed not to require an auxiliary protein (E) Paramyxovirus RNA editing is a co-translational process in which nucleotide insertion occurs through polymerase slipping on the template strand within regions of nucleotide repeats. Bold arrowheads indicate examples of the consequence of editing, which are shown as C→U (C to U), A→I (A to I), U (uridine insertion), and G (guanidine insertion).

250

Table 14.1. Examples where RNA is edited.

Type	Organism	Edited Transcript (or Genome)	Mechanism
U insertion/deletion	Kinetoplastids, *Trypanosoma, Leishmani, Crithidia, Bodonids*	mRNAs (m)	gRNA targeting site, U insertion or deletion and ligation
C insertion (also U, AA, CU, GU, GC, UA)	*Physarum*	mRNAs (m), rRNAs, tRNAs (m)	Co-transcriptional C insertion
G insertion	Paramyxoviruses (SV5, Sendai), mumps, measles	P mRNA	Co-transcriptional G insertion
A insertion	Ebola viruses	Glycoprotein mRNA	Unknown
GA deletion	Rats	vasopressin mRNA (n)	Unknown
C to U	Plants	mRNAs (c), (m), numerous mRNAs at multiple sites	C-deamination
	Physarum	cox1 mRNA (m)	C-deamination
	Mammals	Gly→Asp tRNA anticodon	C-deamination
	Mammals	ApoB mRNA (n), Gln→STOP NF-1 mRNA (n), Arg→STOP	C-deamination
	Mammals	tRNA^{Asp} (n) (adjacent to the anticodon loop)	C-deamination
U to C	Land plants	mRNAs (c), (m)	U-amination
	Mammals	WT-1 mRNA (n), Leu→Pro	Unknown
		tRNA^{Asp} (n) (adjacent to the anticodon loop)	Unknown
		C18 ORF 1 mRNA 5′ UTR	Unknown
A to I	Vertebrates, fly	GluR-B,5,6 (n), Gln→Arg	A-deamination
		GluR-B,C,D (n), Arg→Gly	

(Continued)

251

Table 14.1.   (Continued)

Type	Organism	Edited Transcript (or Genome)	Mechanism
		GluR-6 (n), Tyr→Cys (n), Ile→Val	
		5-HT$_{2C}$R (n), Ile→Val, Asn→Ser	
		PTPN6 phosphatase, ablates splicing branch site	A-deamination
		Endothelin B receptor Glu→Arg	A-deamination
		5′ and 3′ UTRs alu sequences	A-deamination
	Hepatitis delta virus	Antigenome, STOP→Trp	A-deamination
	Mammals, squid, fly	Kv2 K$^+$ channel mRNA	A-deamination
	Rats	α-2,6-Sialyltransferase, Tyr→Cys	Unknown
	Bee, fly, moth, worm	Numerous exon and intron sequences	A-deamination
C to A, A to G, U to G, U to A	*Acanthomoeba*	tRNAs (m)	Unknown

Examples of RNA editing in organisms and viruses were taken from the cited literature in the text (refs 1–11, 15, 17, 21, 25–27, 33, 43). (c), chloroplast; (m), mitochondria; (n), nucleus.

The discovery of plant mRNA editing is of particular note as it brought to light tens to hundreds of editing sites within mRNAs from chloroplasts and mitochondria, respectively. So extensive were these editing events that prior to the discovery of mRNA editing, the disparity between mRNA (cDNA) and organelle genomic sequences had led researchers to speculate that plant organelles used a different genetic code. With the discovery of RNA editing, comparisons of expressed homologous mRNA sequences from either chloroplasts or mitochondria in different species suggested that editing frequently served to generate amino acid substitutions necessary for functional proteins.[2,6–8]

The discovery of mRNA editing in mammalian tissues had the additional effect of establishing that protein expression could be regulated not only through the control of transcription, translation, and mRNA half-life but also through mRNA editing. Perhaps most remarkable was the example of the glutamate-gated calcium channels of the central nervous system (controlling virtually all levels of human cognitive and motor activity).[9,10] Each receptor protein serves as one of the five subunits that interact to establish a transmembrane channel for calcium within the postsynaptic membrane. These channels are regulated (gated) by the neurotransmitter glutamate. A direct translation of the genomic DNA sequence for the subunits positions a glutamine at a key position within the channel. Channels with glutamine in this position are leaky to calcium even in the absence of glutamate signaling. A to I (inosine) editing (changing CAG to CIG, which is read as CGG) changes this glutamine to arginine, thereby placing a positive charge in the channel. A positive charge in this position helps to exclude calcium, thus closing the channel. Signaling by glutamate during synaptic activity opens the channel by inducing appropriate conformational changes.[1,9–11] Other sites of receptor subunit mRNA editing have been identified that affect the rate with which membranes returned to their resting potential following an action potential.

Flies and worms also require A to I mRNA editing of their homologous channel receptor subunits for the neuronal activity necessary to coordinate motor functions and food foraging.[12,13] Discoveries such as these underscored the underappreciated dependence of organisms on mRNA editing for appropriate protein function.

At about the same time as the discovery of A to I mRNA editing, the mRNA encoding apolipoprotein B (apoB) in mammals was discovered to be C to U edited.[14,15] Virtually 100% of all apoB mRNA is edited within the epithelial cells (enterocytes) that line the small intestines of all mammals and a variable and regulated proportion of apoB mRNA is edited in the liver (hepatocytes) of some species.[16] Editing converts a cytidine at nucleotide 6666 of a CAA glutamine codon to a UAA stop codon, thereby enabling both full-length (apoB100) and truncated (apoB48) variants of apoB protein to be expressed from a single gene.

ApoB48 is stored in the enterocyte and assembled with dietary lipids as the structural protein core of chylomicrons. These are secreted into the lymphatic ducts draining the small intestine and enter the bloodstream, from which they are rapidly taken up by the liver. Chylomicron derived lipids are reassembled in the liver as very low density lipoproteins (VLDLs) on apoB100 protein, which are secreted into the circulation for peripheral tissue utilization. In several mammals, apoB mRNA editing also occurs in liver[16] where, unlike intestine, apoB mRNA editing is regulated to determine the proportion of edited apoB mRNA as well as the amount of secreted B48 VLDLs.[17,18] B48- and B100-containing particles differ greatly in the amount of lipid that they can transport (B48-containing particles have a significantly higher capacity and hence it is the protein of choice for transporting dietary lipid from the intestine). Hepatic VLDLs are assembled and secreted only with B100 protein cores in humans (or with B100 and B48 in other species).[16] B48 lacks a low density lipoprotein (LDL) receptor binding domain, and therefore the body "manages" VLDLs that contain B48 differently than those containing B100. VLDLs assembled with B100 have a longer half-life in the blood stream and as a consequence are digested by liver and bloodstream lipases, rendering them to protein and cholesterol rich LDL. Elevated abundance of LDL in the blood is an atherosclerotic risk factor. ApoB48 VLDL is cleared from the blood more rapidly than apoB100 VLDL and is not metabolized to LDL.[19] For this reason, hepatic apoB mRNA editing has been considered as a means of reducing the risk of atherogenic disease.

ApoB mRNA editing catalytic subunit 1 (APOBEC-1) is the sole cytidine deaminase responsible for editing apoB mRNA.[20,21] Although APOBEC-1 can bind and deaminate free cytidine nucleoside or nucleotide substrates, as well as bind weakly to AU-rich RNA sequences, it cannot bind specifically to, nor under physiological temperature and salt concentrations edit, *apoB* RNA.[22,23] In cells, site-specific apoB mRNA editing requires an editing complex (or C to U editosome) consisting minimally of an APOBEC-1 homodimer[24] interacting with a single-stranded-RNA binding protein known as "APOBEC-1 complementation factor" (ACF), which binds to an RNA editing site recognition motif 3′ of C6666 (figure 14.1).[25–26]

Throughout the 1990s, many more examples of base modification, substitution, insertion, and deletion RNA editing were brought to light. In addition, the protein coding capacity of some viruses with RNA genomes were found to be altered by RNA editing.[1,2,11,28-30] Table 14.1 lists a few examples of RNA editing in organisms and tissues (described in greater detail below and in reviews cited in this chapter). The term "RNA editing" was extended conceptually to include those modifications of nucleotides in tRNA that resulted in a change in the amino acid coded during translation. In some instances editing modified nucleotides in the acceptor stem of tRNAs and thereby changed the specificity of amino acylation (amino acid charging of the tRNA), while in other instances amino acylation remained the same after editing but the anticodon sequence was edited resulting in altered codon recognition.

Distinguished from editing events that result in a change in protein translation are mechanistically related processes that result in nucleotide modifications in mRNA, rRNA, and tRNA. Modification typically affects RNA stability, processing, secondary structure, interaction with other RNAs or proteins and/or affects RNA subcellular localization. As these are generally considered as RNA modification, not editing (i.e., they do not change the protein-coding specificity of the mRNA), they will be discussed only to a limited extent in this chapter. The interested reader is referred to recent texts that broadly address RNA and nucleoside/nucleotide modifications.[1,2,11]

mRNA Editing throughout Time

A simple statement concerning the selective forces acting on mRNA editing is unlikely to be accurate because of the diverse mechanisms involved and the breadth of species with one or more forms of mRNA editing. However, it seems likely that mRNA editing mechanisms must have their roots at the very origin of life due to their apparent relationship to nucleotide modification, which some believe dates back to the "RNA World." (The "RNA world" hypothesis,[31] originally proposed by Gilbert, is that RNA preceded DNA as the genetic material.) A case will be made, from the findings described below, that the machinery for mRNA editing emerged through gene duplication and divergence of preexisting purine and pyrimidine base and nucleoside/nucleotide modifying enzymes.[1,2,32]

Genomic mutations that result in impaired protein function will either be deleterious to an organism and become limited within populations, or will be tolerated because the function, while decreased, remains in an acceptable range or is compensated by redundant pathways. It might be speculated that mRNA editing would render these mutations neutral by "correcting" them at the level of the transcriptome. The possibility also has to be considered that editing is not specifically a 'repair' process, but instead enables protein variants to be expressed with a range of activities and, as discussed below, this might have conferred a more robust phenotype to organisms during their evolution.

The bias inherent in each mRNA-editing enzyme for substrate recognition and site-selective editing that we see today (e.g., nearest neighbor nucleotide preferences of editing enzymes)[33-35] may have been acquired by mutation of modification enzymes and selection of nascent activities with the capacity to edit specific RNA sequences. Presumed in this discussion is that mRNA editing provided a selective advantage in the face of ongoing mutagenesis. In this model, orphan editing activities may have emerged spontaneously and were in some instances, maintained through positive selection. The emergence of genomic mutations that could be corrected by mRNA editing and their propagation throughout populations would have fixed some forms of mRNA editing (or tRNA editing) in modern-day organisms.

Consistent with this "environmental selection pressure" hypothesis, C to U mRNA editing has not been observed in aquatic plants but is evident in the organelles of all dry land based plants. An enriched oxygen environment, desiccation, enhanced radiation exposure, or other changes associated with dry land may have promoted or permitted mutations that could have selected for mRNA editing as a "corrective" capacity. Homologous mRNAs from modern-day monocots and dicots, however, are not edited in all species.[1,2,7,36] A high frequency of C to U editing sites found in the mRNA of one species were genomically encoded as T in other species, while some discrepancies in editing site utilization involved cytidines at the third nucleotide position within codons (wobble base pairing).[7]

The findings suggested a high incidence of genomic nucleotide transitions at positions corresponding to plant organellar editing sites. It is uncertain whether these discrepancies are the result of forward (T to C) or back (C to T) mutations. It is likely, however, that the mutability of these sites maintains selection pressure on the maintenance of an mRNA editing capacity and its evolution as new targets for editing emerge. In this regard, editing site recognition in plants requires unique sequences immediately 5' of the edited C (mammalian C to U mRNA editing requires the 11 nucleotide mooring sequence immediately 3' of the edited C) (figure 14.1). This editing site recognition sequence is not the same for chloroplast mRNAs and mitochondrial mRNAs nor can the mRNAs from one organelle be edited when expressed in the other. Additional studies evaluating editing enzyme–substrate relationships will be necessary if we are to understand the selection pressures driving the evolution of C to U editing within plant chloroplasts and mitochondria.

Nucleotide Modifications in RNA

A considerable number and diversity of enzymes are dedicated to modifying nucleotides in tRNA and rRNA in bacteria, archae, and eukaryotes, as is apparent in the observation that there nearly 100 different posttranscriptional RNA modifications of pyrimidines, purines, and of the 2' hydroxyl moieties of ribose.[1,2] Over half of the different types of RNA modifications found in bacteria also are observed in various organisms found in the Archaea and Eukarea kingdoms. While the catalytic domains of the enzymes carrying out similar modifications of RNA within these kingdoms show considerable homology and structural conservation, the RNA substrates and the positions of modification within homologous substrates are only occasionally similar. Some RNA modifications of tRNAs affect codon sense (now considered tRNA editing) (table 14.1), while other modifications may stabilize RNA secondary structure and are required for tRNA processing or improve translation efficiency and fidelity.[1,8,32,37–39] There are examples in both modification and insertion editing where one type of modification is a prerequisite for additional modifications (frequently of a different type) at either the same nucleotide or at a another site within the RNA.[1,2,8,40]

Given that each RNA modifying enzyme interacts with unique sites within a limited range of substrates, there must have been multiple occasions during evolution where new catalytic activities emerged (or diverged from existing enzymes) with the capacity for different or broader substrate specificities. Grosjean has observed that over half of the types of RNA modifications in various organisms in Eukarea are unique to this kingdom,[1] suggesting that new modification activities have emerged or that unique selection pressure led to the retention of activities, now lost in organisms from other kingdoms. Genomic mutations that affected the structure and/or function of an essential RNA(s) would have provided selection pressure for emerging modification activities or the maintenance of activities that were until that point under neutral selection. The capacity of all life forms to carry out so many diverse RNA modifications with such a large number of enzymes remains a topic of controversy and discovery.

C/U and A/I Base Modification mRNA Editing Is Unique to Eukaryotes

C to U RNA Editing

Have organisms become more or less dependent upon RNA editing over the course of evolution? As editing of mRNA has not yet been reported in Archaea or in Bacteria, it is possible that it is a unique characteristic to eukaryotes and so might have emerged rather recently (either from RNA modification processes or as a new function). Although mRNA editing mechanisms in different organisms are, in some instances, very different reactions, involving unique enzymes and auxiliary factors and occurring in different organelles (figure 14.1), we can ask, for a given mechanism of mRNA editing, how broadly across classes and orders can one find this activity, and in which species?

C to U modification mRNA editing activity has been demonstrated (or implicated by comparisons of DNA and protein sequences) in both lower and higher eukaryotes, including yeast, *Physarum*, all dry-land plants, *Caenorhabditis elegans*, mammals, and marsupials. These editing events have been shown (or are postulated) to be catalyzed by cytidine deaminases active on RNA (CDARs). Phylogenetic studies[41] and structural modeling studies[24] have suggested that CDARs are related to cytosine and cytidine deaminase, which are found in all forms of life that use free pyrimidines or nucleoside/nucleotide as substrates.

These enzymes have homologous domains for the coordination of zinc, which is used as a Lewis acid for hydrolytic deamination of cytidine to form uridine, and a glutamic acid residue for proton shuttling during the reaction.[24,42] Crystal structure analysis suggested that nucleotide deaminases must function as dimers or tetramers because each catalytic center is composed of the residues for deamination in one subunit and a substrate coordination "flap" contributed by the other subunit.[24] Whereas deaminases that are active only on free nucleosides or nucleotides have long and inflexible flap domains, those that have the capacity to deaminate nucleotides within RNA (or DNA) have short flexible flaps. Structural studies have suggested that the evolution of CDARs must have involved changes in the flap domain that enable these enzymes to accommodate nucleic acids as substrates.

CDD1, an orphan C to U mRNA editing enzyme from yeast,[43] bears striking structural and functional homology to the mammalian C to U mRNA editing APOBEC-1, which carries out RNA editing of apoB (the major structural protein of low density lipoproteins) mRNA (table 14.1, figure 14.1). Several APOBEC-1 related proteins (ARPs)[42,44] functioning in mammalian cells as deoxycytidine deaminases on genomic[45] and viral[34,46] ssDNAs have catalytic domain folds homologous to APOBEC-1 (figure 14.2). Based upon sequence and structural homology it has been predicted that ARPs also have a flexible flap domains and a distribution of charged and hydrophobic residues within the catalytic cleft that accommodate either single-stranded RNA or single-stranded DNA substrates. A role for ARPs in mRNA editing, while highly likely, remains hypothetical.[42,45,47]

A to I RNA Editing

A to I mRNA editing is catalyzed by a family of zinc-dependent enzymes known as "adenosine deaminases active on RNA" (ADARs), which edit a large variety of mRNAs expressed in *Xenopus, Drosophila, C. elegans*, squid, and all mammals[9] (table 14.1, figure 14.1), and may function as interferon-inducible antiviral deaminases.[30] ADARs may have evolved from a primordial cytosine/cytidine deaminase.[41] The catalytic domains of both CDARs and ADARs bear greater homology to that of *E. coli* cytidine deaminase and C to U mRNA editing enzymes than they do to adenosine deaminases (figure 14.2). Similarly, adenosine deaminases active on tRNA (ADATs) have been identified in yeast and mammals.[9,48] These enzymes are homologous in their catalytic domains to ADARs and therefore also may have evolved divergently from a primordial cytidine deaminase.

As previously mentioned, ADAR editing can change the sense of codons because I base-pairs to C, and it can generate new open reading frames through the generation of translation start codons (by editing ATA to ATI) or the alteration of mRNA splice site and branch point signals.[9] ADARs have domains that bind to require double-stranded RNA, which restricts these enzymes to the targeting of adenosines within RNA secondary structure (figure 14.2). When ADAR editing sites occur within protein-coding regions of primary transcripts, the editing site typically is situated within an exon sequence that forms a duplex with its 3' intron sequence (figure 14.1).

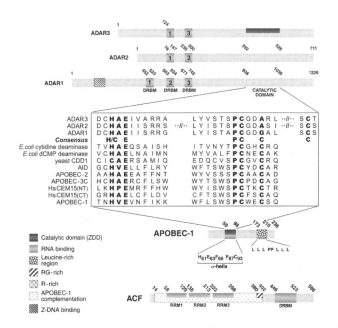

Figure 14.2. Structure-based alignments and the distribution of functional domains of mammalian editing factors. Conserved residues within the zinc-dependent deaminase domain (catalytic domain) are shown for the ADARs (adenosine deaminases active on RNA), APOBEC-1, ARPs (APOBEC-1 related proteins), and *Escherichia coli* deaminases. The catalytic domain of APOBEC-1 is characterized by three zinc coordinating amino acids (each of which can be either histidine or cysteine), a glutamic acid, a proline residue, and a conserved primary sequence spacing (key amino acids, shown in bold type as the "consensus"). The spacing of the terminal cysteine in the primary sequence of ADARs is greater than that seen in cytidine (shown within the consensus as a fourth C in bold type, but note that C to U and A to I deaminases each coordinate zinc with only three amino acids). Shown in comparison to APOBEC-1 and the consensus sequence are: the catalytic domains of deaminases that use free nucleosides/nucleotides as substrates (*E. coli* cytidine deaminase and dCMP deaminase), nucleosides/nucleotides and/or RNA as substrates (CDD1); and those of the ARPs that currently are only know to act as DNA editing enzymes (AID, CEM15) or have no known substrates (APOBEC-2 and APOBEC-3C). ADARs bind to their editing sites through double-stranded RNA binding domains (DRBMs). The indicated residues in the catalytic site of APOBEC-1 bind AU-rich RNA with weak affinity. The leucine-rich region of APOBEC-1 has been implicated in APOBEC-1 dimerization and shown to be required for editing and may be involved in interactions with APOBEC-1 complementation factor (ACF). ACF is an ssRNA binding protein that is required biologically for APOBEC-1 to find and edit apoB mRNA. The three RNA recognition motifs (RRMs) are required for mooring sequence-specific RNA binding, and these domains plus the sequences flanking them are required for APOBEC-1 interaction and complementation. APOBEC-1 complementation activity minimally depends on ACF binding to both APOBEC-1 and ACF binding to the mooring sequence. A broad APOBEC-1 complementation region is indicated on ACF that is inclusive of all regions implicated in this activity. The complete protein sequence is modeled with numbering to indicate key amino acid positions at the borders of domains and at the end of each structure, indicating the total number of amino acids in each protein.

258

In the case of ion channels and transmembrane receptors, A to I editing enables these complexes to form with variable ratios of subunits translated from edited and unedited mRNAs. This is because each channel is composed of five receptor subunits, which may be from either edited or unedited mRNAs and which function together to modulated calcium flux through the pore. This regulation is more like a rheostat than a switch. Editing of these receptors at other sites affects their rate of repolarization following a wave of depolarization. This results in a broader range in an organism's ability to modulate the level of response to signaling and the rate of recovery following a change in membrane potential and thus would be more robust; hence editing would be selected for and spread throughout populations (as long as the genomic sequence corresponded to the unedited version). In this regard, A to I editing of the mRNA encoding glutamate gated calcium channel receptors is ubiquitous in land animals and insects. ADAR gene knockout studies in mice and flies demonstrated that these organisms have become dependent on A to I editing activity not only for central nervous system function but also for the development of several other organ systems.[10,49–52] Interestingly, the editing site within glutamate gated calcium channel receptors are genomically encoded as G in fish but the sodium ion channels in squid require A to I editing. Editing in invertebrates, fish, and amphibians has not been well studied, but it is tempting to speculate (as appears to be the case for plant C to U mRNA editing) that in some instances genomic mutations may have selected for A to I mRNA editing activity as organisms occupied dry land.

The RNA sequences within the RNA secondary structure forming ADAR editing sites are not generic to all edited mRNAs. Further, once mRNA is spliced, little may remain of the ADAR recognition sequence element. This has made the prediction of novel ADAR-edited mRNAs difficult. However, based upon comparisons of cDNA and genomic DNA sequences, it has been estimated that there may be over 12,000 A to I editing sites in as many as 1600 different genes in the human genome.[53] This is a conservative estimate as mRNAs were not scored as "hits" in this study unless they contained minimally three A to G discrepancies. Many presumptive editing sites were within coding regions, but the majority were within *Alu* repeats within 5′ and 3′ untranslated regions where they may have a function in the control of mRNA secondary structure and stability (see chapter 8 for more discussion of *Alu*-containing sequences). Numerous A to I mRNA editing sites also have been predicted in coding and noncoding regions of *Drosophila* mRNAs.[54] One study that compared editing sites from homologous mRNAs showed that some editing sites, including the flanking sequences that contributed to the secondary structures used by ADARs to bind to the editing sites, were highly conserved in two species of *Drosophila* separated by 61–65 million years, whereas other editing sites differed as to whether they were genomically encoded as G or A.[49]

ADAR binding to RNA secondary structure and A to I editing enables these enzymes to act as double-stranded RNA helicases (unwinding enzymes) due to their ability to alter base pairing within RNA secondary structure.[9,48] As ADARs edit adenosines within duplex regions they can disrupt local RNA secondary structure.[55] Aside from the aforementioned mRNA editing activity, ADAR ability to disrupt RNA duplexes has important implications for the regulation of gene expression through RNA interference-mediated (RNAi) mRNA depletion.[56,57] Double-stranded RNA is required as the substrate for the enzyme "dicer," which generates through cleavage small interfering RNAs (siRNAs) (see also chapters 8 and 13).

siRNAs must form perfect duplexes with select mRNAs in order to target them for nucleolytic degradation. ADAR editing and unwinding activity can reduce or eliminate RNAi regulation of gene expression and thereby affect dicer activity and perhaps alter the targeting-specificity of siRNA for mRNAs.[9,58] RNAi has been described in eukaryotes ranging from *Tetrahymena*, *Drosophila*, *C. elegans*, and humans, and is involved in tissue differentiation and organism development. The functions of siRNAs now include chromatin remodeling and genomic DNA sequence deletion and reorganization.[59] RNAi is likely an ancient process and as such is another example of a function that could have contributed selection pressure on emergence and maintenance of ADAR activity in the evolution of organisms.

Nucleotide Insertion mRNA Editing: Here to Stay or Gone Tomorrow?

Uridine insertion and deletion editing of mRNAs in Trypanosome mitochondria can be traced to divergence within the phylum Euglenozoa (one of the earliest groups of organisms with mitochondria).[5] Subsequently, the requirement for editing at some sites has been lost within laboratory strains (which have been maintained as stocks for around 60 years). In these instances, the uridines have become encoded genomically and the genomic regions encoding the gRNAs responsible for targeting these editing sites have, in many instances, been selectively lost. For mRNAs that are generated through numerous editing events (pan edited mRNAs), editing of one site frequently (but not always) generates the anchor sequence for the next gRNA and a subsequent editing event. Consequently, uridine insertion and deletion editing often proceeds with a 3′ to 5′ polarity. Interestingly, the regions of edited mRNAs that have become genomically encoded in laboratory strains are mostly derived from the 3′ end of mRNAs, making possible the remaining editing events that will generate the 5′ end of these mRNAs.

It is apparent in these organisms that selection pressure can be exerted at the level of an individual editing site. Ultimately, it is only the gRNAs that are unique to each editing site, while the editosomal enzymes and structural proteins responsible for either uridine insertion or deletion are assembled (depending on the particular process) at multiple editing sites (figure 14.1).[40] On the other hand, the selection pressure that maintains the proteins involved in nucleotide insertion or deletion editing must manifest at the level of the collective requirement to establish functional mRNAs (i.e., as long as editing is required for the activity of an essential protein there will be positive selection for the genes encoding the editosomal components) (see discussions of "second-order selection" in chapters 3 and 4). While in laboratory strains that have incorporated the As that encode Us into the genome, proteins are encoded explicitly, these proteins are only implied in the genome of natural strains, through the combination of the DNA sequence, the guide RNAs, and the recognition specificities of the editing enzymes.

Insertion RNA editing of tRNA in the mitochondria of amoeba and *Physarum*, and cytidine insertion editing of all forms of mitochondrial RNA from slime molds, were discovered shortly after gRNA-dependent and base-modification mRNA editing.[1] Guanidine insertion mRNA editing of the mumps and measles virus (Paramyxo viruses) also is well documented.[60] In contrast to Trypanosome's use of guide RNAs for U insertion/deletion editing, guanidine and cytidine insertion editing are cotranscriptional processes. Polymerase slippage relative to the template strand during transcription appears to account for the insertion of additional G and/or C nucleotides in nascent transcripts. Edited mRNAs contain frameshifts that enable the expression of essential proteins integral to the virus's and organism's ability to encode essential proteins.

In the case of Paramyxo viruses, proteins expressed from edited and unedited mRNAs are believed to contribute to different stages of the viral life cycle. Perhaps more importantly, there is considerable selection pressure to maintain the capacity for editing as viral replication potentially generates irregular genome lengths that cannot be encapsulated into virions were it not for nucleotide insertion editing that restores genome lengths. This genome length restriction stems from the fact that the nucleocapsid protein has a six ribonucleotide binding capacity and viral genomes that are not exact multiples of six do not assemble functional virions (known as "the rule of six").

DNA Mutational Editing

Homologs of APOBEC-1 and ARPs are expressed in fish, *Xenopus*, birds, and all mammals.[42,44,61] At the turn of the Millennium, some of the enzymes in the ARP family were shown to induce genomic mutations by deamination of deoxycytidine to deoxyuridine.[62] This activity will be referred to as DNA editing because, unlike spontaneous dC to dU deamination, DNA editing is regulated and targeted to regions within genes. The mechanism for DNA editing is nucleotide deamination (although there may be other mechanisms yet-to-be described) largely occurring on singlestranded DNA at the sites of transcription. DNA editing is in this sense a mutationinitiating mechanism with sequence selectivity, and thus distinctly different from environmental factors that give rise to mutations due to DNA damage (see chapter 2).

Up to 1999, C deaminating activity was known only for APOBEC-1, which functions physiologically as an mRNA editing enzyme. Subsequently, under experimental conditions, APOBEC-1 itself was shown to have DNA deaminase activity.[63] This stimulated speculation that APOBEC-1 overexpression might, in addition to promiscuous RNA editing,[63] lead to DNA mutation and thereby induce neoplasias. Indeed, liver-specific transgenic overexpression of APOBEC-1 had previously been shown to induce liver carcinoma and dysplastic disease.[36,42,64] At the time, this effect was proposed to be due to hyper mRNA editing and consequent activation of oncogenic activities, but in light of the discovery of ARP DNA editing, APOBEC-1-induced genomic mutations cannot be ruled out.[42] There has been a recent flurry of interest in the possibility that ARP DNA editing may be morewidespread. For example, it recently has been observed that unregulated expression of the DNA editing ARP known as "activation induced deaminase"

(AID) (figure 14.2),which normally targets immunoglobulin genes (see below and chapters 10 and 11), can lead to oncogene activation[65] and the dysfunctional regulation of antibody expression seen in leukemias.[66] Dysregulation of AID activity also has been suggested as the mechanism by which hepatitis C virus induces genomic mutations and neoplasia.[67]

APOBEC-1 and ARPs have distinct substrates. AID does not edit apoB mRNA and overexpression of APOBEC-1 does not induce DNA mutations in immunoglobulin (Ig) genes of mammalian B lymphocytes. These findings suggest that if APOBEC-1 mutates DNA in mammalian cells, its activity on the genome is nonrandom. Given that genomic mutation frequencies are rare, it is believed that the DNA editing activity of all of the ARPs is highly regulated. In fact ARPs are expressed tissue-specifically and their activity is restricted to nucleic acids within select subcellular compartments. In the case of AID, expression is restricted to activated B lymphocytes within germinal centers (in the spleen and lymph nodes) and AID activity is focused on Ig genes in the cell nucleus (see chapters 10 and 11),[45,68] whereas APOBEC-3G/CEM15 and APOBEC-3F are expressed in T lymphocytes and have activity on HIV-1 and HIV-2 during minus strand DNA reverse transcription in the cytoplasm.[34,46]

That AID is not absolutely selective for Ig genes was seen when exogenous non-Ig reporter genes, recombined randomly throughout the genome, were found to be mutated by AID.[69] The specter of what might happen if AID activity were not regulated properly also is raised by APOBEC-1's and other ARPs' abilities to deaminate deoxycytidine in a wide variety of DNAs under experimental conditions. When APOBEC-1 or ARPs were expressed in *E. coli* (under selection for a DNA mutator phenotype), or reacted with single-stranded DNA in vitro, numerous deoxycytidines were deaminated at a variety of DNA sequences. Although the sites of DNA modification were abundant, their distribution was unique to each enzyme as assessed by nearest neighbor sequence preferences for nucleotides immediately 5' of the target cytidine (dT for APOBEC-1, dA/dG for AID, and dG for CEM15).[33–35] Not all dCs with appropriate flanking sequences were deaminated, suggesting a broader flanking region recognition requirement and/or that although site selectivity of DNA editing may be determined by the intrinsic bias of each enzyme, other factors determine which deoxycytidines are deaminated in vivo (as discussed in chapter 10).

Current hypotheses propose that targeting specificity results from ARP association with other macromolecular assemblies, such as those involved in DNA recombination, repair, transcription, reverse transcription, and chromatin remodeling.[47,70–73] This model is consistent with the known mechanism for apoB mRNA editing and the role the single-stranded RNA binding protein ACF[74] in determining site-specific C to U editing. It is ACF's ability to bind APOBEC-1 and specific sequences 3' of the editing site (the mooring sequence) that restrict editing to a specific mRNA substrate (figure 14.1 and figure 14.2).[36]

The requirement for a robust immune defense system may have selected for AID and APOBEC-3G/CEM15 DNA editing. As described in chapters 10 and 11, AID is essential for increasing the repertoire and affinity of the adaptive immune response through somatic hypermutation and class switch recombination. In mammals with no or low AID, both of these processes are impaired, leading to life-threatening immunodeficiency (an autosomal recessive condition known as "hyper IgM2" which affects one in 106 people).[45,75] Class switch recombination involves not only AID deaminase activity but also the C-terminal domain of the AID protein. The noncatalytic C-terminal domain is thought to be essential for interaction of AID with proteins such as those involved in nonhomologous DNA recombination[76] and transcription[73,77] that facilitate targeting of AID's DNA editing activity and in turn DNA repair and recombination activity to select regions of the genome for CSR and SHM.

APOBEC-3G/CEM15, APOBEC-3F, and possibly APOBEC-3B, previously referred to as "phorbolins,"[42,44] are coexpressed in human lymphoid and myeloid cells and, as is the case for APOBEC-1, can form homodimers but also heterodimers.[34] Our current understanding is that these proteins serve in host defense as antiviral deaminases, although their potential for other activities within the cell has not been explored. For example, APOBEC-3G/CEM15 and 3F deaminate deoxycytidine on HIV-1 and HIV-2 minus strand cDNA that satisfies nearest neighbor nucleotide requirements (as discussed previously). These dC to dU modifications template dG to dA mutations on the positive strand during replication, which inactivate multiple proteins essential for viral infectivity.[46,78] Unlike APOBEC-1 and other ARPs, APOBEC-3G/CEM15 and 3F establish a close proximity with viral genomes by becoming integrated within virions during their assembly.[34,79–81] With regard to the deaminase activity, homodimers of APOBEC-1 or AID are predicted to contain two catalytic centers.[24] APOBEC-3G/CEM15 3F and 3B each have two catalytic centers (both of which have activity).[70,82,83] Homo- and heterodimers of ARPs like APOBEC-3G/CEM15 and 3F are likely to have four catalytic domains and therefore considerable combinatorial substrate targeting potential. This could provide the host cell with an adaptive advantage against a broad spectrum of viruses.

Six phorbolin genes (as well as phorbolin pseudogenes) are clustered on human chromosome 22 (mice have only one phorbolin gene on chromosome 15). Presumably these are the result of gene duplication from a primordial phorbolin gene followed by divergent evolution. The phorbolins known as APOBEC-3A, 3C, 3D, and 3E may be partial gene duplications as they each have single catalytic domain and partial C-terminal sequences like that seen in homologous regions of APOBEC-3G, 3F, 3B, APOBEC-1, APOBEC-2, and AID. The function(s) of phorbolins with one catalytic domain and of APOBEC-2 (and of the predicted APOBEC-4)[84] remains to be determined.

HIV-1 and HIV-2 use the accessory protein known as "viral infectivity factor" (Vif) to defeat the ARP host defense. Vif binds to both APOBEC-3G/CEM15 and 3F and targets them to ubiquitination and proteolytic degradation via the proteasome.[85] Vif's interaction with APOBEC-3G/CEM15 occurs in a noncatalytic region that lies C-terminal to first catalytic domain. Interestingly, a single amino acid within this region (an aspartic acid in humans and a lysine in monkeys) provides the essential charge for the interaction of APOBEC-3G/CEM15 with Vif.[81,86] Site-directed mutagenesis has shown that this single amino acid change in an ARP changes host range of a retroviruses.[81,86,87] Due to this single amino acid difference, Vif derived from simian virus (SIV) cannot bind to human APOBEC-3G/CEM15 and vise versa, and consequently there is species-specific exclusion of APOBEC-3G/CEM15 from the virion. Consequently, this region of APOBEC-3G/CEM15, perhaps more than any other, may have constrained the extent to which Vif can mutate and still protect the virus from the ARP-based host defense.

Conclusions

The discovery of RNA editing appeared at first to be esoteric; uncovered in a small number of organisms and thought not to be mechanistically related. The development of in vitro systems helped identify the proteins involved in mRNA and DNA editing. Sequence alignments and structural comparisons of enzymes, together with molecular approaches and transgenic or gene knockout model organisms, have facilitated the identification of additional enzymes and substrates, and established the biological requirements for mRNA editing. Persistent efforts by labs across the globe validated not only that mRNAs were edited but that editing was required to add diversity, and in many cases specific functionalities, to the proteome of many organisms and viruses.

Striking examples of the importance of editing are gRNA-dependent uridine insertion and deletion, the requirement of AID for immunoglobulin gene diversification and class switch recombination, the multiplicity of C to U modifications in plant organelle mRNAs, and C insertion editing of all RNAs within *Physarum* mitochondria. In some cases, such as gRNA editing, it is very clear how the genome implies the informational content through selection of U insertion sites and their lengths, while in other systems, such as C insertion editing, we do not understand the form of the genomic information that is required to direct the assembly of the final sequence. It appears that RNA modification systems have been a part of biology from its origin and mRNA editing has played an important role.

The past five years have shown dramatic progress in the areas of editosomal component identification, discerning mechanisms, and identification of novel substrates. Deaminases with genomic DNA editing capacity have taken center stage and are being carefully evaluated for the possibility that they, like APOBEC-1, may also have mRNA editing activity. Discovery of the requirement for editing enzymes in viral life cycles on the one hand, and on the other as host antiviral defense factors, and as the agent required for somatic hypermutation and class switch recombination of immunoglobulin genes during the development of the immune system, provided long sought answers in immunology. We need to learn how to identify genomic sequences that may be substrates for RNA and DNA editing systems, and to understand the selection pressures under which they have evolved. Even given our limited knowledge at this time, it is apparent that at least some C to U editing systems, such as those involved in immunity and viral defense, increase the hardiness of organisms.

The discovery of the APOBEC-1 related protein family and their DNA editing activities also brought new insight, with broad implications, as to how genomic instability can be selectively activated in certain cells to regulate diversity in the proteome or prevent viral proteomes from being expressed. It is likely that additional editing mechanisms and novel substrates will be revealed and that these too will prove to increase our appreciation for the extent to which information is implicitly, in addition to explicitly encoded, in the genome.

Acknowledgments   The author acknowledges the many contributions of investigators in the field whose specific work may not have been referenced due to space limitations. The author is grateful to Lynn Caporale and Joseph E. Wedekind for their many helpful suggestions in the preparation of this chapter and to Jenny M.L. Smith for the preparation of the figures. The author's efforts on this chapter and contributions to the field of RNA editing have been supported in part by grants from the National Institutes of Health, the Air Force Office of Scientific Research, the Alcoholic Beverage Medical Research Foundation, the Council for Tobacco Research, and the Office of Naval Research.

15

Alternative Splicing:
One Gene, Many Products

Brenton R. Graveley

Overview

The information content of eukaryotic genes is typically organized in an interrupted, nonlinear fashion. Alternative splicing is a process in which the information encoded by a single stretch of DNA is extracted in a way that results in the production of multiple mRNAs, which can encode related, but distinct, proteins. In addition to containing protein-coding information, eukaryotic RNA transcripts also contain the information that specifies how the protein-coding portions are cut and pasted together to generate messenger RNAs.

Introduction

When students first learn how information is encoded in DNA, they typically are taught that the amino acid sequence of an encoded protein can be translated directly from the linear sequence of DNA. Examples are given from bacteria in which the nucleotide sequence of a gene directly specifies the amino acid sequence of the encoded protein. In other words, there are no breaks in the DNA sequence that codes for the protein, and each codon specifies the amino acid that eventually becomes incorporated into the protein. While this is a useful way to begin to teach students about the flow of genetic information, it is not the whole story. In fact, very few genes in higher eukaryotes are organized in a manner similar to bacterial genes where the protein is encoded in a continuous stretch of DNA. Instead, in higher eukaryotes, the DNA sequence encoding the protein is nearly invariably interrupted by other sequences, usually multiple times. This chapter discusses the phenomenon of split genes—genes that are not organized in a continuous manner and which must contain additional information to specify their assembly—and the consequences of this organization with regards to protein diversity and protein function, as well as how this sort of discontinuous information can be identified and stitched together to specify a correct protein.

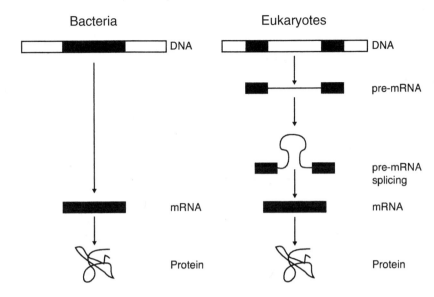

Figure 15.1.   Differences in gene organization between bacteria and eukaryotes. In bacteria, the coding information for proteins is contained in a continuous stretch within the genome. In contrast, the protein coding sequence in eukaryotic genes is frequently interrupted by intron sequences. After transcription, intron sequences are removed by a process called pre-mRNA splicing to generate an mRNA that is translationally competent.

Pre-messenger RNA Splicing

Most bacterial genes contain a continuous open reading frame that codes for a protein in a linear fashion (figure 15.1). For years this was thought to be universally true. However, in 1977, while studying gene structure of Adenovirus, the laboratories of Phillip Sharp at Massachusetts Institute of Technology[1] and Richard Roberts at Cold Spring Harbor Laboratories[2] simultaneously made what at the time was a stunning discovery—that some genes are actually discontinuous: the genetic information that specifies a protein can be contained in separate and discrete segments that are separated by other DNA sequences. The protein-coding portions of the genes are referred to as "exons," while the intervening sequences are referred to as "introns" (figure 15.2). In order for a protein to be synthesized from such a split gene, the gene is first transcribed into a pre-messenger RNA (pre-mRNA) from which the intron is removed and the exons joined together by a process called splicing. Subsequent to this stunning discovery, many other researchers discovered that this type of gene organization was common to most eukaryotic genes, especially in higher eukaryotes. In fact, as we will discuss later, the majority of human genes contain introns, and most contain multiple introns. The consequences of this gene organization are profound in that it may facilitate the evolution of new functional proteins and allow for an entirely new mode of gene regulation that is completely absent in organisms that lack introns. At the same time, this gene organization

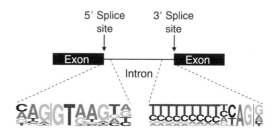

Figure 15.2.   Features of introns. The locations of the 5′ and 3′ splice sites within an intron are shown. Additionally, the consensus sequences of human 5′ and 3′ splice sites are shown as pictograms in which the height of the letters indicates the frequency of that nucleotide at that position.

raises the issue of what types of information enable a cell to "know" which sequences within a pre-mRNA correspond to exons and, more importantly, how it is that the correct exons are joined together.

Splice Sites: Sequence Elements That Direct Pre-mRNA Splicing

There are several conserved sequence elements within the pre-mRNA that serve to direct the splicing reaction. The main sequence elements are located at the 5′ and 3′ ends of an intron, extending just across the intron–exon boundary (figure 15.2). The sequence element located at the 5′ splice site is the most highly conserved of these elements and has the consensus sequence of AG/GURAGU in humans (R represents purine and "/" denotes the location of the exon–intron boundary). The 3′ end of the intron contains two conserved sequence elements—the branchpoint (YNYURAC) (Y represents pyrimidine) and the 3′ splice site, which consists of a pyrimidine tract and an AG dinucleotide at the intron–exon junction (figure 15.2). These sequences are recognized by the spliceosome—the macromolecular complex that catalyzes intron removal. However, these sequence elements are rather degenerate and do not contain sufficient information on their own to distinguish authentic splice sites from pseudo or cryptic splice sites—sequences in the pre-mRNA that have the same nucleotide sequence as a splice site but are never used in the splicing reaction. Additional sequence elements that act to enhance the use of authentic splice sites and/or suppress the use of aberrant splice sites are discussed later.

The Splicing Reaction

The chemical mechanism of pre-mRNA splicing is fairly well understood, although as of this writing it is not yet absolutely clear whether the reaction is protein- or RNA-catalyzed.[3] Introns are removed from a pre-mRNA in two steps (figure 15.3). In the first step, the 2′ hydroxyl of the branchpoint adenosine attacks the phosphodiester bond at the 5′ splice site. This reaction yields two intermediate products—the 5′ exon, which contains both a 2′ and 3′ hydroxyl group, and the

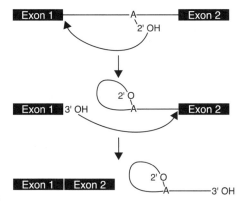

Figure 15.3.   The chemistry of the pre-mRNA splicing reaction. Splicing occurs via a two-step reaction. In the first step, the 2' hydroxyl group of the branchpoint adenosine attacks the 5' splice site. In the second step, the 3' hydroxyl group of exon 1 attacks the 3' splice site to generate the ligated exons and the released lariat intron.

intron linked to the 3' exon. The second product is called a "lariat" intron because the RNA forms a lariat-like structure. The lariat intron contains the phosphate group at the 5' end of the intron covalently linked to the 2' hydroxyl group of an adenosine within the middle of the intron. Because this adenosine contains covalent bonds at both the 3' (to the downstream intron sequence) and the 2' (to the 5' end of the intron) groups on the ribose, it is referred to as the "branchpoint" adenosine. In the second step of the reaction, the 3' hydroxyl group of the first exon attacks the phosphodiester bond at the 3' splice site. This reaction results in the ligation of the two exons and the liberation of the lariat intron.

The Spliceosome

The splicing reaction is carried out by the spliceosome, a large and rather remarkable macromolecular machine. Like the other complex cellular machinery, such as the ribosome or DNA replication machinery, the spliceosome is composed of multiple, separable components. The spliceosome consists of 5 small nuclear RNAs (snRNAs) called U1, U2, U4, U5, and U6 and approximately 200 proteins.[4] The main components of the spliceosome are the small nuclear ribonuclear protein particles (snRNPs). Each snRNP is composed of one or more snRNAs as well as several proteins, some of which are common to all snRNPs and some of which are unique to each snRNP. For example, U1 snRNP contains U1 snRNA, three U1-specific proteins (U1-70K, U1A, and U1C) and 7 Sm proteins (B/B', D1, D2, D3, E, F, and G) which also bind to the U2, U4, and U5 snRNAs.[5]

The spliceosome forms on each intron in a step-wise manner demarcated by a series of discreet intermediate complexes known as E, A, B, and C complexes (figure 15.4).[6] The early, or E, complex forms in the absence of ATP and contains U1 snRNP bound to the 5' splice site via base pairing interactions between the 5' end of U1 snRNA and the pre-mRNA. In addition, the splicing factors SF1,

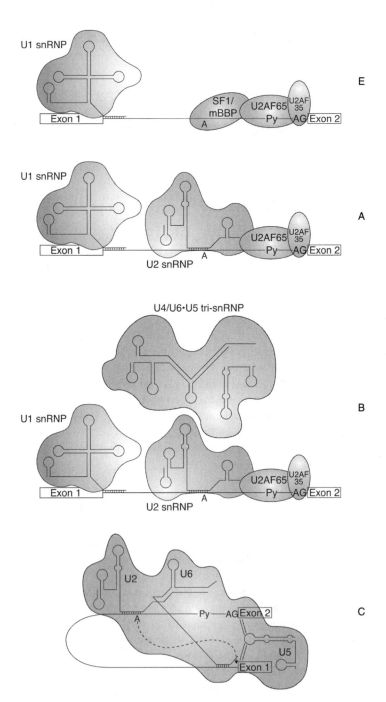

Figure 15.4. Spliceosome assembly pathway. U1 snRNP, SF1, and U2AF first bind to
the pre-mRNA to form E complex. Next, U2 snRNP replaces SF1 to form A complex. The
U4/U6•U5 tri-snRNP then enters the assembling spliceosome to form B complex. Finally, a
massive structural rearrangement occurs to form the catalytically active, or C, complex.

U2AF[65], and U2AF[35] bind to the branchpoint, pyrimidine tract, and YAG, respectively, in E complex. E complex is subsequently converted to A complex when SF1 is displaced from the branchpoint and replaced by U2 snRNP in a step that requires ATP hydrolysis. Next, A complex is converted to B complex when the U4/U6•U5 tri-snRNP joins the assembling spliceosome. Finally, C complex forms after a series of extensive rearrangements of RNA–RNA interactions. This involves the unwinding of the U4 and U6 snRNAs, the interaction of the U2 and U6 snRNAs, and the replacement of U1 snRNA at the 5′ splice site by U6 snRNA. It is the C complex that represents the catalytically active spliceosome.

Variations on a Theme: Alternative Splicing

Eukaryotic genes usually contain multiple exons and introns. In addition, exons can be joined together in different patterns to generate multiple mRNAs from a single gene. Importantly, different mRNA isoforms that are generated from the same gene can encode functionally distinct proteins. This process of alternative splicing is extremely prevalent in higher eukaryotes and is often regulated and therefore plays critical roles in the generation of protein diversity and the regulation of gene expression.[7]

There are many varieties of alternative splicing, as shown in figure 15.5. For example, different 5′ or 3′ splice sites contained within a single exon can be differentially utilized, and this can result in the insertion or deletion of different sequences within the encoded protein. Additionally, entire exons can be differentially included or excluded from a pre-mRNA. This can result from the initiation of transcription at different start sites, which cause different first exons to be present on the pre-mRNA. Likewise, this can result from differential splicing to alternative 3′ exons. This can result from the inclusion of entire exons contained in the internal portion of the pre-mRNA. These so called cassette exons are quite interesting because they almost always result in changes in the encoded protein. Cassette exons can either occur one at a time or in multiples. When multiple cassette exons are present, these can either be individually regulated or included in a mutually exclusive manner—only one of the multiple exons is included. Finally, there are a few remarkable cases, which will be discussed in detail later, in which exons from different pre-mRNA molecules are spliced in trans to one another.

Importantly, a single gene may encode a pre-mRNA that uses more than one of these modes of alternative splicing. Given this tremendous variety in the types and extent of alternative splicing that can occur, the nucleotide sequence of a gene must not only specify the amino acid sequence of the final protein products but also contain all the information necessary to accurately splice the pre-mRNA into the entire collection of mRNA isoforms that can be synthesized from the gene.

Alternative Splicing Is Extraordinarily Prevalent in Higher Eukaryotes

In 1986 it was estimated that perhaps as many as 5% of human genes encode alternatively spliced pre-mRNAs.[8] The recent glut of genomic information and technologies

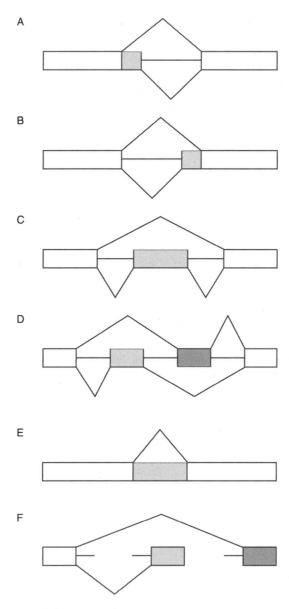

Figure 15.5.   Types of alternative splicing. (A) Alternative 5′ splice sites. (B) Alternative 3′ splice sites. (C) Alternative cassette exon. (D) Mutually exclusive exons. (E) Retained intron. (F) Alternative *trans*-splicing.

over the past few years has allowed for a reevaluation of the extent of alternative splicing. Estimates based on expressed sequence tag (EST) and cDNA sequence analyses have suggested that 40–60% of human genes are alternatively spliced.[9] However, this value is likely to be an underestimate due to the fact that these databases are not fully representative of all expressed sequences. This is because the sequences contained in the EST databases are highly biased towards the 5′ and, in particular, the 3′ ends of transcripts, as well as the fact that low abundance isoforms will likely not be represented in the EST databases. More recently, DNA microarray experiments have been conducted that suggest that nearly 50% and 75% of *Drosophila*[10] and human[11] genes, respectively, encode alternatively spliced mRNAs. Regardless of the precise number of genes that encode alternatively spliced transcripts, it is now abundantly clear that the majority of genes in higher eukaryotes are subject to alternative splicing.

Although many genes encode pre-mRNAs that are alternatively spliced to generate only two different mRNAs, several genes encode many more isoforms. For example, the vertebrate *slo* gene, which encodes a calcium-activated potassium channel gene, encodes at least 500 different mRNA isoforms.[12] Similarly, the three human neurexin genes, which function as neuropeptide receptors and synaptic adhesion molecules,[13] together encode at least 2250 different isoforms.[14] The most dramatic example of this is the *Drosophila Down syndrome cell adhesion molecule* (*Dscam*) gene, which potentially encodes 38,016 different isoforms due to the alternative splicing of 95 of its 115 exons.[15] Thus, alternative splicing plays a dramatic role in increasing protein diversity. In fact, alternative splicing may contribute significantly to the apparent discrepancy between gene number and organismal complexity.[7,16,17]

The Biological Relevance of Alternative Splicing

Alternative splicing is clearly ubiquitous in higher eukaryotes and has the capacity to expand the coding capacity of the genome to a rather amazing extent. But, we are left with the question of the biological significance of alternative splicing, and the challenge of understanding the form in which information that specifies the alternative splice forms is encoded. It turns out that in addition to its role in increasing protein diversity, alternative splicing plays a pivotal role in regulating gene expression in a variety of important biological processes, including sex determination, development, intracellular signaling, and neural wiring. Below several examples of alternative splicing events, and their biological relevance, are described in detail. (Chapter 11 describes the immunoglobulin class switch, in which distinct exons are deleted in a regulated manner when two introns are joined together at the DNA level, thus constraining the choice of alternatively spliced exons in that cell lineage.)

Sex Determination

One of the best examples of the important role of alternative splicing in biological regulation is the sex determination system in the fruit fly *Drosophila melanogaster*. Fruit flies contain five genes that are at the core of the sex determination system: *Sex-lethal* (*Sxl*), *transformer* (*tra*), *male-specific lethal-2* (*msl-2*), *doublesex* (*dsx*), and *fruitless* (*fru*). Together, they regulate the physical and behavioral sexual characteristics of the flies (figure 15.6).[18,19] The pre-mRNAs from each of these genes are differentially spliced in male and female flies. For example, in male flies, a 5′ exon that contains a stop codon remains in the mRNA. In fact, both the *Sxl* and *tra*

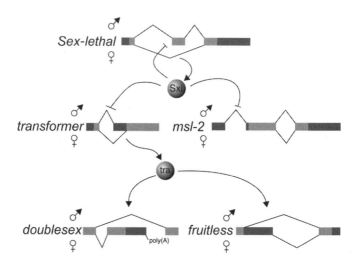

Figure 15.6.   The Drosophila *sex determination pathway.* Five genes—*sex-lethal* (*Sxl*), *transformer, male-specific lethal-2* (*msl-2*), *doublesex,* and *fruitless*—are all spliced one-way in male flies and differently in female flies, in a self-perpetuating pattern that initially is established by the ratio of X chromosomes to autosomes and the splicing of *Sxl*. The splicing of these genes controls all aspects of the physical and behavioral sexual traits of the flies.

pre-mRNAs are spliced in male flies such that no protein is produced from these genes. In contrast, the *msl-2, dsx,* and *fru* pre-mRNAs are all spliced in males such that functional proteins are produced from each gene. The MSL2 protein plays a central role in dosage compensation by increasing transcription from the single X chromosome in males. The DSX^M and FRU^M proteins each function as transcription factors. DSX^M regulates the expression of genes involved in specifying the male physical traits of the fly. In contrast, FRU^M is expressed in a small subset of neurons in the adult brain, where it functions to regulate the expression of genes that are required for male-specific sexual behaviours; *fru* mutant males display defects in courting female flies.[20]

Each of these genes is spliced differently in females. First, *Sxl* is spliced such that the internal exon containing a stop codon is excluded. This results in the synthesis of an mRNA that encodes a functional SXL protein. SXL functions to regulate the expression of *Sxl, tra,* and *msl-2*. Importantly, SXL acts as a negative regulator of the splicing of the *Sxl* internal exon itself, creating a feedback loop that ensures the stable production of SXL protein in females, with the following chain of consequences. SXL represses the splicing of the first intron of *msl-2* and represses of *msl-2* translation (i.e., MSL2 is not produced in females). Finally, SXL represses the inactivating male 3′ splice site in the *tra* mRNA, thus enabling the synthesis of an mRNA that encodes a functional TRA protein. TRA in turn regulates the splicing of *dsx* and *fru*.[20] TRA functions to activate an alternative 3′ splice site in the *dsx* pre-mRNA and results in the synthesis of DSX^F, which regulates the expression of genes required for the development of the female-specific physical traits, rather than DSX^M. TRA also functions to activate an alternative 5′ splice site

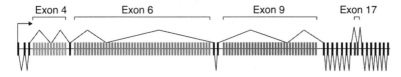

Figure 15.7. *The* Drosophila Dscam *gene.* The *Dscam* gene contains 95 alternatively spliced exons, which are organized into four clusters of mutually exclusive exons. The exon 4, 6, 9, and 17 clusters contain 12, 48, 33, and 2 variable exons, respectively. Due to this organization, *Dscam* can potentially encode 38,016 different isoforms.

in the *fru* pre-mRNA resulting in the synthesis of an mRNA that does not encode a functional FRU protein.[20] As a result, the genes required for male-specific sexual behaviors are not expressed in the females, and the females are therefore interested in courting males. Thus, an initial female-specific splicing of *Sxl* leads to, as it were, a "switch," maintained by feedback of the SXL protein on the splicing of *Sxl* transcript, generating a distinct set of alternative splice variants and stabilizing the female rather than male phenotype.

Dscam: One Gene, Potentially 38,016 Proteins

The most dramatic example to date of the role of alternative splicing in enhancing protein diversity is provided by the *Drosophila Dscam* gene. This single gene may generate 38,016 different mRNAs,[15] nearly three times the estimated number of genes in the entire *Drosophila* genome. This amazing feat results from four clusters of mutually exclusively spliced *Dscam* exons (figure 15.7), i.e., each mRNA contains only one exon selected from each cluster. *Dscam* encodes an axon guidance receptor critical in the wiring of the *Drosophila* nervous system. Alternative splicing of the exon 4, 6, and 9 clusters alter the protein sequence of three extracellular domains. Each distinct DSCAM protein isoform interacts preferentially with itself.[21] Thus, *Dscam* appears to encode a large set of molecular zip codes that determine the specificity of neural wiring in insects.[22] DSCAM also is expressed in hemocytes and fat body cells, which are important in the fly immune system.[23] Because individual DSCAM isoforms bind to, and lead to phagocytic uptake of, different bacteria, and can be secreted, this generation of protein diversity through alternative splicing in this insect "antibody-like" system is reminiscent of the efficient storage of information that enables the generation of vast diversity from gene segments combined (albeit at the DNA level) in the vertebrate immune system, as described in chapter 9, and in pathogen surface antigens, described in chapter 5. Dissecting the molecular details of how the alternative splicing of this attention-grabbing gene is regulated represents an active area of research; dozens of proteins that regulate its splicing have been identified.[24]

mod(mdg4)

Although not rivaling *Dscam* with regards to the number of isoforms it can generate, another *Drosophila* gene, *modifier of mdg4*, or *mod(mdg4)*,[25,26] is known to engage

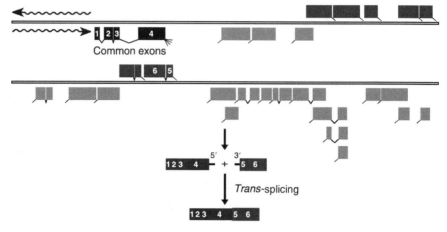

Figure 15.8.   Alternative trans-*splicing.* The *Drosophila mod(mdg4)* gene encodes 26 isoforms. Each mRNA isoform contains four common exons, which are spliced to any of 26 sets of variable exons. Seven of the variable exons are actually encoded on the opposite DNA strand. Expression of these isoforms involves *trans*-splicing.

in its own splicing acrobatics: *trans*-splicing. The *mod(mdg4)* gene encodes a set of 26 protein isoforms involved in chromatin maintenance.[27] The most amazing aspect of this gene, though, is its organization. In contrast to all other currently known genes, even split genes, the information that codes for the MOD(MDG4) proteins is located on different strands of the DNA (figure 15.8). (Chapter 13 describes splicing of information contained on different strands at the DNA level.)

There are four exons that encode the N-terminal portion of the protein and are common to each of the isoforms, while the variable C-terminal domains are each encoded by separately transcribed alternative exons. Several of these are encoded on the same strand of DNA. However, the alternative exons for seven of the isoforms are encoded on the opposite DNA strand. For example, in the example shown in figure 15.8 the common exons 1 thru 4 are expressed on a single pre-mRNA and the three *cis*-introns are spliced out, leaving a 5′ *trans*-splicing precursor containing a free 5′ splice site at its 3′ end. Exons 5 and 6 of this isoform are expressed as a separate pre-mRNA from the opposite DNA strand. Again, the *cis*-intron separating exons 5 and 6 is spliced out, leaving a 3′ *trans*-splicing precursor containing a free 3′ splice site. Next, these two *trans*-splicing precursors are spliced together to produce an intact mRNA. This mechanism is supported by three types of experimental evidence. First, and most obvious, is the fact that several of the alternative exons are encoded on the opposite DNA strand.[25,26] Second, genetic experiments have shown that two different *mod(mdg4)* alleles that contain point mutations in the *trans*-spliced alternative exons can complement one another. In other words, *trans*-splicing can occur not only between pre-mRNAs encoded on two different strands of a single DNA chain, but also between pre-mRNAs encoded on two different DNA molecules.[27] Finally, if a DNA fragment containing some of the *mod(mdg4)* alternative exons is placed on a different chromosome than the common *mod(mdg4)* exons, the *trans*-spliced mRNA products containing these exons can be readily detected on the fly.[26] Although *trans*-splicing has been previously shown in vitro,[28,29] and organisms such as *Caenorhabditis elegans* contain *trans*-spliced leader sequences on the 5' ends of their mRNAs,[30,31] how the spliceosome negotiates the *trans*-splicing of *mod(mdg4)* is currently a complete mystery, as is the nature of the information that constrains this. In particular, nothing is known about the mechanism that prevents the *mod(mdg4)* *trans*-spliced precursors from being randomly spliced to other pre-mRNAs in the cell, a situation that undoubtedly would be disastrous. It is also completely unknown how widespread this phenomenon is in eukaryotes.

Interpreting the Genome: Mechanisms of Splice Site Recognition and Alternative Splicing

As mentioned earlier, one of the greatest challenges we are faced with in regards to pre-mRNA splicing is: how are authentic splice sites recognized and cryptic or aberrant splice sites suppressed? To put this problem into its proper context, human exons are typically about 150 nt while introns are approximately 3500 nt on average.[32] However, human introns can be extraordinarily large. For example, the human *NRNX3* gene contains an intron that is 479,087 nt long.[14] Given the degeneracy of the 5' and 3' splice site sequences, cryptic splice sites that are found within introns and yet are never used significantly outnumber authentic splice sites.[33] How, then, does the spliceosome identify real splice sites amidst the plethora of cryptic splice sites?

The key to this problem is the presence of auxiliary splicing elements; in other words, additional information within the DNA sequence that interacts with proteins that enhance or repress the ability of the spliceosome to recognize the splice sites. In addition to aiding in the discrimination between authentic and cryptic splice sites, the information contained in these auxiliary sequence elements—and the proteins that recognize these elements—are at the heart of the mechanisms involved in regulating alternative splicing.[34] These splicing regulatory proteins are referred to as auxiliary splicing factors because they are not required for the general aspects of the splicing reaction. Rather, they function to control whether the general splicing machinery (the molecules required for the chemical steps of the splicing reaction) bind to the splice sites. The sequences that these auxiliary splicing factors recognize in the pre-mRNA are known as exonic splicing enhancers (ESEs), exonic splicing silencers (ESSs), intronic splicing enhancers (ISEs), and intronic splicing silencers (ISSs).

Exonic Splicing Enhancers

ESEs are sequence elements that enhance the splicing of exons that contain them.[35] ESEs were originally identified in a few exons, most of which were alternatively spliced.[36-38] However, it is now thought that several ESEs are present in most, if not all, constitutively spliced exons.[39] The requirement for instructions that direct splicing (i.e., ESEs) to be contained within coding regions places tremendous constraint upon the choices between otherwise "synonymous" codons.[40]

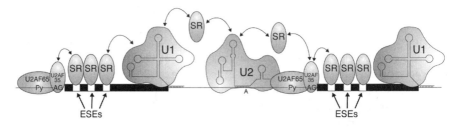

Figure 15.9.   Exon-definition model. Exons are defined by interactions among factors that bind to the splice sites and exons. SR proteins interact with exonic splicing enhancers (ESEs) and interact with other SR proteins, as well as U2AF35 and U1-70K. These interactions are thought to stabilize the binding of U2AF and U1 snRNP at the 3′ and 5′ splice sites, respectively, and therefore help to define authentic exons.

ESEs typically are recognized by members of the SR protein family that function to enhance the binding of components of the splicing machinery to the adjacent splice sites.[35] SR proteins are highly conserved in metazoans and play important roles in both splice site recognition and alternative splicing.[36] All SR proteins contain one or two N-terminal RNA binding motifs, and a C-terminal domain rich in arginine and serine resides called the RS domain. The RNA binding domains of SR proteins have rather degenerate RNA binding specificities.[35] In other words, each SR protein has the ability to interact with a wide range of RNA sequences. This property can explain the dilemma of how ESEs can coexist within protein-coding sequence. The RS domains have been shown to participate in protein–protein interactions with other RS domain-containing proteins, such as SR proteins,[41] U2AF65,[42] U2AF35,[41] and U1-70K.[41,43] Based on these interactions, a model has been proposed to explain how SR proteins contribute to exon definition (figure 15.9).[44] Specifically, SR proteins have been proposed to recognize and bind directly to ESEs and, through interactions with U2AF35 and U1-70K, promote the binding of U2AF and U1 snRNP to the splice sites flanking the exon. SR proteins also may contribute to splice site recognition by interacting directly with the splice sites in a way that somehow facilitates spliceosome assembly.[45,46] Thus, exons contain additional sequence information that marks them as exons, and it is the exon, as opposed to merely individual splice sites, that is the unit of recognition.

SR proteins also participate in the regulation of alternative splicing. This is achieved by the differential expression of individual SR proteins in different tissues or cell types.[47,48] Exons containing ESEs recognized by a specific SR protein are included in tissues or cells that express high levels of that SR protein, but are skipped in cells that express low levels of the same SR protein.

One of the best examples of this type of regulation is the *Drosophila dsx* pre-mRNA. The exon of *dsx* that is expressed in females contains six 13-nt sequence elements that each serve as the binding site for a heterotrimeric complex consisting of the transformer (TRA), transformer 2 (TRA2), and an SR protein.[38] Although TRA and TRA2 are not SR proteins per se, they are functionally related to them because they contain extensive RS domains.[35] The presence of all three of these proteins is required for the entire complex to bind.[49] Because, as described earlier, TRA is not expressed in male flies, the female-specific *dsx* exon is skipped in male flies. In contrast, in females, which do express TRA, this heterotrimeric complex assembles on each 13-nt element,[50] thus promoting the binding of U2AF to the 3′ splice site upstream of the female-specific *dsx* exon.[51,52]

Exonic Splicing Silencers

ESSs are a second class of regulatory elements that reside in exons, but they have effects opposite to those of ESEs—they interact with proteins that inhibit the inclusion of an exon. Most ESSs that have been characterized have been found to interact with members of the hnRNP family.[53] These proteins function by decreasing the binding of components of the splicing machinery to neighboring splice sites, resulting in repression of the exon to which they bind. These effects are thought to be mediated primarily through steric hindrance and hnRNPs have been shown to interfere with the binding of SR proteins, U1 snRNP, and U2AF.[34] However, in most cases the detailed mechanisms by which hnRNPs and ESSs function is not yet clear. Interestingly, ESSs are more prevalent in pseudoexons than in authentic exons, suggesting that these elements may primarily function to prevent the aberrant inclusion of intronic sequences.[54,55] As to why pseudoexons exist at all, there currently is no satisfying answer.

Intronic Splicing Enhancers and Silencers

Splicing regulatory elements are also located in introns flanking the regulated exon. These can function to either enhance (ISE) or suppress (ISS) the inclusion of the adjacent exon. Several auxiliary splicing factors have been identified that bind to these elements. The mammalian Nova proteins (Nova-1 and Nova-2)[56] are splicing regulatory proteins that function by binding to intron sequences. Nova-1 and Nova-2 bind specifically to the sequence YCAY,[57] which is present in all of the pre-mRNAs that are known Nova-regulated splicing targets.[58] Interestingly, Nova can bind either upstream or downstream of the regulated exon and can function as either a positive or negative splicing regulatory factor. To date, the precise mechanism by which the Nova proteins activate or repress splicing is not known. Several other splicing regulatory proteins that function by binding to intronic sequences have been characterized; excellent reviews are available for readers who would like additional information on this topic.[34,53]

Combinatorial Control

It is important to note that the splicing regulatory elements described above rarely work alone. Rather, any given exon typically contains both ESEs and ESSs, and also usually is flanked by ISEs and ISSs (figure 15.10). Together, these sequence elements comprise a splicing code that is interpreted by the spliceosome and its regulatory proteins to aid in the tissue-appropriate definition of authentic exons and prevent the inappropriate inclusion of intronic sequences as exons.[59]

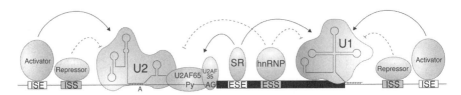

Figure 15.10.   Combinatorial control of pre-mRNA splicing. Individual exons typically contain both exonic splicing silencers (ESSs) and exonic splicing enhancers (ESEs) and are flanked by introns containing intronic splicing silencers (ISSs) and intronic splicing enhancers (ISEs). The frequency with which an exon will be included in a particular cell type is determined by the identity, number, and location of these regulatory elements as well as the relative expression levels of the proteins that bind to each of the element

Throughout this chapter, exons have been referred to as being either included or excluded. This is not meant to imply that a particular exon is either always included or excluded in a certain cell or tissue type. While there are many examples where this "all or not" type of splicing is observed, most often things are less clear cut. Rather, in most cases a particular exon is included some of the time and excluded other times in the same cell. The extent to which any individual exon is included in the mRNA is determined by the number and identity of the entire collection of these regulatory elements in the vicinity of the exon, as well as the relative expression levels of all the splicing regulatory proteins that bind to these elements. Thus, as is discussed in chapter 16 for imprinting, the genomic context in which a sequence lies has a dramatic effect on when and whether it is expressed. RNA editing, described in chapter 14, provides additional examples of the genomic context implying a protein sequence that is different from that directly translated from the linear sequence of the RNA transcript.

Mutually Exclusive Splicing

An additional level of splicing control achieved by the spliceosome are cases in which two or more adjacent exons are included in the mRNA in a strictly mutually exclusive manner. Despite the fact that many genes contain mutually exclusive alternative exons, very little is known about the mechanism that allows only one exon to be included in the mRNA. Two of the best-characterized examples of mutually exclusive exons are from the mammalian α-actinin[60] and α-tropomyosin[61] genes. In each case, the distance between the upstream exon and the branchpoint adenosine 5′ to the downstream exon is critical for the mutually exclusive alternative splicing. In both genes, the branchpoint adenosine of the downstream exon is located near the 5′ splice site of the upstream alternative exon. As a result, if the 5′ splice site of the upstream exon is bound by U1 snRNP, this will prevent the binding of U2 snRNP to the downstream 3′ splice site, thus preventing inclusion of the downstream exon. Although this mechanism can nicely explain how the splicing of the thousands of known pairs of mutually exclusive exons is controlled, it will almost certainly not be involved for genes, such as *Dscam*, that contain clusters with more than two mutually exclusive alternative exons. Although in these situations steric hinderance could prevent adjacent alternative exons from being spliced together, it would not be able to prevent nonadjacent exons from being spliced together. Thus, it appears that a completely novel and unexplored mechanism, involving context-dependent information, which ensures that splicing of the multiple alternative exons is mutually exclusive, remains to be discovered.

Conclusions

As we have seen, alternative splicing is a powerful mechanism for regulating gene expression and increasing protein diversity. This process is a major exception to the idea that information is explicitly encoded in a linear fashion within the genome. In fact, information is contained within the DNA sequence of a gene that implies, through the regulated mechanism of alternative splicing, the sequence of multiple alternative proteins. A major hurdle for the next few years will be the rigorous determination of how it is that the spliceosome manages to find the relatively few and small exons among the sea of intronic sequence.

Acknowledgments   I would like to thank my many colleagues and lab members over the years for their thoughtful discussions and insights into this fascinating area of research. My work on alternative splicing is supported by grants from the National Institutes of Health (R01 GM62516 and R01 GM67842).

16

Imprinting: The Hidden Genome

Alyson Ashe and Emma Whitelaw

Overview

Even in diploid organisms, many genes are expressed in a monallelic fashion, as if the organism were haploid at that locus. In some cases, such as the expression of one allele at the immunoglobulin locus, monoallelic expression enables a cell to experience direct selection based upon the properties of the expressed allele. Perhaps the most intriguing example of monoallelic expression is the silencing of many alleles in a parent-of-origin specific manner. Parental imprinting effectively leaves the affected loci "hidden" for a generation. The understanding of parental imprinting will require a more thorough understanding of the role of chromosomal context and of epigenetics in control of gene expression.

Introduction

Genomic imprinting refers to the transcriptional regulation of alleles in manner that is specific to the parent of origin. In classic cases of genomic imprinting, this regulation leads to the silencing of one parental allele, causing functional hemizygosity at the locus in the offspring. The silencing is maintained during early development, but is occasionally relaxed in adult tissues. In other words, genetic information inherited from one parent may be "hidden" for up to a generation, yet passed on to progeny intact. This process, also known as "parental imprinting," has been discovered in a broad range of organisms, including humans, rodents, marsupials, and plants. The number of classically imprinted genes is not known, but current estimates suggest approximately 100 genes may be subject to imprinting in mammals. This estimate is based on the known number of genes in regions of chromosomes that must be present from both parents for offspring to be viable. There are a number of excellent reviews of genomic imprinting.[1-4]

More recently another set of genes has emerged that are monoallelicly expressed but where the decision as to which copy is silenced is not based on its parent of origin.[5] It is likely that these two phenomena are mechanistically related.

One of the most popular explanations for the existence of parental imprinting is that it reflects a parental conflict with respect to growth of the offspring: the male parent wants his offspring big, while the female needs to preserve her resources for future reproduction.[6] However, examples of silencing phenomena have been found in organisms historically thought to lack genomic imprinting, such as zebrafish, *Drosophila*, and mealybugs, and for which arguments based upon parental conflict are less compelling.[7–9] Some authors question the validity of placing these results under the same heading as classic genomic imprinting, but there is no doubt that silencing does occur in these organisms, and in a parent-of-origin-specific fashion. These cases of parental imprinting have forced us to reconsider the teleological parental-conflict explanation for imprinting. This chapter reviews the molecular mechanisms by which the silencing is established and maintained, the nature of the genes that are parentally imprinted, and the relationship of parental imprinting to other forms of monoallelic expression. In addition, we discuss several different hypotheses regarding the origins of imprinting.

Mechanisms of Epigenetic Silencing

Parent-of-origin-specific silencing is carried out by what are now referred to as epigenetic modifications. Epigenetic modifications include methylation of the cytosine residues of DNA, and methylation and acetylation of histone molecules that package the DNA.[10] These epigenetic mechanisms also are involved in other situations in which gene expression is silenced in a manner that is mitotically heritable, such as X-inactivation.[11–13] X-inactivation is the silencing process by which dosage-compensation for X-linked genes is achieved in mammals. The molecular mechanisms of genomic imprinting are not fully understood, but methylation of cytosine residues seems to be crucial for the process, at least in embryonic tissues in mammals. Most imprinted genes, or gene clusters, contain a differentially methylated domain (DMD) in which the cytosines at CpG dinucleotides are methylated on only one chromosome of the chromosome pair. In many cases, cytosine methylation correlates with the silencing of a gene, for example the DMD located just upstream of the *H19* promoter is paternally methylated and *H19* is paternally silenced. In some cases, however, methylation of the DMD correlates with expression at an adjacent gene.[14] This differential methylation between the maternal and paternal alleles is crucial for the maintenance of silencing: mice with loss-of-function knockouts of any of the DNA methyltransferase enzymes *Dnmt1*, *Dnmt3a*, or *Dnmt3L* produce embryos without the appropriate maternal methylation imprint and show aberrant expression of imprinted genes.[15–17]

The detailed processes that link DNA methylation with silencing are under intense scrutiny (for a detailed review of the literature on regulation of imprinted clusters see reference 14). In particular, it is not yet certain whether DNA methylation is causative or correlative with silencing, although recent evidence suggests DNA methylation may in some cases be linked with the maintenance of silencing, rather than its establishment.[18] One well-characterized locus is the *H19* subdomain of the imprinted cluster on human chromosome 11, mentioned above. A plausible and widely accepted model of how parental imprinting occurs at this locus has emerged. The DMD just upstream from the *H19* promoter is unmethylated on the maternal allele, which correlates with expression of *H19* (figure 16.1). Additionally,

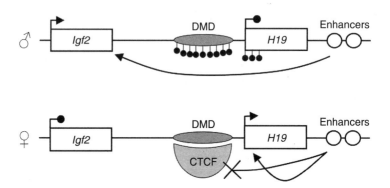

Figure 16.1.   A model of imprinting control of transcription at the H19/Igf2 *locus.* At the paternal locus, the differentially methylated domain (DMD) is heavily methylated (black lollypops). This methylation spreads to the promoter region of *H19* and prevents transcription. Upstream enhancers cause transcription of *Igf2*. At the maternal locus, the DMD is unmethylated and CTCF binds. This blocks the action of the enhancers on *Igf2*, preventing its transcription. The enhancers instead activate transcription of *H19*. (Adapted from reference 14).

the DMD contains binding sites for an insulator factor, CTCF.[19] When CTCF binds to the unmethylated DMD it prevents *Igf2* from accessing downstream enhancers, and hence prevents its expression. On the paternally methylated allele, *H19* is silenced. CTCF is unable to bind to the methylated DMD and can no longer act as an insulator between *Igf2* and its enhancers, allowing *Igf2* expression.[14] For more complicated clusters containing many imprinted genes, the mechanism of silencing is more challenging to dissect.

Nonprotein-coding RNAs are also thought to play an important role in the regulation of some imprinted regions. For example, at the *Igf2r* locus, paternal expression of *Air* RNA, which partially overlaps *Igf2r*, causes paternal silencing of *Igf2r*, and two other genes in the cluster, *Slc22a1* and *Slc22a3*.[20] The mechanism of this silencing is unclear, but histone modifications probably play an important role (see below and reference 21).

Chromatin structure plays a critical role in the silencing of genes, but our ability to study this at any particular locus in the genome has been limited by the available technologies. Chromosomes, or chromosome regions, containing predominantly unmethylated DNA usually have an open chromatin structure, associated with DNase-I-hypersensitivity, indicating a region where nucleosomes are absent or disrupted.[22] Such open regions are characterized by acetylated histones H3 and H4, and methylation of histone H3 lysine 4. Methylation of histone H3 lysine 9 correlates with heterochromatin assembly and silenced genes.[23,24] Our detailed knowledge of the histone modifications present in the chromatin of active (or inactive) regions of the genome has increased greatly over the past few years with the development of chromatin immunoprecipitation (ChIP) assays.[25] Some progress has been made in our understanding of how the histone modifications direct the methylation state of the DNA, and vice versa,[26–29] but it is still not completely clear how cytosine methylation and histone modifications work together to establish the transcriptional activity of a locus. Studies at the *Kcnq1* locus show how complicated this interplay can be. *Kcnq1* contains a DMD, which drives expression of an antisense, noncoding transcript, *Kcnq1ot1 (Lit1)*. Expression of *Kcnq1ot1* is crucial to both silencing of genes in the cluster and spreading of DNA methylation in the embryo.[30] However, silencing of many of the genes in this cluster occurs in placenta, in the absence of DNA methylation.

In some cases of parental imprinting, the DNA methylation mark is re-set in the germline, in others it seems to be established later (see below), perhaps indicating that it is not the primary imprint.[14] When compared with other genes, imprinted regions are characterized by a large number of matrix attachment regions.[31] Matrix attachment regions usually are associated with heterochromatic domains, so this finding strengthens the idea that chromatin structure is intricately involved in parental imprinting. Imprinted loci also are characterized by asynchronous replication timing. At imprinted loci replication timing differs depending on the parent of origin. Asynchronous replication does not depend upon prior DNA methylation as the distinct times of replication are maintained in the absence of DNA methylation.[32]

Establishment of Parental Marks

In order for the correct set of imprinted marks to be passed on to the next generation, complete erasure of existing marks and resetting of the new imprint must occur in the germline of each individual. In mice, primordial germ cells start to differentiate at about embryonic day 7, and migrate to the genital ridges between day 10.5 and day 11.5.[33] Mitotic arrest in male gonads and meiotic prophase in female gonads occurs around day 13.5. Clearing of the imprinted marks, at least with respect to cytosine methylation, appears to occur during, and just after, the migration stage, with marks being partially cleared at day 11.5 and fully cleared by day 12.5.[34,35] The timing of clearing is similar in both male and female primordial germ cells.

The resetting of the appropriate parental mark occurs later in development. In females, establishment of cytosine methylation imprints occurs during the maturation of oocytes.[36] The imprinted loci in male gametes undergo methylation at an earlier developmental stage than the equivalent for oocytes, with most methylation occurring during the gonocyte stage and essentially completed by the time they mature to spermatogonia.[37] In both cases, methylation of imprinted genes relies on the presence of at least two members of the DNA methyltransferase family, Dnmt3a and Dnmt3L.[15,17,38]

After fertilization, the imprinted marks are actively maintained in the developing embryo. Prior to fertilization, both oocyte and sperm DNA have high levels of methylation genome-wide, in a specific pattern different from that in somatic tissue.[39] In the 6–12 hours directly after fertilization, the paternal contribution to the genome undergoes an extremely rapid, active wave of demethylation.[40,41] This demethylation appears to affect most of the genome, with the exception of imprinted genes and possibly some retrotransposons.[39,42] It is not known how these sequences avoid demethylation.

The maternal contribution to the genome also undergoes demethylation, but this demethylation phase is believed to be a passive process resulting from continued replication of DNA in the absence of the maintenance methyltransferase Dnmt1.[39] As replication proceeds, the methylation marks are not established on the new strand, and so the amount of methylation in the genome decreases in proportion to the amount of replication. The marks on imprinted genes are somehow maintained and avoid the clearing process. While this maintenance is, again, not completely understood, we do know that it involves Dnmt1o, a maternally expressed oocyte-specific form of Dnmt1.[43]

Which Genes Are Imprinted?

One of the first indications that some genes in the embryo are not expressed equally from the maternal and paternal contributions came from studies of androgenetic (bipaternal) and gynogenetic (bimaternal) embryos. While these embryos had the correct amount of DNA, they never developed to term.[44] These experiments showed that the contribution of a haploid genome from each parent was necessary for proper development, suggesting the possibility that not all genes were biallelicly expressed. Cattanach and colleagues, using mice with translocations, identified specific chromosomal regions in which paternal or maternal disomy led to embryonic lethality.[45] Many genes that show imprinted patterns of expression within these regions have now been identified. Currently, there are around 100 genes in mammals that are known to be parentally imprinted,[46] but this number may well increase somewhat as further studies are carried out. A number of relatively rare human diseases, including Beckwith–Wiedemann syndrome, Angelman syndrome and Prader–Willi syndrome, have been identified which result from defects in imprinted genes.[47]

There still is no definitive way of predicting, whether by function or by examining upstream sequences, which genes will turn out to be imprinted. While many of the imprinted genes identified so far seem to be involved in growth and behavior, the significance of this observation is unclear since changes in these two traits are relatively easy to detect.[48] It also should be noted that the number of growth and growth-related genes that are not imprinted vastly outweighs the number that are. Nonetheless, it is still possible that there is an overrepresentation of growth-related genes among those genes that are imprinted. In addition to these growth-related genes, there are some imprinted genes (identified by comparing expression patterns from the maternal and the paternal genomes) that show no obvious phenotype when knocked out, and whose function remains unknown.[49] So, at present there is no way of looking at the function of a gene and deducing whether or not it is parentally imprinted.

Since function is not a clear indicator of novel imprinted genes, scientists have been using laboratory screens to identify them. One approach has been to carry out genome-wide searches for DMDs, those regions that are differentially methylated on the paternal and maternal alleles.[50] Another approach is to compare cDNA from normal, parthenogenetic, and androgenetic embryos.[51–53] This has been relatively successful in the mouse, having identified at least six previously unknown parentally imprinted genes or regions.[54,55]

Attempts have been made to see if it is possible to distinguish the two subgroups of imprinted genes, paternal and maternal, from each other on the basis of sequence analysis.[54,55] In both mice and humans, maternally expressed genes are associated with a higher GC content than paternally expressed genes.[54,55] In humans, paternally expressed genes are also associated with a significantly greater L1 content compared with maternally expressed genes and their control regions.[55] However, attempts to design predictive methods for imprinted gene discovery using the above sequence characteristics remain problematic, although for genes known to be imprinted they are very accurate at telling whether the imprinting is maternal or paternal.[56] Because broader chromatin context is likely to play a critical role in imprinting, local sequence analysis in the absence of experimentation is unlikely to be completely effective at imprinted gene discovery.

Other Examples of Monoallelic Expression

Parental imprinting is not the only in vivo situation where monoallelic expression of genes occurs. There are a number of other autosomal genes that are expressed in some cells in a monoallelic fashion but not in a parent-of-origin-specific fashion. These include immunoglobulins,[57] T-cell receptors,[58] natural killer cell receptors,[59] olfactory receptors,[60] and pheromone receptors.[61] The monoallelic expression of the immunoglobulin and T-cell receptor genes correlates with mutational events,[62] but at least in the case of the immunoglobulin κ locus,[63] the decision as to which allele undergoes DNA rearrangement is associated with epigenetic events. Similarly, for the olfactory receptors, the mechanism causing random monoallelic expression appears to be epigenetic.[64] There are a number of other genes (e.g., fibrinogen beta chain, 5-hydroxytryptamine-2A receptor, interleukin-1a, and klotho[65]) that appear to be monoallelicly expressed, and, although it has not been shown, it is likely their expression is controlled by epigenetic mechanisms.

There are some genes, first identified in plants (reviewed in reference 66) and more recently in mammals,[67,68] where the expression state is variable from cell to cell, among cells of the same cell type. At these alleles there is a certain probability, independent of the expression state of the same gene on the homologous chromosome, that the gene will be transcriptionally active. So, even in a homozygous organism, a cell of the relevant tissue may have none, one, or both alleles active. These are known as metastable epialleles and are instances of epigenetic silencing causing variation in patterns of gene expression and a variegated phenotype. One example of a murine metastable epiallele is *agouti viable yellow* (A^{vy}). An active locus, which correlates with hypomethylation of the cytosine residues, results in a yellow coat, while epigenetic silencing of the allele, which correlates with hypermethylation, results an agouti coat. In many cases, however, the mice display variegated coats, with patches of yellow and agouti fur.[68–70] A similar result is seen in mice carrying the $Axin^{Fu}$ allele, but in these mice the visible phenotype is the presence or absence of tail kinks.[71] The presence of kinks again correlates with a lack of cytosine methylation at the $Axin^{Fu}$ allele.[72] The epigenetic silencing at metastable epialleles is established during early development of the embryo. Once the mark is established in a cell, it is inherited mitotically (i.e., through subsequent rounds of cell division). Thus the variegated coat color in the A^{vy} mice mentioned above results from the epigenetic mark being established stochastically in founder cells in the early embryo. Parent-of-origin effects have also been reported at metastable alleles, with some being expressed more strongly following female transmission and others more strongly following male transmission.[73] For example, offspring of a parent carrying the A^{vy} locus are more likely to be yellow (active allele) if that parent is female than if that parent is a male carrying the locus.[68]

Neither monoallelicly expressed genes, such as the immunoglobulin locus, nor metastable epialleles would fall into the category of classic parentally imprinted genes, but there are many similarities, from the inheritance of the mark mitotically to the epigenetic basis behind the silencing of the alleles in question. In each case, something in addition to the sequence of the bases in the gene's regulatory regions is affecting its expression.

Why Have Imprinting?

At first glance, imprinting seems to be contradictory to the central tenet of diploidy: that diploidy is advantageous due to the masking of deleterious recessive alleles. Imprinting causes the silencing of one parental allele, therefore causing functional haploidy at the loci involved. Why should this occur? Many have argued that understanding the function of imprinted genes should help us to understand why parental imprinting occurs. Indeed, many of the hypotheses put forward for the evolution of imprinting have been based on an analysis of the function of imprinted genes. For example, the large number of imprinted genes involved in fetal/placental growth has led to the development of the intergenomic conflict hypothesis (see below). More recently, models have emerged where the function of the imprinted genes is not the primary driver in the evolution of the process (but rather a side effect of a chromatin-related parental mark). The following sections review some of the more prominent theories for the evolution of imprinting.

Conflict Theory

The conflict theory of genomic imprinting is the most widely accepted explanation for the existence of parental imprinting. It has received more attention in the literature than any of the other hypotheses, perhaps because the first few imprinted genes to be discovered seemed to fit the hypothesis well. As more imprinted genes have been discovered, the hypothesis has grown less attractive, and its validity now is hotly debated.[48,49,74] The conflict theory of genomic imprinting centers around the conflict between maternal and paternal interests in the partitioning of resources to offspring. Ideally, the paternal genome "wants" all the resources to be concentrated in his offspring, while the maternal genome "wants" her resources to be spread equally amongst all her offspring, both current and future. Thus, this conflict is predicted to result in the maternal silencing of growth enhancers, and the paternal silencing of growth repressors. This is nicely demonstrated in the silencing pattern of the growth factor *Igf2* (maternally silenced) and its repressor, *Igf2R* (paternally silenced), in mice. The idea also is supported by examining defects in the development of androgenetic and gynogenetic embryos. The inner cell mass of androgenetic embryos does not grow well, while the trophoblast development is essentially normal, supporting the theory that the paternal genome "wants" to get as many resources from the mother as possible by promoting placental growth and hence the transfer of nutrients from mother to offspring. In contrast, gynogenetic embryos have trophoblasts that fail to thrive, but the inner cell mass is reasonably normal (the maternal genome "wants" each of her offspring to do well, without one individual monopolizing all her resources).[75]

The conflict theory also is supported by the discovery of imprinting in angiosperms; indeed, the original theory was developed with the results from maize polyploid experiments in mind.[6] In angiosperms, the embryo is surrounded by endosperm, much as the mammalian embryo is surrounded by extraembryonic membranes. Endosperm size varies among angiosperms, from large and full of starch in monocots, such as maize, to small and inconspicuous in most dicots, such as *Arabidopsis*. The endosperm is involved in directing nutrients to the developing embryo, a similar function to that of mammalian extraembryonic tissues. The endosperm is not derived from the same tissue as the embryo, but from the fusion of two maternal haploid nuclei with one sperm nucleus. It is thought to be genetically identical to the embryo except that it has two copies of the maternal genome. This 1p:2m ratio is crucial for proper endosperm development. Polyploid embryos are viable, but only if this ratio is maintained: for example 2p:4m ratios in the endosperm are viable, but 3p:3m or 1p:5m ratios cause embryo abortion, presumably due to endosperm failure.[6,76] In addition, when the endosperm has a paternal excess it undergoes greater mitotic activity, resulting in larger seed. This contrasts with maternal excess endosperms, which have less mitotic activity and a drastically reduced seed size.[76] These parallels between the development of extraembryonic tissue and parental imprinting in such widely divergent species led to the development of the conflict theory.

A major criticism of the conflict hypothesis is that there are now large numbers of imprinted genes whose function does not appear to be related to prenatal growth. In an attempt to account for this discrepancy, the concept of resource constraints underlying the conflict theory has been expanded into the kinship theory of genomic imprinting.[74,77] In its new form, this theory incorporates more-general effects on kin, in the manner of inclusive fitness, as first modeled by Hamilton.[78] Hamilton theorized that a gene can expand in a population not only by increasing the fitness of its bearer but also by increasing the fitness of relatives that are likely to share copies of this gene by common descent. The combination of these two fitness parameters is called "inclusive fitness." By incorporating inclusive fitness into the conflict theory, genes that have an effect on more-distantly related relatives (through, for example, behavior) can be suggested to fit the hypothesis, despite the fact that they seem to have no effect on parental partitioning of resources to the offspring. The kinship theory also helps to justify the fact that imprinting persists after weaning, when the offspring are no longer using resources directly from the mother and as such should not be reducing the fitness of subsequent litters.[48]

The conflict hypothesis requires that parental imprinting exists only in organisms in which the offspring have the potential to affect the acquisition of resources from the maternal parent. As such, it should occur in viviparous mammals (and angiosperms) but not in oviparous animals. Parental imprinting has been reported in all eutherian mammals studied to date and in marsupials, but not in monotremes.[79,80] As monotremes are oviparous, the lack of imprinting has been used to support the conflict hypothesis. However, only the IGF2R and IGF2 loci have been studied in monotremes. As imprinted loci are not completely conserved between humans and mice, which are much more closely related to each other than either is to monotremes, the possibility remains that there are loci that are imprinted in monotremes.

The Ovarian Time Bomb Hypothesis

The lack of development of the trophoblast in parthenogenetic and gynogenetic embryos (see above) is proposed by Varmuza and Mann[81] to be the driving force behind the evolution of imprinting. They propose that the silencing of maternal genes crucial for trophoblast development exists to prevent the formation of malignant teratomas from parthenogenetically activated oocytes. Another early hypothesis that is closely related to the ovarian time bomb hypothesis is Solter's hypothesis[82] that imprinting evolved to prevent parthenogenetic reproduction. One of the criticisms of the ovarian time bomb hypothesis is that while humans have a fairly high risk of ovarian cancers, the frequency is much reduced in rodents, and in other mammals that lack an invasive trophoblast (e.g., sheep).[74] However, if there was a high risk of ovarian cancer in the ancestral organism, this argument does not invalidate the hypothesis.

One of the major problems with the ovarian time bomb hypothesis is the existence of parentally imprinted genes that appear to have no function in trophoblast development. Varmuza and Mann suggest these genes are "innocent bystanders" that have become imprinted due to their chromosomal proximity to genes involved in trophoblast development.[81] However, as the number of known imprinted genes increases, the number of genes that do not appear to be involved in trophoblast development also increases.[49]

An intriguing development is the birth of the first parthenogenetic mouse, Kaguya, from the fusion of nongrowing oocytes harboring a 13 kb deletion in the H19 gene and fully grown oocytes.[83] Although only 0.6% of embryos transferred to pseudopregnant females survived in this experiment, a surprising 8.2% of pups, albeit all but two showing developmental retardation, were recovered at 19.5 days of gestation. Perhaps the most interesting observation was that the expression levels of almost all imprinted genes studied was normal in these embryos (i.e., levels of expression were similar to control embryos). Since the knocked-out region was only 13 kb at one locus, this raises the question of how appropriate expression of one imprinted gene can lead to the appropriate expression of most other imprinted genes scattered throughout the genome, although it is possible that normal imprinting at other loci was rare and related to the rarity of survival. If in fact all that was needed to allow parthenogenetic development of Kaguya was a modification at one locus, then the imprinting at 100 additional loci cannot be argued to have evolved as a mechanism to prevent parthenogenetic development of ovaries (and hence ovarian teratomas).

Rheostat Model for a Rapid Form of Imprinting-dependent Evolution

In this model, Beaudet and Jiang[84] propose that genomic imprinting evolved as a mechanism to maximize the variation in gene expression at dosage-sensitive loci among individuals. As mentioned above, imprinting creates functional hemizygosity at the loci in question, which at first glance appears risky considering the advantages of diploidy, such as the effective elimination of deleterious mutations. So what counterbalancing selective advantage might there be of haploidy at certain loci?

The rheostat model of imprinting-dependent evolution[84] proposes that imprinted genes are highly dosage-dependent, such that a small change in the amount of gene product available can result in a change in phenotype: when a small percentage of cells are silenced, say 10%, there will be a visible effect. Alleles that display this quantitative hypervariability are effectively the same as metastable epialleles (see above). Thus both metastable epialleles and imprinted alleles will result in a spectrum of phenotypes in the population simply from epigenetic variability. Additionally, the imprinted alleles are out of circulation in some of the individuals in a population, allowing mutation to occur without immediate selection.[7] These two characteristics combined should give much more variation in the population, due not just to allelic variation but also to epiallelic variation. The selective advantage would accrue primarily to the group, not the individual. Obviously, this would be an advantage to the population as a whole in times of environmental change and/or catastrophe.

The authors suggest that the "reason" most genes are not imprinted is because they are not sensitive enough to dosage variation, and thus imprinting would not be positively selected at these loci. These ideas are supported by the observation that most null mutations are recessive, indicating that a large change in amount of gene product (50%) does not result in a significant change in phenotype.

Parental Imprinting as a By-product of Heterochromatin

Parent-of-origin effects actually occur in many more organisms than just mammals and angiosperms, including, for example, insects. Although for some organisms, such as the fruit fly *Drosophila*, researchers have not been able to find any phenotypic effects in the wild-type fly, there are numerous reports of genes which, following translocation to pericentric heterochromatin, display parent-of-origin dependent patterns of gene expression.[8] After translocation to pericentric heterochromatin, but not at their evolved location, the genes are also expressed in a variegated manner, termed "position effect variegation" (PEV).[85,86] Parental imprinting is observed for marker genes only when they are translocated next to specific regions of heterochromatin (it is important to note that it is only certain regions of heterochromatin that produce this effect). Because no phenotypic consequences have been observed for parent-of-origin effects in the wild-type fly, imprinting in *Drosophila* is mostly ignored in discussions of classic parental imprinting. Lloyd argues that these findings imply that the evolution of imprinting, at least in *Drosophila*, is unlikely to have been driven by the function of specific imprinted genes.[8]

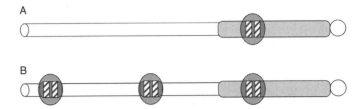

Figure 16.2.   A model for imprinting in Drosophila *and mammals.* (A) A *Drosophila* chromosome with an imprinting centre (striped boxes) in heterochromatin (gray). The imprinted region (shaded oval) remains in heterochromatin and gene expression is not influenced. (B) A similar mammalian chromosome. The imprinting centers have become scattered throughout the genome, and induced localized areas of heterochromatin-like regions into euchromatin (white). Genes lying near these regions exhibit imprinted expression. (Adapted from reference 8.)

An especially interesting example of parental imprinting occurs in the mealybug (lecanoid Coccids). In the male mealybug, the entire paternally derived chromosome set is heterochromatized during development so that the male is functionally haploid, solely expressing genes inherited from its mother.[9] This does not occur in female mealybugs, and they remain diploid. In some tissues of the adult male, such as some cells of the mid- and hind-gut and some skeletal muscle cells, this imprint is relaxed and the paternal chromosome set reverts to euchromatin. This relaxation of imprinting in some tissues is reminiscent of parental imprinting in mammals. What is even more interesting is that meiosis in males lacks independent assortment, and the paternal (heterochromatic) chromosomes assort separately from the maternal (euchromatic) chromosomes and the spermatids that inherit the heterochromatic set degenerate so that the paternal genome cannot be transmitted through male offspring.[9]

Thus in two insects, *Drosophila* and Coccids, the association between parental imprinting and heterochromatin is obvious. Lloyd[8] proposes an ancient mechanism of unknown function (but see next paragraph) in which regions within heterochromatin are parentally marked in the gametes. In *Drosophila*, these regions have remained exclusively within heterochromatin and therefore have no obvious phenotypic effect. Chromosomal rearrangement during the course of evolution has led to juxtaposition of these parentally marked heterochromatic regions next to euchromatin in mammals (figure 16.2). Consistent with this theory is the observation that imprinted genes are not evenly distributed throughout the genome, and that imprinting spreads out from imprinting centers (see above and reference 14). Furthermore, when these imprinting centers (from mammals) are inserted into the *Drosophila* genome they are recognized by it, and this results in silencing of adjacent genes.[87,88]

Sapienza and colleagues[89,90] propose that parental marks exist on chromosomes in order to facilitate pairing during meiosis and play the essential role of maintaining the distinction between homologues during DNA repair and recombination.

The basis of this theory is that the imprinting mark is chromatin structure, and in this respect the model is similar to that proposed by Lloyd.[8]

Conclusions

Obviously there is a great deal more to understanding the genome than translating isolated sequences, and it certainly seems that simply looking at the sequence of bases, without understanding their genomic context, is not going to be enough to fully understand the expression patterns of genes. The silencing of certain alleles in a parent-of-origin-specific manner interferes with the classic Mendelian notions of dominant and recessive genes, and means that individuals in the population are carrying genetic information that is "hidden" in the sense that only the allele inherited from one parent contributes to the phenotype of that individual. But from the point of view of the extended family (over a number of generations), the genetic information does contribute to phenotype and hence "fitness" of the group. Leaky silencing (or metastable epialleles) and haploidization can result in a much more variable range of phenotypes in the population than would otherwise be possible.

The discovery of more imprinted genes and the analysis of their function, along with a better understanding of what focuses imprinting on certain sequence regions, and the mechanisms of imprinting establishment and maintenance, are likely to lead to a better understanding of the evolutionary origins of, and selective forces acting upon, the process by which some genes "remember" their parent of origin.

Epilogue

An Engineering Perspective: The Implicit Protocols

John Doyle, Marie Csete, and Lynn Caporale

Introduction

While the sequencing of the human genome is a remarkable feat of technology, it is increasingly clear that the true complexity of the genome has little to do with the number and heterogeneity of its list of "parts," such as genes. It is not so much the ability to manufacture a diversity of parts as the ability to regulate and respond, often rapidly but on a wide range of time scales, that is the hallmark of life, and indeed of all sophisticated technologies. As evidenced in this book, the apparent linear structure of genomic DNA in highways of A, T, G and C compresses a variety of remarkable evolved mechanisms of recognition and control.

The Implicit Genome makes it clear that we need to enrich our conventional view, both of how information is encoded in the genome and of what information a genome might contain. In fact, the functioning genome challenges us even to examine what we mean by "information" and by "code." At first glance, the diverse biological examples in this book may appear to be a collection of disconnected special cases that do not fit conveniently into any existing conceptual framework. As discussed in the overview, we anticipate that new conceptual frameworks will emerge from careful analyses of these and other examples; in this brief epilogue we suggest one such framework.

To start, the word "code" already is heavily overloaded with meaning for biologists and engineers, with genetic code (biology), error-correcting and data compression codes (communications), object, source, and machine codes (computers), encryption codes (cryptography), etcetera, in wide use. While it might be possible to formulate a unified definition of "code" that captures these diverse examples, we think it more useful to introduce a distinct concept that does not assume a linear relationship between sequences and "meaning." Engineering offers the notion of "protocol," which is more general and richer than "code" and lets "code" preserve its existing meanings particular to specialized disciplines. To engineers, the term "protocol" is the set of rules by which components interact to create a new (system) level of functionality. In the next few pages, we describe the significance of protocols in engineering and what this concept might offer to biologists.

Layers of Interacting Protocols

As biologists dig deeper past the superficiality of sequence data, they unearth layers of control that are, at a fundamental level, organized in ways that are surprisingly similar to those in advanced engineered technologies. The most consistent, coherent, and salient features of all complex technologies are their protocols. Advanced technologies, such as the Internet and aerospace systems and, we suggest, biology, are organized based on suites and layers of protocols. For example, the "layering" of allostery and covalent modifications on top of transcription for fast responses to perturbations is strikingly analogous to the "layering" of transmission control protocols (TCP) to control for congestion and loss of packets on top of Internet protocols (IP) that route packets.[1-3]

Indeed, in advanced technologies, and, we maintain, in the organization of cells and organisms, the protocols are more fundamental than the modules that follow protocol "rules" and whose interconnections protocols facilitate. That is, in any given system it is more crucial to understand the protocols than the modules as the former are the more permanent and invariant features of complex system function. A central design feature of efficient, protocol-based systems is that they are very flexible with respect to what modules are allowed. Modules can easily be rearranged and interchanged and new modules introduced, provided that they conform to the protocols and provide the desired functionality. A major challenge in understanding the role of protocols is that often they are obscured by the overwhelming diversity and details that now characterize both experimental biology and advanced technologies. Indeed, the very nature of protocol-based systems contributes to this obscurity.

When engineers speak of the TCP/IP protocol suite, they are describing a huge collection of protocols, with acronyms usually ending in "P" for protocol, organized around the core of TCP/IP and layered to form what is called a "protocol stack." The details are less important here than the consequences, which are the system-level robustness and evolvability that these protocols facilitate.[3]

It stretches the notion of a "code" past the breaking point to think of TCP/IP as a code, and the notion of a protocol suite, which may initially seem vague and ill-defined compared with all the examples of a "code" given above, is nevertheless the right way to understand and describe TCP/IP. It is true that implementation of the protocols always involves multiple kinds of code, including a source code and a machine code, and that data may be encrypted and compressed, so that all the different notions of "code" may be relevant in structuring protocols. But the notion of protocol is something distinct, and beyond the "codes" that help to underpin the protocols. That is, TCP/IP exists as an abstract protocol, independent of any and all the specific codes associated with its implementation. One analogous biological example might be the conserved feature of high mutability at sites at which the DNA sequences differ, as described on page 119 for *mutS*.

Protocols and Adaptability, Interchangeability, and Interoperability

Ironically, in engineering as in biology, it often is the most transient and superficial elements of hardware and application software that are most visible to the user. We touch and see the computer screens and keyboards, and are constantly using the email, browser, computational, and word processing applications that run on them. The far more critical and persistent infrastructure facilitating these uses is the core protocols, which by design remain largely hidden to the point that most consumers are unaware even that they exist. Because of the protocols, new (and even radically different) hardware is easily incorporated at the lowest physical layers, and even more radically varying applications are enabled at the highest layers.

In engineering, protocol-based organization facilitates control, coordination, and integration. Thus, the system using the protocols has a coherent and global adaptation to variations in system components and in the environment, despite implementation mechanisms that are largely decentralized and asynchronous.[1] The TCP/IP protocol suite enables adaptation and control on time scales from the sub-microsecond changes in physical media, to the millisecond-to-second changes in traffic flow, to the daily fluctuations in user interactions, to evolving hardware and application modules that take place over years and decades. The remarkable robustness to changing circumstances, and to evolution of Internet-related technologies could only have come about as the result of a highly structured and organized suite of a relatively invariant set of universally shared, well-engineered protocols.

Similarly in biology, protocol-based architectures and their control mechanisms by their very nature facilitate both robustness and evolvability, despite massive impinging pressures and variation in the environment. The most familiar (and least complex) protocols could be thought of as codes. For example, a string of base pairs is given the abstract label of a "gene" simply because we identify a particular kind of modularity that is recognized through the rules of codon usage. To engineers, it is natural to think of biology's universally shared set of protocols as being more fundamental and invariant than the modules, such as individual genes, whose function these protocols enable and whose control and evolution they facilitate.

Other sets of protocols, or a protocol suite, are harder to describe as codes. Allostery, a huge suite of posttranslational modifications, and the rapid changes in location of macromolecular modules all enable adaptive responses to environmental signals or alterations on rapid time scales. Translational and transcriptional control and regulation of alternative splicing and editing act on somewhat longer time scales. On still longer time scales within and across generations, as highlighted in this book, the sequences of the DNA itself can change, not only through random mutation but also through highly structured and evolved mechanisms that facilitate the generation of adaptive diversity. All of these mechanisms follow specific rules that are becoming increasingly well understood.

The familiar example of lateral gene transfer in bacteria is possible because bacteria have a shared set of protocols, quite appropriately described as the "bacterial Internet".[4] As described in chapter 11, bacteria can simply grab DNA encoding new genes from other bacteria and incorporate them into their genome, just like consumers can plug a new computer into a home or office network. This "plug and play" modularity works because a shared set of protocols allows even novel genetic material to function in an entirely new cellular setting. Plug and play DNA mobility and expression are further facilitated by protocols involving, for example, integrons[5] or plasmids (chapter 7). This protocol-based organization allows bacteria to acquire antibiotic resistance on time scales that would be vanishingly improbable via point mutations, an example of how rapid evolution of complexity is possible by Darwinian mechanisms. Multiple examples in this book illustrate that natural selection can favor the evolution of whole protocol suites and their interactions, which in turn massively accelerates the acquisition and sharing of functional adaptive change.

Reverse Engineering of Protocols

The most explicit, though transient, features of the Internet are hardware and application software. The more permanent and fundamental hidden protocols are only implicit when viewed externally. A radical extension of this engineering perspective is that biological protocols, which may at first glance appear to be severe abstractions only tenuously connected with real biology, are in fact the most coherent and consistent locus of biological function. Thus protocols become an important substrate for evolution, and their appearance under natural selection is inevitable.

TCP/IP would be very hard to identify simply by examining all the hardware and software making up the Internet. However, once the architecture of the Internet is studied in detail, it is clear that TCP/IP is the essence of the Internet. It is far more vital and fundamental than the hardware and software that is its "code," interacting with other "codes" that facilitate additional functionality through TCP/IP. Even within engineering the understanding of what constitutes good protocols and how to design them, especially for new and novel problems, is on the frontiers of research.[6,7] Fortunately, biology and engineering together provide a diversity of examples that can help us create a collection of case studies to inform further research.

This book likely highlights just the tip of an iceberg. Ingenious and unexpected mechanisms, many described in this book and more to be discovered, are likely to be as systematic, pervasive, and protocol-based as more familiar forms of coding and regulation, giving additional flexibility, robustness, and evolvability to organisms and their successful lineages.

That protocols can be highly coded and compressed at the sequence level, however, makes their discovery challenging. They create functionality and flexibility that is only "implicit" in the genome, in the same way that the TCP/IP protocol suite is only "implicit" in the hardware and application software. The explicit genome codes for RNA and proteins. To gain a greater understanding of the functioning, and evolving genome, we must learn to translate the implicit genome into its critical protocol suites.

Developing a Shared Language

One immediate significance of this apparent convergence between biology and technology is that the types of mathematical (analytic) tools used in engineering complex systems are proving valuable for dissecting biological complexity.[1,6–11] With feedback between engineers and biologists, these tools may facilitate the further extraction of the protocols to increase understanding of how mechanisms are built up and interact with their environment and each other to create increasingly higher-level network functionality.

In bringing methods of mathematical analysis developed for engineering to bear on biological systems, we are encouraged by the fact that at the highest levels within the best-understood parts of biology, the global architecture of advanced technologies and biological systems are conspicuously similar.[11,12] For example, manufacturing of all sorts is facilitated by what we have described as a bowtie architecture.[13] An increasingly wide variety of raw materials can be drawn upon to construct a limited number of standardized building blocks. These building blocks then can be incorporated into a wide and evolving variety of complex appliances and machines using factory-based protocols. Similarly, apparently any sequence surrounded by the appropriate recombination signals can become part of an immunoglobulin variable region, a spirochete, or trypanosome coat protein, or can be moved by a transposase (chapters 9, 5, and 8, respectively). Some sequences contain information that marks them for regulated change, whether at the RNA level, during alternative splicing (chapter 15), or at the DNA level, such as in the immunoglobulin class switch (chapter 11). Under certain circumstances, such as the loss of mismatch repair (chapter 6), repetitive sequences throughout the genome (chapter 4) will increase the rate at which they sample the implied "species genome" (the term introduced in chapter 3).

Much as in advanced technologies, where viruses, worms, and parasites can be wildly disruptive, and even deadly if they hijack control protocols, biological protocols are vulnerable, reflected both in devastating diseases (e.g., immunodeficiency from the loss of a single enzyme involved in the generation of antibody diversity)[14] and destruction of ecologies sculpted as organisms coevolved.[13] This "robust yet fragile" nature is an inherent feature of evolved or designed complex systems.[13,15] From an engineering point of view, the complexity that has evolved under the pressure of natural selection is not merely reminiscent of today's complex technological "design" but also appears to be as much or more advanced than even our most sophisticated technologies. Thus, it seems likely that as the structure and protocols involved in regulation of cellular- and organism-level biology are better understood, biology is likely to inform engineering.[16–18]

An important take-home message from this book is that identification and description of protocols is likely to deepen our understanding of biological organization. Progress towards understanding could be furthered by an integrated scientific education for interested biologists and engineers who then can develop and share a common language. We anticipate that this book will inspire the development of new approaches that will enable us more fully to perceive how living organisms emerge from the diverse levels of recognition, interactions and control that are implied in a genome.

References

This list of references for the overview is only to supplement the more detailed information regarding the subjects discussed in the overview that can be found in the reference lists for the relevant chapters.

Overview

Epigraph: Wallace wrote these words after arriving in the South Pacific, but prior to his independent realization that evolution can proceed by survival of those who are "most fitted."

1. Nirenberg, M. W. and Matthaei, J. H. The dependence of cell-free protein synthesis in *E. coli* upon naturally occurring or synthetic polyribonucleotides. *Proc. Natl. Acad. Sci. U. S. A.* **47**(10): 1588–602 (1961).
2. Crick, F. H., Barnett, L., Brenner, S., and Watts-Tobin R. J. General nature of the genetic code for proteins. *Nature* **192**: 1227–32 (1961).
3. International Human Genome Sequencing Consortium. Finishing the euchromatic sequence of the human genome. *Nature* **431**: 931–45 (2004).
4. Levine M. and Davidson E. H. Gene Regulatory Networks Special Feature: Gene regulatory networks for development *Proc. Natl. Acad. Sci. U. S. A.* **102**: 4936–42 (2005).
5. Ambros, V. The functions of animal microRNAs. *Nature* **431**: 350–5 (2004).
6. Arber, W. Concluding remarks. *Ann. N.Y. Acad. Sci.* **870**: 344–345 (1998).
7. Gerhart, J. and Kirschner, M. *Cells Embryos and Evolution*. Blackwell (1997).
8. West-Eberhard, M. J. *Developmental Plasticity and Evolution*. Oxford University Press (2003).
9. Caporale, L. H. Is there a higher-level genetic code? *Mol. Cell. Biochem.* **64**: 5–13 (1984).
10. Liao G. C., Rehm E. J., and Rubin G. M. Insertion site preferences of the P transposable element in *Drosophila melanogaster*. *Proc. Natl. Acad. Sci. U. S. A.* **97**: 3347–51 (2000).
11. Kolter R. and Yanofsky, C. Genetic analysis of the tryptophan operon regulatory region using site-directed mutagenesis. *J. Mol. Biol.* **175**: 299–312 (1984).
12. Mandal, M. and Breaker R. R. Gene regulation by riboswitches. *Nat. Rev. Mol. Cell Biol.* **5**: 451–63 (2004).
13. Herbert, A. The four Rs of RNA-directed evolution. *Nature Genetics* **36**: 19–25 (2004).
14. van Noort V., Worning P., Ussery D. W., Rosche W. A., and Sinden R. R. Strand misalignments lead to quasipalindrome correction. *Trends Genet.* **19**: 365–9 (2003).

15. Rozen, S., Skaletsky, H., Marszalek, J. D., Minx, P. J., Cordum, H. S., Waterston, R. H., Wilson, R. K., and Pagem D. C. Abundant gene conversion between arms of palindromes in human and ape Y chromosomes. *Nature* **423**: 873–6 (2003).
16. Holbrook, S. R. RNA structure: the long and the short of it. *Curr. Opin. Struct. Biol.* **15**: 302–8 (2005).
17. Dickinson, W. J. and Seger, J. Cause and effect in evolution. *Nature* **399**: 30 (1999).
18. Stelling, J., Sauer, U., Szallasi, Z., Doyle F.J. III, and Doyle, J. Robustness of cellular function. *Cell* **118**: 675–85 (2004).
19. Olivera, B. M., Walker, C., Cartier, G. E., Hooper, D., Santos, A. D., Schoenfeld, R., Shetty, R., Watkins, M., Bandyopadhyay, P., and Hillyard, D. R. Speciation of cone snails and interspecific hyperdivergence of their venom peptides. Potential evolutionary significance of introns. *Ann. N. Y. Acad. Sci.* **870**: 223–37 (1999).
20. Darwin, C. *On the Origin of Species* (available online at http://www.talkorigins.org/faqs/origin.html).
21. Caporale, L. H. Mutation is modulated: implications for evolution. *Bioessays* **22**: 388–95 (2000).
22. Balakirev, E. S. and Ayala, F. J. Pseudogenes: are they "junk" or functional DNA? *Annu. Rev. Genet.* **37**: 123–51 (2003).
23. Sabarinadh, C., Subramanian, S., Tripathi, A., and Mishra, R. K. Extreme conservation of noncoding DNA near *HoxD* complex of vertebrates. *BMC Genomics* **5**: 75 (http://www.biomedcentral.com/1471-2164/5/75) (2004).
24. Arnheim, N., Calabrese, P., and Nordborg, M. Hot and cold spots of recombination in the human genome: the reason we should find them and how this can be achieved. *Am. J. Hum. Genet.* **73**: 5–16 (2003).
25. Watanabe, Y., Ikemura, T., and Sugimura, H. Amplicons on human chromosome 11q are located in the early/late-switch regions of replication timing. *Genomics,* **84**: 796–805 (2004).
26. Murphy, W.J., Larkin, D.M., Everts-van der Wind, A., Bourque, G., Tesler, G., Auvil, L., et al. Dynamics of mammalian chromosome evolution inferred from multispecies comparative maps. *Science* **309**: 613–617 (2005).
27. Caporale, L. H. Natural selection and the emergence of a mutation phenotype: an update of the evolutionary synthesis considering mechanisms that affect genome variation. *Annu. Rev. Microbiol.* **57**: 467–85 (2003).
28. Alberts, B., Johnson, A., Lewis, J., Raff, M., Roberts, K., and Walter, P. *Molecular Biology of the Cell,* 4th edn. Garland Science (2002).
29. Fedoroff, N. and Botstein, D. (eds). *The Dynamic Genome: Barbara McClintock's ideas in the century of genetics.* Cold Spring Harbor Laboratory Press, Cold Spring Harbor, NY (1992).

Chapter 1

1. Watson, J. D. and Crick, F. H. Molecular structure of nucleic acids; a structure for deoxyribose nucleic acid. *Nature* **171**: 737–8 (1953).
2. Wilkins, M. H. F., Stokes, A. R., and Wilson, H. R. Molecular structure of deoxypentose nucleic acids, *Nature* **171**: 738–40 (1953)
3. Franklin, R. E. and Gosling, R. G. Molecular configuration in sodium thymonucleate, *Nature* **171**: 740–1 (1953).
4. Arnott, S. Poly nucleotide secondary structures: an historical perspective, in S. Neidle (ed.), *Oxford Handbook of Nucleic Acid Structures*, pp. 1–38, Oxford University Press, Oxford (1999).

5. Marmur, J. and Doty, P. Heterogeneity in deoxyribonucleic acids. I. Dependence on composition of the configurational stability of deoxyribonucleic acids. *Nature* **183**: 1427–9 (1959).

6. Lavery, R. and Zarkrzewska, K. Base and base pair morphologies, helical parameters, and definitions, in S. Neidle (ed.), *Oxford Handbook of Nucleic Acid Structures*, pp. 39–76, Oxford University Press, Oxford (1999).

7. Olson, W. K. DNA higher order structures, in S. Neidle (ed.), *Oxford Handbook of Nucleic Acid Structures*, pp. 499–532, Oxford University Press, Oxford (1999).

8. Wang, E. and Feigon, J. Structures of nucleic acid triplexes, in S. Neidle (ed.), *Oxford Handbook of Nucleic Acid Structures*, pp. 355–88, Oxford University Press, Oxford (1999).

9. Patel, D. J., Bouazia, S., Kettani, A., and Wang, Y. Structures of guanine-rich and cytosine rich quadruplexes formed *in vitro* by telomeric, centromeric, and triplet disease DNA sequences, in S. Neidle (ed.), *Oxford Handbook of Nucleic Acid Structures*, pp. 389–454, Oxford University Press, Oxford (1999).

10. Lilly, D. M. J. Structures and interactions of helical junctions in nucleic acids, in S. Neidle (ed.), *Oxford Handbook of Nucleic Acid Structures*, pp. 471–498, Oxford University Press, Oxford (1999).

11. Neidle, S. (ed.), *Oxford Handbook of Nucleic Acid Structures*, Oxford University Press, Oxford (1999).

12. Berman, H. M., Olson, W. K., Beveridge, D. L., Westbrook, J., Gelbin, A., Demeny, T., Hsieh, S.-H., Srinivasan, A. R., and Schneider, B. The Nucleic Acid Database: a comprehensive relational database of three-dimensional structures of nucleic acids. *Biophys. J.* **63**: 751–9 (1992).

13. Berman, H. M., Zardecki, C., and Westbrook, J. The nucleic acid database: a research and teaching tool, in S. Neidle (ed.), *Oxford Handbook of Nucleic Acid Structures*, pp. 77–94, Oxford University Press, Oxford (1999).

14. Gorin, A. A., Zhurkin, V. B., and Olson, W. K. B-DNA twisting correlates with base-pair morphology. *J. Mol. Biol.* **247**: 34–48 (1995).

15. Zhurkin, V. B., Lysov, Y. P., and Ivanov, V. I. Anisotropic flexibility of DNA and the nucleosomal structure. *Nucleic Acids Res.* **6**: 1081–96 (1979).

16. Trifonov, E. N. Curved DNA. *CRC Crit. Rev. Biochem.* **19**: 89–106 (1985).

17. Schellman, J. A. and Harvey, S. C. Static contributions to the persistence length of DNA and dynamic contributions to DNA curvature. *Biophys. Chem.* **55**: 95–114 (1995).

18. Bednar, J., Furrer, P., Katritch, V., Stasiak, A. Z., Dubochet, J., and Stasiak A. Determination of DNA persistence length by cryoelectron microscopy. Separation of the static and dynamic contributions to the apparent persistence length of DNA. *J. Mol. Biol.* **254**: 579–94 (1995).

19. Marini, J. C., Levene, S. D., Crothers, D. M., and Englund, P. T. Bent helical structure in kinetoplast DNA. *Proc. Nat. Acad. Sci. U.S.A.* **79**: 7664–8 (1982).

20. Wu, H.-M. and Crothers, D. M. The locus of sequence-directed and protein-induced DNA bending. *Nature* **308**: 509–13 (1984).

21. Perez-Martin, J. and de Lorenzo, V. Clues and consequences of DNA bending in transcription. *Annu. Rev. Microbiol.* **51**: 593–628 (1997).

22. Gartenberg, M. R. and Crothers, D. M. Synthetic DNA bending sequences increase the rate of *in vitro* transcription initiation at the *Escherichia coli lac* promoter *J. Mol. Biol.* **219**: 217–30 (1991).

23. Bracco, L., Kotlarz, D., Kolb, A., Diekmann, S., and Buc, H. Synthetic curved DNA sequences can act as transcriptional activators in *Escherichia coli*. *EMBO J.* **8**: 4289–96 (1989).

24. Koo, H. S., Wu, H.-M and Crothers, D. M. DNA bending at adenine-thymine tracts. *Nature* **320**: 501–6 (1986).

25. Matsumoto, A. and Olson, W. K. Sequence-dependent motions of DNA: a normal mode analysis at the base pair level. *Biophys. J.* **83**: 22–41 (2002).
26. Crothers, D. M. and Shakked, Z. DNA bending by adenine-thymine tracts, in S. Neidle (ed.), *Oxford Handbook of Nucleic Acid Structures*, pp. 455–69, Oxford University Press, Oxford (1999).
27. Beutel, B. A. and Gold, L. *In vitro* evolution of intrinsically bent DNA. *J. Mol. Biol.* **228**: 803–12 (1992).
28. Goodsell, D. S., Kopka, M. L., Cascio, D., and Dickerson, D. Crystal structure of CATGGCCATG and its implications for A-tract bending models. *Proc. Nat. Acad. Sci. U.S.A.* **90**: 2930–4 (1993).
29. Dickerson, R. E., Goodsell, D., and Kopka, M. L. MPD and DNA bending in crystals and in solution. *J. Mol. Biol.* **256**: 108–25 (1996).
30. Brukner, I., Susic, S., Dlakic, M., Savic, A., and Pongor, S. Physiological concentration of magnesium ions induces a strong macroscopic curvature in GGGCCC-containing DNA. *J. Mol. Biol.* **236**: 26–32 (1994).
31. Zinkel, S. S. and Crothers, D. M. DNA bend direction by phase sensitive detection. *Nature* **328**: 178–81 (1987).
32. Calladine, C. R., Drew, H. R., and McCall, M. J. The intrinsic curvature of DNA in solution. *J. Mol. Biol.* **201**: 127–37 (1988).
33. Ulanovsky, L. E., Bodner, M., Trifonov, E. N., and Choder, M. Curved DNA: design, synthesis, and circularization. *Proc. Natl. Acad. Sci. U.S.A.* **83**: 862–6 (1986).
34. Levene, S. D., Wu, H.-M., and Crothers, D. M. bending and flexibility of kinetoplast DNA. *Biochemistry* **25**: 3988–96 (1986).
35. Koo, H.-S., Drak, J., Rice, J. A., and Crothers, D. M. Determination of the extent of DNA bending by an adenine-thymine tract. *Biochemistry* **29**: 4227–34 (1990).
36. Haran, T. E., Kahn, J. D., and Crothers, D. M. Sequence elements responsible for DNA curvature. *J. Mol. Biol.* **244**: 135–43 (1994).
37. Marini, J. C., Effron, P. N., Goodman, T. C., Singleton, C. K., Wells, R. D., Wartell, R. M. and Englund, P. T. Physical characterization of a kinetoplast DNA fragment with unusual properties. *J. Biol. Chem.* **259**: 8974–9 (1984).
38. Diekmann, S. Temperature and salt dependence of the gel migration anomaly of curved DNA fragments. *Nucl. Acids Res.* **15**: 247–65 (1987).
39. Arnott, S. and Selsing, E. Structures for the polynucleotide complexes poly(dA) with poly(dT) and poly(dT) with poly(dA) with poly(dT). *J. Mol. Biol.* **88**: 509–21 (1974).
40. Alexeev, D. G., Lipanov, A. A., and Skuratovskii, I. Y. Poly(dA) poly(dT) is a B-type double helix with a distinctively narrow minor groove, *Nature* **325**: 821–3 (1987).
41. Alexeev, D. G., Lipanov, A. A., and Skuratovskii, I. Y. The structure of poly(dA): Epoly(dT) as revealed by an X-ray fibre diffraction. *J. Biomol. Struct. Dynam.* **4**: 989–1011 (1987).
42. Aymami, J., Coll, M., Frederick, C. A., Wang, A. H., and Rich, A. The propeller DNA conformation of poly(dA).poly(dT). *Nucleic Acids Res.* **17**: 3229–45 (1989).
43. Burkhoff, A. M. and Tullius, T. D. The unusual conformation adopted by the adenine tracts in kinetoplast DNA. *Cell* **48**: 935–43 (1987).
44. Hagerman, P. J. Sequence-directed curvature of DNA. *Nature* **321**: 449–50 (1986).
45. Nelson, H. C. M., Finch, J. T., Luisi, B. F., and Klug, A. The structure of an oligo(dA)-oligo(dT) tract and its biological implications. *Nature* **330**: 221–6 (1987).
46. Shatzky-Schwartz, M., Arbuckle, N. D., Eisenstein, M., Rabinovich, D., Bareket-Samish, A., Haran, T. E., Luisi, B. F., and Shakked, Z. X-ray and solution studies of DNA oligomers and implications for the structural basis of A-tract-dependent curvature. *J. Mol. Biol.* **267**: 595–623 (1997).

47. Chan, S. S., Austin, R. H., Mukerji, I., and Spiro, T. G. Temperature-dependent ultraviolet resonance Raman spectroscopy of the premelting state of dA.dT DNA. *Biophys. J.* **72**: 1512–20 (1997).

48. Zhurkin, V. B. (1985) Sequence-dependent bending of DNA and phasing of nucleosomes. *J. Biomol. Struct. Dynam.* **2**: 785–804 (1985).

49. Maroun, R. C. and Olson, W. K. Base sequence effects in double helical DNA. III. Curved DNA. *Biopolymers* **27**: 585–603 (1988).

50. Hizver, J., Rozenberg, H., Frolow, F., Rabinovich, D., and Shakked, Z. DNA bending by an adenine–thymine tract and its role in gene regulation. *Proc. Natl. Acad. Sci. U. S. A.* **98**: 8490–5 (2001).

51. Koo, H. S. and Crothers D. M. Calibration of DNA curvature and a unified description of sequence-directed bending. *Proc. Natl. Acad. Sci. U. S. A.* **85**: 1763–7 (1988).

52. Barbic, A., Zimmer, D. P., and Crothers, D. M.. Structural origins of adenine-tract bending. *Proc. Natl. Acad. Sci. U. S. A.* **100**: 2369–73 (2003).

53. MacDonald, D., Herbert, K., Zhang, X., Polgruto, T., and Lu, P. Solution structure of an A-tract DNA bend. *J. Mol. Biol.* **306**: 1081–98 (2001).

54. Merling, A., Sagaydakova, N., and Haran, T. E. A-tract polarity dominates the curvature in flanking sequences. *Biochemistry* **42**: 4978–84 (2003).

55. Haran, T. E. and Crothers, D. M. Cooperativity in A-tract structure and bending properties of composite TnAn blocks. *Biochemistry* **28**: 2763–7 (1989).

56. Shore, D., Langowski, J., and Baldwin, R. L. DNA flexibility studied by covalent closure of short fragments into circles. *Proc. Natl. Acad. Sci. U. S. A.* **78**: 4833–7 (1981).

57. Shore, D. and Baldwin, R. L. Energetics of DNA twisting. I. Relation between twist and cyclization probability. *J. Mol. Biol.* **170**: 957–81 (1983).

58. Jacobson, H. and Stockmayer, W. H. Intramolecular reaction in polycondensations. I. The theory of linear systems. *J. Chem. Phys.* **18**: 1600–1606 (1950).

59. Crothers, D. M., Drak, J., Kahn, J. D., and Levene, S. D. DNA bending, flexibility and helical repeat by cyclization kinetics. *Methods Enzymol.* **212**: 3–71 (1992).

60. Roychoudhury, M., Sitlani, A., Lapham, J., and Crothers, D. M. Global structure and mechanical properties of a 10-bp nucleosome positioning motif. *Proc. Natl. Acad. Sci. U. S. A.* **97**: 13608–13 (2000).

61. Widlund, H. R., Cao, H., Simonsson, S., Magnusson, E., Simonsson, T., Nielsen, P. E., Kahn, J. D., Crothers, D. M., and Kubista, M. Identification and characterization of genomic nucleosome-positioning sequences. *J. Mol. Biol.* **267**: 807–18 (1997).

62. Zhang, Y. and Crothers, D. M. Statistical mechanics of sequence-dependent circular DNA and its application for DNA cyclization. *Biophys. J.* **84**: 136–53 (2003).

63. Hogan, M. E. and Austin, R. H. Importance of DNA stiffness in protein-DNA binding specificity. *Nature* **329**: 263–6 (1987).

64. Zhang, Y., Xi, Z., Hegde, R. S., Shakked, Z., and Crothers, D. M. Predicting indirect readout effects in protein-DNA interactions. *Proc. Natl. Acad. Sci. U. S. A.* **101**: 8337–41 (2004).

65. Anselmi, C., Bocchinfuso, G., De Santis, P., Savino, M., and Scipioni, A. A theoretical model for the prediction of sequence-dependent nucleosome thermodynamic stability. *Biophys. J.* **79**: 601–13 (2000).

66. Virstedt, J., Berge, T., Henderson, R. M., Waring, M. J., and Travers, A. A. The influence of DNA stiffness upon nucleosome formation. *J. Struct. Biol.* **148**: 66–85 (2004).

67. Cloutier, T. E. and Widom, J. Spontaneous sharp bending of double-stranded DNA. *Mol. Cell.* **14**: 355–62 (2004).

68. Isaacs, R. J. and Spielmann, H. P. NMR evidence for mechanical coupling of phosphate B(I)-B(II) transitions with deoxyribose conformational exchange in DNA. *J. Mol. Biol.* **311**: 149–60 (2001).

69. Lynch, T. W, Read, E. K., Mattis, A. N., Gardner, J. F., and Rice, P. A. Integration host factor: putting a twist on protein-DNA recognition. *J. Mol. Biol.* **330**: 493–502 (2003).

70. Koudelka, G. B., Harrison, S. C., and Ptashne, M. Effect of non-contacted bases on the affinity of 434 operator for 434 repressor and Cro. *Nature* **326**: 886–8 (1987).

71. Kim, S. S., Tam, J. K., Wang, A. F., and Hegde, R. S. The structural basis of DNA target discrimination by papillomavirus E2 proteins. *J. Biol. Chem.* **275**: 31245–54 (2000).

72. Mauro, S. A., Pawlowski, D., and Koudelka, G. B. The role of the minor groove substituents in indirect readout of DNA sequence by 434 repressor. *J. Biol. Chem.* **278**: 12955–60 (2003).

Chapter 2

1. Friedberg, E. C., Walker, G., and Siede, W. *DNA Repair and Mutagenesis*, ASM Press, Washington, DC (1995).

2. Friedberg, E. C., Walker, G. C., Siede, W., Wood, R. D., Schultz, R. A. and Ellenberger, T. *DNA Repair and Mutagenesis and Other Cellular Responses to DNA Damage*, ASM Press, Washington DC (in press) (2005).

3. Michaels, M. L. and Miller, J. H. The GO system protects organisms from the mutagenic effect of the spontaneous lesion 8-hydroxyguanine (7, 8-dihydro-8-oxoguanine). *J. Bacteriol.* **174**: 6321–5 (1992).

4. Maki, H. and Sekiguchi, M. MutT protein specifically hydrolyses a potent mutagenic substrate for DNA synthesis. *Nature* **355**: 273–5 (1992).

5. Mo, J. Y., Maki, H., and Sekiguchi, M. Hydrolytic elimination of a mutagenic nucleotide, 8-oxodGTP, by human 18-kilodalton protein: sanitization of nucleotide pool. *Proc. Natl. Acad. Sci. U. S. A.* **89**: 11021–5 (1992).

6. Sakumi, K., Fuichi, M., Tsuzuki, T., Kakuma, T., Kawabata, S., Maki, H., and Sekiguchi, M. Cloning and expression of cDNA of a human enzyme that hydrolyzes 8-oxo-dGTP, a mutagenic substrate for DNA synthesis. *J. Biol. Chem.* **268**: 23524–30 (1993).

7. Kornberg, A. and Baker, T. A. *DNA Replication* W, H. Freeman, New York, NY (1991).

8. Echols, H. and Goodman, M. F. Fidelity mechanisms in DNA replication. *Annu. Rev. Biochem.* **60**: 477–511 (1991).

9. Kunkel, T. A. and Bebenek, K. DNA replication fidelity. *Annu. Rev. Biochem.* **69**: 497–529 (2000).

10. Sawaya, M. R., Prasad, R., Wilson, S. H., Kraut, J, and Pelletier, H. Crystal structures of human DNA polymerase beta complexed with gapped and nicked DNA: evidence for an induced fit mechanism. *Biochemistry* **36**: 11205–15 (1997).

11. Wong, I., Patel, S. S., and Johnson, K. A. An induced-fit kinetic mechanism for DNA replication fidelity: direct measurement by single-turnover kinetics. *Biochemistry* **30**: 526–37 (1991).

12. Johnson, K. A. Conformational coupling in DNA polymerase fidelity. *Annu. Rev. Biochem.* **62**: 685–713 (1993).

13. Petruska, J., Sowers, L. C., and Goodman, M. F. Comparison of nucleotide interactions in water, proteins, and vacuum: model for DNA polymerase fidelity. *Proc. Natl. Acad. Sci. U. S. A.* **83**: 1559–62 (1986).

14. Lindahl, T. Instability and decay of the primary structure of DNA. *Nature* **362**: 709–15 (1993).

15. Lindahl, T. An N-glycosidase from *Escherichia coli* that releases free uracil from DNA containing deaminated cytosine residues. *Proc. Natl. Acad. Sci. U. S. A.* **71**: 3649–53 (1974).

16. Kunkel, T. A. DNA replication fidelity. *J. Biol. Chem.* **267**: 18251–18254 (1992).

17. Kunkel, T. A. Misalignment mediated DNA synthesis errors. *Biochemistry* **29**: 8003–8011 (1990).

18. Papanicolaou, C. and Ripley, L. S. An *in vitro* approach to identifying specificity determinants of mutagenesis mediated by DNA misalignments. *J. Mol. Biol.* **221**: 805–821 (1991).

19. Friedberg, E. C., Wagner, R., and Radman, M. Specialized DNA polymerases, cellular survival, and the genesis of mutations. *Science* **296**: 1627–1630 (2002).

20. Goodman, M. F. Error-prone repair DNA polymerases in prokaryotes and eukaryotes. *Annu. Rev. Biochem.* **71**: 17–50 (2002).

21. Plosky, B. S. and Woodgate, R. Switching from high-fidelity to low-fidelity lesion-bypass polymerases. *Curr. Opin. Genet. Dev.* **14**: 113–119 (2004).

22. Woodgate, R. A plethora of lesion-replicating DNA polymerases. *Genes Dev.* **13**: 2191–2195 (1999).

23. Rajagopalan, M., Lu, C., Woodgate, R., O'Donnell, M., Goodman, M. F, and Echols, H. Activity of the purified mutagenesis proteins UmuC, UmuD′, and RecA in replicative bypass of an abasic DNA lesion by DNA polymerase III. *Proc. Natl. Acad. Sci. U. S. A.* **89**: 10777–10781 (1992).

24. Friedberg, E. C., Fischhaber, P. L., and Kisker, C. Error-prone DNA polymerases: unexpected structures and the benefits of infidelity. *Cell* **107**: 9–12 (2001).

25. Lingh, H., Boudsocq, F., Woodgate, R., and Yang, W. Snapshots of replication through an abasic lesion; structural basis for base substitutions and frameshifts. *Mol. Cell.* **12**: 751–62 (2004).

26. Lingh, H., Boudsocq, F., Plosky, B. S., Woodgate, R., and Yang, W. Replication of a *cis-syn* thymine dimer at atomic resolution. *Nature* **424**: 1083–1087 (2003).

27. Elledge, S. J. and Walker, G. C. Proteins required for ultraviolet light and chemical mutagenesis: Identification of the products of the *umuC* locus of *Escherichia coli*. *J. Mol. Biol.* **164**: 175–192 (1983).

28. Shinagawa, H., Kato, T., Ise, T., Makino, K., and Nakata, A. Cloning and characterization of the *umu* operon responsible for inducible mutagenesis in *Escherichia coli*. *Gene* **23**: 167–174 (1983).

29. Masutani, C., Kusumoto, R., Yamada, A., Dohmae, N., Yokoi, M., Yuasa, M, Araki, M., Iwai, S., Takio K., and Hanaoka, F. The XPV (xeroderma pigmentosum variant) gene encodes human DNA polymerase eta. *Nature* **399**: 700–704 (1999).

30. Masutani, C., Kusumoto, R., Yamada, A., Yuasa, M., Araki, M., Nogimori, T., Yokoi, M., Eki, T., Iwai S., and Hanaoka, F. Xeroderma pigmentosum variant: from a human genetic disorder to a novel DNA polymerase. *Cold Spring Harb. Symp. Quant. Biol.* **65**: 71–80 (2000).

31. Sancar, A. DNA excision repair. *Annu. Rev. Biochem.* **65**: 43–81 (1996).

32. Wood, R. D. DNA repair in eukaryotes. *Annu. Rev. Biochem.* **65**: 135–167 (1996).

33. Guo, C., Fischhaber, P. L., Luk-Paszyc, M., Masuda, Y., Zhou, J., Kisker, K., and Friedberg. E. C. Mouse Rev1 protein interacts with multiple DNA polymerases involved in translesion DNA synthesis. *EMBO J.* **22**: 6621–6630 (2003).

34. Nelson, J. R., Lawrence, C. W., and Hinkle, D. C. Deoxycytidyl transferase activity of yeast REV1 protein. *Nature* **382**: 729–731 (1996).

35. Lawrence, C. W. Cellular roles of DNA polymerase zeta and Rev1 protein. *DNA Repair* **1**: 425–435 (2002).

36. Kannouche, P. L., Wing, J., and Lehmann, A. R. Interaction of human DNA polymerase h with monoubiquitinated PCNA: a possible mechanism for the polymerase switch in response to DNA damage. *Mol. Cell* **14**: 491–500 (2004).

37. Lebecque, S.G. and Gearhart, P.J. Boundaries of somatic mutation in rearranged immunoglobulin genes: 5′ boundary is near the promoter, and 3′ boundary is approximately 1 kb from V(D)J gene. *J. Exp. Med.* **172**: 1717–1727 (1990).

38. Muramatsu, M., Kinoshita, K., Fagarasan, S., Yamada, S., Shinkai, Y., and Honjo, T. Class switch recombination and hypermutation require activation-induced cytidine deaminase (AID), a potential RNA editing enzyme. *Cell* **102**: 553–63 (2000).

39. Martin, A., Bardwell, P. D., Woo, C. J., Fan, M., Shulman, M. J., and Scharff, M. D. Activation-induced cytidine deaminase turns on somatic hypermutation in hybridomas. *Nature* **415**: 802–6 (2002).

40. Martin, A. and Scharff, M.D. Somatic hypermutation of the AID transgene in B and non-B cells. *Proc. Natl. Acad. Sci. U. S. A.* **99**: 12304–8 (2002).

41. Yoshikawa, K., Okazaki, I.M., Eto, T., Kinoshita, K., Muramatsu, M., Nagaoka, H., and Honjo, T. AID enzyme-induced hypermutation in an actively transcribed gene in fibroblasts. *Science* **296**: 2033–6 (2002).

42. Caporale, L. H. Natural selection and the emergence of a mutator phenotype: an update of the evolutionary synthesis considering mechanisms that affect genome variation. *Annu. Rev. Microbiol.* **57**: 467–485 (2003).

43. Glassner, B. J., Rasmussen, L. J., Najarian, M. T., Posnick L. M., and Samson, L. D. Generation of a strong mutator phenotype in yeast by imbalanced base excision repair. *Proc. Natl. Acad. Sci U. S. A.* **95**: 9997–10002 (1998).

44. Witkin, E. Ultraviolet mutagenesis and the SOS response in *Escherichia coli*: a personal perspective. *Environ. Mol. Mutagen.* **14**: Suppl. **16**, 30–34 (1989).

Chapter 3

1. Caporale, L. H. Natural selection and the emergence of a mutation phenotype: an update of the evolutionary synthesis considering mechanisms that affect genome variation. *Annu. Rev. Microbiol* **57**: 467–85 (2003).

2. Fleischmann, R. D., Alland, D., Eisen, J. A., Carpenter, L., White, O., Peterson, J., DeBoy, R., Dodson, R., Gwinn, M., Haft, D., Hickey, E., Kolonay, J. F., Nelson, W. C., Umayam, L. A., Ermolaeva, M., Salzberg, S. L., Delcher, A., Utterback, T., Weidman, J., Khouri, H., Gill, J., Mikula, A., Bishai, W., Jacobs Jr, W. R., Venter, J. C., and Fraser, C. M. Whole-genome comparison of *Mycobacterium tuberculosis* clinical and laboratory strains. *J. Bacteriol.* **184**: 5479–90 (2002).

3. Lindsay, J. A. and Holden, M. T. *Staphylococcus aureus*: superbug, super genome? *Trends Microbiol.* **12**: 378–85 (2004).

4. Dawkins, R. and Krebs, J. R. Arms races within and between species. *Proc. R. Soc. Lond. [Biol.]* **205**: 489–511 (1979).

5. Van Valen, L. A new evolutionary law. *Evol. Theory* **1**: 1–30 (1973).

6. Tonjum, T. and Seeberg, E. Microbial fitness and genome dynamics. *Trends Microbiol.* **9**: 356–8 (2001).

7. Brookes, M. Day of the mutators. *New Scientist* **157**: 38 (1998).

8. Anderson, R. M. and May, R. M. The invasion, persistence and spread of infectious diseases within animal and plant communities. *Philos. Trans. R. Soc. Lond. B Biol. Sci.* **314**: 533–70 (1986).

9. Nomura, M. and Benzer, S. The nature of the "deletion" mutants in the rII region of phage T4. *J. Mol. Biol.* **3**: 684–92 (1961).

10. Moxon, E. R., Rainey, P. B., Nowak, M. A., and Lenski, R. Adaptive evolution of highly mutable loci in pathogenic bacteria. *Curr. Biol.* **4**: 24–33 (1994).

11. Barry, J. D., Ginger, M. L., Burton, P., and McCulloch, R. Why are parasite contingency genes often associated with telomeres? *Int. J. Parasitol.* **33**: 29–45 (2003).

12. Kashi, Y., King, D., and Soller, M. Simple sequence repeats as a source of quantitative genetic variation. *Trends Genet.* **13**: 74–8 (1997).

13. Koch, A. L. Catastrophe and what to do about it if you are a bacterium: the importance of frameshift mutants. *Crit. Rev. Microbiol.* **30**: 1–6 (2004).

14. Ritz, D., Lim, J., Reynolds, C. M., Poole, L. B., and Beckwith, J. Conversion of a peroxiredoxin into a disulfide reductase by a triplet repeat expansion. *Science* **294**: 158–60 (2001).

15. Hopwood, D. A. The *Streptomyces* genome—be prepared! *Nature Biotechnol.* **21**: 505–6 (2003).

16. Farabaugh, P. J., Schmeissner, U., Hofer, M., and Miller, J. H. Genetic studies of the lac repressor. VII. On the molecular nature of spontaneous hotspots in the lacI gene of *Escherichia coli. J. Mol. Biol.* **126**: 847–57 (1978).

17. Bennett, P. Demystified ... Microsatellites. *Mol. Pathol.* **53**: 177–83 (2000).

18. Gur-Arie, R., Cohen, C. J., Eitan, Y., Shelef, L., Hallerman, E. M., and Kashi, Y. Simple sequence repeats in *Escherichia coli*: abundance, distribution, composition, and polymorphism. *Genome Res.* **10**: 62–71 (2000).

19. Karlin, S., Mrazek, J., and Campbell, A. M. Compositional biases of bacterial genomes and evolutionary implications. *J. Bacteriol.* **179**: 3899–3913 (1997).

20. Bayliss, C. D., Field, D., and Moxon, E. R. The simple sequence contingency loci of *Haemophilus influenzae* and *Neisseria meningitidis. J. Clin. Invest.* **107**: 657–662 (2001).

21. Hood, D. W., Deadman, M. E., Jennings, M. P., Bisercic, M., Fleischmann, R. D., Venter, J. C., and Moxon, E. R. DNA repeats identify novel virulence genes in *Haemophilus influenzae. Proc. Natl. Acad. Sci. U. S. A.* **93**: 11121–11125 (1996).

22. Saunders, N. J., Jeffries, A. C., Peden, J. F., Hood, D. W., Tettelin, H., Rappuoli, R., and Moxon, E. R. Repeat-associated phase variable genes in the complete genome sequence of *Neisseria meningitidis* strain MC58. *Mol. Microbiol.* **37**: 207–215 (2000).

23. Martin, P., van de Ven, T., Mouchel, N., Jeffries, A. C., Hood, D. W., and Moxon E. R. Experimentally revised repertoire of putative contingency loci in *Neisseria meningitidis* strain MC58: evidence for a novel mechanism of phase variation. *Mol. Microbiol.* **50**: 245–57 (2003).

24. Wegrzyn, G. and Thomas, M. S. Modulation of the susceptibility of intestinal bacteria to bacteriophages in response to Ag43 phase variation – a hypothesis. *Med. Sci. Monit.* **8**: HY15–8 (2002).

25. Bayliss, C. D., Dixon, K. M., and Moxon, E. R. Simple sequence repeats (microsatellites): mutational mechanisms and contributions to bacterial pathogenesis. A meeting review. *FEMS Immunol. Med. Microbiol.* **40**: 11–9 (2004).

26. Nassif, X., Pujol, C., Morand, P., and Eugene, E. Interactions of pathogenic *Neisseria* with host cells. Is it possible to assemble the puzzle? *Mol. Microbiol.* **32**: 1124–1132 (1999).

27. Hammerschmidt, S., Muller, A., Sillmann, H., Muhlenhoff, M., Borrow, R., Fox, A., van Putten, J., Zollinger, W. D., Gerardy-Schahn, R., and Frosch, M. Capsule phase variation in *Neisseria meningitidis* serogroup B by slipped-strand mispairing in the

polysialytransferase gene (*siaD*): correlation with bacterial invasion and the outbreak of meningococcal disease. *Mol. Microbiol.* **20**: 1211–1220 (1996).

28. Hauck, C. R. and Meyer, T. F. "Small" talk: Opa proteins as mediators of *Neisseria*-host-cell communication. *Curr. Opin. Microbiol.* **6**: 43–9 (2003).

29. Flick, K. and Chen, Q. var genes, PfEMP1 and the human host. *Mol. Biochem. Parasitol.* **134**: 3–9 (2004).

30. Snyder, L. A. S., Butcher, S. A., and Saunders, N. J. Comparative whole-genome analyses reveal over 100 putative phase-variable genes in the pathogenic *Neisseria* spp. *Microbiology* **147**: 2321–2332 (2001).

31. Hood, D. W. and Moxon, E. R., Lipopolysaccharide phase variation in *Haemophilus* and *Neisseria*, in H. Brade, S. M. Opal, S. N. Vogel, and D. C. Morrison (eds.), *Endotoxin in Health and Disease*, pp. 39–54, Marcel Dekker Inc., New York (1999).

32. Wang, G., Ge, Z., Rasko, D. A., and Taylor, D. E. Lewis antigens in *Helicobacter pylori*: biosynthesis and phase variation. *Mol. Microbiol.* **36**: 1187–96 (2000).

33. Hood, D. W., Makepeace, K., Deadman, M. E., Rest, R. F., Thibault, P., Martin, A., Richards, J. C., and Moxon, E. R. Sialic acid in the lipopolysaccharide of *Haemophilus influenzae*: strain distribution, influence on serum resistance and structural characterization. *Mol. Microbiol.* **33**: 679–692 (1999).

34. Lysenko, E., Richards, J. C., Cox, A. D., Stewart, A., Martin, A., Kapoor, M., and Weiser, J. N. The position of phosphorylcholine on the lipopolysaccharide of *Haemophilus influenzae* affects binding and sensitivity to C-reactive protein mediated killing. *Mol. Microbiol.* **35**: 234–245 (2000).

35. Lysenko, E. S., Gould, J., Bals, R., Wilson, J. M., and Weiser, J. N. Bacterial phosphorylcholine decreases susceptibility to the antimicrobial peptide LL-37/hCAP18 expressed in the upper respiratory tract. *Infect. Immunol.* **68**: 1664–1671 (2000).

36. Srikhanta Y. N., Maguire, T. L., Stacey, K. J., Grimmond, S. M., and Jennings, M. P. The phasevarion: a genetic system controlling coordinated, random switching of expression of multiple genes. *Proc. Natl. Acad. Sci. U. S. A.* **102**: 5547–51. (2005)

37. Gumulak-Smith, J., Teachman, A., Tu, A. H., Simecka, J. W., Lindsey, J. R., and Dybvig, K. Variations in the surface proteins and restriction enzyme systems of *Mycoplasma pulmonis* in the respiratory tract of infected rats. *Mol. Microbiol.* **40**: 1037–44 (2001).

38. Aras, R. A., Small, A. J., Ando, T., and Blaser, M. J. *Helicobacter pylori* interstrain restriction-modification diversity prevents genome subversion by chromosomal DNA from competing strains. *Nucleic Acids Res.* **30**: 5391–7 (2002).

39. Jablonska, E., Kauc, L., and Piekarowicz, A. An *Haemophilus influenzae* mutant which inhibits the growth of HP1c1 phage. *Mol. Gen. Genet.* **139**: 157–66 (1975).

40. Raleigh, E. A. Restricting parasites: why are there so many restriction endonucleases? *NEB transcript.* **9**: 6–7 (1998).

41. Corn, P. G., Anders, J. A., Takala, A. K., Kayhty, H., and Hoiseth, S. K. Genes involved in *Haemophilus influenzae* type b capsule expression are frequently amplified. *J. Infect. Dis.* **167**: 356–364 (1993).

42. van Ham, S. M., van Alphen, L., Mooi, F. R., and van Putten, J. P. Phase variation of *H.influenzae* fimbriae: transcriptional control of two divergent genes through a variable combined promoter region. *Cell* **73**: 1187–1196 (1993).

43. Martin, P., Makepeace, K., Hill, S. A., Hood, D. W., Moxon, E. R. Microsatellite instability regulates transcription factor binding and gene expression. *Proc. Natl. Acad. Sci. U. S. A.* **102**: 3800–4. (2005)

44. Liu, L., Panangala, V. S., and Dybvig, K. Trinucleotide GAA repeats dictate pMGA gene expression in *Mycoplasma gallisepticum* by affecting spacing between flanking regions. *J. Bacteriol.* **184**: 1335–9 (2002).

45. De Bolle, X., Bayliss, C. D., Field, D., van de Ven, T., Saunders, N. J., Hood, D. W., and Moxon E. R. The length of a tetranucleotide repeat tract in *Haemophilus influenzae* determines the phase variation rate of a gene with homology to type III DNA methyltransferases. *Mol. Microbiol.* **35**: 211–222 (2000).

46. Richardson, A. R., Yu, Z., Popovic, T., and Stojiljkovic, I. Mutator clones of *Neisseria meningitidis* in epidemic serogroup A disease. *Proc. Natl. Acad. Sci. U. S. A.* **99**: 6103–6107 (2002).

47. Bayliss, C. D., van de Ven, T., and Moxon, E. R. Mutations in *polI* but not *mutSLH* destabilize *Haemophilus influenzae* tetranucleotide repeats. *EMBO J.* **21**: 1465–1476 (2002).

48. Bayliss, C. D., Sweetman, W. A., and Moxon, E. D. Tetranucleotide repeats are destabilised in *Haemophilus influenzae* mutants lacking RnaseHI and the Klenow domain of PolI. *Nucleic Acids Res.* **33**: 400–408 (2005).

49. Martin, P., Sun, L., Hood, D. W., and Moxon, E. R. Involvement of genes of genome maintenance in the regulation of phase variation frequencies in *Neisseria meningitidis. Microbiology* **150**: 3001–12 (2004).

50. Morel, P., Reverdy, C., Michel, B., Ehrlich, S. D., and Cassuto, E. The role of SOS and flap processing in microsatellite instability in *Escherichia coli. Proc. Natl. Acad. Sci. U. S. A.* **95**: 10003–10008 (1998).

51. Yamamura, E., Nunoshiba, T., Nohmi, T., and Yamamoto, K. Hydrogen peroxide-induced microsatellite instability in the *Escherichia coli* K-12 endogenous tonB gene. *Biochem. Biophys. Res. Commun.* **306**: 570–6 (2003).

52. Schumacher, S., Pinet, I., and Bichara, M. Modulation of transcription reveals a new mechanism of triplet repeat instability in *Escherichia coli. J. Mol. Biol.* **307**: 39–49 (2001).

53. Bayliss, C. D., Sweetman, W. A., and Moxon, E. R. Mutations in *Haemophilus influenzae* mismatch repair genes increase mutation rates of dinucleotide repeat tracts but not dinucleotide repeat-driven pilin phase variation rates. *J. Bacteriol.* **186**: 2928–35 (2004).

54. Blomfield, I. C. DNA rearrangements and regulation of gene expression, in D. A. Hodgson and C. M. Thomas (eds.) *Signals, Switches, Regulons and Cascades: control of bacterial gene expression*, pp. 57–72 Cambridge University Press, Cambridge (2002).

55. El-Labany, S., Sohanpal, B. K., Lahooti, M., Akerman, R., and Blomfield, I. C. Distant cis-active sequences and sialic acid control the expression of fimB in *Escherichia coli* K-12. *Mol. Microbiol.* **49**: 1109–18 (2003).

56. Sniegowski, P. D., Gerrish, P. J., Johnson, T., and Shaver, A. The evolution of mutation rates: separating causes from consequences. *Bioessays* **22**: 1057–66 (2000).

57. Saunders, N. J., Moxon, E. R., and Gravenor, M. B. Mutation rates: estimating phase variation rates when fitness differences are present and their impact on population structure. *Microbiology* **149**: 485–95 (2003).

58. Webb, G. F. and Blaser, M. J. Dynamics of bacterial phenotype selection in a colonized host. *Proc. Natl. Acad. Sci. U. S. A.* **99**: 3135–3140 (2002).

59. Leigh Jr, E. G. Natural selection and mutability. *American Naturalist* **104**: 301–335 (1970).

Chapter 4

1. Nikitina, T. V. and Nazarenko, S. A. Human microsatellites: mutation and evolution. *Russ. J. Genet.* **40**: 1065–1079 (2004).

2. International Human Genome Sequencing Consortium. Initial sequencing and analysis of the human genome. *Nature* **409**: 860–921 (2001).

3. Rockman, M. V. and Wray, G. A. Abundant raw material for *cis*-regulatory evolution in humans. *Mol. Biol. Evol.* **19**: 1991–2004 (2002).

4. Chambers, G. K. and MacAvoy, E. S. Microsatellites: consensus and controversy. *Comp. Biochem. Physiol.* B **126**: 455–476 (2000).

5. Tautz, D. and Schlötterer, C. Simple sequences. *Curr. Opin. Genet. Dev.* **4**: 832–837 (1994).

6. Edwards, Y. J. K., Elgar, G., Clark, M. S., and Bishop, M. J. The identification and characterization of microsatellites in the compact genome of the Japanese pufferfish, *Fugu rubripes*: Perspectives in functional and comparative genomic analyses. *J. Mol. Biol.* **278**: 843–854 (1998).

7. Gur-Arie, R., Cohen, C. J., Eitan, Y., Shelef, L., Hallerman, E. M., and Kashi, Y. Simple sequence repeats in *Escherichia coli*: Abundance, distribution, composition, and polymorphism. *Genome Res.* **10**: 62–71 (2000).

8. Tóth, G., Gáspári, Z., and Jurka, J. Microsatellites in different eukaryotic genomes: Survey and analysis. *Genome Res.* **10**: 967–981 (2000).

9. Young, E. T., Sloan, J. S., and Van Riper, K. Trinucleotide repeats are clustered in regulatory genes in *Saccharomyces cerevisiae*. *Genetics* **154**: 1053–1068 (2000).

10. Gentles, A. J. and Karlin, S. Genome-scale compositional comparisons in eukaryotes. *Genome Res.* **11**: 540–546 (2001).

11. Katti, M. V., Ranjekar, P. K., and Gupta, V. S. Differential distribution of simple sequence repeats in eukaryotic genome sequences. *Mol. Biol. Evol.* **18**: 1161–1167 (2001).

12. Hancock, J. M. Genome size and the accumulation of simple sequence repeats: implications of new data from genome sequencing projects. *Genetica* **115**: 93–103 (2002).

13. Subramanian, S., Mishra, R. K., and Singh, L. Genome-wide analysis of microsatellite repeats in humans: their abundance and density in specific genomic regions. *Genome Biol.* **4**: R13, online at http://genomebiology.com/2003/4/2/R13 (2003).

14. Eichinger, L. et al. The genome of the social amoeba *Dictyostelium discoideum*. *Nature* **435**: 43–57 (2005).

15. Künzler, P., Matsuo, K., and Schaffner, W. Pathological, physiological, and evolutionary aspects of short unstable DNA repeats in the human genome. *Biol. Chem. Hoppe Seyler* **376**: 201–211 (1995).

16. Kashi, Y., King, D. G., and Soller, M. Simple sequence repeats as a source of quantitative genetic variation. *Trends Genet.* **13**: 74–78 (1997).

17. Nakamura, Y., Koyama, K., and Matsushima, M. VNTR (variable number of tandem repeat) sequences as transcriptional, translational, or functional regulators. *J. Hum. Genet.* **43**: 149–152 (1998).

18. Kashi, Y. and Soller, M. Functional roles of microsatellites and minisatellites, in D. B. Goldstein and C. Schlötterer (eds.), *Microsatellites Evolution and Applications* pp. 10–23, Oxford University Press, Oxford (1999).

19. King, D. G. and Soller, M. in Wasser, S. P. (ed.), Variation and fidelity: The evolution of simple sequence repeats as functional elements in adjustable genes, *Evolutionary Theory and Processes: Modern Perspectives*, pp. 65–82, Kluwer Academic Publishers, Dordrecht (1999).

20. Pisarchik, A. V. and Kartel, N. A. Simple repetitive sequences and gene expression. *Mol. Biol.* **34**: 303–307 (2000).

21. Karlin, S., Brocchieri, L., Bergman, A., Mrázek, J., and Gentles, A. J. Amino acid runs in eukaryotic proteomes and disease associations. *Proc. Natl. Acad. Sci. U. S. A.* **99**: 333–338 (2002).

22. Li, Y.-C., Korol, A. B., Fahima, T., Beiles, A., and Nevo, E. Microsatellites: genomic distribution, putative functions and mutational mechanisms: a review. *Mol. Ecol.* **11**: 2453–2465 (2002).

23. Li, Y.-C., Korol, A. B., Fahima, T., and Nevo, E. Microsatellites within genes: Structure, function, and evolution. *Mol. Biol. Evol.* **21**: 991–1007 (2004).

24. Queller, D. C., Strassmann, J. E., and Hughes, C. R. Microsatellites and kinship. *Trends Ecol. Evol.* **8**: 285–288 (1993).

25. Goldstein, D. B. and Schlötterer, C. (eds.), *Microsatellites Evolution and Applications*, Oxford University Press, Oxford (1999).

26. Cohen, H., Danin-Poleg, Y., Cohen, C. J., Sprecher, E., Darvasi, A., and Kashi, Y. Mono-nucleotide repeats (MNRs): A neglected polymorphism for generating high density genetic maps *in silico*. *Hum. Genet.* **115**: 213–220 (2004).

27. Rubinsztein, D. C., Trinucleotide expression mutations cause diseases which do not conform to classical Mendelian expectations, in D. B. Goldstein and C. Schlötterer (eds.), *Microsatellites Evolution and Applications*, pp. 80–97, Oxford University Press, Oxford (1999).

28. Cummings, C. J. and Zoghbi, H. Y. Fourteen and counting: unravelling trinucleotide repeat diseases. *Hum. Mol. Genet.* **9**: 906–916 (2000).

29. Brown, L. Y. and Brown, S. A. Alanine tracts: the expanding story of human illness and trinucleotide repeats. *Trends Genet.* **20**: 51–58 (2004).

30. Ranum, L. P. W. and Day, J. W. Pathogenic RNA repeats: an expanding role in genetic disease. *Trends Genet.* **20**: 506–512 (2004).

31. Morell, V. The puzzle of the triple repeats. *Science* **260**: 1422–1423 (1993).

32. Drake, J. W., Glickman, B. W., and Ripley, L. S. Updating the theory of mutation. *Am. Sci.* **71**: 621–630 (1983).

33. Hamada, H., Seidman, M., Howard, B. H., and Gorman, C. M. Enhanced gene expression by the poly(dT-dG) · poly(dC-dA) sequence. *Mol. Cellular Biol.* **4**: 2622–2630 (1984).

34. Trifonov, E. N. The multiple codes of nucleotide sequences. *Bull. Math. Biol.* **51**: 417–432 (1989).

35. Gerber, H.-P., Seipel, K., Georgiev, O., Höfferer, M., Hug, M., Rusconi, S., and Schaffner, W. Transcriptional activation modulated by homopolymeric glutamine and proline stretches. *Science* **263**: 808–811 (1994).

36. Karlin, S. and Burge, C. Trinucleotide repeats and long homopeptides in genes and proteins associated with nervous system disease and development. *Proc. Natl. Acad. Sci. U. S. A.* **93**: 1560–1565 (1996).

37. Lavoie, H., Debeane, F., Trinh, Q.-D., Turcotte, J.-F., Corbeil-Girard, L.-P., Dicaire, M.-J., Saint-Denis, A., Pagé, M., Rouleau, G. A., and Brais, B. Polymorphism, shared functions and convergent evolution of genes with sequences coding for polyalanine domains. *Hum. Mol. Genet.* **12**: 2967–2979 (2003).

38. Rothenburg, S., Koch-Nolte, F., and Haag, F. DNA methylation and Z-DNA formation as mediators of quantitative differences in the expression of alleles. *Immunol. Rev.* **184**: 286–298 (2001).

39. Godde, J. S. and Wolffe, A. P. Nucleosome assembly on CTG triplet repeats. *J. Biol. Chem.* **271**: 15222–15229 (1996).

40. Cao, H., Widlund, H. R., Simonsson, T., and Kubista, M. TGGA repeats impair nucleosome formation. *J. Mol. Biol.* **281**: 253–260 (1998).

41. Nelson, H. C. M., Finch, J. T., Luisi, B. F., and Klug, A. The structure of an oligo(dA)·oligo(dT) tract and its biological implications. *Nature* **330**: 221–226 (1987).

42. Morgante, M., Hanafey, M., and Powell, W. Microsatellites are preferentially associated with nonrepetitive DNA in plant genomes. *Nature Genet.* **30**: 194–200 (2002).

43. Yeap, B. B., Wilce, J. A., and Leedman, P. J. The androgen receptor mRNA. *Bioessays* **26**: 672–682 (2004).

44. Bayliss, C. D., Field, D., and Moxon, E. R. The simple sequence contingency loci of *Haemophilus influenzae* and *Neisseria meningitidis*. *J. Clin. Invest.* **107**: 657–662 (2001).

45. Strauss, B. S. Frameshift mutation, microsatellites and mismatch repair. *Mutat. Res.* **437**: 195–203 (1999).

46. Gebhardt, F., Zänker, K. S., and Brandt, B. Modulation of epidermal growth factor receptor gene transcription by a polymorphic dinucleotide repeat in intron 1. *J. Biol. Chem.* **274**: 13176–13180 (1999).

47. Albanèse, V., Biguet, N. F., Kiefer, H., Bayard, E., Mallet, J., and Meloni, R. Quantitative effects on gene silencing by allelic variation at a tetranucleotide microsatellite. *Hum. Mol. Genet.* **10**: 1785–1792 (2001).

48. Tian, B., White, R. J., Xia, T., Welle, S., Turner, D. H., Mathews, M. B., and Thornton, C. A. Expanded CUG repeat RNAs form hairpins that activate the double-stranded RNA-dependent protein kinase PKR. *RNA* **6**: 79–87 (2000).

49. Riley, D. E. and Krieger, J. N. Simple repeat replacements support similar functions of distinct repeats in inter-species mRNA homologs. *Gene* **328**: 17–24 (2004).

50. Krasilnikova, M. M. and Mirkin, S. M. Replication stalling at Friedreich's ataxia $(GAA)_n$ repeats in vivo. *Mol. Cellular Biol.* **24**: 2286–2295 (2004).

51. Majewski, J. and Ott, J. GT repeats are associated with recombination on human chromosome 22. *Genome Res.* **10**: 1108–1114 (2000).

52. Gendrel, C.-G., Boulet, A., and Dutreix, M. $(CA/GT)_n$ microsatellites affect homologous recombination during yeast meiosis. *Genes Dev.* **14**: 1261–1268 (2000).

53. Nadir, E., Margalit, H., Gallily, T., and Ben-Sasson, S. A. Microsatellite spreading in the human genome: Evolutionary mechanisms and structural implications. *Proc. Natl. Acad. Sci. U. S. A.* **93**: 6470–6475 (1996).

54. Ramsay, L., Macaulay, M., Cardle, L., Morgante, M., degli Ivanissevich, S., Maestri, E., Powell, W., and Waugh, R. Intimate association of microsatellite repeats with retrotransposons and other dispersed repetitive elements in barley. *Plant J.* **17**: 415–425 (1999).

55. Chang, D. K., Metzgar, D., Wills, C., and Boland, C. R. Microsatellites in the eukaryotic DNA mismatch repair genes as modulators of evolutionary mutation rate. *Genome Res.* **11**: 1145–1146 (2001).

56. Rosenberg, S. M. Evolving responsively: adaptive mutation. *Nat. Rev. Genet.* **2**: 504–515 (2001).

57. Schmidt, A. L. and Mitter, V. Microsatellite mutation directed by an external stimulus. *Mutat. Res.* **568**: 233–243 (2004).

58. Kijas, J. M. H., Moller, M., Plastow, G., and Andersson, L. A frameshift mutation in *MC1R* and a high frequency of somatic reversions cause black spotting in pigs. *Genetics* **158**: 779–785 (2001).

59. Comings, D. E. Polygenic inheritance and micro/minisatellites. *Mol. Psychiatry* **3**: 21–31 (1998).

60. Kühn, C., Thaller, G., Winter, A., Bininda-Emonds, O. R. P., Kaupe, B., Erhardt, G., Bennewitz, J., Schwerin, M., and Fries, R. Evidence for multiple alleles at the *DGAT1* locus better explains a quantitative trait locus with major effect on milk fat content in cattle. *Genetics* **167**: 1873–1881 (2004).

61. Hale, C. S., Herring, W. O., Shibuya, H., Lucy, M. C., Lubahn, D. B., Keisler, D. H., and Johnson, G. S. Decreased growth in Angus steers with a short TG-microsatellite allele in the P1 promoter of the growth hormone receptor gene. *J. Anim. Sci.* **78**: 2099–2104 (2000).

62. Streelman, J. T. and Kocher, T. D. Microsatellite variation associated with prolactin expression and growth of salt-challenged tilapia. *Physiol. Genomics* **9**: 1–4 (2002).

63. Sawyer, L. A., Hennessy, J. M., Peixoto, A. A., Rosato, E., Parkinson, H., Costa, R., and Kyriacou, C. P. Natural variation in a *Drosophila* clock gene and temperature compensation. *Science* **278**: 2117–2120 (1997).

64. Hammock, E. A. D. and Young, L. J. Microsatellite instability generates diversity in brain and sociobehavioral traits. *Science* **308**: 1630–1634 (2005).

65. Fondon III, J. W. and Garner, H. R. Molecular origins of rapid and continuous morphological evolution. *Proc. Natl. Acad. Sci. U. S. A.* **101**: 18058–18063 (2004).

66. Reif, A. and Lesch, K.-P. Toward a molecular architecture of personality. *Behav. Brain Res.* **139**: 1–20 (2003).

67. Comings, D. E., Muhleman, D., Johnson, J. P., and MacMurray, J. P. Parent-daughter transmission of the androgen receptor gene as an explanation of the effect of father absence on age of menarche. *Child Dev.* **73**: 1046–1051 (2002).

68. Choong, C. S., Kemppainen, J. A., and Wilson, E. M. Evolution of the primate androgen receptor: A structural basis for disease. *J. Mol. Evol.* **47**: 334–342 (1998).

69. Amos, W., A comparative approach to the study of microsatellite evolution, in D. B. Goldstein and C. Schlötterer (eds.), *Microsatellites Evolution and Applications*, pp. 66–79, Oxford University Press, Oxford (1999).

70. Tan, Q., Bellizzi, D., Rose, G., Garasto, S., Franceschi, C., Kruse, T., Vaupel, J. W., De Benedictis, G., and Yashin, A. I. The influences on human longevity by HUMTHO1.STR polymorphism (Tyrosine Hydroxylase gene): A relative risk approach. *Mech. Ageing Dev.* **123**: 1403–1410 (2002).

71. Caspi, A., Sugden, K., Moffitt, T. E., Taylor, A., Craig, I. W., Harrington, H., McClay, J., Mill, J., Martin, J., Braithwaite, A., and Poulton, R. Influence of life stress on depression: Moderation by a polymorphism in the 5-HTT gene. *Science* **301**: 386–389 (2003).

72. Trefilov, A., Berard, J., Krawczak, M., and Schmidtke, J. Natal dispersal in rhesus macaques is related to serotonin transporter gene promoter variation. *Behav. Genet.* **30**: 295–301 (2000).

73. Rockman, M. V., Hahn, M. W., Soranzo, N., Loisel, D. A., Goldstein, D. B., and Wray, G. A. Positive selection on *MMP3* regulation has shaped heart disease risk. *Curr. Biol.* **14**: 1531–1539 (2004).

74. Hancock, J. M. Microsatellites and other simple sequences: genomic context and mutational mechanisms, in D. B. Goldstein and C. Schlötterer (eds.), *Microsatellites Evolution and Applications*, pp. 1–9, Oxford University Press, Oxford (1999).

75. Primmer, C. R., Saino, N., Møller, A. P., and Ellegren, H. Unravelling the processes of microsatellite evolution through analysis of germ line mutations in barn swallows *Hirundo rustica*. *Mol. Biol. Evol.* **15**: 1047–1054 (1998).

76. Bacon, A. L., Dunlop, M. G., and Farrington, S. M. Hypermutability at a poly(A/T) tract in the human germline. *Nucleic Acids Res.* **29**: 4405–4413 (2001).

77. Armour, J. A. L., Alegre, S. A., Miles, S., Williams, L. J., and Badge, R. M. Minisatellites and mutation processes in tandemly repetitive DNA, in D. B. Goldstein and C. Schlötterer (eds.), *Microsatellites Evolution and Applications*, pp. 24–33, Oxford University Press, Oxford (1999).

78. Eisen, J. A. Mechanistic basis for microsatellite instability, in D. B. Goldstein and C. Schlötterer (eds.), *Microsatellites Evolution and Applications*, pp. 34–48, Oxford University Press, Oxford (1999).

79. Richard, G.-F. and Pâques, F. Mini- and microsatellite expansions: the recombination connection. *EMBO Rep.* **1**: 122–126 (2000).

80. Vergnaud, G. and Denoeud, F. Minisatellites: Mutability and genome architecture. *Genome Res.* **10**: 899–907 (2000).

81. Bois, P. R. J. Hypermutable minisatellites, a human affair? *Genomics* **81**: 349–355 (2003).

82. Nichol, K. and Pearson, C. E. CpG methylation modifies the genetic stability of cloned repeat sequences. *Genome Res.* **12**: 1246–1256 (2002).

83. Bayliss, C. D., van de Ven, T., and Moxon, E. R. Mutations in *poll* but not *mutSLH* destabilize *Haemophilus influenzae* tetranucleotide repeats. *EMBO J.* **21**: 1465–1476 (2002).

84. Webster, M. T., Smith, N. G. C., and Ellegren, H. Microsatellite evolution inferred from human-chimpanzee genomic sequence alignments. *Proc. Natl. Acad. Sci. U. S. A.* **99**: 8748–8753 (2002).

85. Calabrese, P. and Durrett, R. Dinucleotide repeats in the Drosophila and human genomes have complex, length-dependent mutation processes. *Mol. Biol. Evol.* **20**: 715–725 (2003).

86. Trifonov, E. N. Elucidating sequence codes: Three codes for evolution. *Ann. N. Y. Acad. Sci.* **870**: 330–338 (1999).

87. Caporale, L. H. Mutation is modulated: implications for evolution. *Bioessays* **22**: 388–395 (2000).

88. Laken, S. J., Petersen, G. M., Gruber, S. B., Oddoux, C., Ostrer, H., Giardiello, F. M., Hamilton, S. R., Hampel, H., Markowitz, A., Klimstra, D., Jhanwar, S., Winawer, S., Offit, K., Luce, M. C., Kinzler, K. W., and Vogelstein, B. Familial colorectal cancer in Ashkenazim due to a hypermutable tract in *APC*. *Nature Genet.* **17**: 79–83 (1997).

89. Zhu, Y., Strassmann, J. E., and Queller, D. C. Insertions, substitutions, and the origin of microsatellites. *Genet. Res.* **76**: 227–236 (2000).

90. Taylor, J. S. and Breden, F. Slipped-strand mispairing at noncontiguous repeats in *Poecilia reticulata*: A model for minisatellite birth. *Genetics* **155**: 1313–1320 (2000).

91. Boán, F., Blanco, M. G., Quinteiro, J., Mouriño, S., and Gómez-Márquez, J. Birth and evolutionary history of a human minisatellite. *Mol. Biol. Evol.* **21**: 228–235 (2004).

92. Wilder, J. and Hollocher, H. Mobile elements and the genesis of microsatellites in dipterans. *Mol. Biol. Evol.* **18**: 384–392 (2001).

93. Zhang, D.-X. Lepidopteran microsatellite DNA: redundant but promising. *Trends Ecol. Evol.* **19**: 507–509 (2004).

94. Wilke, K., Jung, M., Chen, Y., and Geldermann, H. Porcine $(GT)_n$ sequences: Structure and association with dispersed and tandem repeats. *Genomics* **21**: 63–70 (1994).

95. Arcot, S. S., Wang, Z., Weber, J. L., Deininger, P. L., and Batzer, M. A. *Alu* repeats: A source for the genesis of primate microsatellites. *Genomics* **29**: 136–144 (1995).

96. Yandava, C. N., Gastier, J. M., Pulido, J. C., Brody, T., Sheffield, V., Murray, J., Buetow, K., and Duyk, G. M. Characterization of *Alu* repeats that are associated with trinucleotide and tetranucleotide repeat microsatellites. *Genome Res.* **7**: 716–724 (1997).

97. Sharma, V. K., B-Rao, C., Sharma, A., Brahmachari, S. K., and Ramachandran, S. $(TG/CA)_n$ repeats in human housekeeping genes. *J. Biomol. Str. Dyn.* **21**: 303–310 (2003).

98. Katti, M. V., Sami-Subbu, R., Ranjekar, P. K., and Gupta, V. S. Amino acid repeat patterns in protein sequences: Their diversity and structural-functional implications. *Protein Sci.* **9**: 1203–1209 (2000).

99. Tompa, P. Intrinsically unstructured proteins evolve by repeat expansion. *Bioessays* **25**: 847–855 (2003).

100. Sturtevant, A. H. Essays on evolution. I. On the effects of selection on mutation rate. *Q. Rev. Biol.* **12**: 464–467 (1937).

101. Williams, G. *Adaptation and Natural Selection*, Princeton University Press, Princeton (1966).

102. Dickinson, W. J. and Seger, J. Cause and effect in evolution. *Nature* **399**: 30 (1999).

103. Sniegowski, P. D., Gerrish, P. J., Johnson, T., and Shaver, A. The evolution of mutation rates: Separating causes from consequences. *Bioessays* **22**: 1057–1066 (2000).

104. Fisher, R. A. *The Genetical Theory of Natural Selection*, Oxford University Press, Oxford (1930).

105. Crow, J. F. and Kimura, M. *An Introduction to Population Genetics Theory*, Harper and Row, New York (1970).

106. Orr, H. A. The population genetics of adaptation: The distribution of factors fixed during adaptive evolution. *Evolution* **52**: 935–949 (1998).

107. King, D. G., Soller, M., and Kashi, Y. Evolutionary tuning knobs. *Endeavour* **21**: 36–40 (1997).

108. Trifonov, E. N. The tuning function of tandemly-repeating sequences: A molecular device for fast adaptation, in S. P. Wasser (ed.) *Evolutionary Theory and Processes: Modern Horizons*, pp. 115–138, Kluwer Academic Publishers, Dordrecht (2004).

109. Moxon, E. R. and Wills, C. DNA microsatellites: Agents of evolution? *Sci. Am.* **280**(1): 94–99 (1999).

110. Endler, J. A. *Natural Selection in the Wild*, Princeton University Press, Princeton (1986).

111. Dufresne, F., Bourget, E., and Bernatchez, L. Differential patterns of spatial divergence in microsatellite and allozyme alleles: further evidence for locus-specific selection in the acorn barnacle, *Semibalanus balanoides? Mol. Ecol.* **11**: 113–123 (2002).

112. Nevo, E., Beharav, A., Meyer, R. C., Hackett, C. A., Forster, B. P., Russell, J. R., and Powell, W. Genomic microsatellite adaptive divergence of wild barley by microclimatic stress in 'Evolution Canyon', Israel. *Biol. J. Linn. Soc.* **84**: 205–224 (2005).

113. Fahima, T., Röder, M. S., Wendehake, K., Kirzhner, V. M., and Nevo, E. Microsatellite polymorphism in natural populations of wild emmer wheat, *Triticum dicoccoides*, in Israel. *Theor. Appl. Genet.* **104**: 17–29 (2002).

114. Levins, R. *Evolution in Changing Environments*, Princeton University Press, Princeton (1968).

115. Weber, M., Blot, M., and Arber, W. On the origin of genetic diversity. *Gaia* **4**: 191–198 (1995).

116. Harr, B., Zangerl, B., and Schlötterer, C. Removal of microsatellite interruptions by DNA replication slippage: Phylogenetic evidence from *Drosophila. Mol. Biol. Evol.* **17**: 1001–1009 (2000).

117. Metzgar, D., Bytof, J., and Wills, C. Selection against frameshift mutations limits microsatellite expansion in coding DNA. *Genome Res.* **10**: 72–80 (2000).

118. King, D. G. Modeling selection for adjustable genes based on simple sequence repeats. *Ann. N. Y. Acad. Sci.* **870**: 396–399 (1999).

119. Albà, M. M., Santibáñez-Koref, M. F., and Hancock, J. M. Conservation of polyglutamine tract size between mice and humans depends on codon interruption. *Mol. Biol. Evol.* **16**: 1641–1644 (1999).

120. Tei, H., Okamura, H., Shigeyoshi, Y., Fukuhara, C., Ozawa, R., Hirose, M., and Sakaki, Y. Circadian oscillation of a mammalian homologue of the *Drosophila period* gene. *Nature* **389**: 512–516 (1997).

121. Galant, R. and Carroll, S. B. Evolution of a transcriptional repression domain in an insect Hox protein. *Nature* **415**: 910–913 (2002).

122. Wray, G. A., Hahn, M. W., Abouheif, E., Balhoff, J. P., Pizer, M., Rockman, M. V., and Romano, L. A. The evolution of transcriptional regulation in eukaryotes. *Mol. Biol. Evol.* **20**: 1377–1419 (2003).

123. West-Eberhard, M. J. Evolution in the light of developmental and cell biology, and *vice versa. Proc. Natl. Acad. Sci. U. S. A.* **95**: 8417–8419 (1998).

124. Kirschner, M. and Gerhart, J. Evolvability. *Proc. Natl. Acad. Sci. U. S. A.* **95**: 8420–8427 (1998).

Chapter 5

1. Moxon, E. R., Rainey, P. B., Nowak, M. A., and Lenski, R. E. Adaptive evolution of highly mutable loci in pathogenic bacteria. *Curr. Biol.* **4**: 24–33 (1994).
2. Augusto-Pinto, L., Teixeira, S. M., Pena, S. D., and Machado, C. R. Single-nucleotide polymorphisms of the *Trypanosoma cruzi* MSH2 gene support the existence of three phylogenetic lineages presenting differences in mismatch-repair efficiency. *Genetics* **164**: 117–126 (2003).
3. Deitsch, K. W., Moxon, E. R., and Wellems, T. E. Shared themes of antigenic variation and virulence in bacterial, protozoal, and fungal infections. *Microbiol. Mol. Biol. Rev.* **61**: 281–294 (1997).
4. Barry, J. D., Ginger, M. L., Burton, P., and McCulloch, R. Why are parasite contingency genes often associated with telomeres? *Int. J. Parasitol* **33**: 29–45 (2003).
5. Borst, P. Mechanisms of antigenic variation: an overview, in A. Craig and A. Scherf (eds.), *Antigenic Variation*, pp. 1–15, Academic Press, Amsterdam (2003).
6. Barry, J. D. and McCulloch, R. Antigenic variation in trypanosomes: Enhanced phenotypic variation in a eukaryotic parasite. *Adv. Parasitol.* **49**: 1–70 (2001). www.vsgdb.org
7. Mehlert, A., Bond, C. S., and Ferguson, M. A. The glycoforms of a *Trypanosoma brucei* variant surface glycoprotein and molecular modeling of a glycosylated surface coat. *Glycobiology* **12**: 607–612 (2002).
8. Turner, C. M. R. and Barry, J. D. High frequency of antigenic variation in *Trypanosoma brucei rhodesiense* infections. *Parasitology* **99**: 67–75 (1989).
9. Miller, E. N., Allan, L. M., and Turner, M. J. Topological analysis of antigenic determinants on a variant surface glycoprotein of *Trypanosoma brucei*. *Mol. Biochem. Parasitol.* **13**: 67–81 (1984).
10. Hsia, R. C., Beals, T., and Boothroyd, J. C. Use of chimeric recombinant polypeptides to analyze conformational, surface epitopes on trypanosome variant surface glycoproteins. *Mol. Microbiol.* **19**: 53–63 (1996).
11. Barry, J. D. Antigenic variation during *Trypanosoma vivax* infections of different host species. *Parasitology* **92**: 51–65 (1986).
12. Barry, J. D., Graham, S.V., Fotheringham, M., Graham,V.S., Kobryn, K., and Wymer, B. VSG gene control and infectivity strategy of metacyclic stage *Trypanosoma brucei*. *Mol. Biochem. Parasitol.* **91**: 93–105 (1998).
13. Wain-Hobson, S. HIV variation – a question of signal-to-noise, in A. Craig and A. Scherf (eds.), *Antigenic Variation*, pp. 16–32, Academic Press, Amsterdam (2003).
14. Oxford, J., Eswarasaran, R., Mann, A., and Lambkin, R. Influenza – the chameleon virus, in A. Craig and A. Scherf (eds.), *Antigenic Variation*, pp. 52–83, Academic Press, Amsterdam (2003).
15. Melville, S. E., Leech, V., Navarro, M., and Cross, G. A. M. The molecular karyotype of the megabase chromosomes of *Trypanosoma brucei* stock 427. *Mol. Biochem. Parasitol.* **111**: 261–273 (2000).
16. Dickin, S. K. and Gibson, W. C. Hybridization with a repetitive DNA probe reveals the presence of small chromosomes in *Trypanosoma vivax*. *Mol. Biochem. Parasitol.* **33**: 135–142 (1989).
17. Ersfeld, K. and Gull, K. Partitioning of large and minichromosomes in *Trypanosoma brucei*. *Science* **276**: 611–614 (1997).
18. Gull, K., Alsford, S., and Ersfeld, K. Segregation of minichromosomes in trypanosomes: implications for mitotic mechanisms. *Trends Microbiol.* **6**: 319–323 (1998).

19. Aline, R., Macdonald, G., Brown, E., Allison, J., Myler, P., Rothwell, V., and Stuart, K. (TAA)$_n$ within sequences flanking several intrachromosomal variant surface glycoprotein genes in *Typanosoma brucei*. *Nucleic Acids Res.* **13**: 3161–3177 (1985).

20. Michels, P. A. M., Liu, A. Y. C., Bernards, A., Sloof, P., Vanderbijl, M. M. W., Schinkel, A. H., Menke, H. H., Borst, P., Veeneman, G. H., Tromp, M. C., and Vanboom, J. H. Activation of the genes for variant surface glycoprotein-117 and glycoprotein–118 in *Trypanosoma brucei*. *J. Mol. Biol.* **166**: 537–556 (1983).

21. Pays, E., Lips, S., Nolan, D., Vanhamme, L., and Perez-Morga, D. The VSG expression sites of *Trypanosoma brucei*: multipurpose tools for the adaptation of the parasite to mammalian hosts. *Mol. Biochem. Parasitol.* **114**: 1–16 (2001).

22. Borst, P. and Ulbert, S. Control of VSG gene expression sites. *Mol. Biochem. Parasitol.* **114**: 17–27 (2001).

23. Gunzl, A., Bruderer, T., Laufer, G., Schimanski, B., Tu, L. C., Chung, H. M., Lee, P. T., and Lee, M. G. RNA polymerase I transcribes procyclin genes and variant surface glycoprotein gene expression sites in *Trypanosoma brucei*. *Eukaryot. Cell* **2**: 542–551 (2003).

24. Robinson, N. P., Burman, N., Melville, S. E., and Barry, J. D. Predominance of duplicative *VSG* gene conversion in antigenic variation in African trypanosomes. *Mol. Cell. Biol.* **19**: 5839–5846 (1999).

25. Borst, P. Antigenic variation and allelic exclusion. *Cell* **109**: 5–8 (2002).

26. Navarro, M. and Gull, K. A pol I transcriptional body associated with VSG monoallelic expression in *Trypanosoma brucei*. *Nature* **414**: 759–763 (2001).

27. Ulbert, S., Chaves, I., and Borst, P. Expression site activation in *Trypanosoma brucei* with three marked variant surface glycoprotein gene expression sites. *Mol. Biochem. Parasitol.* **120**: 225–235 (2002).

28. Gerrits, H., Mussmann, R., Bitter, W., Kieft, R, and Borst, P. The physiological significance of transferrin receptor variations in *Trypanosoma brucei*. *Mol. Biochem. Parasitol.* **119**: 237–247 (2002).

29. Barry, J.D. and McCulloch, R. Trypanosome antigenic variation – a heavy investment in the evasion of immunity, in A. Craig and A. Scherf (eds.), *Antigenic Variation*, pp. 224–242, Academic Press, Amsterdam (2003).

30. Kim, K. S. and Donelson, J. E. Co-duplication of a variant surface glycoprotein gene and its promoter to an expression site in African trypanosomes. *J. Biol. Chem.* **272**: 24637–24645 (1997).

31. Ginger, M. L., Blundell, P. A., Lewis, A. M., Browitt, A., Gunzl, A., and Barry, J. D. *Ex vivo* and *in vitro* identification of a consensus promoter for *VSG* genes expressed by metacyclic-stage trypanosomes in the tsetse fly. *Euk. Cell* **1**: 1000–1009 (2002).

32. Graham, S. V., Terry, S., and Barry, J. D. A structural and transcription pattern for variant surface glycoprotein gene expression sites used in metacyclic stage *Trypanosoma brucei*. *Mol. Biochem. Parasitol.* **103**: 141–154 (1999).

33. Bringaud, F., Biteau, N., Donelson, J. E., and Baltz, T. Conservation of metacyclic variant surface glycoprotein expression sites among different trypanosome isolates. *Mol. Biochem. Parasitol.* **113**: 67–78 (2001).

34. de Lange, T., Kooter, J. M., Michels, P. A. M., and Borst, P. Telomere conversion in trypanosomes. *Nucleic Acids Res.* **11**: 8149–8165 (1983).

35. McCulloch, R. and Barry, J. D. A role for RAD51 and homologous recombination in *Trypanosoma brucei* antigenic variation. *Genes Dev.* **13**: 2875–2888 (1999).

36. Pays, E. Gene conversion in trypanosome antigenic variation. *Prog. Nucleic Acid Res. Mol. Biol.* **32**: 1–26 (1985).

37. Barry, J. D. The relative significance of mechanisms of antigenic variation in African trypanosomes. *Parasitol. Today* **13**: 212–218 (1997).

38. Liu, A. Y. C., Michels, P. A. M., Bernards, A., and Borst, P. Trypanosome variant surface glycoprotein genes expressed early in infection. *J. Mol. Biol.* **182**: 383–396 (1985).

39. Timmers, H. T. M., de Lange, T., Kooter, J. M., and Borst, P. Coincident multiple activations of the same surface antigen gene in *Trypanosoma brucei*. *J. Mol. Biol.* **194**: 81–90 (1987).

40. Laroche, T., Martin, S. G., Tsai-Pflugfelder, M., and Gasser, S. M. The dynamics of yeast telomeres and silencing proteins through the cell cycle. *J. Struct. Biol.* **129**: 159–174 (2000).

41. Pays, E. Pseudogenes, chimaeric genes and the timing of antigen variation in african trypanosomes. *Trends Genet.* **5**: 389–391 (1989).

42. Thon, G., Baltz, T., Giroud, C., and Eisen, H. Trypanosome variable surface glycoproteins: composite genes and order of expression. *Genes Dev.* **4**: 1374–1383 (1990).

43. Kamper, S. M. and Barbet, A. F. Surface epitope variation via mosaic gene formation is potential key to long-term survival of *Trypanosoma brucei*. *Mol Biochem. Parasitol.* **53**: 33–44 (1992).

44. Meyer, T. F. and Hill, S. A. Genetic variation in the pathogenic *Neisseria* species, in A. Craig and A. Scherf (eds.), *Antigenic Variation*, pp. 142–164, Academic Press, Amsterdam (2003).

45. Howell-Adams, B. and Seifert, H. S. Molecular models accounting for the gene conversion reactions mediating gonococcal pilin antigenic variation. *Mol. Microbiol.* **37**: 1146–1158 (2000).

46. Haas, R., Veit, S., and Meyer, T. F. Silent pilin genes of *Neisseria gonorrhoeae* MS11 and the occurrence of related hypervariant sequences among other gonococcal isolates. *Mol. Microbiol* **6**: 197–208 (1992).

47. Hamrick, T. S., Dempsey, J. A., Cohen, M. S., and Cannon, J. G. Antigenic variation of gonococcal pilin expression *in vivo*: analysis of the strain FA1090 pilin repertoire and identification of the *pilS* gene copies recombining with *pilE* during experimental human infection. *Microbiology* **147**: 839–849 (2001).

48. Zhang, J. R. and Norris, S. J. Genetic variation of the *Borrelia burgdorferi* gene *vlsE* involves cassette-specific, segmental gene conversion. *Infect. Immun.* **66**: 3698–3704 (1998).

49. Barbour, A. G. Antigenic variation in *Borrelia*: relapsing fever and Lyme borreliosis, in A. Craig and A. Scherf (eds.), *Antigenic Variation*, pp. 319–356, Academic Press, Amsterdam (2003).

50. Plasterk, R. H. A., Simon, M. I., and Barbour, A. G. Transposition of structural genes to an expression sequence on a linear plasmid causes antigenic variation in the bacterium *Borrelia hermsii*. *Nature* **318**: 257–263 (1985).

51. Futse, J. E., Brayton, K. A., Knowles, D. P., and Palmer, G. H. Structural basis for segmental gene conversion in generation of *Anaplasma marginale* outer membrane protein variants. *Mol. Microbiol.* **57**: 212–221 (2005).

52. Mahan, S. M. Antigenic variation in *Anaplasma marginale* and *Ehrlichia (Cowdria) ruminantium*, in A. Craig and A. Scherf (eds.), *Antigenic Variation*, pp. 243–272, Academic Press, Amsterdam (2003).

53. Schaffzin, J. K., Sunkin, S. M., and Stringer, J. R. A new family of *Pneumocystis carinii* genes related to those encoding the major surface glycoprotein. *Curr. Genet.* **35**: 134–143 (1999).

54. Stringer, J. R. The MSG gene family and antigenic variation in the fungus *Pneumocystis carinii*, in A. Craig and A. Scherf (eds.), *Antigenic Variation*, pp. 202–223, Academic Press, Amsterdam (2003).

55. Allred, D. R. and Al Khedery, B. Antigenic variation and cytoadhesion in *Babesia bovis* and *Plasmodium falciparum*: different logics achieve the same goal. *Mol. Biochem. Parasitol.* **134**: 27–35 (2004).

56. Allred, D. R., Al Khedery, B., and O'Connor, R. M. Antigenic variation and its significance to *Babesia*, in A. Craig and A. Scherf (eds.), *Antigenic Variation*, pp. 273–290, Academic Press, Amsterdam (2003).

57. Conway, C., Proudfoot, C., Burton, P., Barry, J. D., and McCulloch, R. Two pathways of homologous recombination in *Trypanosoma brucei*. *Mol. Microbiol.* **45**: 1687–1700 (2002).

58. Tetley, L., Turner, C. M. R., Barry, J. D., Crowe, J. S., and Vickerman, K. Onset of expression of the variant surface glycoproteins of *Trypanosoma brucei* in the tsetse fly studied using immunoelectron microscopy. *J. Cell Sci.* **87**: 363–372 (1987).

59. Barry, J. D., Crowe, J. S., and Vickerman, K. Instability of the *Trypanosoma brucei rhodesiense* metacyclic variable antigen repertoire. *Nature* **306**: 699–701 (1983).

60. Graham, V.S. and Barry, J. D. Is point mutagenesis a mechanism for antigenic variation in *Trypanosoma brucei*? *Mol. Biochem. Parasitol.* **79**: 35–45 (1996).

61. Flick, K. and Chen, Q. *var* genes, PfEMP1 and the human host. *Mol. Biochem. Parasitol.* **134**: 3–9 (2004).

62. Wyatt, R., Kwong, P. D., Desjardins, E., Sweet, R. W., Robinson, J., Hendrickson, W. A., and Sodroski, J. G. The antigenic structure of the HIV gp120 envelope glycoprotein. *Nature* **393**: 705–711 (1998).

63. Galinski, M. R. and Corredor, V. Variant antigen expression in malaria infections: post-transcriptional gene silencing, virulence and severe pathology. *Mol. Biochem. Parasitol.* **134**: 17–25 (2004).

64. Duraisingh, M. T., Voss, T. S., Marty, A. J., Duffy, M. F., Good, R. T., Thompson, J. K., Freitas-Junior, L. H., Scherf, A., Crabb, B. S., and Cowman, A. F. Heterochromatin silencing and locus repositioning linked to regulation of virulence genes in *Plasmodium falciparum*. *Cell* **121**: 13–24 (2005).

65. Freitas-Junior, L. H., Hernandez-Rivas, R., Ralph, S. A., Montiel-Condado, D., Ruvalcaba-Salazar, O. K., Rojas-Meza, A. P., Mancio-Silva, L., Leal-Silvestre, R. J., Gontijo, A. M., Shorte, S., and Scherf, A. Telomeric heterochromatin propagation and histone acetylation control mutually exclusive expression of antigenic variation genes in malaria parasites. *Cell* **121**: 25–36 (2005).

66. Blum, M. L., Down, J. A., Gurnett, A. M., Carrington, M., Turner, M. J., and Wiley, D. C. A structural motif in the variant surface glycoproteins of *Trypanosoma brucei*. *Nature* **362**: 603–609 (1993).

67. Meeus, P. F. and Barbet, A. F. Ingenious gene generation. *Trends Microbiol.* **9**: 353–355 (2001).

Chapter 6

1. Wilson, D. S. What is wrong with absolute individual fitness? *TREE* **19**: 245–48 (2004).

2. Cooper, V. S. and Lenski, R. E. The population genetics of ecological specialization in evolving *Escherichia coli* populations. *Nature* **407**: 736–739 (2000).

3. Elena, S. F. and Lenski, R. E. Evolution experiments with microorganisms: the dynamics and genetic bases of adaptation. *Nat Rev. Genet.* **4**: 457–469 (2003).

4. Vasi, F., Travisano, M., and Lenski, R. E. Long-term experimental evolution in Escherichia coli. II. Changes in life history traits during adaptation to a seasonal environment. *Am. Nat.* **144**: 432–456 (1994).

5. Funchain, P., Yeung, A., Stewart, J.L., Lin, R., Slupska, M.M., and Miller, J.H. The consequences of growth of a mutator strain of *Escherichia coli* as measured by loss of function among multiple gene targets and loss of fitness. *Genetics* **154**: 959–970 (2000).

6. Giraud, A., Matic, I., Tenaillon, O., Clara, A., Radman, M., Fons, M., and Taddei, F. Costs and benefits of high mutation rates: adaptive evolution of bacteria in the mouse gut. *Science* **291**: 2606–2608 (2001).

7. Cooper, V. S., Schneider, D., Blot, M., and Lenski, R. E. Mechanisms causing rapid and parallel losses of ribose catabolism in evolving populations of *Escherichia coli* B. *J. Bacteriol.* **183**: 2834–2841 (2001).

8. Feldman, D. E. and Frydman, J. Protein folding *in vivo*: the importance of molecular chaperones. *Curr. Opin. Struct. Biol.* **10**: 26–33 (2000).

9. Maisnier-Patin, S., Nordstrom, K., and Dasgupta, S. Replication arrests during a single round of replication of the *Escherichia coli* chromosome in the absence of DnaC activity. *Mol. Microbiol.* **42**: 1371–1382 (2001).

10. McGlynn, P. and Lloyd, R. G. Recombinational repair and restart of damaged replication forks. *Nat. Rev. Mol. Cel. Biol.* **3**: 859–870 (2002).

11. Kimura, M. Evolutionary rate at the molecular level. *Nature* **217**: 624–626 (1968).

12. Elowitz, M. B., Levine, A. J., Siggia, E. D., and Swain, P. S. Stochastic gene expression in a single cell. *Science* **297**: 1183–1186 (2002).

13. Caporale, L. H. Natural selection and the emergence of a mutation phenotype: an update of the evolutionary synthesis considering mechanisms that affect genome variation. *Annu. Rev. Microbiol* **57**: 467–485 (2003).

14. Rainey, P. B. Evolutionary genetics : The economics of mutation. *Curr. Biol.* **9**: R371–R373 (1999).

15. Watt, W. B. and Dean, A. M. Molecular-functional studies of adaptive genetic variation in prokaryotes and eukaryotes. *Annu. Rev. Genet.* **34**: 593–622 (2000).

16. Médigue, C., Rouxel, T., Vigier, P., Henaut, A., and Danchin, A. Evidence for horizontal gene transfer in *E. coli* speciation. *J. Mol. Biol.* **222**: 851–856 (1991).

17. Ochman, H., Lawrence, J. G., and Groisman, E. A. Lateral gene transfer and the nature of bacterial innovation. *Nature* **405**: 299–304 (2000).

18. Lawrence, J. G. and Hendrickson, H. Lateral gene transfer: when will adolescence end? *Mol. Microbiol.* **50**: 739–749 (2003).

19. Rowe-Magnus, D. A., Guerout, A. M., Biskri, L., Bouige, P., and Mazel, D. Comparative analysis of superintegrons: engineering extensive genetic diversity in the *Vibrionaceae*. *Genome Res.* **13**: 428–442 (2003).

20. Michel, B. Illegitimate recombination bacteria, in R. L. Charlebois (ed.), *Organization of the Prokaryotic Genome*, pp. 129–150, ASM Press, Washington DC (1999).

21. Lovett, S. T. Encoded errors: mutations and rearrangements mediated by misalignment at repetitive DNA sequences. *Mol. Microbiol.* **52**: 1243–1253 (2004).

22. Levinson, G. and Gutman, G. A. Slipped-strand mispairing: a major mechanism for DNA sequence evolution. *Mol. Biol. Evol.* **4**: 203–221 (1987).

23. Achaz, G., Rocha, E. P. C., Netter, P., and Coissac, E. Origin and fate of repeats in bacteria. *Nucleic Acids Res.* **30**: 2987–2994. (2002).

24. Roth, J. R. et al. Rearrangements of the bacterial chromosome: formation and applications, in F. Neidhardt, F. et al. (eds.), *Escherichia coli and Salmonella: cellular and molecular biology*, pp. 2256–2276, ASM Press, Washington DC (1996).

25. Hughes, D. Transformation and recombination, in R. L. Charlebois (ed.), *Organization of the Prokaryotic Genome*, pp. 109–128, ASM Press, Washington DC (1999).

26. Rocha, E. P. C. DNA repeats lead to the accelerated loss of gene order in Bacteria. *Trends Genet.* **19**: 600–604 (2003).

27. Rayssiguier, C., Thaler, D. S., and Radman, M. The barrier to recombination between *E. coli* and *S. typhimurium* is disrupted in mismatch-repair mutants. *Nature* **342**: 396–401 (1989).

28. Heale, S. M. and Petes, T. D. The stabilization of repetitive tracts of DNA by variant repeats requires a functional DNA mismatch repair system. *Cell* **83**: 539–545 (1995).

29. Bayliss, C. D., van de Ven, T., and Moxon, E. R. Mutations in polI but not *mutSLH* destabilize *Haemophilus influenzae* tetranucleotide repeats. *EMBO J.* **21**: 1465–1476. (2002).

30. Bayliss, C. D., Sweetman, W. A., and Moxon, E. R. Mutations in *Haemophilus influenzae* mismatch repair genes increase mutation rates of dinucleotide repeat tracts but not dinucleotide repeat-driven pilin phase variation rates. *J. Bacteriol.* **186**: 2928–2935 (2004).

31. Matic, I., Rayssiguier, C., and Radman, M. Interspecies gene exchange in bacteria: the role of SOS and mismatch repair systems in evolution of species. *Cell* **80**: 507–515 (1995).

32. LeClerc, J. E., Li, B., Payne, W. L., and Cebula, T. A. High mutation frequencies among *Escherichia coli* and *Salmonella* pathogens. *Science* **274**: 1208–1211 (1996).

33. Matic, I., Radman, M., Taddei, F., Picard, B., Doit, C., Bingen, E., Denamur, E., and Elion, J. Highly variable mutation rates in commensal and pathogenic *Escherichia coli*. *Science* **277**: 1833–1834 (1997).

34. Rocha, E. P. C., Matic, I., and Taddei, F. Over-representation of close repeats in stress response genes: a strategy to increase versatility under stressful conditions? *Nucleic Acids Res.* **30**: 1886–1894 (2002).

35. Oliver, A., Baquero, F., and Blázquez, J. The Mismatch Repair System (*mutS, mutL,* and *uvrD* genes) in *Pseudomonas aeruginosa*: molecular characterisation of naturally occurring mutants. *Mol. Microbiol.* **43**: 1641–1650 (2002).

36. Kotewicz, M. L., Brown, E. W., Eugene LeClerc, J., and Cebula, T. A. Genomic variability among enteric pathogens: the case of the mutS-rpoS intergenic region. *Trends Microbiol.* **11**: 2–6 (2003).

37. Caporale, L. H. Mutation is modulated: implications for evolution. *Bioessays* **22**: 388–395 (2000).

38. Weinbauer, M. G. Ecology of prokaryotic viruses. *FEMS Microbiol. Rev.* **28**: 127–181 (2004).

39. Capy, P., Gasperi, G., Biémont, C., and Bazin, C. Stress and transposable elements: co-evolution of useful parasites? *Heredity* **85**: 101–106 (2000).

40. Dubnau, D. and Lovett, C. M., Transformation and recombination, in A. L. Sonenshein, J. A. Hoch, and R. Losick (eds.), Bacillus subtilis *and Its Closest Relatives*, pp. 453–471, ASM Press, Washington DC (2002).

41. Muela, A., Pocino, M., Arana, I., Justo, J.I., Iriberri, J., and Barcina, I. Effect of growth phase and parental cell survival in river water on plasmid transfer between *Escherichia coli* strains. *Appl. Environ. Microbiol.* **60**: 4273–4278 (1994).

42. Feng, G., Tsui, H. C., and Winkler, M. E. Depletion of the cellular amounts of the MutS and MutH methyl-directed mismatch repair proteins in stationary-phase *Escherichia coli* K-12 cells. *J. Bacteriol.* **178**: 2388–2396 (1996).

43. Harris, R. S. et al. Mismatch repair protein MutL becomes limiting during stationary-phase mutation. *Genes Dev.* **11**: 2426–2437 (1997).

44. Bjedov, I., Tenaillon, O., Gerard, B., Souza, V., Denamur, E., Radman, M., Taddei, F., and Matic, I. Stress-induced mutagenesis in bacteria. *Science* **300**: 1404–1409 (2003).

45. Shen, P. and Huang, H. V. Homologous recombination in *Escherichia coli*: dependence on substrate length and homology. *Genetics* **112**: 441–457 (1986).

46. Rocha, E. P. C., Danchin, A., and Viari, A. Analysis of long repeats in bacterial genomes reveals alternative evolutionary mechanisms in *Bacillus subtilis* and other competent prokaryotes. *Mol. Biol. Evol.* **16**: 1219–1230 (1999).

47. Alokam, S., Liu, S. L., Said, K., and Sanderson, K. E. Inversions over the terminus region in *Salmonella* and *Escherichia coli*: IS200s as the sites of homologous recombination inverting the chromosome of *Salmonella enterica* serovar typhi. *J. Bacteriol.* **184**: 6190–6197 (2002).

48. Hill, C. W. and Harnish, B. Inversions between ribosomal RNA genes of *E. coli*. *Proc. Natl. Acad. Sci. U. S. A.* **78**: 7069–7072 (1981).

49. Mandal, M., Boese, B., Barrick, J. E., Winkler, W. C., and Breaker, R. R. Riboswitches control fundamental biochemical pathways in *Bacillus subtilis* and other bacteria. *Cell* **113**: 577–586 (2003).

50. Rocha, E. P. C. An appraisal of the potential for illegitimate recombination in bacterial genomes and its consequences: from duplications to genome reduction. *Genome Res.* **13**: 1123–1132 (2003).

51. Lovett, S. T., Gluckman, T. J., Simon, P. J., Sutera, V. A., and Drapkin, P. T. Recombination between repeats in *E. coli* by a recA-independent, proximity-sensitive mechanism. *Mol. Gen. Genet.* **245**: 294–300 (1994).

52. Chédin, F., Dervyn, E., Ehrlich, S. D., and Noirot, P. Frequency of deletion formation decreases exponentially with distance between short direct repeats. *Mol. Microbiol.* **12**: 561–569 (1994).

53. Peeters, B. P., de Boer, J. H., Bron, S., and Venema, G. Structural plasmid instability in *Bacillus subtilis*: effect of direct and inverted repeats. *Mol. Gen. Genet.* **212**: 450–458 (1988).

54. Pierce, J. C., Kong, D., and Masker, W. The effect of the length of direct repeats and the presence of palindromes on deletion between directly repeated DNA sequences in bacteriophage T7. *Nucleic Acids Res.* **19**: 3901–3905 (1991).

55. Albertini, A. M., Hofer, M., Calos, M. P., and Miller, J. H. On the formation of spontaneous deletions: the importance of short sequence homologies in the generation of large deletions. *Cell* **29**: 319–328 (1982).

56. Singer, B. S. and Westlye, J. Deletion formation in bacteriophage T4. *J. Mol. Biol.* **202**: 233–243 (1988).

57. Moxon, E. R., Rainey, P. B., Nowak, M. A., and Lenski, R. E. Adaptive evolution of highly mutable loci in pathogenic bacteria. *Curr. Biol.* **4**: 24–33 (1994).

58. van Belkum, A., Scherer, S., van Alphen, L., and Verbrugh, H. Short-sequence DNA repeats in prokaryotic genomes. *Microbiol. Mol. Biol. Rev.* **62**: 275–293 (1998).

59. Tautz, D., Trick, M., and Dover, G. A. Cryptic simplicity in DNA is a major source of genetic variation. *Nature* **322**: 652–656 (1986).

60. Jansen, R., van Embden, J. D., Gaastra, W., and Schouls, L. M. Identification of a novel family of sequence repeats among prokaryotes. *OMICS* **6**: 23–33 (2002).

61. Neidhardt, F., Curtiss, R., Ingraham, J.L., Lin, E.C.C., Low, K.B., Magasanik, B., Reznikoff, W.S., Riley, M., Schaechter, M., and Umbarger, H.E. (eds.), Escherichia coli *and* Salmonella: *Cellular and Molecular Biology*, ASM Press, Washington DC (1996).

62. Horst, J.-P., Wu, T.-H., and Marinus, M. G. *Escherichia coli* mutator genes. *Trends Microbiol.* **7**: 29–36 (1999).

63. Yang, H., Wolff, E., Kim, M., Diep, A., and Miller, J. H. Identification of mutator genes and mutational pathways in *Escherichia coli* using a multicopy cloning approach. *Mol. Microbiol.* **53**: 283–295 (2004).

64. Jankovic, M., Kostic, T., and Savic, D. J. DNA sequence analysis of spontaneous histidine mutations in a polA1 strain of *Escherichia coli* K12 suggests a specific role of the GTGG sequence. *Mol. Gen. Genet.* **223**: 481–486 (1990).

65. Canceill, D. and Ehrlich, S. D. Copy-choice recombination mediated by DNA polymerase III holoenzyme from *Escherichia coli. Proc. Natl. Acad. Sci. U. S. A.* **93**: 6647–6652 (1996).

66. Hebert, M. L., Spitz, L. A., and Wells, R. D. DNA double-strand breaks induce deletion of CTG.CAG repeats in an orientation-dependent manner in *Escherichia coli. J. Mol. Biol.* **336**: 655–672 (2004).

67. Bowater, R. P., Jaworski, A., Larson, J. E., Parniewski, P., and Wells, R. D. Transcription increases the deletion frequency of long CTG.CAG triplet repeats from plasmids in *Escherichia coli. Nucleic Acids Res.* **25**: 2861–2868 (1997).

68. Hood, D.W., Deadman, M.E., Jennings, M.P., Bisercic, M., Fleischmann, R.D., Venter, J.C., and Moxon, R. DNA repeats identify novel virulence genes in *Haemophilus influenzae. Proc. Natl. Acad. Sci. U. S. A.* **93**: 11121–11125 (1996).

69. Saunders, N. J., Peden, J. F., Hood, D. W., and Moxon, E. R. Simple sequence repeats in the *Helicobacter pylori* genome. *Mol. Microbiol.* **27**: 1091–1098 (1998).

70. Bayliss, C. D., Field, D., and Moxon, E. R. The simple sequence contingency loci of *Haemophilus influenzae* and *Neisseria meningitidis. J. Clin. Invest.* **107**: 657–662. (2001).

71. Rocha, E. P. C., Pradillon, O., Bui, H., Sayada, C., and Denamur, E. A new family of highly variable proteins in the *Chlamydophila pneumoniae* genome. *Nucleic Acids Res.* **30**: 4351–4360 (2002).

72. Tenaillon, O., Taddei, F., Radmian, M., and Matic, I. Second-order selection in bacterial evolution: selection acting on mutation and recombination rates in the course of adaptation. *Res. Microbiol.* **152**: 11–16. (2001).

73. Drake, J. W., Charlesworth, B., Charlesworth, D., and Crow, J. F. Rates of spontaneous mutation. *Genetics* **148**: 1667–1686 (1998).

74. Richardson, A. R., Yu, Z., Popovic, T., and Stojiljkovic, I. Mutator clones of *Neisseria meningitidis* in epidemic serogroup A disease. *Proc. Natl. Acad. Sci. U. S. A.* **99**: 6103–6107 (2002).

75. Negri, M.-A. et al. Very Low Cefotaxime concentrations select for hypermutable *Streptococcus pneumoniae* populations. *Antimicrob. Agents Chemother.* **46**: 528–530 (2002).

76. Vulic, M., Dionisio, F., Taddei, F., and Radman, M. Molecular keys to speciation: DNA polymorphism and the control of genetic exchange in enterobacteria. *Proc. Natl. Acad. Sci. U. S. A.* **94**: 9763–9767 (1997).

77. Denamur, E., Lecointre, G., Darlu, P., Tenaillon, O., Acquaviva, C., Sayada, C., Sunjevaric, I., Rothstein, R., Elion, J., Taddei, F., Radman, M., and Matic I. Evolutionary implications of the frequent horizontal transfer of mismatch repair genes. *Cell* **103**: 711–721. (2000).

78. Brown, E. W., LeClerc, J. E., Li, B., Payne, W. L., and Cebula, T. A. Phylogenetic evidence for horizontal transfer of *mutS* alleles among naturally occurring *Escherichia coli* strains. *J. Bacteriol.* **183**: 1631–1644. (2001).

79. Davidsen, T., Bjoras, M., Seeberg, E.C., and Tonjum, T. Biased distribution of DNA uptake sequences towards genome maintenance genes. *Nucleic Acids Res.* **32**: 1050–1058 (2004).

80. Scocca, J. J., Poland, R. L., and Zoon, K. C. Specificity in deoxyribonucleic acid uptake by transformable *Haemophilus influenzae. J. Bacteriol.* **118**: 369–373 (1974).

81. Smith, H. O., Tomb, J.-F., Dougherty, B. A., Fleischmann, R. D., and Venter, J. C. Frequency and distribution of DNA uptake signal sequences in the *Haemophilus influenzae* Rd genome. *Science* **269**: 538–540 (1995).

82. Baumann, P., Baumann, L., Lai, C.-H., and Rouhbakhsh, D. Genetics, physiology, and evolutionary relationships of the genus *Buchnera*: intracellular symbionts of aphids. *Annu. Rev. Microbiol.* **49**: 55–94 (1995).

83. Andersson, J. O. and Andersson, S. G. E. Insights into the evolutionary process of genome degradation. *Curr. Opin. Genet. Dev.* **9**: 664–671 (1999).

84. Citti, C., Kim, M. F. and Wise, K. S. Elongated versions of Vlp surface lipoproteins protect Mycoplasma hyorhinis escape variants from growth-inhibiting host antibodies. *Infect. Immun.* **65**: 1773–1785. (1997).

85. Zambrano, M. M., Siegele, D. A., Almiron, M., Tormo, A., and Kolter, R. Microbial competition: *Escherichia coli* mutants that take over stationary phase cultures. *Science* **259**: 1757–1760 (1993).

86. Kurusu, Y., Narita, T., Suzuki, M., and Watanabe, T. Genetic analysis of an incomplete *mutS* gene from *Pseudomonas putida*. *J. Bacteriol.* **182**: 5278–5279 (2000).

87. Wu, T.-H. and Marinus, M. G. Deletion mutation analysis of the *mutS* gene in *Escherichia coli*. *J Biol. Chem.* **274**: 5948–5952 (1999).

88. Young, D. M. and Ornston, L. N. Functions of the mismatch repair gene *mutS* from *Acinetobacter* sp. strain ADP1. *J. Bacteriol.* **183**: 822–6831 (2001).

89. Hastings, P. J. and Rosenberg, S. M. In pursuit of a molecular mechanism for adaptive gene amplification. *DNA Repair (Amst.)* **1**: 111–123 (2002).

90. Petit, M.-A., Mesas, J. M., Noirot, P., Morel-Deville, F., and Ehrlich, S. D. Induction of DNA amplification in the *Bacillus subtilis* chromosome. *EMBO J.* **11**: 1317–1326 (1992).

91. Normark, S., Edlund, T., Grundstrom, T., Bergstrom, S., and Wolf-Watz, H. *Escherichia coli* K-12 mutants hyperproducing chromosomal beta-lactamase by gene repetitions. *J. Bacteriol.* **132**: 912–922 (1977).

92. Edlund, T. and Normark, S. Recombination between short DNA homologies causes tandem duplication. *Nature* **292**: 269–271 (1981).

93. Fisher, R. A. *The Genetical Theory of Natural Selection*, Oxford University Press, Oxford (1930).

94. Caporale, L. H. Is there a higher level genetic code that directs evolution? *Mol. Cell Biochem.* **64**: 5–13 (1984).

95. Himmelreich, R., Hilbert, H., Plagens, H., Pirki, E., Li, B.-C., and Herrmann, R. Complete sequence analysis of the genome of the bacterium *Mycoplasma pneumoniae*. *Nucleic Acids Res.* **24**: 4420–4449 (1996).

96. Kenri, T., Taniguchi, R., Sasaki, Y., Okazaki, N., Narita, M, Izumikawa, K., Umetsu, M., and Sasaki, T. Identification of a new variable sequence in the P1 cytadhesin gene of *Mycoplasma pneumoniae*: evidence for the generation of antigenic variation by DNA recombination between repetitive sequences. *Infect. Immun.* **67**: 4557–4562. (1999).

97. Rocha, E. P. C. and Blanchard, A. Genomic repeats, genome plasticity and the dynamics of *Mycoplasma* evolution. *Nucleic Acids Res.* **30**: 2031–2042 (2002).

Chapter 7

1. Caporale, L. H. Natural selection and the emergence of a mutation phenotype: An update of the evolutionary synthesis considering mechanisms that affect genome variation. *Ann. Rev. Microbiol.* **57**: 467–85 (2003).

2. Avery, O. T., MacLeod, C., and McCarty, M. Studies on the chemical nature of the substance inducing transformation of pneumoccal types: Induction of transformation by a deoxyribonucleic acid fraction isolated from *Pneumococcus* type III. *J. Exp. Med.* **79**: 137–58 (1944).

3. Blattner, F. R. et al. The complete genome sequence of *Escherichia coli* K-12. *Science* **277**: 1453–74 (1997).

4. Welch, R. A. et al. Extensive mosaic structure revealed by the complete genome sequence of uropathogenic *Escherichia coli*. *Proc. Natl. Acad. Sci. U. S. A.* **99**: 17020–4 (2002).

5. Day, W. A., Jr., Fernandez, R. E., and Maurelli, A. T. Pathoadaptive mutations that enhance virulence: genetic organization of the cadA regions of *Shigella* spp. *Infect. Immun.* **69**: 7471–80 (2001).

6. Giraud, A., Radman, M., Matic, I., and Taddei, F. The rise and fall of mutator bacteria. *Curr. Opin. Microbiol.* **4**: 582–5 (2001).

7. Taddei, F., Matic, I., Godelle, B., and Radman, M. To be a mutator, or how pathogenic and commensal bacteria can evolve rapidly. *Trends Microbiol.* **5**: 427–8; discussion 428–9 (1997).

8. Oliver, A., Canton, R., Campo, P., Baquero, F., and Blazquez, J. High frequency of hypermutable *Pseudomonas aeruginosa* in cystic fibrosis lung infection. *Science* **288**: 1251–4 (2000).

9. Taddei, F., Radman, M., Maynard-Smith, J., Toupance, B., Gouyon, P. H., and Godelle, B. Role of mutator alleles in adaptive evolution. *Nature* **387**: 700–2 (1997).

10. Romero, D. and Palacios, R. Gene amplification and genomic plasticity in prokaryotes. *Annu. Rev. Genet.* **31**: 91–111 (1997).

11. Achaz, G., Rocha, E. P., Netter, P., and Coissac, E. Origin and fate of repeats in bacteria. *Nucleic Acids Res.* **30**: 2987–94 (2002).

12. Levinson, G. and Gutman, G. A. Slipped-strand mispairing: a major mechanism for DNA sequence evolution. *Mol. Biol. Evol.* **4**: 203–21 (1987).

13. Fraser, C. M. et al. The minimal gene complement of *Mycoplasma genitalium*. *Science* **270** (1995).

14. Gieffers, J., Durling, L., Ouellette, S. P., Rupp, J., Maass, M., Byrne, G. I., Caldwell, H. D., and Belland, R. J. Genotypic differences in the *Chlamydia pneumoniae* tyrP locus related to vascular tropism and pathogenicity. *J. Infect. Dis.* **188**: 1085–93 (2003).

15. van Ham, S. M., van Alphen, L., Mooi, F. R., and van Putten, J. P. The fimbrial gene cluster of *Haemophilus influenzae* type b. *Mol. Microbiol.* **13**: 673–84 (1994).

16. Eisen, J. A., Heidelberg, J. F., White, O., and Salzberg, S. L. Evidence for symmetric chromosomal inversions around the replication origin in bacteria. *Genome Biol.* **1** (2000).

17. Suyama, M. and Bork, P. Evolution of prokaryotic gene order: genome rearrangements in closely related species. *Trends Genet.* **17**: 10–3 (2001).

18. Tillier, E. R. and Collins, R. A. Genome rearrangement by replication-directed translocation. *Nature Genet.* **26**: 195–7 (2000).

19. Read, T. D et al. Genome sequence of *Chlamydophila caviae* (*Chlamydia psittaci* GPIC): Examining the role of niche-specific genes in the evolution of the Chlamydiaceae. *Nucl. Acids Res.* **31**: 2134–47 (2003).

20. Read, T. D. et al. The genome sequence of *Bacillus anthracis* Ames and comparison to closely related bacteria. *Nature* **423**: 81–6 (2003).

21. Salama, N., Guillemin, K., McDaniel, T. K., Sherlock, G., Tompkins, L., and Falkow, S. A whole-genome microarray reveals genetic diversity among *Helicobacter pylori* strains. *Proc. Natl. Acad. Sci. U. S. A.* **97**: 14668–73 (2000).

22. Lecompte, O., Ripp, R., Puzos-Barbe, V., Duprat, S., Heilig, R., Dietrich, J., Thierry, J. C., and Poch, O. Genome evolution at the genus level: comparison of three complete genomes of hyperthermophilic archaea. *Genome Res.* **11**: 981–93 (2001).

23. Brunder, W. and Karch, H. Genome plasticity in Enterobacteriaceae. *Int. J. Med. Microbiol.* **290**: 153–65 (2000).

24. Arber, W. Involvement of gene products in bacterial evolution. *Ann. N. Y. Acad. Sci.* **870**: 36–44 (1999).

25. Tettelin, H. et al. Complete genome sequence of a virulent isolate of *Streptococcus pneumoniae*. *Science* **293**: 498–506. (2001).

26. Tobiason, D. M., Lenich, A. G., and Glasgow, A. C. Multiple DNA binding activities of the novel site-specific recombinase, Piv, from *Moraxella lacunata*. *J Biol. Chem.* **274**: 9698–706 (1999).

27. Krinos, C. M., Coyne, M. J., Weinacht, K. G., Tzianabos, A. O., Kasper, D. L., and Comstock, L. E. Extensive surface diversity of a commensal microorganism by multiple DNA inversions. *Nature* **414**: 555–8 (2001).

28. Coyne, M. J., Weinacht, K. G., Krinos, C. M., and Comstock, L. E. Mpi recombinase globally modulates the surface architecture of a human commensal bacterium. *Proc. Natl. Acad. Sci. U. S. A.* **100**: 10446–51 (2003).

29. Belfort, M., Reaban, M. E., Coetzee, T., and Dalgaard, J. Z. Prokaryotic introns and inteins: a panoply of form and function. *J. Bacteriol.* **177**: 3897–903 (1995).

30. Pietrokovski, S. Intein spread and extinction in evolution. *Trend Genet.* **17**: 465–472 (2001)

31. Lampson, B., Inouye, M., and Inouye, S. The msDNAs of bacteria. *Prog. Nucleic Acid. Res. Mol. Biol.* **67**: 65–91 (2001).

32. Liu, S. V., Saunders, N. J., Jeffries, A., and Rest, R. F. Genome analysis and strain comparison of *correia* repeats and *correia* repeat-enclosed elements in pathogenic *Neisseria*. *J. Bacteriol.* **184**: 6163–73 (2002).

33. Correia, F. F., Inouye, S., and Inouye, M. A 26-base-pair repetitive sequence specific for *Neisseria gonorrhoeae* and *Neisseria meningitidis* genomic DNA. *J. Bacteriol.* **167**: 1009–15 (1986).

34. Correia, F. F., Inouye, S., and Inouye, M. A family of small repeated elements with some transposon-like properties in the genome of *Neisseria gonorrhoeae*. *J. Biol. Chem.* **263**: 12194–8 (1988).

35. Meyer, B. J. and Schottel, J. L. Characterization of cat messenger RNA decay suggests that turnover occurs by endonucleolytic cleavage in a 3′ to 5′ direction. *Mol. Microbiol.* **6**: 1095–104 (1992).

36. Hoskins, J. et al. Genome of the bacterium *Streptococcus pneumoniae* strain R6. *J. Bacteriol.* **183**: 5709–17 (2001).

37. Martin, B., Humbert, O., Camara, M., Guenzi, E., Walker, J., Mitchell, T., Andrew, P., Prudhomme, M., Alloing, G., and Hakenbeck, R. A highly conserved repeated DNA element located in the chromosome of *Streptococcus pneumoniae*. *Nucleic Acids Res.* **20**: 3479–83 (1992).

38. Oggioni, M. R. and Claverys, J. P. Repeated extragenic sequences in prokaryotic genomes: a proposal for the origin and dynamics of the RUP element in *Streptococcus pneumoniae*. *Microbiology* **145**(10): 2647–53 (1999).

39. Mahillon, J. and Chandler, M. Insertion sequences. *Microbiol. Mol. Biol. Rev.* **62**: 725–74 (1998).

40. Craig, N. L. Transposition, in F. C. Neidhardt et al. (eds), Escherichia coli *and* Salmonella typhimurium, pp. 2339–62, ASM Press, Washington DC (1996).

41. Lawrence, J. G., Ochman, H., and Hartl, D. L. The evolution of insertion sequences within enteric bacteria. *Genetics* **131**: 9–20. (1992).

42. Paulsen, I. T. et. al. Role of mobile DNA in the evolution of vancomycin-resistant *Enterococcus faecalis*. *Science* **299**: 2071–4 (2003).

43. Zerbib, D., Gamas, P., Chandler, M., Prentki, P., Bass, S., and Galas, D. Specificity of insertion of IS1. *J. Mol. Biol.* **185**: 517–24 (1985).

44. Peters, J. E. and Craig, N. L. Tn7: smarter than we thought. *Nat. Rev. Mol. Cell Biol.* **2**: 806–14 (2001).

45. Olasz, F., Kiss, J., Konig, P., Buzas, Z., Stalder, R., and Arber, W. Target specificity of insertion element IS30. *Mol. Microbiol.* **28**: 691–704 (1998).

46. Hall, R. M. and Collis, C. M. Mobile gene cassettes and integrons: capture and spread of genes by site-specific recombination. *Mol. Microbiol.* **15**: 593–600 (1995).

47. Stokes, H. W., O'Gorman, D. B., Recchia, G. D., Parsekhian, M., and Hall, R. M. Structure and function of 59-base element recombination sites associated with mobile gene cassettes. *Mol. Microbiol.* **26**: 731–45 (1997).

48. Recchia, G. D. and Hall, R. M. Gene cassettes: a new class of mobile element. *Microbiology* **141**(12): 3015–27 (1995).

49. Holmes, A. J., Gillings, M. R., Nield, B. S., Mabbutt, B. C., Nevalainen, K. M., and Stokes, H. W. The gene cassette metagenome is a basic resource for bacterial genome evolution. *Environ. Microbiol.* **5**: 383–94 (2003).

50. Dubnau, D. and Lovett, C. M. Transformation and recombination, in A. L. Sonenshein et al. (eds.), Bacillus subtilis *and Its Closest Relatives: from Genes to Cells*, pp. 453–71, ASM Press, Washington DC (2002).

51. Saunders, N. J., Hood, D. W., and Moxon, E. R. Bacterial evolution: bacteria play pass the gene. *Curr. Biol.* **9**: R180–3 (1999).

52. Mandel, M. and Higa, A. Calcium-dependent bacteriophage DNA infection. *J. Mol. Biol.* **53**: 159–62 (1970).

53. Zechner, E. L., de la Cruz, F., Eisenbrandt, R., Grahn, A. M., Koraimann, G., Lanka, E., Muth, G., Pansegrau, W., Thomas, C. M., Wilkins, B. M., and Zatyka, M. Conjugative DNA transfer processes, in C. M. Thomas (ed.), *The Horizontal Gene Pool: Bacterial Plasmids and Gene Spread*, pp. 87–174, Harwood Academic Publishers, Amsterdam (2000).

54. Bates, S., Cashmore, A. M., and Wilkins, B. M. IncP plasmids are unusually effective in mediating conjugation of *Escherichia coli* and *Saccharomyces cerevisiae*: involvement of the tra2 mating system. *J. Bacteriol.* **180**: 6538–43 (1998).

55. Szpirer, C., Top, E., Couturier, M., and Mergeay, M. Retrotransfer or gene capture: a feature of conjugative plasmids, with ecological and evolutionary significance. *Microbiology* **145**(12): 3321–9 (1999).

56. Ronchel, M. C., Ramos-Diaz, M. A., and Ramos, J. L. Retrotransfer of DNA in the rhizosphere. *Environ. Microbiol.* **2**: 319–23 (2000).

57. Von Bodman, S. B., Bauer, W. D., and Coplin, D. L. Quorum sensing in plant-pathogenic bacteria. *Annu. Rev. Phytopathol.* **41**: 455–82 (2003).

58. Kado, C. I. Origin and evolution of plasmids. *Antonie Van Leeuwenhoek* **73**: 117–26 (1998).

59. Klare, I., Werner, G., and Witte, W. Enterococci. Habitats, infections, virulence factors, resistances to antibiotics, transfer of resistance determinants. *Contrib. Microbiol* **8**: 108–22 (2001).

60. Dunny, G. M., Antiporta, M. H., and Hirt, H. Peptide pheromone-induced transfer of plasmid pCF10 in *Enterococcus faecalis*: probing the genetic and molecular basis for specificity of the pheromone response. *Peptides* **22**: 1529–39 (2001).

61. Dunny, G. M. and Leonard, B. A. Cell–cell communication in Gram-positive bacteria. *Annu. Rev. Microbiol* **51**: 527–64 (1997).

62. Parkhill, J. et al. Genome sequence of *Yersinia pestis*, the causative agent of plague. *Nature* **413**: 523–7. (2001).

63. Wei, J. et al. Complete genome sequence and comparative genomics of *Shigella flexneri* serotype 2a strain 2457T. *Infect Immun* **71**: 2775–86 (2003).

64. Jin, Q. et al. Genome sequence of *Shigella flexneri* 2a: insights into pathogenicity through comparison with genomes of *Escherichia coli* K12 and O157. *Nucleic Acids Res.* **30**: 4432–41 (2002).

65. Lindler, L. E., Plano, G. V., Burland, V., Mayhew, G. F., and Blattner, F. R. Complete DNA sequence and detailed analysis of the *Yersinia pestis* KIM5 plasmid encoding murine toxin and capsular antigen. *Infect. Immun.* **66**: 5731–42 (1998).

66. Tobe, T., Hayashi, T., Han, C. G., Schoolnik, G. K., Ohtsubo, E., and Sasakawa, C. Complete DNA sequence and structural analysis of the enteropathogenic *Escherichia coli* adherence factor plasmid. *Infect. Immun.* **67**: 5455–62 (1999).

67. Schmidt, H. Shiga-toxin-converting bacteriophagesphages. *Res. Microbiol.* **152**: 687–95 (2001).

68. Ehrbar, K., Mirold, S., Friebel, A., Stender, S., and Hardt, W. D. Characterization of effector proteins translocated via the SPI1 type III secretion system of *Salmonella typhimurium*. *Int. J. Med. Microbiol.* **291**: 479–85 (2002).

69. Allison, G. E., Angeles, D., Tran-Dinh, N., and Verma, N. K. Complete genomic sequence of SfV, a serotype-converting temperate bacteriophage of *Shigella flexneri*. *J. Bacteriol.* **184**: 1974–87 (2002).

70. Adams, M. M., Allison, G. E., and Verma, N. K. Type IV O antigen modification genes in the genome of *Shigella flexneri* NCTC 8296. *Microbiology* **147**: 851–60 (2001).

71. Lawrence, J. G. and Hendrickson, H. Lateral gene transfer: when will adolescence end? *Mol. Microbiol.* **50**: 739–49 (2003).

72. Gogarten, J. P., Doolittle, W. F., and Lawrence, J. G. Prokaryotic evolution in light of gene transfer. *Mol. Biol. Evol.* **19**: 2226–38 (2002).

73. Nakamura, Y., Itoh, T., Matsuda, H., and Gojobori, T. Biased biological functions of horizontally transferred genes in prokaryotic genomes. *Nature Genet.* **36**: 760–6 (2004).

74. Lander, E. S. et al. Initial sequencing and analysis of the human genome. *Nature* **409**, 860–921 (2001).

75. Roelofs, J. and Van Haastert, P. J. Genes lost during evolution. *Nature* **411**: 1013–4 (2001).

76. Salzberg, S. L., White, O., Peterson, J., and Eisen, J. A. Microbial genes in the human genome: lateral transfer or gene loss? *Science* **292**: 1903–6 (2001).

77. Stanhope, M. J., Lupas, A., Italia, M. J., Koretke, K. K., Volker, C., and Brown, J. R. Phylogenetic analyses do not support horizontal gene transfers from bacteria to vertebrates. *Nature* **411**: 940–4 (2001).

78. Eisen, J. A. Horizontal gene transfer among microbial genomes: new insights from complete genome analysis. *Curr. Opin. Genet. Dev.* **10**: 606–11 (2000).

79. Hacker, J., Bender, L., Ott, M., Wingender, J., Lund, B., Marre, R., and Goebel, W. Deletions of chromosomal regions coding for fimbriae and hemolysins occur *in vitro* and *in vivo* in various extraintestinal *Escherichia coli* isolates. *Microb. Pathog.* **8**: 213–25 (1990).

80. Brussow, H., Canchaya, C., and Hardt, W. D. Phage and the evolution of bacterial pathogens: from genomic rearrangements to lysogenic conversion. *Microbiol. Mol. Biol. Rev.* **68**: 560–602 (2004).

81. Karaolis, D. K., Somara, S., Maneval, D. R., Jr., Johnson, J. A., and Kaper, J. B. A bacteriophage encoding a pathogenicity island, a type-IV pilus and a phage receptor in cholera bacteria. *Nature* **399**: 375–9 (1999).

82. Faruque, S. M., Zhu, J., Asadulghani, Kamruzzaman, M., and Mekalanos, J. J. Examination of diverse toxin-coregulated pilus-positive *Vibrio cholerae* strains fails to demonstrate evidence for *Vibrio* pathogenicity island phage. *Infect. Immun.* **71**: 2993–9 (2003).

83. Vokes, S. A., Reeves, S. A., Torres, A. G., and Payne, S. M. The aerobactin iron transport system genes in *Shigella flexneri* are present within a pathogenicity island. *Mol. Microbiol.* **33**: 63–73 (1999).

Chapter 8

1. Labrador, M. and Corces, V. Interactions between transposable elements and the host genome, in N. L. Craig, Craigie, R., Gellert, M., Lambowitz, A.M. (eds.), *Mobile DNA II*, pp. 1008–23, ASM Press, Washington DC (2002).
2. Feschotte, C., Jiang, N., and Wessler, S. R. Plant transposable elements: where genetics meets genomics. *Nat. Rev. Genet.* **3**: 329–41 (2002).
3. Kazazian Jr., H. H. Mobile elements: drivers of genome evolution. *Science* **303**: 1626–32 (2004).
4. Kidwell, M. G. Transposable elements and the evolution of genome size in eukaryotes. *Genetica* **115**: 49–63 (2002).
5. McClintock, B. The origin and behavior of mutable loci in maize. *Proc. Natl. Acad. Sci. U. S. A.* **36**: 344–55 (1950).
6. Craig, N. L., Craigie, R., Gellert, M., and Lambowitz, A. M. *Mobile DNA II*, ASM Press, Washington DC (2002).
7. Luan, D. D., Korman, M. H., Jakubczak, J. L., and Eickbush, T. H. Reverse transcription of R2Bm RNA is primed by a nick at the chromosomal target site: a mechanism for non-LTR retrotransposition. *Cell* **72**: 595–605 (1993).
8. McClintock, B. Chromosome organization and genic expression. *Cold Spg Hbr Symp. Quant. Biol.* **16**: 13–47 (1951).
9. McClintock, B. The significances of responses of the genome to challenge. *Science* **226**: 792–801 (1984).
10. Bowen, N. J. and Jordan, I. K. Transposable elements and the evolution of eukaryotic complexity. *Curr. Issues Mol. Biol.* **4**: 65–76 (2002).
11. Doolittle, W. F. and Sapienza, C. Selfish genes, the phenotype paradigm and genome evolution. *Nature* **284**: 601–3 (1980).
12. Orgel, L. E. and Crick, F. H. C. Selfish DNA: the ultimate parasite. *Nature* **284**: 604–7 (1980).
13. McClintock, B. The stability of broken ends of chromosomes in *Zea mays*. *Genetics* **26**: 234–82 (1941).
14. Gray, Y. H. It takes two transposons to tango. *Trends Genet.* **16**: 461–8 (2000).
15. McClintock, B. The *Suppressor-mutator* system of control of gene action in maize. *Carnegie Institution of Washington Yearbook* **57**: 415–29 (1958).
16. Deininger, P. L. and Batzer, M. A. Mammalian retroelements. *Gen. Res.* **12**: 1455–65 (2002).
17. Kumar, A. and Bennetzen, J. L. Plant retrotransposons. *Ann. Rev. Genet.* **33**: 479–532 (1999).
18. Caceres, M., Ranz, J. M., Barbadilla, A., Long, M., and Ruiz, A. Generation of a widespread *Drosophila* inversion by a transposable element. *Science* **285**: 415–18 (1999).
19. Puig, M., Caceres, M., and Ruiz, A. Silencing of a gene adjacent to the breakpoint of a *Drosophila* inversion by a transposon-induced antisense RNA. *Proc. Natl. Acad. Sci. U. S. A.* **101**: 9013–18 (2004).
20. Dunham, M. J., Badrane, H., Ferea, T., Adams, J., Brown, P. O., Rosenzweig, F., and Botstein, D. Characteristic genome rearrangements in experimental evolution of *Saccharomyces cerevisiae*. *Proc. Natl. Acad. Sci. U. S. A.* **99**: 16144–9 (2002).

21. Bailey, J. A., Gu, Z., Clark, R. A., Reinert, K., Samonte, R. V., Schwartz, S., Adams, M. D., Myers, E. W., Li, P. W., and Eichler, E. E. Recent segmental duplication in the human genome. *Science* **297**: 1003–7 (2002).

22. Babcock, M., Pavlicek, A., Spiteri, E., Kashork, D. D., Ioshikhes, I., Shaffer, L. G., Jurka, J., and Morrow, B. E. Shuffling of genes within low-copy repeats on 22q11 (*LCR22*) by Alu-mediated recombination events during evolution. *Gen. Res.* **13**: 2519–32 (2003).

23. Deininger, P. L. and Batzer, M. A. *Alu* repeats and human genetic disease. *Mol. Genet. Metab.* **67**: 183–93 (1999).

24. Bailey, J. A., Liu, G., and Eichler, E. E. An *Alu* transposition model for the origin and expansion of human segmental duplications. *Am. J. Hum. Genet.* **73**: 823–34 (2003).

25. Gibbs, R. A., Weinstock, G. M., Metzker, M. L., Muzny, D. M., Sodergren, E.J., Scherer, S., et al. Genome sequence of the Brown Norway rat yields insights into mammalian evolution. *Nature* **428**: 493–521 (2004).

26. McDonald, J. F. Transposable elements: possible catalysts of organismic evolution. *Trends Ecol. Evol.* **10**: 123–6 (1995.).

27. Weil, C. F. and Wessler, S. R. The effects of plant transposable element insertion on transcription initiation and RNA processing. *Ann. Rev. Plant Phys. Mol. Biol.* **41**: 527–52 (1990).

28. Kidwell, M. G. and Lisch, D. Transposable elements as sources of genomic variation, in N. L. Craig, Craigie, R., Gellert, M., Lambowitz, A. M. (eds.), *Mobile DNA II*, pp. 59–90, ASM Press, Washington DC (2002).

29. Gibbs, R. A., Weinstock, G. M., Metzker, M. L., Muzny, D. M., Sodergren, E. J., and Scherer, S. Transposable elements in mammals promotes regulatory variation and diversification of genes with specialized functions. *Trends Genet.* **19**(10): 530–6 (2003).

30. Feschotte, C., Zhang, X., and Wessler, S. Miniature inverted-repeat transposable elements (MITEs) and their relationship with established DNA transposons, in N. L. Craig, Craigie, R., Gellert, M., Lambowitz, A. M. (eds.), *Mobile DNA II*, pp. 1147–58, ASM Press, Washington DC (2002).

31. Jiang, N., Bao, Z., Zhang, X., McCouch, S. R., Eddy, S. R., and Wessler, S. R. An active DNA transposon in rice. *Nature* **421**: 163–7 (2003).

32. Kikuchi, K., Terauchi, K., Wada, M., Hirano, H. Y. The plant MITE *mPing* is mobilized in anther culture. *Nature* **421**: 167–70 (2003).

33. Nakazaki, T., Okumoto, Y., Horibata, A., Yamahira, S., Teraishi, M., Nishida, H., Inoue, H., Tanisaka, T. Mobilization of a transposon in the rice genome. *Nature* **421**: 170–2 (2003).

34. Goff, S. A., Ricke, D., Lan, T. H., Presting, G., Wang, R. et al. A draft sequence of the rice genome (*Oryza sativa L.* ssp. *japonica*). *Science* **296**: 92–100 (2002).

35. Yu, J., Hu, S., Wang, J., Wong, G. K., Li, S., et al. A draft sequence of the rice genome (*Oryza sativa L.* ssp. *indica*). *Science* **296**: 79–92 (2002).

36. Hans, J. S., Szak, S. T., and Boeke, J.D. Transcriptional disruption by the *L1* retrotransposon and implications for mammalian transcriptomes. *Nature* **429**: 268–74 (2004).

37. Perepelistsa-Belancio, B. and Deininger, P. L. RNA truncation by premature polyadenylation attenuates human mobile element activity. *Nature Genet.* **35**: 363–6 (2003).

38. Kreahling, J. and Graveley, B. R. The origin and implications of Aluternative splicing. *Trends Genet.* **20**: 1–4 (2004).

39. Nekrutenko, A. and Li, W. H. Transposable elements are found in a large number of human protein-coding genes. *Trends Genet.* **17**: 619–21 (2001).

40. Lev-Maor, G., Sorek, R., Shomron, N., and Ast, G. The birth of an alternatively spliced exon: 3' splice-site selection in *Alu* exons. *Science* **300**: 1288–91 (2003).

41. Gilbert, W. Why genes in pieces? *Nature* **271**: 501 (1978).

42. Moran, J. V., DeBerardinis, R. J., and Kazazian Jr., H. H. Exon shuffling by *L1* retrotransposition. *Science* **283**: 1530–4 (1999).

43. Goodier, J. L., Ostertag, E. M., and Kazazian Jr., H. H. Transduction of 3'-flanking sequences is common in *L1* retrotransposition. *Hum. Mol. Genet.* **9**: 653–7 (2000).

44. Pickeral, O. K., Makalowski, W., Boguski, M. S., and Boeke, J. D. Frequent human genomic DNA transduction driven by *LINE-1* retrotransposition. *Gen. Res.* **10**: 411–5 (2000).

45. Talbert, L. E. and Chandler, V. L.Characterization of a highly conserved sequence related to *mutator* transposable elements in maize. *Mol. Biol. Evol.* **5**: 519–29 (1988).

46. Jiang, N., Bao, Z., Zhang, X., Eddy, S. R., and Wessler, S. R. Pack-Mule transposable elements mediate gene evolution in plants. *Nature* **431**: 569–73 (2004).

47. Fedoroff, N. V. and Chandler, V. Inactivation of maize transposable elements, in J. Paskowski (ed.), *Homologous Recombination and Gene Silencing in Plants*, pp. 349–85, Kluwer Academic Publishers (1994).

48. Vastenhouw, N. L. and Plasterk, R. H. RNAi protects *Caenorhabditis elegans* germline transcription against transposition. *Trends Genet.* **20**: 314–19 (2004).

49. Vongs, A., Kakutani, T., Martienssen, R. A., and Richards, E. J., *Arabidopsis thaliana* DNA methylation mutants. *Science* **260**: 1926–8 (1993).

50. Lippman, Z. G., Black, A.-V., Vaughn, M. W., Dedhia, N., McCombie, R. W. et al. Role of transposable elements in heterochromatin and epigenetic control. *Nature* **430**: 471–6 (2004).

51. Jeddeloh, J. A., Stokes, T. L., and Richards, E. J. Maintenance of genomic methylation requires a SWI2/SNF2-like protein. *Nature Genet.* **22**: 94–7 (1999).

52. Volpe, T., Schramke, V., Hamilton, G. L., White, S. A., Teng, G., Martienssen, R. A., and Allshire, R. C. RNA interference is required for normal centromere function in fission yeast. *Chrom. Res.* **11**: 137–46 (2003).

53. Volpe, T. A., Kidner, C., Hall, I. M., Teng, G., Grewal, S. I., and Martienssen, R. A. Regulation of heterochromatic silencing and histone H3 lysine-9 methylation by RNAi. *Science* **297**: 1833–7 (2002).

54. Adams, K. L. and Wendel, J. F. Polyploidy and genome evolution in plants. *Curr. Opin. Plant Biol.* **8**: 135–41 (2005).

55. Kashkush, K., Feldman, M., and Levy, A. Transcriptional activation of retrotransposons alters the expression of adjacent genes in wheat. *Nature Genet.* **33**: 102–6 (2003).

56. Robertson, K. D. DNA methylation and human disease. *Nat. Rev. Genet.* **6**: 597–610 (2005).

57. Galagan, J. and Selker, E. RIP: the evolutionary cost of genome defense. *Trends Genet.* **20**: 417–23 (2004).

58. Galagan, J. E., Calvo, S. E., Borkovich, K. A., Selker, E. U., Read, N. D., Jaffe, D. et al. The genome sequence of the filamentous fungus *Neurospora crassa*. *Nature* **422**: 859–68 (2003).

Chapter 9

1. Lewis, S. M. The mechanism of V(D)J joining: lessons from molecular, immunological, and comparative analyses. *Adv. Immunol.* **56**: 27–150 (1994).

2. Fugmann, S. D., Lee, A. I., Shockett, P. E., Villey, I. J., and Schatz D. G. The RAG proteins and V(D)J recombination: complexes, ends, and transposition. *Annu. Rev. Immunol.* **18**: 495–527 (2000).

3. Gellert, M. V(D)J recombination: RAG proteins, repair factors, and regulation. *Annu. Rev. Biochemistry* **71**: 101–32 (2002).

4. Schlissel, M. S. Regulating antigen-receptor gene assembly. *Nat. Rev. Immunol.* **3**: 890–9 (2003).

5. Lieber, M. R., Ma Y., Pannicke, U., and Schwarz, K. Mechanism and regulation of human non-homologous DNA end-joining. *Nat. Rev. Mol. Cell Biol.* **4**: 712–20 (2003).

6. Tonegawa, S. Somatic generation of antibody diversity. *Nature* **302**: 575–81 (1983).

7. Matsuda, F. Human immunoglobulin heavy chain locus, in T. Honjo, Alt, F., and Neuberger, M. (eds.), *Molecular Biology of B cells*, Elsevier Academic Press, Amsterdam (2004).

8. Bengten, E., Wilson. M., Miller, N., Clem, L. W., Pilstrom, L., and Warr, G. W. Immunoglobulin isotypes: structure, function, and genetics. *Curr. Top. Microbiol. Immunol.* **248**: 189–219 (2000).

9. Schatz, D. G., Oettinger, M. A., and Baltimore, D. The V(D)J recombination activating gene, RAG-1. *Cell* **59**: 1035–48 (1989).

10. Oettinger, M. A., Schatz, D. G., Gorka, C., and Baltimore, D. RAG-1 and RAG-2, adjacent genes that synergistically activate V(D)J recombination. *Science* **248**: 1517–23 (1990).

11. McBlane, J. F., van Gent, D. C., Ramsden, D. A., Romeo, C., Cuomo, C. A., Gellert, M., and Oettinger, M. A. Cleavage at a V(D)J recombination signal requires only RAG1 and RAG2 proteins and occurs in two steps. *Cell* **83**: 387–95 (1995).

12. Sakano, H., Huppi, K., Heinrich, G., and Tonegawa S. Sequences at the somatic recombination sites of immunoglobulin light-chain genes. *Nature* **280**: 288–94 (1979).

13. Early, P., Huang, H., Davis, M., Calame, K., and Hood, L. An immunoglobulin heavy chain variable region gene is generated from three segments of DNA: V_H, D and J_H. *Cell* **19**: 981–92 (1980).

14. Krangel, M. S. Gene segment selection in V(D)J recombination: accessibility and beyond. *Nature Immunol.* **4**: 624–30 (2003).

15. Aidinis, V., Bonaldi, T., Beltrame, M., Santagata, S., Bianchi, M. E., and Spanopoulou, E. The RAG1 homeodomain recruits HMG1 and HMG2 to facilitate recombination signal sequence binding and to enhance the intrinsic DNA-bending activity of RAG1-RAG2. *Mol. Cell Biol.* **19**: 6532–42 (1999).

16. van Gent, D. C., McBlane, J. F., Ramsden, D. A., Sadofsky, M. J., Hesse, J. E., and Gellert, M. Initiation of V(D)J recombination in a cell-free system. *Cell* **81**: 925–34 (1995).

17. van Gent, D. C., Mizuuchi, K., and Gellert, M. Similarities between initiation of V(D)J recombination and retroviral integration. *Science* **271**: 1592–4 (1996).

18. Moshous, D., Callebaut, I., de Chasseval, R., Corneo, B., Cavazzana-Calvo, M., Le Deist, F., Tezcan, I., Sanal, O., Bertrand, Y., Philippe, N., Fischer, A., and de Villartay, J. P. Artemis, a novel DNA doublestrand break repair/V(D)J recombination protein, is mutated in human severe combined immune deficiency. *Cell* **105**: 177–86 (2001).

19. Ma, Y., Pannicke, U., Schwarz, K., and Lieber, M. R. Hairpin opening and overhang processing by an artemis DNA-dependent protein kinase complex in nonhomologous end joining and V(D)J recombination. *Cell* **108**: 781–94 (2002).

20. Lafaille, J. J., DeCloux, A., Bonneville, M., Takagaki, Y., and Tonegawa, S. Junctional sequences of T cell receptor gamma delta genes: implications for gamma delta T cell lineages and for a novel intermediate of V-(D)-J joining. *Cell* **59**: 859–70 (1989).

21. Sawchuk, D. J., Mansilla-Soto, J., Alarcon, C., Singha, N. C., Langen, H., Bianchi, M. E., Lees-Miller, S. P., Nussenzweig, M. C., and Cortes, P. Ku70/Ku80 and DNA-dependent protein kinase catalytic subunit modulate RAG-mediated cleavage. *J. Biol. Chem.* **279**: 29821–31 (2004).

22. Lewis, S. M., Hesse, J. E., Mizuuchi, K., and Gellert, M. Novel strand exchanges in V(D)J recombination. *Cell* **55**: 1099–107 (1988).

23. Brandt, V. L. and Roth, D. B. A recombinase diversified: new functions of the RAG proteins. *Curr. Opin. Immunol.* **14**: 224–9 (2002).

24. Gilfillan, S., Benoist, C., and Mathis, D. Mice lacking terminal deoxynucleotidyl transferase: adult mice with a fetal antigen receptor repertoire. *Immunol. Rev.* **148**: 201–19 (1995).

25. de Villartay, J. P., Poinsignon, C., de Chasseval, R., Buck, D., Le Guyader, G., and Villey, I. Human and animal models of V(D)J recombination deficiency. *Curr. Opin. Immunol.* **15**: 592–8 (2003).

26. Hiom, K. and Gellert, M. Assembly of a 12/23 paired signal complex: a critical control point in V(D)J recombination. *Mol. Cell* **1**: 1011–9 (1998).

27. Eastman, Q. M. and Schatz, D. G. Nicking is asynchronous and stimulated by synapsis in 12/23 rule-regulated V(D)J cleavage. *Nucleic Acids Res.* **25**: 4370–8 (1997).

28. van Gent, D. C., Ramsden, D. A., and Gellert, M. The RAG1 and RAG2 proteins establish the 12/23 rule in V(D)J recombination. *Cell* **85**: 107–13 (1996).

29. West, R. B. and Lieber, M. R. The RAG-HMG1 complex enforces the 12/23 rule of V(D)J recombination specifically at the double-hairpin formation step. *Mol. Cell Biol.* **18**: 6408–15 (1998).

30. Corbett, S. J., Tomlinson, I. M., Sonnhammer, E. L. L., Buck, D., and Winter, G. Sequence of the human immunoglobulin diversity (D) segment locus: A systematic analysis provides no evidence for the use of DIR segments, inverted D segments, "minor" D segments or D-D recombination. *J. Mol. Biol.* **270**: 587–597 (1997).

31. Kokubu, F., Litman, R., Shamblott, M. J., Hinds, K., and Litman, G. W. Diverse organization of immunoglobulin V$_H$ gene loci in a primitive vertebrate. *EMBO J.* **7**: 3413–22 (1988).

32. Ramsden, D. A., Baetz, K., and Wu, G. E. Conservation of sequence in recombination signal sequence spacers. *Nucleic Acids Res.* **22**: 1785–96 (1994).

33. Hesse, J. E., Lieber, M. R., Mizuuchi, K., and Gellert, M. V(D)J recombination: a functional definition of the joining signals. *Genes Dev.* **3**: 1053–61 (1989).

34. Timsit, Y., Vilbois, E., and Moras, D. Base-pairing shift in the major groove of (CA)n tracts by B-DNA crystal structures. *Nature* **354**: 167–70 (1991).

35. Cuomo, C. A., Mundy, C. L., and Oettinger, M. A. DNA sequence and structure requirements for cleavage of V(D)J recombination signal sequences. *Mol. Cell Biol.* **16**: 5683–90 (1996).

36. Ramsden, D. A., McBlane, J. F., van Gent, D. C., and Gellert, M. Distinct DNA sequence and structure requirements for the two steps of V(D)J recombination signal cleavage. *EMBO J.* **15**: 3197–206 (1996).

37. Posnett, D. N., Vissinga, C. S., Pambuccian, C., Wei, S., Robinson, M. A., Kostyu, D., and Concannon, P. Level of human TCRBV3S1 (V beta 3) expression correlates with allelic polymorphism in the spacer region of the recombination signal sequence. *J. Exp. Med.* **179**: 1707–11 (1994).

38. Feeney, A., J., Tang, A., and Ogwaro, K. M. B-cell repertoire formation: role of the recombination signal sequence in non-random V segment utilization. *Immunol. Rev.* **175**: 59–69 (2000).

39. Montalbano, A., Ogwaro, K. M., Tang, A., Matthews, A., G., Larijani, M., Oettinger, M. A., and Feeney, A. J. V(D)J recombination frequencies can be profoundly affected by changes in the spacer sequence. *J. Immunol.* **171**: 5296–304 (2003).

40. Bassing, C. H., Alt, F. W., Hughes, M. M., D'Auteuil, M., Wehrly, T. D., Woodman, B. B., Gartner, F., White, J. M., Davidson, L., and Sleckman, B. P. Recombination signal sequences restrict chromosomal V(D)J recombination beyond the 12/23 rule. *Nature* **405**: 583–6 (2000).

41 Jung, D., Bassing, C. H., Fugmann, S. D., Cheng, H. L., Schatz, D. G., and Alt, F. W. Extrachromosomal recombination substrates recapitulate beyond 12/23 restricted VDJ recombination in nonlymphoid cells. *Immunity* **18**: 65–74 (2003).

42. Olaru, A., Patterson, D. N., Villey, I., and Livak, F. DNA-Rag protein interactions in the control of selective D gene utilization in the TCR beta locus. *J. Immunol.* **171**: 3605–11 (2003).

43. Ramsden, D. A. and Wu, G. E. Mouse kappa light-chain recombination signal sequences mediate recombination more frequently than do those of lambda light chain. *Proc. Natl. Acad. Sci. U. S. A.* **88**: 10721–5 (1991).

44. Hassanin, A., Golub, R., Lewis, S. M., and Wu, G. E. Evolution of the recombination signal sequences in the Ig heavy-chain variable region locus of mammals. *Proc. Natl. Acad. Sci. U. S. A.* **97**: 11415-20 (2000). Erratum in: *Proc. Natl. Acad. Sci. U. S. A.* **97**: 14015 (2000).

45. Weigert, M., Gatmaitan, L., Loh, E., Schilling, J., and Hood, L. Rearrangement of genetic information may produce immunoglobulin diversity. *Nature* **276**: 785–90 (1978).

46. Janeway, C.A., Travers, P., Walport, M., and Shlomchik, M. *Immunobiology*, Garland Publishing, New York, NY (2001), pp. 138.

47. Edmundson, A. B., Ely, K. R., Abola, E. E., Schiffer, M., and Panagiotopoulos, N. Rotational allomerism and divergent evolution of domains in immunoglobulin light chains. *Biochemistry* **14**: 3953–61 (1975).

48. Poljak, R. J., Amzel, L. M., Avey, H. P., Chen, B. L., Phizackerley, R. P., and Saul, F. Three-dimensional structure of the Fab' fragment of a human immunoglobulin at 2,8-A resolution. *Proc. Natl. Acad. Sci. U. S. A.* **70**: 3305–10 (1973).

49. Lewis, S. M., Wu, G. E., and Hsu, E., The origin of V(D)J diversification, in T. Honjo, Alt, F. W., and Neuberger, M. (eds.), *Molecular Biology of B Cells*, Elsevier, Amsterdam (2004).

50. Agrawal, A., Eastman, Q. M., and Schatz, D. G. Transposition mediated by RAG1 and RAG2 and its implications for the evolution of the immune system. *Nature* **394**: 744–51 (1998).

51. Hiom, K., Melek, M., and Gellert, M. DNA transposition by the RAG1 and RAG2 proteins: a possible source of oncogenic translocations. *Cell* **94**: 463–70 (1998).

52. Clatworthy, A. E., Valencia, M. A., Haber, J. E., and Oettinger, M. A. V(D)J recombination and RAG-mediated transposition in yeast. *Mol. Cell* **12**: 489–99 (2003).

53. Thompson, C. B. New insights into V(D)J recombination and its role in the evolution of the immune system. *Immunity* **3**: 531–9 (1995).

54. Lee, S. S., Tranchina, D. S., Ohta, Y., Flajnik, M. F., and Hsu, E. Hypermutation in shark immunoglobulin light chain genes results in contiguous substitutions. *Immunity* **16**: 571–82 (2002).

55. Lee, S. S., Fitch, D., Flajnik, M. F., and Hsu, E. Rearrangement of immunoglobulin genes in shark germ cells. *J. Exp. Med.* **191**: 1637–48 (2000).

56. Lewis, S. M. and Wu, G. E. The old and the restless. *J. Exp. Med.* **191**: 1631–6 (2000).

57. Rumfelt, L. L., Avila, D., Diaz, M., Bartl, S., McKinney, E. C., and Flajnik, M. F. A shark antibody heavy chain encoded by a nonsomatically rearranged VDJ is preferentially expressed in early development and is convergent with mammalian IgG. *Proc. Natl. Acad. Sci. U. S. A.* **98**: 1775–80 (2001).

Chapter 10

1. Ehrlich, P. Chemotherapeutics: Scientific principles, methods and results. *Lancet* **2**: 445–451 (1913).

2. Janeway, C. A. J. and Medzhitov, R. Innate immune recognition. *Annu. Rev. Immunol* **20**: 197–216 (2002).

3. Li, Z., Woo, C. J., Iglesias-Ussel, M. D., Ronai, D., and Scharff, M. D. The generation of antibody diversity through somatic hypermutation and class switch recombination. *Genes Dev* **18**(1): 1–11 (2004).

4. Muramatsu, M., Kinoshita, K., Fagarasan, S., Yamada, S., Shinkai, Y., and Honjo, T. Class switch recombination and hypermutation require activation-induced cytidine deaminase (AID), a potential RNA editing enzyme. *Cell* **102**(5): 553–563 (2000).

5. Arakawa, H., Hauschild, J., and Buerstedde, J.-M. Requirement of the activation-induced deaminase (AID) gene for immunoglobulin gene conversion. *Science* **295**(5558): 1301–1306 (2002).

6. Harris, R. S., Sale, J. E., Petersen-Mahrt, S. K., and Neuberger, M. S. AID is essential for immunoglobulin V gene conversion in a cultured B cell line. *Curr. Biol.* **12**(5): 435–438 (2002).

7. Petersen-Mahrt, S. K., Harris, R. S., and Neuberger, M. S. AID mutates *E. coli* suggesting a DNA deamination mechanism for antibody diversification. *Nature* **418**(6893): 99–103 (2002).

8. Bransteitter, R., Pham, P., Scharff, M. D., and Goodman, M. F. Activation-induced cytidine deaminase deaminates deoxycytidine on single-stranded DNA but requires the action of RNase. *Proc. Natl. Acad. Sci. U. S. A.* **100**(7): 4102–4107 (2003).

9. Takata, M., Sasaki, M. S., Tachiiri, S., Fukushima, T., Sonoda, E., Schild, D., Thompson, L. H. and Takeda, S. Chromosome instability and defective recombinational repair in knockout mutants of the five Rad51 paralogs. *Mol. Cell Biol.* **21**(8): 2858–2866 (2001).

10. Sale, J. E., Calandrini, D. M., Takata, M., Takeda, S., and Neuberger, M. S. Ablation of XRCC2/3 transforms immunoglobulin V gene conversion into somatic hypermutation. *Nature* **412**(6850): 921–926 (2001).

11. Di Noia, J. and Neuberger, M. S. Altering the pathway of immunoglobulin hypermutation by inhibiting uracil-DNA glycosylase. *Nature* **419**(6902): 43–48 (2002).

12. Rada, C., Williams, G. T., Nilsen, H., Barnes, D. E., Lindahl, T., and Neuberger, M. S. Immunoglobulin isotype switching is inhibited and somatic hypermutation perturbed in UNG-deficient mice. *Curr. Biol.* **12**(20): 1748–1755 (2002).

13. Imai, K., Slupphaug, G., Lee, W.-I., Revy, P., Nonoyama, S., Catalan, N., Yel, L., Forveille, M., Kavli, B., Krokan, H. E., Ochs, H. D., Fischer, A., and Durandy, A. Human uracil-DNA glycosylase deficiency associated with profoundly impaired immunoglobulin class-switch recombination. *Nature Immunol.* **4**(10): 1023–1028 (2003).

14. Pham, P., Bransteitter, R., Petruska, J., and Goodman, M. F. Processive AID-catalysed cytosine deamination on single-stranded DNA simulates somatic hypermutation. *Nature* **424**(6944): 103–107 (2003).

15. Yu, K., Huang, F.-T., and Lieber, M. R. DNA substrate length and surrounding sequence affect the activation-induced deaminase activity at cytidine. *J. Biol. Chem.* **279**(8): 6496–6500 (2004).

16. Beale, R. C. L., Petersen-Mahrt, S. K., Watt, I. N., Harris, R. S., Rada, C., and Neuberger, M. S. Comparison of the differential context-dependence of DNA deamination by APOBEC enzymes: correlation with mutation spectra *in vivo*. *J. Mol. Biol.* **337**(3): 585–596 (2004).

17. Brenner, S. and Milstein, C. Origin of antibody variation. *Nature* **211**(46): 242–243 (1966).

18. Martin, A. and Scharff, M. D. AID and mismatch repair in antibody diversification. *Nat. Rev. Immunol* **2**(8): 605–614 (2002).

19. Rada, C., Ehrenstein, M. R., Neuberger, M. S., and Milstein, C. Hot spot focusing of somatic hypermutation in MSH2-deficient mice suggests two stages of mutational targeting. *Immunity* **9**(1): 135–141 (1998).

20. Bardwell, P. D., Woo, C. J., Wei, K., Li, Z., Martin, A., Sack, S. Z., Parris, T., Edelmann, W., and Scharff, M. D. Altered somatic hypermutation and reduced class-switch recombination in exonuclease 1-mutant mice. *Nature Immunol.* **5**(2): 224–229 (2004).

21. Zeng, X., Winter, D. B., Kasmer, C., Kraemer, K. H., Lehmann, A. R., and Gearhart, P. J. DNA polymerase eta is an A-T mutator in somatic hypermutation of immunoglobulin variable genes. *Nature Immunol.* **2**(6): 537–541 (2001).

22. Rada, C., Di Noia, J. M., and Neuberger, M. S. Mismatch recognition and uracil excision provide complementary paths to both Ig switching and the A/T-focused phase of somatic mutation. *Mol. Cell.* **16**(2): 163–171 (2004).

23. Neuberger, M. S., Noia, J. M. D., Beale, R. C. L., Williams, G. T., Yang, Z., and Rada, C. Opinion: Somatic hypermutation at A.T pairs: polymerase error versus dUTP incorporation. *Nat. Rev. Immunol.* **5**(2): 171–178 (2005).

24. Yelamos, J., Klix, N., Goyenechea, B., Lozano, F., Chui, Y. L., Gonzalez Fernandez, A., Pannell, R., Neuberger, M. S., and Milstein, C. Targeting of non-Ig sequences in place of the V segment by somatic hypermutation. *Nature* **376**(6537): 225–229 (1995).

25. Rada, C. and Milstein, C. The intrinsic hypermutability of antibody heavy and light chain genes decays exponentially. *EMBO J.* **20**(16): 4570–4576 (2001).

26. Shen, H. M. and Storb, U. Activation-induced cytidine deaminase (AID) can target both DNA strands when the DNA is supercoiled. *Proc. Natl. Acad. Sci. U. S. A.* **101**(35): 12997–13002 (2004).

27. Chaudhuri, J., Khuong, C., and Alt, F. W. Replication protein A interacts with AID to promote deamination of somatic hypermutation targets. *Nature* **430**(7003): 992–998 (2004).

28. MacLennan, I., de Vinuesa, C. G., and Casamayor-Palleja, M. B-cell memory and the persistence of antibody responses. *Philos. Trans. R. Soc. Lond. B Biol. Sci.* **355**: 345–350 (2000).

29. MacLennan, I. Germinal centers. *Annu. Rev. Immunol.* **12**: 117–139 (1994).

30. Palmer, E. Negative selection-clearing out the bad apples from the T-cell repertoire. *Nat. Rev. Immunol.* **3**(5): 383–391 (2003).

31. Matsumoto, M., Lo, S. F., Carruthers, C. J., Min, J., Mariathasan, S., Huang, G., Plas, D. R., Martin, S. M., Geha, R. S., Nahm, M. H., and Chaplin, D. D. Affinity maturation without germinal centres in lymphotoxin-alpha-deficient mice. *Nature* **382**(6590): 462–466 (1996).

32. Liu, Y., Zhang, J., Chan, E., and MacLennan, I. Sites of specific B cell activation in primary and secondary responses to T-cell-dependent and T-cell-independent antigens. *Eur. J. Immunol.* **21**: 2951–2962 (1991).

33. Tew, J. G., Wu, J., Qin, D., Helm, S., Burton, G. F., and Szakal, A. K. Follicular dendritic cells and presentation of antigen and costimulatory signals to B cells. *Immunol. Rev.* **156**: 39–52 (1997).

34. Kosco-Vilbois, M. H. Are follicular dendritic cells really good for nothing? *Nat. Rev. Immunol.* **3**(9): 764–769 (2003).

35. Koni, P. A. and Flavell, R. A. Lymph node germinal centers form in the absence of follicular dendritic cell networks. *J. Exp. Med.* **189**(5): 855–864 (1999).

36. Arpin, C., Dechanet, J., Kooten, C. V., Merville, P., Grouard, G., Briere, F., Banchereau, J., and Liu, Y. Generation of memory B cells and plasma cells *in vitro*. *Science* **268**: 720–722 (1995).

37. Liu, Y. J., de Bouteiller, O., and Fugier-Vivier, I. Mechanisms of selection and differentiation in germinal centers. *Curr. Opin. Immunol.* **9**(2): 256–262 (1997).

38. Kepler, T. and Perelson, A. Somatic hypermutation in B cells: an optimal control treatment. *J. Theor. Biol.* **164**: 37–64 (1993).

39. Dal Porto, J. M., Haberman, A. M., Kelsoe, G., and Shlomchik, M. J. Very low affinity B cells form germinal centers, become memory B cells, and participate in secondary immune responses when higher affinity competition is reduced. *J. Exp. Med.* **195**(9): 1215–1221 (2002).

40. Shih, T.-A. Y., Meffre, E., Roederer, M., and Nussenzweig, M. C. Role of BCR affinity in T cell dependent antibody responses *in vivo*. *Nat. Immunol.* **3**(6): 570–575 (2002).

41. Siskind, G. W. and Benacerraf, B. Cell selection by antigen in the immune response. *Adv. Immunol.* **10**: 1–50 (1969).

42. Tarlinton, D. and Smith, K. Dissecting affinity maturation: a model explaining selection of antibody-forming cells and memory B cells in the germinal centre. *Immunol Today* **9**: 436–441 (2000).

43. Iber, D. and Maini, P. A mathematical model of germinal centre kinetics and affinity maturation. *J. Theor. Biol.* **219**: 153–175 (2002).

44. Haberman, A. and Shlomchik, M. Reassessing the function of immune-complex retention by follicular dendritic cells. *Nat. Rev. Immunol.* **3**: 757–764 (2003).

45. Manser, T. Textbook germinal centers? *J. Immunol.* **172**(6): 3369–3375 (2004).

46. Hannum, L., Haberman, A., Anderson, S., and Shlomchik, M. Germinal center initiation, variable gene region hypermutation, and mutant B cell selection without detectable immune complexes on follicular dendritic cells. *J. Exp. Med.* **192**: 931–942 (2000).

47. Kouskoff, V., Famiglietti, S., Lacaud, G., Lang, P., Rider, J. E., Kay, B. K., Cambier, J. C., and Nemazee, D. Antigens varying in affinity for the B cell receptor induce differential B lymphocyte responses. *J. Exp. Med.* **188**(8): 1453–1464 (1998).

48. Batista, F. and Neuberger, M. Affinity dependencs of the B cell response to antigen: A threshold, a ceiling and the importance of off-rate. *Immunity* **8**: 751–759 (1998).

49. Harding, C. V. and Unanue, E. R. Quantitation of antigen-presenting cell MHC class II/peptide complexes necessary for T-cell stimulation. *Nature* **346**(6284): 574–576 (1990).

50. Grakoui, A., Bromley, S., Sumen, C., Davis, M., Shaw, A., Allen, P., and Dustin, M. The immunological synapse: A molecular machine controlling T cell activation. *Science* **285**: 221–227 (1999).

51. Clark, M. R., Massenburg, D., Zhang, M., and Siemasko, K. Molecular mechanisms of B cell antigen receptor trafficking. *Ann. N. Y. Acad. Sci.* **987**: 26–37 (2003).

52. Kunichika, K., Hashimoto, Y., and Imoto, T. Robustness of hen egg lysozyme monitored by random mutations. *Protein Engng* **15**: 805–9 (2002).

53. Berek, C., Berger, A., and Apel, M. Maturation of the immune response in germinal centers. *Cell* **67**: 1121–1129 (1991).

54. Scrutton, N. S., Berry, A., and Perham, R. N. Redesign of the coenzyme specificity of a dehydrogenase by protein engineering. *Nature* **343**(6253): 38–43 (1990).

55. Kauffman, S. *The Origins of Order*. Oxford University Press, Oxford (1993).

56. Takahashi, Y., Cerasoli, D. M., Dal Porto, J. M., Shimoda, M., Freund, R., Fang, W., Telander, D. G., Malvey, E. N., Mueller, D. L., Behrens, T. W., and Kelsoe, G. Relaxed

negative selection in germinal centers and impaired affinity maturation in bcl-xL transgenic mice. *J. Exp. Med.* **190**(3): 399–410 (1999).

57. Takahashi, Y., Ohta, H., and Takemori, T. Fas is required for clonal selection in germinal centers and the subsequent establishment of the memory B cell repertoire. *Immunity* **14**(2): 181–192 (2001).

58. Cumbers, S. J., Williams, G. T., Davies, S. L., Grenfell, R. L., Takeda, S., Batista, F. D., Sale, J. E., and Neuberger, M. S. Generation and iterative affinity maturation of antibodies *in vitro* using hypermutation B-cell lines. *Nat. Biotechnol* **20**(11): 1129–1134 (2002).

Chapter 11

1. Sen, D. and Gilbert, W. Formation of parallel four-stranded complexes by guanine rich motifs in DNA and its implications for meiosis. *Nature* **334**: 364–366 (1988).

2. Gellert, M., Lipsett, M. N., and Davies, D. R. Helix formation by guanylic acid. *Proc. Natl. Acad. Sci. U. S. A.* **48**: 2014–2018 (1962).

3. Gilbert, D. E. and Feigon, J. Multistranded DNA structures. *Curr. Opin. Struct. Biol.* **9**: 305–314 (1999).

4. Keniry, M. A. Quadruplex structures in nucleic acids. *Biopolymers* **56**: 123–146 (2000).

5. Sen, D. and Gilbert, W. Novel DNA superstructures formed by telomere-like oligomers. *Biochemistry* **31**: 65–70 (1992).

6. Parkinson, G. N., Lee, M. P., and Neidle, S. Crystal structure of parallel quadruplexes from human telomeric DNA. *Nature* **417**: 876–880 (2002).

7. Duquette, M. L., Handa, P., Vincent, J. A., Taylor, A. F., and Maizels, N. Intracellular transcription of G-rich DNAs induces formation of G-loops, novel structures containing G4 DNA. *Genes Dev.* **18**: 1618–1629 (2004).

8. Chaganti, R. S., Schonberg, S., and German, J. A manyfold increase in sister chromatid exchanges in Bloom's syndrome lymphocytes. *Proc. Natl. Acad. Sci. U. S. A.* **71**: 4508–4512 (1974).

9. Watt, P. M., Louis, E. J., Borts, R. H., and Hickson, I. D. Sgs1: a eukaryotic homolog of *E. coli* RecQ that interacts with topoisomerase II *in vivo* and is required for faithful chromosome segregation. *Cell* **81**: 253–260 (1995).

10. Mills, M., Lacroix, L., Arimondo, P. B., Leroy, J. L., Francois, J. C., Klump, H., and Mergny, J. L. Unusual DNA conformations: implications for telomeres. *Curr. Med. Chem. Anti-Canc. Agents* **2**: 627–644 (2002).

11. Neidle, S. and Parkinson, G. N. The structure of telomeric DNA. *Curr. Opin. Struct. Biol.* **13**: 275–283 (2003).

12. Osheim, Y., Mougey, E. B., Windle, J., Anderson, M., O'Reilly, M., Miller, O. L., Beyer, A., and Sollner-Webb, B. Metazoan rDNA enhancer acts by making more genes transcriptionally active. *J. Cell Biol.* **133**: 943–954 (1996).

13. Hanakahi, L. A., Sun, H., and Maizels, N. High affinity interactions of nucleolin with G-G paired rDNA. *J. Biol. Chem.* **274**: 15908–15912 (1999).

14. Wong, Z., Wilson, V., Patel, I., Povey, S., and Jeffreys, A. J. Characterization of a panel of highly variable minisatellites cloned from human DNA. *Ann. Hum. Genet.* **51**(4): 269–288 (1987).

15. Jeffreys, A. J., Royle, N. J., Wilson, V., and Wong, Z. Spontaneous mutation rates to new length alleles at tandem-repetitive hypervariable loci in human DNA. *Nature* **332**: 278–281 (1988).

16. Fry, M. and Loeb, L. A. The fragile X syndrome d(CGG)n nucleotide repeats form a stable tetrahelical structure. *Proc. Natl. Acad. Sci. U. S. A.* **91**: 4950–4954 (1994).

17. Weitzmann, M. N., Woodford, K. J., and Usdin, K. DNA secondary structures and the evolution of hypervariable tandem arrays. *J. Biol. Chem.* **272**: 9517–9523 (1997).

18. Hagerman, P. J. and Hagerman, R. J. The fragile-X premutation: a maturing perspective. *Am. J. Hum. Genet.* **74**: 805–816 (2004).

19. Darnell, J. C., Jensen, K. B., Jin, P., Brown, V., Warren, S. T., and Darnell, R. B. Fragile X mental retardation protein targets G quartet mRNAs important for neuronal function. *Cell* **107**: 489–499 (2001).

20. Sun, H., Yabuki, A., and Maizels, N. A. human nuclease specific for G4 DNA. *Proc. Natl. Acad. Sci. U. S. A.* **89**: 12444–12449 (2001).

21. Reaban, M. E. and Griffin, J. A. Induction of RNA-stabilized DNA conformers by transcription of an immunoglobulin switch region. *Nature* **348**: 342–344 (1990).

22. Mizuta, R., Iwai, K., Shigeno, M., Mizuta, M., Ushiki, T., and Kitamura, D. Molecular visualization of immunoglobulin switch region RNA/DNA complex by atomic force microscope. **278**(7): 4431–4434 (2003).

23. Sugimoto, N., Nakano, S., Katoh, M., Matsumura, A., Nakamuta, H., Ohmichi, T., Yoneyama, M., and Sasaki, M. Thermodynamic parameters to predict stability of RNA/DNA hybrid duplexes. *Biochemistry* **34**: 11211–11216 (1995).

24. Yin, Y. W. and Steitz, T. A. Structural basis for the transition from initiation to elongation transcription in T7 RNA polymerase. *Science* **298**: 1387–1395 (2002).

25. Westover, K. D., Bushnell, D. A., and Kornberg, R. D. Structural basis of transcription: separation of RNA from DNA by RNA polymerase II. *Science* **303**: 1014–1016 (2004).

26. Chapados, B. R. et al. Structural biochemistry of a type 2 RNase H: RNA primer recognition and removal during DNA replication. *J. Mol. Biol.* **307**: 541–556 (2001).

27. Ohtani, N. et al. Identification of the genes encoding Mn2+-dependent RNase HII and Mg2+-dependent RNase HIII from *Bacillus subtilis*: classification of RNases H into three families. *Biochemistry* **38**: 605–618 (1999).

28. Qiu, J., Qian, Y., Chen, V., Guan, M. X., and Shen, B. Human exonuclease 1 functionally complements its yeast homologues in DNA recombination, RNA primer removal, and mutation avoidance. *J. Biol. Chem.* **274**: 17893–17900 (1999).

29. Strasser, K., Masuda, S., Mason, P., Pfannstiel, J., Oppizzi, M., Rodriguez-Navarro, S., Rondon, A. G., Aguilera, A., Struhl, K., Reed, R., and Hurt, E. TREX is a conserved complex coupling transcription with messenger RNA export. *Nature* **417**: 304–308 (2002).

30. Huertas, P. and Aguilera, A. Cotranscriptionally formed DNA: RNA hybrids mediate transcription elongation impairment and transcription-associated recombination. *Mol. Cell* **12**: 711–721 (2003).

31. Manis, J. P., Tian, M., and Alt, F. W. Mechanism and control of class-switch recombination. *Trends Immunol.* **23**: 31–39 (2002).

32. Muramatsu, M., Kinoshita, K., Fagarasan, S., Yamada, S., Shinkai, Y., and Honjo, T. Class switch recombination and hypermutation require activation-induced cytidine deaminase (AID), a potential RNA editing enzyme. *Cell* **102**: 553–563 (2000).

33. Revy, P., Muto, T., Levy, Y., Geissmann, F., Plebani, A., Sanal, O., Catalan, N., Forveille, M., Dufourcq-Labelouse, R., Gennery, A., Tezcan, I., Ersoy, F., Kayserili, H., Ugazio, A. G., Brousse, N., Muramatsu, M., Notarangelo, L. D., Kinoshita, K., Honjo, T., Fischer, A., and Durandy, A. Activation-induced cytidine deaminase (AID) deficiency causes the autosomal recessive form of the Hyper-IgM syndrome (HIGM2). *Cell* **102**: 565–575 (2000).

34. Neuberger, M. S., Harris, R. S., Di Noia, J., and Petersen-Mahrt, S. K. Immunity through DNA deamination. *Trends Biochem. Sci.* **28**: 305–312 (2003).

35. Maizels, N. and Scharff, M. D., Molecular mechanisms of hypermutation, in T. Honjo, M. Neuberger, and F. W. Alt (eds.), *Molecular Biology of B Cells*, pp. 327–338, Elsevier Academic Press, Amsterdam (2004).

36. Okazaki, I. M., Kinoshita, K., Muramatsu, M., Yoshikawa, K., and Honjo, T. The AID enzyme induces class switch recombination in fibroblasts. *Nature* **416**: 340–345. (2002).

37. Yoshikawa, K., Okazaki, I. M., Eto, T., Kinoshita, K., Muramatsu, M., Nagaoka, H., and Honjo, T. AID enzyme-induced hypermutation in an actively transcribed gene in fibroblasts. *Science* **296**: 2033–2036. (2002).

38. Dunnick, W., Hertz, G. Z., Scappino, L., and Gritzmacher, C. DNA sequences at immunoglobulin switch region recombination sites. *Nucleic Acids Res.* **21**: 365–372 (1993).

39. Mussman, R., Courtet, M., Schwaer, J., and Du Pasquier, L. Microsites for immunoglobulin switch recombination breakpoints from *Xenopus* to mammals. *Eur. J. Immunol.* **27**: 2610–2619 (1997).

40. Zarrin, A. A. et al. An evolutionarily conserved target motif for immunoglobulin class-switch recombination. *Nature Immunol.* **5**: 1275–1281 (2004).

41. Snapper, C. M., Marcu, K. B., and Zelazowski, P. The immunoglobulin class switch: beyond accessibility. *Immunity* **6**: 217–223 (1997).

42. Shinkura, R. et al. The influence of transcriptional orientation on endogenous switch region function. *Nature Immunol.* **4**: 435–441 (2003).

43. Chaudhuri, J., Khuong, C., and Alt, F. W. Replication protein A interacts with AID to promote deamination of somatic hypermutation targets. *Nature*, 992–998 (2004).

44. Buermeyer, A. B., Deschenes, S. M., Baker, S. M., and Liskay, R. M. Mammalian DNA mismatch repair. *Annu. Rev. Genet.* **33**: 533–564 (1999).

45. Ehrenstein, M. R. and Neuberger, M. S. Deficiency in Msh2 affects the efficiency and local sequence specificity of immunoglobulin class-switch recombination: parallels with somatic hypermutation. *EMBO J.* **18**: 3484–3490 (1999).

46. Li, Z., Scherer, S. J., Ronai, D., Iglesias-Ussel, M. D., Peled, J. U., Bardwell, P. D., Zhuang, M., Lee, K., Martin, A., Edelmann, W., and Scharff, M. D. Examination of Msh6- and Msh3-deficient mice in class switching reveals overlapping and distinct roles of MutS homologues in antibody diversification. *J. Exp. Med.* **200**: 47–59 (2004).

47. Martomo, S. A., Yang, W. W., and Gearhart, P. J. A role for msh6 but not msh3 in somatic hypermutation and class switch recombination. *J. Exp. Med.* **200**: 61–68 (2004).

48. Larson, E. D., Duquette, M. L., Cummings, W. J., Streiff, R. J., and Maizels, N. MutSα binds to and promotes synapsis of transcriptionally activated immunoglobulin S regions. *Curr. Biol.* **15**: 470–474 (2005).

49. Duquette, M. L., Pham, P., Goodman, M. F., and Maizels, N. AID binds to transcription-induced structures in c-MYC that map to regions associated with translocation and hypermutation. *Oncogene.* **24**: 5791–5798 (2005).

50. Simonsson, T., Pecinka, P., and Kubista, M. DNA tetraplex formation in the control region of c-myc. *Nucleic Acids Res* **26**: 1167–1172 (1998).

51. Rangan, A., Fedoroff, O. Y., and Hurley, L. H. Induction of duplex to G-quadruplex transition in the c-myc promoter region by a small molecule. *J. Biol. Chem.* **276**: 4640–4646 (2001).

52. Siddiqui-Jain, A., Grand, C. L., Bearss, D. J., and Hurley, L. H. Direct evidence for a G-quadruplex in a promoter region and its targeting with a small molecule to repress c-MYC transcription. *Proc. Natl. Acad. Sci. U. S. A.* **99**: 11593–11598 (2002).

53. Kuppers, R. and Dalla-Favera, R. Mechanisms of chromosomal translocations in B cell lymphomas. *Oncogene* **20**: 5580–5594 (2001).

54. Blackburn, E. H. Switching and signaling at the telomere. *Cell* **106**: 661–673 (2001).

55. de Lange, T. Protection of mammalian telomeres. *Oncogene* **21**: 532–540 (2002).

56. Vega, L. R., Mateyak, M. K., and Zakian, V. A. Getting to the end: telomerase access in yeast and humans. *Nat. Rev. Mol. Cell. Biol.* **4**: 948–959 (2003).

57. Shay, J. W., Zou, Y., Hiyama, E., and Wright, W. E. Telomerase and cancer. *Hum. Mol. Genet.* **10**: 677–685 (2001).

58. Griffith, J. D., Comeau, L., S., R., Stansel, R. M., Bianchi, A., Moss, H., and de Lange, H. Mammalian telomeres end in a large duplex loop. *Cell* **97**: 503–514 (1999).

59. Dunham, M. A., Neumann, A. A., Fasching, C. L., and Reddel, R. R. Telomere maintenance by recombination in human cells. *Nature Genet.* **26**: 447–450 (2000).

60. Neumann, A. A. and Reddel, R. R. Telomere maintenance and cancer – look, no telomerase. *Nat. Rev. Cancer* **2**: 879–884 (2002).

61. Cohen, H. and Sinclair, D.A. Recombination-mediated lengthening of terminal telomeric repeats requires the Sgs1 DNA helicase. *Proc. Natl. Acad. Sci. U. S. A.* **98**: 3174–3179 (2001).

62. Huang, P., Pryde, F. E., Lester, D., Maddison, R. L., Borts, R., H., Hickson, I. D., and Louis, E. J. SGS1 is required for telomere elongation in the absence of telomerase. *Curr. Biol.* **11**: 125–129 (2001).

63. Johnson, F. B., Marciniak, R. A., McVey, M., Stewart, S. A., Hahn, W. C., and Guarente, L. The *Saccharomyces cerevisiae* WRN homolog Sgs1p participates in telomere maintenance in cells lacking telomerase. *EMBO J.* **20**: 905–913 (2001).

64. Sun, H., Karow, J. K., Hickson, I. D., and Maizels, N. The Bloom's syndrome helicase unwinds G4 DNA. *J. Biol. Chem.* **273**: 27587–27592 (1998).

65. Hickson, I. D. RecQ helicases: caretakers of the genome. *Nat. Rev. Cancer* **3**: 169–178 (2003).

66. Sun, H., Bennett, R. J., and Maizels, N. The *S. cerevisiae* Sgs1 helicase efficiently unwinds G-G paired DNAs. *Nucleic Acids Res.* **27**: 1978–1984 (1999).

67. Fry, M. and Loeb, L. A. Human Werner syndrome DNA helicase unwinds tetrahelical structures of the fragile X syndrome repeat sequence d(CGG)n. *J. Biol. Chem.* **274**: 12797–12802 (1999).

68. Huber, M. D., Lee, D. C., and Maizels, N. G4 DNA unwinding by BLM and Sgs1p: substrate specificity and substrate-specific inhibition. *Nucleic Acids Res.* **30**: 3954–3961 (2002).

69. Wu, X. and Maizels, N. Substrate-specific inhibition of RecQ helicase. *Nucleic Acids Res.* **29**: 1765–1771 (2001).

70. Cobb, J. A., Bjergbaek, L., Shimada, K., Frei, C., and Gasser, S. M. DNA polymerase stabilization at stalled replication forks requires Mec1 and the RecQ helicase Sgs1. *EMBO J.* **22**: 4325–4336 (2003).

71. Mohaghegh, P., Karow, J. K., Brosh Jr, R. M., Jr., Bohr, V. A., and Hickson, I. D. The Bloom's and Werner's syndrome proteins are DNA structure-specific helicases. *Nucleic Acids Res.* **29**: 2843–2849 (2001).

72. Bachrati, C. Z. and Hickson, I. D. RecQ helicases: suppressors of tumorigenesis and premature aging. *Biochem. J.* **374**: 577–606 (2003).

73. Sinclair, D.A. and Guarente, L. Extrachromosomal rDNA circles—a cause of aging in yeast. *Cell* **91**: 1033–1042 (1997).

74. Sinclair, D. A., Mills, K., and Guarente, L. Accelerated aging and nucleolar fragmentation in yeast sgs1 mutants. *Science* **277**: 1313–1316 (1997).

75. Lee, S. K., Johnson, R. E., Yu, S. L., Prakash, L., and Prakash, S. Requirement of yeast SGS1 and SRS2 genes for replication and transcription. *Science* **286**: 2339–2342 (1999).

76. Versini, G., Comet, I., Wu, M., Hoopes, L., Schwob, E., and Pasero, P. The yeast Sgs1 helicase is differentially required for genomic and ribosomal DNA replication. *EMBO J.* **22**: 1939–1949 (2003).

77. Shiratori, M., Suzuki, T., Itoh, C., Goto, M., Furuichi, Y., and Matsumoto, T. WRN helicase accelerates the transcription of ribosomal RNA as a component of an RNA polymerase I-associated complex. *Oncogene* **21**: 2447–2454 (2002).

78. Kondo, N., Motoyoshi, F., Mori, S., Kuwabara, N., Orii, T., and German, J. Long-term study of the immunodeficiency of Bloom's syndrome. *Acta Pediatr.* **81**: 86–90 (1992).

79. Hu, P. et al. Evidence for BLM and Topoisomerase IIIalpha interaction in genomic stability. *Hum. Mol. Genet.* **10**: 1287–1298 (2001).

80. Opresko, P. L., von Kobbe, C., Laine, J. P., Harrigan, J., Hickson, I. D., and Bohr, V. A. Telomere-binding protein TRF2 binds to and stimulates the Werner and Bloom syndrome helicases. *J. Biol. Chem.* **277**: 41110–41119 (2002).

81. Bai, Y. and Murnane, J. P. Telomere instability in a human tumor cell line expressing a dominant-negative WRN protein. *Hum. Genet.* **113**: 337–347 (2003).

82. Ding, H., Schertzer, M., Wu, X., Gertsenstein, M., Selig, S., Kammori, M., Pourvali, R., Poon, S., Vulto, I., Chavez, E., Tam, P. P., Nagy, A., and Lansdorp, P. M. Regulation of murine telomere length by Rtel: an essential gene encoding a helicase-like protein. *Cell* **117**: 873–886 (2004).

83. Blasco, M. A., Lee, H. W., Hande, M. P., Samper, E., Lansdorp, P. M., DePinho, R. A., and Greider, C. W. Telomere shortening and tumor formation by mouse cells lacking telomerase RNA. *Cell* **91**: 25–34 (1997).

84. Cheung, I., Schertzer, M., Rose, A., and Lansdorp, P. M. Disruption of dog-1 in Caenorhabditis elegans triggers deletions upstream of guanine-rich DNA. *Nature Genet.* **31**: 405–409 (2002).

85. Kusano, A., Staber, C., and Ganetzky, B. Nuclear mislocalization of enzymatically active RanGAP causes segregation distortion in *Drosophila. Dev. Cell* **1**: 351–361 (2001).

86. Hollingsworth, N.M., Goetsch, L., and Byers, B. The HOP1 gene encodes a meiosis-specific component of yeast chromosomes. *Cell* **61**: 73–84 (1990).

87. Muniyappa, K., Anuradha, S., and Byers, B. Yeast meiosis-specific protein Hop1 binds to G4 DNA and promotes its formation. *Mol. Cell. Biol.* **20**: 1361–1369 (2000).

88. Anuradha, S. and Muniyappa, K. Meiosis-specific yeast Hop1 protein promotes synapsis of double-stranded DNA helices via the formation of guanine quartets. *Nucleic Acids Res.* **32**: 2378–2385 (2004).

89. Venczel, E. A. and Sen, D. Synapsable DNA. *J. Mol. Biol.* **257**: 219–224 (1996).

90. Gilbert, W. The exon theory of genes. *Cold Spring Harb. Symp. Quant. Biol.* **52**: 901–905 (1987).

Chapter 12

1. Szostak, J. W., Orr-Weaver, T. L., Rothstein, R. J., and Stahl, F. W. The double-strand break model for recombination. *Cell* **33**: 25–35 (1983).

2. Merker, J. D., Dominska, M., and Petes, T. D. Patterns of heteroduplex formation associated with the initiation of meiotic recombination in the yeast *Saccharomyces cerevisiae. Genetics* **165**: 47–63 (2003).

3. Hoffmann, E. R. and Borts, R. H. Meiotic recombination intermediates and mismatch repair proteins. *Cytogenet Genome Research* **107**; 232–248 (2004).

4. Sun, H., Treco, D., and Szostak, J. W. Extensive 3′-overhanging, single-stranded DNA associated with the meiosis-specific double-strand breaks at the *ARG4* recombination initiation site. *Cell* **64**: 1155–1161 (1991).

5. Holliday, R. A mechanism for gene conversion in fungi. *Gen. Res.* **5**: 282–304 (1964).

6. Hollingsworth, N. M. and Brill, S. J. The Mus81 solution to resolution: generating meiotic crossovers without Holliday junctions. *Genes Dev* **18**: 117–25 (2004).

7. Borner, G. V., Kleckner, N., and Hunter, N. Crossover/noncrossover differentiation, synaptonemal complex formation, and regulatory surveillance at the leptotene/zygotene transition of meiosis. *Cell* **117**: 29–45 (2004).

8. Allers, T. and Lichten, M. Differential timing and control of noncrossover and crossover recombination during meiosis. *Cell* **106**: 47–57 (2001).

9. Stahl, F. W., Foss, H. M., Young, L. S., Borts, R. H., Abdullah, M. F., and Copenhaver, G. P. Does crossover interference count in *Saccharomyces cerevisiae*? *Genetics* **168**: 35–48 (2004).

10. Kleckner, N., Zickler, D., Jones, G. H., Dekker, J., Padmore, R., Henle, J., and Hutchinson, J. A mechanical basis for chromosome function. *Proc. Natl. Acad. Sci. U. S. A.* **101**, 12592–7 (2004).

11. Alani, E., Reenan, R. A. G., and Kolodner, R. D. Interaction between mismatch repair and genetic recombination in *Saccharomyces cerevisiae*. *Genetics* **137**: 19–39 (1994).

12. Kirkpatrick, D. T. and Petes, T. D. Repair of DNA loops involves DNA mismatch and nucleotide excision repair proteins. *Nature* **387**: 929–931 (1997).

13. Kirkpatrick, D. T. Roles of the DNA mismatch repair and nucleotide excision repair proteins during meiosis. *Cell Mol. Life Sci.* **55**: 437–449 (1999).

14. Borts, R. H., Chambers, S. R., and Abdullah, M. F. The many faces of mismatch repair in meiosis. *Mutat. Res.* **451**: 129–50 (2000).

15. Nicolas, A. and Petes, T. D. Polarity of meiotic gene conversion in fungi: Contrasting views. *Experientia* **50**: 242–252 (1994).

16. Detloff, P., White, M. A., and Petes, T. D. Analysis of a gene conversion gradient at the *HIS4* locus in *Saccharomyces cerevisiae*. *Genetics* **132**: 113–123 (1992).

17. Schultes, N. P. and Szostak, J. Decreasing gradients of gene conversion on both sides of the initiation site for meiotic recombination at the *ARG4* locus in yeast. *Genetics* **126**: 813–822 (1990).

18. Fan, Q.-Q., Xu, F., and Petes, T. D. Meiosis-specific double-strand DNA breaks at the *HIS4* recombination hot spot in the yeast *Saccharomyces cerevisiae*: Control in *cis* and *trans*. *Mol. Cell Biol.* **15**: 1679–1688 (1995).

19. Xu, L. and Kleckner, N. Sequence non-specific double-strand breaks and interhomolog interactions prior to double-strand break formation at a meiotic recombination hot spot in yeast. *EMBO J.* **14**: 5115–5128 (1995).

20. Sun, H., Treco, D., Schultes, N. P., and Szostak, J. W. Double-strand breaks at an initiation site for meiotic gene conversion. *Nature* **338**, 87–90 (1989).

21. Baudat, F. and Nicolas, A. Clustering of meiotic double-strand breaks on yeast chromosome III. *Proc. Natl. Acad. Sci. U. S. A.* **94**: 5213–5218 (1997).

22. Gerton, J. L., DeRisi, J., Shroff, R., Lichten, M., Brown, P. O., and Petes, T. D. Global mapping of meiotic recombination hotspots and coldspots in the yeast *Saccharomyces cerevisiae*. *Proc. Natl. Acad. Sci. U. S. A.* **97**, 11383–11390 (2000).

23. Liu, J., Wu, T. C., and Lichten, M. The location and structure of double-strand DNA breaks induced during yeast meiosis: evidence for a covalently linked DNA-protein intermediate. *EMBO J.* **14**: 4599–4608 (1995).

24. de Massy, B., Rocco, V., and Nicolas, A. The nucleotide mapping of DNA double-strand breaks at the *CYS3* initiation site of meiotic recombination in *Saccharomyces cerevisiae*. *EMBO J.* **14**: 4589–4598 (1995).

25. Xu, F. and Petes, T. D. Fine-structure mapping of meiosis-specific double-strand DNA breaks at a recombination hotspot associated with an insertion of telomeric sequences upstream of the *HIS4* locus in yeast. *Genetics* **143**: 1115–1125 (1996).

26. Keeney, S., Giroux, C. N., and Kleckner, N. Meiosis-specific DNA double-strand breaks are catalyzed by Spo11, a member of a widely conserved protein family. *Cell* **88**: 375–384 (1997).

27. Wu, T.-C. and Lichten, M. Meiosis-induced double-strand break sites determined by yeast chromatin structure. *Science* **263**: 515–518 (1994).

28. Fan, Q.-Q. and Petes, T. D. Relationship between nuclease-hypersensitive sites and meiotic recombination hot spot activity at the *HIS4* locus of *Saccharomyces cerevisiae*. *Mol. Cell Biol.* **16**: 2037–2043 (1996).

29. Ohta, K., Shibata, T., and Nicolas, A. Changes in chromatin structure at recombination initiation sites during yeast meiosis. *EMBO J.* **13**: 5754–5763 (1994).

30. Eggleston, A. K. and West, S. C. Recombination initiation: easy as A, B, C, D... chi? *Curr. Biol.* **7**: R745–R749 (1997).

31. Blumental-Perry, A., Zenvirth, D., Klein, S., Onn, I., and Simchen, G. DNA motif associated with meiotic double-strand break regions in *Saccharomyces cerevisiae*. *EMBO Rep.* **1**: 232–238 (2000).

32. Haring, S. J., Lautner, L. J., Comeron, J. M., and Malone, R. E. A test of the CoHR motif associated with meiotic double-strand breaks in *Saccharomyces cerevisiae*. *EMBO Rep.* **5**: 41–46 (2004).

33. Diaz, R. L., Alcid, A. D., Berger, J. M., and Keeney, S. Identification of residues in yeast Spo11p critical for meiotic DNA double-strand break formation. *Mol. Cell Biol.* **22**: 1106–1115 (2002).

34. Kirkpatrick, D. T., Fan, Q.-Q., and Petes, T. D. Maximal stimulation of meiotic recombination by a yeast transcription factor requires the transcription activation domain and a DNA-binding domain. *Genetics* **152**: 101–115 (1999).

35. Petes, T. D. Meiotic recombination hot spots and cold spots. *Nat. Rev. Genet.* **2**: 360–369 (2001).

36. Abdullah, M. F. and Borts, R. H. Meiotic recombination frequencies are affected by nutritional states in *Saccharomyces cerevisiae*. *Proc. Natl. Acad. Sci. U. S. A.* **98**: 14524–14529. (2001).

37. White, M. A., Wierdl, M., Detloff, P., and Petes, T. D. DNA-binding protein *RAP1* stimulates meiotic recombination at the *HIS4* locus in yeast. *Proc. Natl. Acad. Sci. U. S. A.* **88**: 9755–9759 (1991).

38. White, M. A., Dominska, M., and Petes, T. D. Transcription factors are required for the meiotic recombination hotspot at the *HIS4* locus in *Saccharomyces cerevisiae*. *Proc. Natl. Acad. Sci. U. S. A.* **90**: 6621–6625 (1993).

39. White, M. A., Detloff, P., Strand, M., and Petes, T. D. A promoter deletion reduces the rate of mitotic, but not meiotic, recombination at the *HIS4* locus in yeast. *Curr. Genet.* **21**: 109–116 (1992).

40. Kon, N., Krawchuk, M. D., Warren, B. G., Smith, G. R., and Wahls, W. P. Transcription factor Mts1/Mts2 (Atf1/Pcr1, Gad7/Pcr1) activates the M26 meiotic recombination hotspot in *Schizosaccharomyces pombe*. *Proc. Natl. Acad. Sci. U. S. A.* **94**: 13765–13770 (1997).

41. Kirkpatrick, D. T., Wang, Y.-H., Dominska, M., Griffith, J. D., and Petes, T. D. Control of meiotic recombination and gene expression in yeast by a simple repetitive DNA sequence that excludes nucleosomes. *Mol. Cell Biol.* **19**: 7661–7671 (1999).

42. Stapleton, A. and Petes, T. D. The Tn3 B-lactamase gene acts as a hotspot for meiotic recombination in yeast. *Genetics* **127**: 39–51 (1991).

43. Borde, V., Wu, T. C., and Lichten, M. Use of a recombination reporter insert to define meiotic recombination domains on chromosome III of *Saccharomyces cerevisiae*. *Mol. Cell Biol.* **19**: 4832–4842 (1999).

44. Jauert, P. A., Edmiston, S. N., Conway, K., and Kirkpatrick, D. T. *RAD1* controls the meiotic expansion of the human *HRAS1* minisatellite in *Saccharomyces cerevisiae*. *Mol. Cell Biol.* **22**: 953–964 (2002).

45. Debrauwere, H., Buard, J., Tessier, J., Aubert, D., Vergnaud, G., and Nicolas, A. Meiotic instability of human minisatellite *CEB1* in yeast requires DNA double-strand breaks. *Nature Genet.* **23**: 367–371 (1999).

46. Fischer, G., James, S. A., Roberts, I. N., Oliver, S. G., and Louis, E. J. Chromosomal evolution in *Saccharomyces*. *Nature* **405**: 451–454 (2000).

47. Bailey, J. A., Baertsch, R., Kent, W. J., Haussler, D., and Eichler, E. E. Hotspots of mammalian chromosomal evolution. *Genome Biol.* **5**: R23 (2004).

48. Murphy, W. J., Larkin, D. M., Everts-van der Wind, A., Bourque, G., Tesler, G., Auvil, L., Beever, J. E., Chowdhary, B. P., Galibert, F., Gatzke, L., Hitte, C., Meyers, S. N., Milan, D., Ostrander, E. A., Pape, G., Parker, H. G., Raudsepp, T., Rogatcheva, M. B., Schook, L. B., Skow, L. C., Welge, M., Womack, J. E., O'Brien S, J., Pevzner, P. A., and Lewin, H. A. Dynamics of mammalian chromosome evolution inferred from multispecies comparative maps. *Science* **309**: 613–617 (2005).

49. Shaw, C. J. and Lupski, J. R. Implications of human genome architecture for rearrangement-based disorders: the genomic basis of disease. *Hum. Mol. Genet.* **13**: Spec No 1, R57–R64 (2004).

50. Casaregola, S., Nguyen, H. V., Lepingle, A., Brignon, P., Gendre, F., and Gaillardin, C. A family of laboratory strains of *Saccharomyces cerevisiae* carry rearrangements involving chromosomes I and III. *Yeast* **14**: 551–564 (1998).

51. Chibana, H., Beckerman, J. L., and Magee, P. T. Fine-resolution physical mapping of genomic diversity in *Candida albicans*. *Genome Res.* **10**: 1865–1877 (2000).

52. Kolodner, R. D., Putnam, C. D., and Myung, K. Maintenance of genome stability in *Saccharomyces cerevisiae*. *Science* **297**: 552–557 (2002).

53. Lambie, E. J. and Roeder, G. S. Repression of meiotic crossing over by a centromere (*CEN3*) in *Saccharomyces cerevisiae*. *Genetics* **114**: 769–789 (1986).

54. Borde, V., Goldman, A. S., and Lichten, M. Direct coupling between meiotic DNA replication and recombination initiation. *Science* **290**: 806–809 (2000).

55. Kupiec, M. and Petes, T. D. Meiotic recombination between repeated transposable elements in *Saccharomyces cerevisiae*. *Mol. Cell Biol.* **8**: 2942–2954 (1988).

56. Ben-Aroya, S., Mieczkowski, P. A., Petes, T. D., and Kupiec, M. The compact chromatin structure of a Ty repeated sequence suppresses recombination hotspot activity in *Saccharomyces cerevisiae*. *Mol. Cell* **15**: 221–231 (2004).

57. Egel, R. Two tightly linked silent cassettes in the mating-type region of *Schizosaccharomyces pombe*. *Curr. Genet.* **8**: 199–203 (1984).

58. Thon, G., Cohen, A., and Klar, A. J. Three additional linkage groups that repress transcription and meiotic recombination in the mating-type region of *Schizosaccharomyces pombe*. *Genetics* **138**: 29–38 (1994).

59. Ekwall, K. and Ruusala, T. Mutations in *rik1*, *clr2*, *clr3* and *clr4* genes asymmetrically derepress the silent mating-type loci in fission yeast. *Genetics* **136**: 53–64 (1994).

60. Fan, Q.-Q., Xu, F., White, M. A., and Petes, T. D. Competition between adjacent meiotic recombination hotspots in the yeast *Saccharomyces cerevisiae*. *Genetics* **145**: 661–670 (1997).

61. Wu, T. C. and Lichten, M. Factors that affect the location and frequency of meiosis-induced double-strand breaks in *Saccharomyces cerevisiae*. *Genetics* **140**: 55–66 (1995).

62. de Massy, B. and Nicolas, A. The control *in cis* of the position and the amount of the *ARG4* meiotic double-strand break of *Saccharomyces cerevisiae. EMBO J.* **12**: 1459–1466 (1993).

63. Jeffreys, A. J. and Neumann, R. Reciprocal crossover asymmetry and meiotic drive in a human recombination hot spot. *Nature Genet.* **31**: 267–271 (2002).

64. Hunter, N., Chambers, S. R., Louis, E. J., and Borts, R. H. The mismatch repair system contributes to meiotic sterility in an interspecific yeast hybrid. *EMBO J.* **15**, 1726–1733 (1996).

65. Borts, R. H. and Haber, J. E. Meiotic recombination in yeast: Alteration by multiple heterozygosities. *Science* **237**: 1459–1465 (1987).

66. Nilsson-Tillgren, T., Gjermansen, C., Kielland-Brandt, M. C., Peterson, J. G. L., and Holmberg, S. Genetic differences between *Saccharomyces carlsbergensis* and *Saccharomyces cerevisiae*; analysis of chromosome III by single chromosome transfer. *Carlsberg Res. Commun.* **46**: 65–71 (1981).

67. Chambers, S. R., Hunter, N., Louis, E. J., and Borts, R. H. The mismatch repair system reduces meiotic homeologous recombination and stimulates recombination-dependent chromosome loss. *Mol. Cell Biol.* **16**: 6110–6120 (1996).

68. Hunter, N. and Kleckner, N. The single-end invasion: An asymmetric intermediate at the double-strand break to double-Holliday junction transition of meiotic recombination. *Cell* **106**: 59–70 (2001).

69. Greig, D., Travisano, M., Louis, E. J., and Borts, R. H. A role for the mismatch repair system during incipient speciation in Saccharomyces. *J. Evol. Biol.* **16**: 429–437 (2003).

70. Sollier, J., Lin, W., Soustelle, C., Suhre, K., Nicolas, A., Geli, V., and De La Roche Saint-Andre, C. Set1 is required for meiotic S-phase onset, double-strand break formation and middle gene expression. *EMBO J.* **23**: 1957–1967 (2004).

71. Yamashita, K., Shinohara, M., and Shinohara, A. Rad6-Bre1-mediated histone H2B ubiquitylation modulates the formation of double-strand breaks during meiosis. *Proc. Natl. Acad. Sci. U. S. A.* **101**(31): 11380–11385 (2004).

72. Borts, R. H., Lichten, M., and Haber, J. E. Analysis of meiosis-defective mutations in yeast by physical monitoring of recombination. *Genetics* **113**: 551–567 (1986).

73. Murakami, H., Borde, V., Shibata, T., Lichten, M., and Ohta, K. Correlation between premeiotic DNA replication and chromatin transition at yeast recombination initiation sites. *Nucleic Acids Res.* **31**: 4085–4090 (2003).

74. Kupiec, M., Byers, B., Esposito, R. E., and Mitchell, A. P. Meiosis and sporulation in *Saccharomyces cerevisiae*, in J. R. Pringle, Broach, J. R., and Jones, E. W. (eds.), *The Molecular and Cellular Biology of the Yeast Saccharomyces: Cell Cycle and Cell Biology*, pp. 889–1036, Cold Spring Harbor Laboratory Press, Cold Spring Harbor (1997).

75. Natarajan, K., Meyer, M. R., Jackson, B. M., Slade, D., Roberts, C., Hinnebusch, A. G., and Marton, M. J. Transcriptional profiling shows that Gcn4p is a master regulator of gene expression during amino acid starvation in yeast. *Mol. Cell Biol.* **21**: 4347–4368 (2001).

76. Pinson, B., Kongsrud, T. L., Ording, E., Johansen, L., Daignan-Fornier, B., and Gabrielsen, O. S. Signaling through regulated transcription factor interaction: mapping of a regulatory interaction domain in the Myb-related Bas1p. *Nucleic Acids Res.* **28**: 4665–4673 (2000).

77. Koren, A., Ben-Aroya, S., and Kupiec, M. Control of meiotic recombination initiation: a role for the environment? *Curr. Genet.* **42**: 129–139 (2002).

78. Ptak, S. E., Hinds, D. A., Koehler, K., Nickel, B., Patil, N., Ballinger, D. G., Przeworski, M., Frazer, K. A., and Paabo, S. Fine-scale recombination patterns differ between chimpanzees and humans. *Nature Genet.* **37**: 429–434 (2005).

79. Winckler, W., Myers, S. R., Richter, D. J., Onofrio, R. C., McDonald, G. J., Bontrop, R. E., McVean, G. A. T., Gabriel, S. B., Reich, D., Donnelly, P., and Altshuler, D. Comparison of fine-scale recombination rates in humans and chimpanzees. *Science* **308**: 107–111 (2005).

80. Birdsell, J. A. Integrating genomics, bioinformatics, and classical genetics to study the effects of recombination on genome evolution. *Mol. Biol. Evol.* **19**: 1181–1197 (2002).

81. Petes, T. D. and Merker, J. D. Context dependence of meiotic recombination hotspots in yeast. The relationship between recombination activity of a reporter construct and base composition. *Genetics* **162**: 2049–2052 (2002).

Chapter 13

1. Raikov, I. B. Evolution of macronuclear organization. *Annu. Rev. Genet.* **10**: 413–40 (1976).

2. Orias, E. Evolution of amitosis of the ciliate macronucleus: gain of the capacity to divide. *J. Protozool.* **38**: 217–21 (1991).

3. Hammerschmidt, B. Schlegel, M., Lynn, D. H., Leipe, D. D., Sogin, M. L., and Raikov, I. B. Insights into the evolution of nuclear dualism in the ciliates revealed by phylogenetic analysis of rRNA sequences. *J. Eukaryot. Microbiol.* **43**: 225–30 (1996).

4. Frankel, J. Cell biology of *Tetrahymena thermophila*. *Methods Cell Biol.* **62**: 27–125 (2000).

5. Jenuwein, T. and Allis, C. D. Translating the histone code. *Science* **293**: 1074–80 (2001).

6. Rice, J. C. and Allis, C. D. Histone methylation versus histone acetylation: new insights into epigenetic regulation. *Curr. Opin. Cell Biol.* **13**: 263–73 (2001).

7. Baroin-Tourancheau, A., Delgado, P., Perasso, R., and Adoutte, A. A broad molecular phylogeny of ciliates: identification of major evolutionary trends and radiations within the phylum. *Proc. Natl. Acad. Sci. U. S. A.* **89**: 9764–8 (1992).

8. Doak, T. G. Cavalcanti, A. R., Stover, N. A., Dunn, D. M., Weiss, R., Herrick, G., and Landweber, L. F. Sequencing the *Oxytricha trifallax* macronuclear genome: a pilot project. *Trends Genet.* **19**: 603–7 (2003).

9. Klobutcher, L. A. and Farabaugh, P. J. Shifty ciliates: frequent programmed translational frameshifting in euplotids. *Cell* **111**: 763–6 (2002).

10. Coyne, R. S., Chalker, D. L., and Yao, M. C. Genome downsizing during ciliate development: nuclear division of labor through chromosome restructuring. *Annu. Rev. Genet.* **30**: 557–78 (1996).

11. Gratias, A. and Betermier, M. Developmentally programmed excision of internal DNA sequences in *Paramecium aurelia*. *Biochimie* **83**: 1009–22 (2001).

12. Jahn, C. L. and Klobutcher, L. A. Genome remodeling in ciliated protozoa. *Annu. Rev. Microbiol.* **56**: 489–520 (2002).

13. Klobutcher, L. A. and Herrick, G. Developmental genome reorganization in ciliated protozoa: the transposon link. *Prog. Nucleic Acid Res. Mol. Biol.* **56**: 1–62 (1997).

14. Prescott, D. M. Genome gymnastics: unique modes of DNA evolution and processing in ciliates. *Nat. Rev. Genet.* **1**: 191–8 (2000).

15. Yao, M. C. Programmed DNA deletions in *Tetrahymena*: mechanisms and implications. *Trends Genet.* **12**: 26–30 (1996).

16. Cech, T. R. Beginning to understand the end of the chromosome. *Cell* **116**: 273–9 (2004).

17. Collins, K. Ciliate telomerase biochemistry. *Annu. Rev. Biochem.* **68**: 187–218 (1999).

18. Melek, M. and Shippen, D. E. Chromosome healing: spontaneous and programmed *de novo* telomere formation by telomerase. *Bioessays* **18**: 301–8 (1996).

19. Karamysheva, Z. Wang, L., Shrode, T., Bednenko, J., Hurley, L. A., and Shippen, D. E. Developmentally programmed gene elimination in *Euplotes crassus* facilitates a switch in the telomerase catalytic subunit. *Cell* **113**: 565–76 (2003).

20. Fan, Q. and Yao, M. C. A long stringent sequence signal for programmed chromosome breakage in *Tetrahymena thermophila*. *Nucleic Acids Res.* **28**: 895–900 (2000).

21. Klobutcher, L. A. Characterization of *in vivo* developmental chromosome fragmentation intermediates in *E. crassus*. *Mol. Cell* **4**: 695–704 (1999).

22. Mollenbeck, M. and Klobutcher, L. A. *De novo* telomere addition to spacer sequences prior to their developmental degradation in *Euplotes crassus*. *Nucleic Acids Res.* **30**: 523–31 (2002).

23. Wuitschick, J. D., Gershan, J. A., Lochowicz, A. J., Li, S., and Karrer, K. M. A novel family of mobile genetic elements is limited to the germline genome in *Tetrahymena thermophila*. *Nucleic Acids Res.* **30**: 2524–37 (2002).

24. Reeves, R. and Nissen, M. S. The A.T-DNA-binding domain of mammalian high mobility group I chromosomal proteins. A novel peptide motif for recognizing DNA structure. *J. Biol. Chem.* **265**: 8573–82 (1990).

25. Bewley, C. A., Gronenborn, A. M., and Clore, G. M. Minor groove-binding architectural proteins: structure, function, and DNA recognition. *Annu. Rev. Biophys. Biomol. Struct.* **27**: 105–31 (1998).

26. Prescott, D. M. The evolutionary scrambling and developmental unscrambling of germline genes in hypotrichous ciliates. *Nucleic Acids Res.* **27**: 1243–50 (1999).

27. Hogan, D. J., Hewitt, E. A., Orr, K. E., Prescott, D. M., and Muller, K. M. Evolution of IESs and scrambling in the actin I gene in hypotrichous ciliates. *Proc. Natl. Acad. Sci. U. S. A.* **98**: 15101–6 (2001).

28. Novina, C. D. and Sharp, P. A. The RNAi revolution. *Nature* **430**: 161–4 (2004).

29. Madireddi, M. T. Coyne, R. S., Smothers, J. F., Mickey, K. M., Yao, M. C., and Allis, C. D. Pdd1p, a novel chromodomain-containing protein, links heterochromatin assembly and DNA elimination in *Tetrahymena*. *Cell* **87**: 75–84 (1996).

30. Nikiforov, M. A., Smothers, J. F., Gorovsky, M. A., and Allis, C. D. Excision of micronuclear-specific DNA requires parental expression of pdd2p and occurs independently from DNA replication in *Tetrahymena thermophila*. *Genes Dev.* **13**: 2852–62 (1999).

31. Nikiforov, M. A., Gorovsky, M. A., and Allis, C. D. A novel chromodomain protein, pdd3p, associates with internal eliminated sequences during macronuclear development in *Tetrahymena thermophila*. *Mol. Cell Biol.* **20**: 4128–34 (2000).

32. Coyne, R. S., Nikiforov, M. A., Smothers, J. F., Allis, C. D., and Yao, M. C. Parental expression of the chromodomain protein Pdd1p is required for completion of programmed DNA elimination and nuclear differentiation. *Mol. Cell* **4**: 865–72 (1999).

33. Chalker, D. L. and Yao, M. C. Nongenic, bidirectional transcription precedes and may promote developmental DNA deletion in *Tetrahymena thermophila*. *Genes Dev.* **15**: 1287–98 (2001).

34. Mochizuki, K., Fine, N. A., Fujisawa, T., and Gorovsky, M. A. Analysis of a piwi-related gene implicates small RNAs in genome rearrangement in tetrahymena. *Cell* **110**: 689–99 (2002).

35. Liu, Y., Mochizuki, K., and Gorovsky, M. A. Histone H3 lysine 9 methylation is required for DNA elimination in developing macronuclei in *Tetrahymena*. *Proc. Natl. Acad. Sci. U. S. A.* **101**: 1679–84 (2004).

36. Taverna, S. D., Coyne, R. S., and Allis, C. D. Methylation of histone h3 at lysine 9 targets programmed DNA elimination in *Tetrahymena*. *Cell* **110**: 701–11 (2002).

37. Yao, M. C., Fuller, P., and Xi, X. Programmed DNA deletion as an RNA-guided system of genome defense. *Science* **300**: 1581–4 (2003).
38. Meyer, E. and Garnier, O. Non-Mendelian inheritance and homology-dependent effects in ciliates. *Adv. Genet.* **46**: 305–37 (2002).
39. Meyer, E. and Duharcourt, S. Epigenetic programming of developmental genome rearrangements in ciliates. *Cell* **87**: 9–12 (1996).
40. Preer, J. R., Jr. Epigenetic mechanisms affecting macronuclear development in *Paramecium* and *Tetrahymena. J. Eukaryot. Microbiol.* **47**: 515–24 (2000).
41. Agrawal, N. Dasaradhi, P. V., Mohmmed, A., Malhotra, P., Bhatnagar, R. K., and Mukherjee, S. K. RNA interference: biology, mechanism, and applications. *Microbiol. Mol. Biol. Rev.* **67**: 657–85 (2003).
42. Vastenhouw, N. L. Fischer, S. E., Robert, V. J., Thijssen, K. L., Fraser, A. G., Kamath, R. S., Ahringer, J., and Plasterk, R. H. A genome-wide screen identifies 27 genes involved in transposon silencing in *C. elegans. Curr. Biol.* **13**: 1311–6 (2003).
43. Tabara, H. Sarkissian, M., Kelly, W. G., Fleenor, J., Grishok, A., Timmons, L., Fire, A., and Mello, C. C. The rde-1 gene, RNA interference, and transposon silencing in *C. elegans. Cell* **99**: 123–32 (1999).
44. Volpe, T. A. Kidner, C., Hall, I. M., Teng, G., Grewal, S. I., and Martienssen, R. A. Regulation of heterochromatic silencing and histone H3 lysine-9 methylation by RNAi. *Science* **297**: 1833–7 (2002).
45. Sherman, J. M. and Pillus, L. An uncertain silence. *Trends Genet.* **13**, 308–13 (1997).
46. Lippman, Z., May, B., Yordan, C., Singer, T., and Martienssen, R. Distinct mechanisms determine transposon inheritance and methylation via small interfering RNA and histone modification. *PLoS Biol.* **1**: E67 (2003).

Chapter 14

1. Grosjean, H. and Benne, R. *Modification and Editing of RNA*, ASM Press, Washington DC (1998).
2. Hoffmann, M., Kuhn, J., Daschner, K., and Binder, S. The RNA world of plant mitochondria. *Prog. Nucleic Acid. Res. Mol. Biol.* **70**: 119–54 (2001).
3. Benne, R., Van den Burg, J., Brakenhoff, J. P., Sloof, P., Van Boom, J. H., and Tromp, M. C Major transcript of the frameshifted coxII gene from trypanosome mitochondria contains four nucleotides that are not encoded in the DNA. *Cell* **46**: 819–26 (1986).
4. Feagin, J. E., Abraham, J. M., and Stuart, K. (1988). Extensive editing of the cytochrome c oxidase III transcript in *Trypanosoma brucei. Cell* **53**: 413–22.
5. Simpson, L., Thiemann, O. H., Savill, N. J., Alfonzo, J. D., and Maslov, D. A. Evolution of RNA editing in trypanosome mitochondria. *Proc. Natl. Acad. Sci. U. S. A.* **97**: 6986–93. (2000).
6. Bock, R. Sense from nonsense: how the genetic information of chloroplasts is altered by RNA editing. *Biochimie* **82**: 549–57. (2000).
7. Shields, D. C. and Wolfe, K. H. Accelerated evolution of sites undergoing mRNA editing in plant mitochondria and chloroplasts. *Mol. Biol. Evol.* **14**: 344–9 (1997).
8. Fey, J., Tomita, K., Bergdoll, M., and Marechal-Drouard, L. Evolutionary and functional aspects of C-to-U editing at position 28 of tRNA(Cys)(GCA) in plant mitochondria. *RNA* **6**: 470–4 (2000).
9. Keegan, L. P., Leroy, A., Sproul, D., and O'Connell, M. A. Adenosine deaminases acting on RNA (ADARs): RNA-editing enzymes. *Genome Biol.* **5**: 209 (2004).

10. Seeburg, P. H. A-to-I editing. New and old sites, functions and speculations. *Neuron* **35**: 17–20 (2002).

11. Bass, B. L. RNA editing by adenosine deaminases that act on RNA. *Annu. Rev. Biochem.* **71**: 817–46 (2002).

12. Palladino, M. J., Keegan, L. P., O'Connell, M. A., and Reenan, R. A. A-to-I pre-mRNA editing in *Drosophila* is primarily involved in adult nervous system function and integrity. *Cell* **102**: 437–49 (2000).

13. Tomita, K., Ueda, T., and Watanabe, K. RNA editing by ADARs is important for normal behavior in *Caenorhabditis elegans*. *EMBO J.* **21**: 6025–35 (2002).

14. Chen, S. H., Habib, G., Yang, C. Y., Gu, Z. W., Lee, B. R., Weng, S. A., Silberman, S. R., Cai, S. J., Deslypere, J. P., Rosseneu, M. et al. Apolipoprotein B-48 is the product of a messenger RNA with an organ-specific in-frame stop codon. *Science* **238**: 363–6. (1987).

15. Powell, L. M., Wallis, S. C., Pease, R. J., Edwards, Y. H., Knott, T. J., and Scott, J. A novel form of tissue-specific RNA processing produces apolipoprotein- B48 in intestine. *Cell* **50**: 831–40. (1987).

16. Greeve, J., Altkemper, I., Dieterich, J. H., Greten, H., and Windler, E. Apolipoprotein B mRNA editing in 12 different mammalian species: hepatic expression is reflected in low concentrations of apoB- containing plasma lipoproteins. *J. Lipid Res.* **34**: 1367–83. (1993).

17. Pan, M., Cederbaum, A. I., Zhang, Y. L., Ginsberg, H. N., Williams, K. J., and Fisher, E. A. Lipid peroxidation and oxidant stress regulate hepatic apolipoprotein B degradation and VLDL production. *J. Clin. Invest.* **113**: 1277–87 (2004).

18. Sparks, C. E. and Marsh, J. B. Metabolic heterogeneity of apolipoprotein B in the rat. *J. Lipid Res.* **22**: 519–27 (1981).

19. Chan, L. Apolipoprotein B, the major protein component of triglyceride-rich and low density lipoproteins. *J. Biol. Chem.* **267**: 25621–4. (1992).

20. Hirano, K. I., Young, S.G., Farese, R.V., Ng, J., Sande, E., Warburton, C., Powell-Braxton, L.M., and Davidson, N.O. Targeted disruption of the mouse apobec-1 gene abolishes apolipoprotein B mRNA editing and eliminates apolipoprotein B48. *J. Biol. Chem.* **271**: 9887–90 (1996).

21. Nakamuta, M., Chang, B.H.J., Zsigmond, E., Kobayashi, K., Lei, H., Ishida, B.Y., Oka, K., Li, E., and Chan, L. Complete phenotypic characterization of the apobec-1 knockout mice with a wild-type genetic background and a human apolipoprotein B transgenic background, and restoration of apolipoprotein B mRNA editing by somatic gene transfer of Apobec-1. *J. Biol. Chem.* **271**: 25981–8 (1996).

22. Chester, A., Weinreb, V., Carter, C. W., Jr., and Navaratnam, N. Optimization of apolipoprotein B mRNA editing by APOBEC1 apoenzyme and the role of its auxiliary factor, ACF. *RNA* **10**: 1399–411 (2004).

23. Teng, B., Burant, C. F., and Davidson, N. O. Molecular cloning of an apolipoprotein B messenger RNA editing protein. *Science* **260**: 1816–9. (1993).

24. Xie, K., Sowden, M. P., Dance, G. S., Torelli, A. T., Smith, H. C., and Wedekind, J. E. The structure of a yeast RNA-editing deaminase provides insight into the fold and function of activation-induced deaminase and APOBEC-1. *Proc. Natl. Acad. Sci. U. S. A.* **101**: 8114–9 (2004).

25. Sowden, M. P., Lehmann, D. M., Lin, X., Smith, C. O., and Smith, H. C. Identification of novel alternative splice variants of APOBEC-1 complementation factor with different capacities to support apolipoprotein B mRNA editing. *J. Biol. Chem.* **279**: 197–206 (2004).

26. Mehta, A. and Driscoll, D. M. (2002). Identification of domains in apobec-1 complementation factor required for RNA binding and apolipoprotein-B mRNA editing. *RNA* **8**: 69–82.

27. Blanc, V., Henderson, J. O., Kennedy, S., and Davidson, N. O. Mutagenesis of apobec-1 complementation factor reveals distinct domains that modulate RNA binding, protein-protein interaction with apobec-1, and complementation of C to U RNA-editing activity. *J. Biol. Chem.* **276**: 46386–93. (2001).

28. Wu, T. T., Netter, H. J., Bichko, V., Lazinski, D., and Taylor, J. RNA editing in the replication cycle of human hepatitis delta virus. *Biochimie* **76**: 1205–8 (1994).

29. Jacques, J. P., Hausmann, S., and Kolakofsky, D. Paramyxovirus mRNA editing leads to G deletions as well as insertions. *EMBO J.* **13**: 5496–503 (1994).

30. George, C. X. and Samuel, C. E. Human RNA-specific adenosine deaminase ADAR1 transcripts possess alternative exon 1 structures that initiate from different promoters, one constitutively active and the other interferon inducible. *Proc. Natl. Acad. Sci. U. S. A.* **96**: 4621–6 (1999).

31. Joyce, G. F. (2002). The antiquity of RNA-based evolution. *Nature* **418**: 214–21.

32. Covello, P. S. and Gray, M. W. On the evolution of RNA editing. *Trends Genet.* **9**: 265–8 (1993).

33. Beale, R. C., Petersen-Mahrt, S. K., Watt, I. N., Harris, R. S., Rada, C., and Neuberger, M. S. Comparison of the differential context-dependence of DNA deamination by APOBEC enzymes: correlation with mutation spectra *in vivo*. *J Mol. Biol.* **337**: 585–96 (2004).

34. Wiegand, H. L., Doehle, B. P., Bogerd, H. P., and Cullen, B. R. A second human anti-retroviral factor, APOBEC3F, is suppressed by the HIV-1 and HIV-2 Vif proteins. *EMBO J.* **23**: 2451–8 (2004).

35. Polson, A. G. and Bass, B. L. Preferential selection of adenosines for modification by double-stranded RNA adenosine deaminase. *EMBO J.* **13**: 5701–11 (1994).

36. Smith, H. C., Gott, J. M., and Hanson, M. R. A guide to RNA editing. *RNA* **3**: 1105–23 (1997).

37. Lonergan, K. M. and Gray, M. W. Editing of transfer RNAs in *Acanthamoeba castellanii* mitochondria. *Science* **259**: 812–6 (1993).

38. Janke, A. and Paabo, S. Editing of a tRNA anticodon in marsupial mitochondria changes its codon recognition. *Nucleic Acids Res.* **21**: 1523–5 (1993).

39. Alfonzo, J. D., Blanc, V., Estevez, A. M., Rubio, M. A., and Simpson, L. C to U editing of the anticodon of imported mitochondrial tRNA(Trp) allows decoding of the UGA stop codon in *Leishmania tarentolae*. *EMBO J.* **18**: 7056–62 (1999).

40. Stuart, K., Panigrahi, A. K., and Schnaufer, A. Identification and characterization of trypanosome RNA-editing complex components. *Methods Mol. Biol.* **265**: 273–91 (2004).

41. Mian, I. S., Moser, M. J., Holley, W. R., and Chatterjee, A. Statistical modelling and phylogenetic analysis of a deaminase domain. *J. Comput. Biol.* **5**: 57–72 (1998).

42. Wedekind, J. E., Dance, G. S., Sowden, M. P., and Smith, H. C. Messenger RNA editing in mammals: new members of the APOBEC family seeking roles in the family business. *Trends Genet.* **19**: 207–16 (2003).

43. Dance, G. S., Beemiller, P., Yang, Y., Mater, D. V., Mian, I. S., and Smith, H. C. Identification of the yeast cytidine deaminase CDD1 as an orphan C→URNA editase. *Nucleic Acids Res.* **29**: 1772–80. (2001).

44. Jarmuz, A., Chester, A., Bayliss, J., Gisbourne, J., Dunham, I., Scott, J., and Navaratnam, N. An anthropoid-specific locus of orphan C to U RNA-editing enzymes on chromosome 22. *Genomics* **79**: 285–96. (2002).

45. Honjo, T., Muramatsu, M., and Fagarasan, S. AID: how does it aid antibody diversity? *Immunity* **20**: 659–68 (2004).

46. Yu, Q., Konig, R., Pillai, S., Chiles, K., Kearney, M., Palmer, S., Richman, D., Coffin, J. M., and Landau, N. R. Single-strand specificity of APOBEC3G accounts for minus-strand deamination of the HIV genome. *Nat. Struct. Mol. Biol.* **11**: 435–42 (2004).

47. Smith, H. C., Bottaro, A., Sowden, M. P., and Wedekind, J. E. Activation induced deaminase: the importance of being specific. *Trends Genet.* **20**: 224–7 (2004).

48. Maas, S., Rich, A., and Nishikura, K. A-to-I RNA editing: recent news and residual mysteries. *J. Biol. Chem.* **278**: 1391–4 (2003).

49. Hanrahan, C. J., Palladino, M. J., Ganetzky, B., and Reenan, R. A. RNA editing of the Drosophila para Na(+) channel transcript. Evolutionary conservation and developmental regulation. *Genetics* **155**: 1149–60 (2000).

50. Beghini, A., Ripamonti, C. B., Peterlongo, P., Roversi, G., Cairoli, R., Morra, E., and Larizza, L. RNA hyperediting and alternative splicing of hematopoietic cell phosphatase (PTPN6) gene in acute myeloid leukemia. *Hum. Mol. Genet.* **9**: 2297–304 (2000).

51. Hartner, J. C., Schmittwolf, C., Kispert, A., Muller, A. M., Higuchi, M., and Seeburg, P. H. Liver disintegration in the mouse embryo caused by deficiency in the RNA-editing enzyme ADAR1. *J. Biol. Chem.* **279**: 4894–902 (2004).

52. Wang, Q., Khillan, J., Gadue, P., and Nishikura, K. Requirement of the RNA editing deaminase ADAR1 gene for embryonic erythropoiesis. *Science* **290**: 1765–8. (2000).

53. Levanon, E. Y., Eisenberg, E., Yelin, R., Nemzer, S., Hallegger, M., Shemesh, R., Fligelman, Z. Y., Shoshan, A., Pollock, S. R., Sztybel, D., Olshansky, M., Rechavi, G., and Jantsch, M. F. Systematic identification of abundant A-to-I editing sites in the human transcriptome. *Nature Biotechnol.* **22**: 1001–5 (2004).

54. Hoopengardner, B., Bhalla, T., Staber, C., and Reenan, R. Nervous system targets of RNA editing identified by comparative genomics. *Science* **301**: 832–6 (2003).

55. Bass, B. L. and Weintraub, H. An unwinding activity that covalently modifies its double-stranded RNA substrate. *Cell* **55**: 1089–98 (1988).

56. Dykxhoorn, D. M., Novina, C. D., and Sharp, P. A. Killing the messenger: short RNAs that silence gene expression. *Nat. Rev. Mol. Cell Biol.* **4**: 457–67 (2003).

57. Paddison, P. J., Caudy, A. A., Bernstein, E., Hannon, G. J., and Conklin, D. S. Short hairpin RNAs (shRNAs) induce sequence-specific silencing in mammalian cells. *Genes Dev.* **16**: 948–58. (2002).

58. Knight, S. W. and Bass, B. L. The role of RNA editing by ADARs in RNAi. *Mol. Cell* **10**: 809–17 (2002).

59. Mochizuki, K. and Gorovsky, M. A. Conjugation-specific small RNAs in *Tetrahymena* have predicted properties of scan (scn) RNAs involved in genome rearrangement. *Genes Dev.* **18**: 2068–73 (2004).

60. Hausmann, S., Jacques, J. P., and Kolakofsky, D. Paramyxovirus RNA editing and the requirement for hexamer genome length. *RNA* **2**: 1033–45 (1996).

61. Saunders, H. L. and Magor, B. G. Cloning and expression of the AID gene in the channel catfish. *Dev. Comp. Immunol.* **28**: 657–63 (2004).

62. Harris, R. S., Petersen-Mahrt, S. K., and Neuberger, M. S. RNA editing enzyme APOBEC1 and some of its homologs can act as DNA mutators. *Mol. Cell* **10**: 1247–53 (2002).

63. Sowden, M., Hamm, J. K., and Smith, H. C. Overexpression of APOBEC-1 results in mooring sequence-dependent promiscuous RNA editing. *J. Biol. Chem.* **271**: 3011–7 (1996).

64. Yamanaka, S., M. Balestra, L. Ferrell, J. Fan, K. S. Arnold, S. Taylor, J. M. Taylor, Innerarity, T. L. Apolipoprotein B mRNA editing protein induces hepatocellular carcinoma and dysplasia in transgenic animals. *Proc. Natl. Acad. Sci. U. S. A.* **92**: 8483–7. (1995).

65. Okazaki, I. M., Hiai, H., Kakazu, N., Yamada, S., Muramatsu, M., Kinoshita, K., and Honjo, T.Constitutive expression of AID leads to tumorigenesis. *J. Exp. Med.* **197**: 1173–81 (2003).

66. Oppezzo, P., Vuillier, F., Vasconcelos, Y., Dumas, G., Magnac, C., Payelle-Brogard, B., Pritsch, O., and Dighiero, G. Chronic lymphocytic leukemia B cells expressing AID

display dissociation between class switch recombination and somatic hypermutation. *Blood* **101**: 4029–32 (2003).

67. Machida, K., Cheng, K. T., Sung, V. M., Shimodaira, S., Lindsay, K. L., Levine, A. M., Lai, M. Y., and Lai, M. M. Hepatitis C virus induces a mutator phenotype: enhanced mutations of immunoglobulin and protooncogenes. *Proc. Natl. Acad. Sci. U. S. A.* **101**: 4262–7 (2004).

68. Brar, S. S., Watson, M., and Diaz, M. Activation-induced cytosine deaminase (AID) is actively exported out of the nucleus but retained by the induction of DNA breaks. *J. Biol. Chem.* **279**: 26395–401 (2004).

69. Wang, C. L., Harper, R. A., and Wabl, M. Genome-wide somatic hypermutation. *Proc. Natl. Acad. Sci. U. S. A.* **101**: 7352–6 (2004).

70. Mangeat, B., Turelli, P., Caron, G., Friedli, M., Perrin, L., and Trono, D. Broad anti-retroviral defence by human APOBEC3G through lethal editing of nascent reverse transcripts. *Nature* **424**: 99–103 (2003).

71. Rada, C., Williams, G. T., Nilsen, H., Barnes, D. E., Lindahl, T., and Neuberger, M. S. Immunoglobulin isotype switching is inhibited and somatic hypermutation perturbed in UNG-deficient mice. *Curr. Biol.* **12**: 1748–55 (2002).

72. Woo, C. J., Martin, A., and Scharff, M. D. Induction of somatic hypermutation is associated with modifications in immunoglobulin variable region chromatin. *Immunity* **19**: 479–89 (2003).

73. Chaudhuri, J., Tian, M., Khuong, C., Chua, K., Pinaud, E., and Alt, F. W. Transcription-targeted DNA deamination by the AID antibody diversification enzyme. *Nature* **422**: 726–30 (2003).

74. Dance, G. S. C., Sowden, M. P., Cartegni, L., Cooper, E., Krainer, A. R., and Smith, H. C. Two proteins essential for apolipoprotein B mRNA editing are expressed from a single gene through alternative splicing. *J. Biol. Chem.* **277**: 12703–9 (2002).

75. Fugmann, S. D. and Schatz, D. G. Immunology. One AID to unite them all. *Science* **295**: 1244–5. (2002).

76. Ta, V. T., Nagaoka, H., Catalan, N., Durandy, A., Fischer, A., Imai, K., Nonoyama, S., Tashiro, J., Ikegawa, M., Ito, S., Kinoshita, K., Muramatsu, M., and Honjo, T. AID mutant analyses indicate requirement for class-switch-specific cofactors. *Nature Immunol.* **4**: 843–8 (2003).

77. Chaudhuri, J., Khuong, C., and Alt, F. W. Replication protein A interacts with AID to promote deamination of somatic hypermutation targets. *Nature* **430**: 992–8 (2004).

78. Sheehy, A. M., Gaddis, N. C., Choi, J. D., Malim, M. H. Isolation of a human gene that inhibits HIV-1 infection and is suppressed by the viral Vif protein. *Nature* **418**: 646–50 (2002).

79. Stopak, K., De Noronha, C., Yonemoto, W., and Greene, W. C. HIV-1 Vif blocks the antiviral activity of APOBEC3G by impairing both its translation and intracellular stability. *Mol. Cell* **12**: 591–601 (2003).

80. Sheehy, A. M., Gaddis, N. C., and Malim, M. H. The antiretroviral enzyme APOBEC3G is degraded by the proteasome in response to HIV-1 Vif. *Nat. Med.* **9**: 1404–7 (2003).

81. Mariani, R., Chen, D., Schrofelbauer, B., Navarro, F., Konig, R., Bollman, B., Munk, C., Nymark-McMahon, H., and Landau, N. R. Species-specific exclusion of APOBEC3G from HIV-1 virions by Vif. *Cell* **114**: 21–31 (2003).

82. Shindo, K., Takaori-Kondo, A., Kobayashi, M., Abudu, A., Fukunaga, K., and Uchiyama, T. The enzymatic activity of CEM15/Apobec-3G is essential for the regulation of the infectivity of HIV-1 virion but not a sole determinant of its antiviral activity. *J. Biol. Chem.* **278**: 44412–6 (2003).

83. Zhang, H. et al. The cytidine deaminase CEM15 induces hypermutation in newly synthesized HIV-1 DNA. *Nature* **424**, 94–8 (2003).

84. Rogozin, I. B., Basu, M. K., Jordan, I. K., Pavlov, Y. I., and Koonin, E. V. APOBEC4, a new member of the AID/APOBEC family of polynucleotide (deoxy)cytidine deaminases predicted by computational analysis. *Cell Cycle* **4**: 1281–5 (2005).

85. Yu, X., Yu, Y., Liu, B., Luo, K., Kong, W., Mao, P., and Yu, X. F. Induction of APOBEC3G ubiquitination and degradation by an HIV-1 Vif-Cul5-SCF complex. *Science* **302**: 1056–60 (2003).

86. Bogerd, H. P., Doehle, B. P., Wiegand, H. L., and Cullen, B. R. A single amino acid difference in the host APOBEC3G protein controls the primate species specificity of HIV type 1 virion infectivity factor. *Proc. Natl. Acad. Sci. U. S. A.* **101**: 3770–4 (2004).

87. Zhang, J. and Webb, D. M. Rapid evolution of primate antiviral enzyme APOBEC3G. *Hum. Mol. Genet.* **13**: 1785–91 (2004).

Chapter 15

1. Berget, S. M., Moore, C., and Sharp, P. A. Spliced segments at the 5' terminus of adenovirus 2 late mRNA. *Proc. Natl. Acad. Sci. U. S. A.* **74**: 3171–5 (1977).

2. Chow, L. T., Gelinas, R. E., Broker, T. R., and Roberts, R. J. An amazing sequence arrangement at the 5' ends of adenovirus 2 messenger RNA. *Cell* **12**: 1–8 (1977).

3. Nilsen, T. W. The case for an RNA enzyme. *Nature* **408**: 782–3 (2000).

4. Jurica, M. S. and Moore, M. J. Pre-mRNA splicing: awash in a sea of proteins. *Mol. Cell* **12**: 5–14 (2003).

5. Hochleitner, E.O., Kastner, B., Frohlich, T., Schmidt, A., Luhrmann, R., Arnold, G., and Lottspeich, F. Protein stoichiometry of a multiprotein complex, the human spliceosomal U1 small nuclear ribonucleoprotein: absolute quantification using isotope-coded tags and mass spectrometry. *J. Biol. Chem.* **280**: 2536–42 (2005).

6. Konarska, M. M. and Sharp, P. A. Interactions between small nuclear ribonucleoprotein particles in formation of spliceosomes. *Cell* **49**: 763–74 (1987).

7. Graveley, B. R. Alternative splicing: increasing diversity in the proteomic world. *Trends Genet.* **17**: 100–7 (2001).

8. Sharp, P. A. Split genes and RNA splicing. *Cell* **77**: 805–15 (1994).

9. Modrek, B. and Lee, C. A genomic view of alternative splicing. *Nature Genet.* **30**: 13–9 (2002).

10. Stolc, V., Gauhar, Z., Mason, C., Halasz, G., van Batenburg, M. F., Rifkin, S. A., Hua, S., Herreman, T., Tongprasit, W., Barbano, P. E., Bussemaker, H. J., and White, K. P. A gene expression map for the euchromatic genome of *Drosophila melanogaster*. *Science* **306**: 655–60 (2004).

11. Johnson, J. M., Castle, J., Garrett-Engele, P., Kan, Z., Loerch, P. M., Armour, C. D., Santos, R., Schadt, E. E., Stoughton, R., and Shoemaker, D. D. Genome-wide survey of human alternative pre-mRNA splicing with exon junction microarrays. *Science* **302**: 2141–4 (2003).

12. Black, D. L. Splicing in the inner ear: a familiar tune, but what are the instruments? *Neuron* **20**: 165–8 (1998).

13. Missler, M. and Südhof, T. C. Neurexins: three genes and 1001 products. *Trends Gen.* **14**: 20–6 (1998).

14. Rowen, L., Young, J., Birditt, B., Kaur, A., Madan, A., Philipps, D. L., Qin, S., Minx, P., Wilson, R. K., Hood, L., and Graveley, B. R. Analysis of the human neurexin genes: alternative splicing and the generation of protein diversity. *Genomics* **79**: 587–97 (2002).

15. Schmucker, D., Clemens, J. C., Shu, H., Worby, C. A., Xiao, J., Muda, M., Dixon, J. E., and Zipursky, S. L. Drosophila Dscam is an axon guidance receptor exhibiting extraordinary molecular diversity. *Cell* **101**: 671–84 (2000).
16. Black, D. L. Protein diversity from alternative splicing: a challenge for bioinformatics and post-genome biology. *Cell* **103**: 367–70 (2000).
17. Maniatis, T. and Tasic, B. Alternative pre-mRNA splicing and proteome expansion in metazoans. *Nature* **418**: 236–43 (2002).
18. Baker, B. S. Sex in flies: the splice of life. *Nature* **340**: 521–4 (1989).
19. Graveley, B. R. Sex, AGility, and the regulation of alternative splicing. *Cell* **109**: 409–12 (2002).
20. Ryner, L. C., Goodwin, S. F., Castrillon, D. H., Anand, A., Villella, A., Baker, B. S., Hall, J. C., Taylor, B. J., and Wasserman, S. A. Control of male sexual behavior and sexual orientation in *Drosophila* by the fruitless gene. *Cell* **87**: 1079–89 (1996).
21. Wojtowicz, W. M., Flanagan, J. J., Millard, S. S., Zipursky, S. L., and Clemens, J. C. Alternative splicing of *Drosophila* Dscam generates axon guidance receptors that exhibit isoform-specific homophilic binding. *Cell* **118**: 619–33 (2004).
22. Graveley, B. R., Kaur, A., Gunning, D., Zipursky, S. L., Rowen, L., and Clemens, J. C. The organization and evolution of the dipteran and hymenopteran Down syndrome cell adhesion molecule (Dscam) genes. *RNA* **10**: 1499–506 (2004).
23. Watson, F. L., Püttmann-Holgado, R., Thomas, F., Lamar, D. L., Hughes, M., Konda, M., Rebel, V. I., and Schumucker, D. Extensive diversity of Ig-superfamily proteins in the immune system of insects. *Science* **309**: 1874–8 (2005).
24. Park, J. W., Parisky, K., Celotto, A. M., Reenan, R. A., and Graveley, B. R. Identification of alternative splicing regulators by RNA interference in *Drosophila*. *Proc. Natl. Acad. Sci. U. S. A.* **101**: 15974–9 (2004).
25. Labrador, M., Mongelard, F., Plata-Rengifo, P., Baxter, E. M., Corces, V. G., and Gerasimova, T. I. Protein encoding by both DNA strands. *Nature* **409**: 1000 (2001).
26. Dorn, R., Reuter, G., and Loewendorf, A. Transgene analysis proves mRNA transsplicing at the complex mod(mdg4) locus in *Drosophila*. *Proc. Natl. Acad. Sci. U. S. A.* **98**: 9724–9 (2001).
27. Mongelard, F., Labrador, M., Baxter, E. M., Gerasimova, T. I., and Corces, V. G. Transsplicing as a novel mechanism to explain interallelic complementation in *Drosophila*. *Genetics* **160**: 1481–7 (2002).
28. Chiara, M. D. and Reed, R. A two-step mechanism for 5′ and 3′ splice-site pairing. *Nature* **375**: 510–3 (1995).
29. Bruzik, J. P. and Maniatis, T. Enhancer-dependent interaction between 5′ and 3′ splice sites in trans. *Proc. Natl. Acad. Sc.i U. S. A.* **92**: 7056–9 (1995).
30. Nilsen, T. W. Evolutionary origin of SL-addition trans-splicing: still an enigma. *Trends Genet.* **17**: 678–680 (2001).
31. Blumenthal, T. Trans-splicing and polycistronic transcription in *Caenorhabditis elegans*. *Trends Genet.* **11**: 132–136 (1995).
32. Deutsch, M. and Long, M. Intron-exon structures of eukaryotic model organisms. *Nucleic Acids Res.* **27**: 3219–3228 (1999).
33. Sun, H. and Chasin, L. A. Multiple splicing defects in an intronic false exon. *Mol. Cell Biol.* **20**: 6414–25 (2000).
34. Smith, C. W. and Valcarcel, J. Alternative pre-mRNA splicing: the logic of combinatorial control. *Trends Biochem Sci.* **25**, 381–8 (2000).
35. Graveley, B. R. Sorting out the complexity of SR protein functions. *RNA* **6**: 1197–211 (2000).
36. Lavigueur, A., La Branche, H., Dornblihtt, A. R., and Chabot, B. A splicing enhancer in the human fibronectin alternate ED1 exon interacts with SR proteins and stimulates U2 snRNP binding. *Genes Dev.* **7**: 2405–17 (1993).

37. Sun, Q., Mayeda, A., Hampson, R. K., Krainer, A. R., and Rottman, F. M. General splicing factor SF2/ASF promotes alternative splicing by binding to an exonic splicing enhancer. *Genes Dev.* **7**: 2598–2608 (1993).

38. Tian, M. and Maniatis, T. A splicing enhancer complex controls alternative splicing of doublesex pre-mRNA. *Cell* **74**: 105–14 (1993).

39. Schaal, T. D. and Maniatis, T. Multiple distinct splicing enhancers in the protein-coding sequences of a constitutively spliced pre-mRNA. *Mol. Cell Biol.* **19**: 261–73 (1999).

40. Cartegni, L., Chew, S. L., and Krainer, A. R. Listening to silence and understanding nonsense: exonic mutations that affect splicing. *Nat. Rev. Genet.* **3**: 285–98 (2002).

41. Wu, J. Y. and Maniatis, T. Specific interactions between proteins implicated in splice site selection and regulated alternative splicing. *Cell* **75**: 1061–70 (1993).

42. Zhang, W.-J. and Wu, J. Y. Functional properties of p54, a novel SR protein active in constitutive and alternative splicing. *Mol. Cell. Biol.* **16**: 5400–8 (1996).

43. Kohtz, J. D., Jamison, S. F., Will, C. L., Zuo, P., Luhrmann, R., Garcia-Blanco, M. A., and Manley, J. L. Protein-protein interactions and 5′ splice site recognition in mammalian mRNA precursors. *Nature* **368**: 119–24 (1994).

44. Berget, S. M. Exon recognition in vertebrate splicing. *J. Biol. Chem.* **270**: 2411–4 (1995).

45. Shen, H., Kan, J. L., and Green, M. R. Arginine-serine-rich domains bound at splicing enhancers contact the branchpoint to promote prespliceosome assembly. *Mol. Cell* **13**: 367–76 (2004).

46. Shen, H. and Green, M. R. A pathway of sequential arginine-serine-rich domain-splicing signal interactions during mammalian spliceosome assembly. *Mol. Cell* **16**: 363–73 (2004).

47. Wang, H. Y., Xu, X., Ding, J. H., Bermingham, J. R., Jr., and Fu, X. D. SC35 plays a role in T cell development and alternative splicing of CD45. *Mol. Cell* **7**: 331–42 (2001).

48. Xu, X., Yang, D., Ding, J. H., Wang, W., Chu, P. H., Dalton, N. D., Wang, H. Y., Bermingham, J. R. Jr., Ye, Z., Liu, F., Rosenfeld, M. G., Manley, J. L., Ross, J. Jr., Chen, J., Xiao, R. P., Cheng, H., and Fu, X. D. ASF/SF2-regulated CaMKIIdelta alternative splicing temporally reprograms excitation-contraction coupling in cardiac muscle. *Cell* **120**: 59–72 (2005).

49. Lynch, K. W. and Maniatis, T. Assembly of specific SR protein complexes on distinct regulatory elements of the *Drosophila* doublesex splicing enhancer. *Genes Dev.* **10**: 2089–101 (1996).

50. Hertel, K. J. and Maniatis, T. The function of multisite splicing enhancers. *Mol. Cell* **1**: 449–55 (1998).

51. Zuo, P. and Maniatis, T. The splicing factor U2AF35 mediates critical protein-protein interactions in constitutive and enhancer-dependent splicing. *Genes Dev.* **10**: 1356–68 (1996).

52. Graveley, B. R., Hertel, K. J., and Maniatis, T. The role of U2AF35 and U2AF65 in enhancer-dependent splicing. *RNA* **7**: 806–18 (2001).

53. Black, D. L. Mechanisms of alternative pre-messenger RNA splicing. *Annu. Rev. Biochem.* **72**: 291–336 (2003).

54. Wang, Z., Rolish, M. E., Yeo, G., Tung, V., Mawson, M., and Burge, C. B. Systematic identification and analysis of exonic splicing silencers. *Cell* **119**: 831–45 (2004).

55. Zhang, X. H. and Chasin, L. A. Computational definition of sequence motifs governing constitutive exon splicing. *Genes Dev.* **18**: 1241–50 (2004).

56. Jensen, K. B., Dredge, B. K., Stefani, G., Zhong, R., Buckanovich, R. J., Okano, H. J., Yang, Y. Y., and Darnell, R. B. Nova-1 regulates neuron-specific alternative splicing and is essential for neuronal viability. *Neuron* **25**: 359–71 (2000).

57. Jensen, K. B., Musunuru, K., Lewis, H. A., Burley, S. K., and Darnell, R. B. The tetranucleotide UCAY directs the specific recognition of RNA by the Nova K-homology 3 domain. *Proc. Natl. Acad. Sci. U. S. A.* **97**: 5740–5 (2000).

58. Ule, J., Jensen, K. B., Ruggiu, M., Mele, A., Ule, A., and Darnell, R. B. CLIP identifies Nova-regulated RNA networks in the brain. *Science* **302**: 1212–5 (2003).
59. Fu, X. D. Towards a Splicing Code. *Cell* **119**: 736–738 (2004).
60. Southby, J., Gooding, C., and Smith, C. W. Polypyrimidine tract binding protein functions as a repressor to regulate alternative splicing of alpha-actinin mutally exclusive exons. *Mol. Cell Biol.* **19**: 2699–711 (1999).
61. Smith, C. W. and Nadal-Ginard, B. Mutually exclusive splicing of alpha-tropomyosin exons enforced by an unusual lariat branch point location: implications for constitutive splicing. *Cell* **56**: 749–58 (1989).

Chapter 16

1. Reik, W., Dean, W., and Walter, J. Epigenetic reprogramming in mammalian development. *Science* **293**: 1089–93 (2001).
2. Latham, K. E. Epigenetic modification and imprinting of the mammalian genome during development. *Curr. Top. Dev. Biol.* **43**: 1–49 (1999).
3. Ferguson-Smith, A. C., and Surani, M. A. Imprinting and the epigenetic asymmetry between parental genomes. *Science* **293**: 1086–9 (2001).
4. Paldi, A. Genomic imprinting: could the chromatin structure be the driving force? *Curr. Top. Dev. Biol.* **53**: 115–38 (2003).
5. Blewitt, M. E., Chong, S., and Whitelaw, E. How the mouse got its spots. *Trends Genet.* **20**: 550–4 (2004).
6. Moore, T. and Haig, D. Genomic imprinting in mammalian development: a parental tug-of-war. *Trends Genet.* **7**: 45–9 (1991).
7. McGowan, R. A. and Martin, C. C. DNA methylation and genome imprinting in the zebrafish, *Danio rerio*: some evolutionary ramifications. *Biochem. Cell Biol.* **75**: 499–506 (1997).
8. Lloyd, V. Parental imprinting in *Drosophila*. *Genetica* **109**: 35–44 (2000).
9. Bongiorni, S. and Prantera, G. Imprinted facultative heterochromatization in mealybugs. *Genetica* **117**: 271–9 (2003).
10. Jenuwein, T. and Allis, C. D. Translating the histone code. *Science* **293**: 1074–80 (2001).
11. Brockdorff, N. X-chromosome inactivation: closing in on proteins that bind Xist RNA. *Trends Genet.* **18**: 352–8 (2002).
12. Ferguson-Smith, A. X inactivation: pre- or post-fertilisation turn-off. *Curr. Biol.* **14**: R323–5 (2004).
13. Riggs, A. D. X chromosome inactivation, differentiation, and DNA methylation revisited, with a tribute to Susumu Ohno. *Cytogenet. Genome Res.* **99**: 17–24 (2002).
14. Verona, R. I., Mann, M. R., and Bartolomei, M. S. Genomic imprinting: intricacies of epigenetic regulation in clusters. *Annu. Rev. Cell Dev. Biol.* **19**: 237–59 (2003).
15. Hata, K., Okano, M., Lei, H., and Li, E. Dnmt3L cooperates with the Dnmt3 family of *de novo* DNA methyltransferases to establish maternal imprints in mice. *Development* **129**: 1983–93 (2002).
16. Li, E., Beard, C., and Jaenisch, R. Role for DNA methylation in genomic imprinting. *Nature* **366**: 362–5 (1993).
17. Kaneda, M., Okano, M., Hata, K., Sado, T., Tsujimoto, N., Li, E., and Sasaki, H. Essential role for *de novo* DNA methyltransferase Dnmt3a in paternal and maternal imprinting. *Nature* **429**: 900–3 (2004).
18. Umlauf, D., Goto, Y., Cao, R., Cerqueira, F., Wagschal, A., Zhang, Y., and Feil, R. Imprinting along the Kcnq1 domain on mouse chromosome 7 involves repressive

histone methylation and recruitment of Polycomb group complexes. *Nature Genet.* **36**: 1296–300 (2004).

19. Pant, V., Mariano, P., Kanduri, C., Mattsson, A., Lobanenkov, V., Heuchel, R., Ohlsson, R. The nucleotides responsible for the direct physical contact between the chromatin insulator protein CTCF and the H19 imprinting control region manifest parent of origin-specific long-distance insulation and methylation-free domains. *Genes Dev.* **17**: 586–90 (2003).

20. Sleutels, F., Zwart, R., and Barlow, D. P. The non-coding Air RNA is required for silencing autosomal imprinted genes. *Nature* **415**: 810–3 (2002).

21. Vu, T. H., Li, T., and Hoffman, A. R. Promoter-restricted histone code, not the differentially methylated DNA regions or antisense transcripts, marks the imprinting status of IGF2R in human and mouse. *Hum. Mol. Genet.* **13**: 2233–45 (2004).

22. Feil, R. and Khosla, S. Genomic imprinting in mammals: an interplay between chromatin and DNA methylation? *Trends Genet.* **15**, 431–5 (1999).

23. Grewal, S. I. S. and Moazed, D. Heterochromatin and epigenetic control of gene expression. *Science* **301**: 798–802 (2003).

24. Li, E. Chromatin modification and epigenetic reprogramming in mammalian development. *Nature Rev. Genet.* **3**: 662–673 (2002).

25. Orlando, V., Strutt, H., and Paro, R. Analysis of chromatin structure by *in vivo* formaldehyde cross-linking. *Methods* **11**: 205–14 (1997).

26. Fuks, F., Hurd, P. J., Deplus, R., and Kouzarides, T. The DNA methyltransferases associate with HP1 and the SUV39H1 histone methyltransferase. *Nucl. Acids Res.* **31**, 2305–12 (2003).

27. Hashimshony, T., Zhang, J., Keshet, I., Bustin, M., and Cedar, H. The role of DNA methylation in setting up chromatin structure during development. *Nature Genet.* **34**: 187–92 (2003).

28. Jones, P. L., Veenstra, G., Wade, P., Vermaak, D., Kass, S., Landsberger, N., Strouboulis, J., and Wolffe, A. Methylated DNA and MeCP2 recruit histone deacetylase to repress transcription. *Nature Genet.* **19**: 187–91 (1998).

29. K, N. H., Chow, M., Baker, E., Pal, S., Bassal, S., Brasacchio, D., Wang, L., Craig, J., Jones, P., Sif, S., and El-Osta, A. Brahma links the SWI/SNF chromatin-remodeling complex with MeCP2-dependent transcriptional silencing. *Nature Genet.* **37**: 254–64 (2005).

30. Thakur, N., Tiwari, V., Thomassin, H., Pandey, R., Kanduri, M., Gondor, A., Grange, T., Ohlsson, R., and Kanduri, C. An antisense RNA regulates the bidirectional silencing property of the Kcnq1 imprinting control region. *Mol. Cell Biol.* **24**: 7855–62 (2004).

31. Greally, J. M., Gray, T., Gabriel, J., Song, L., Zemel, S., and Nicholls, R. Conserved characteristics of heterochromatin-forming DNA at the 15q11-q13 imprinting center. *Proc. Natl. Acad. Sci. U. S. A.* **96**: 14430–5 (1999).

32. Gribnau, J., Hochedlinger, K., Hata, K., Li, E., and Jaenisch, R. Asynchronous replication timing of imprinted loci is independent of DNA methylation, but consistent with differential subnuclear localization. *Genes Dev.* **17**: 759–73 (2003).

33. Molyneaux, K. A., Stallock, J., Schaible, K., and Wylie, C. Time-lapse analysis of living mouse germ cell migration. *Dev. Biol.* **240**: 488–98 (2001).

34. Hajkova, P., Erhardt, S., LAne, N., Haaf, T., El-Maarri, O., Reik, W., Walter, J., and Surani, M. Epigenetic reprogramming in mouse primordial germ cells. *Mech. Dev.* **117**: 15–23 (2002).

35. Lee, J., Inoue, K., Ono, R., Ogonuki, N., Kohda, T., Kaneko-Ishino, T., Ogura, A., and Ishino, F. Erasing genomic imprinting memory in mouse clone embryos produced from day 11.5 primordial germ cells. *Development* **129**: 1807–17 (2002).

36. Lucifero, D., Mann, M. R., Bartolomei, M. S., and Trasler, J. M. Gene-specific timing and epigenetic memory in oocyte imprinting. *Hum. Mol. Genet.* **13**: 839–49 (2004).

37. Ueda, T., Abe, K., Miura, A., Yuzuriha, M., Zubair, M., Noguchi, M., Niwa, K., Kawase, Y., Kono, T., Matsuda, Y., Fujimoto, H., Shibata, H., Hayashizaki, Y., and Sasaki, H. The paternal methylation imprint of the mouse H19 locus is acquired in the gonocyte stage during foetal testis development. *Genes Cells* **5**: 649–59 (2000).

38. Bourc'his, D., Xu, G. L., Lin, C. S., Bollman, B., and Bestor, T. H. Dnmt3L and the establishment of maternal genomic imprints. *Science* **294**: 2536–9 (2001).

39. Dean, W., Santos, F., and Reik, W. Epigenetic reprogramming in early mammalian development and following somatic nuclear transfer. *Semin. Cell Dev. Biol.* **14**: 93–100 (2003).

40. Mayer, W., Niveleau, A., Walter, J., Fundele, R., and Haaf, T. Demethylation of the zygotic paternal genome. *Nature* **403**: 501–2 (2000).

41. Oswald, J., Engemann, S., Lane, N., Mayer, W., Olek, A., Fundele, R., Dean, W., Reik, W., and Walter, J. Active demethylation of the paternal genome in the mouse zygote. *Curr. Biol.* **10**: 475–8 (2000).

42. Lane, N., Dean, W., Erhardt, S., Hajkova, P., Surani, A., Walter, J., and Reik, W. Resistance of IAPs to methylation reprogramming may provide a mechanism for epigenetic inheritance in the mouse. *Genesis* **35**: 88–93 (2003).

43. Dean, W. and Ferguson-Smith, A. Genomic imprinting: mother maintains methylation marks. *Curr. Biol.* **11**: R527–30 (2001).

44. McGrath, J. and Solter, D. Completion of mouse embryogenesis requires both the maternal and paternal genomes. *Cell* **37**: 179–83 (1984).

45. Cattanach, B. M. and Kirk, M. Differential activity of maternally and paternally derived chromosome regions in mice. *Nature* **315**: 496–8 (1985).

46. Beechey, C., Cattanach, B., Blake, A., and Peters, J. World Wide Web Site - Genetic and Physical Imprinting Map of the Mouse. http://www.mgu.har.mrc.ac.uk/imprinting/imprinting.html (2003).

47. Lalande, M. Parental imprinting and human disease. *Annu. Rev. Genet.* **30**: 173–95 (1996).

48. Hurst, L. D. and McVean, G. T. Do we understand the evolution of genomic imprinting? *Curr. Opin. Genet. Dev.* **8**: 701–8 (1998).

49. Tycko, B. and Morison, I. M. Physiological functions of imprinted genes. *J. Cell Physiol.* **192**: 245–58 (2002).

50. Plass, C., Shibata, H., Kalcheva, I., Mullins, L., Kotelevtseva, N., Mullins, J., Kato, R., Sasaki, H., Hirotsune, S., Okazaki, Y., Held, W., Hayashizaki, Y., and Chapman, V. Identification of Grf1 on mouse chromosome 9 as an imprinted gene by RLGS-M. *Nature Genet.* **14**: 106–9 (1996).

51. Kaneko-Ishino, T. Kuroiwa, Y., Miyoshi, N., Kohda, T., Suzuki, R., Yokoyama, M., Viville, S., Barton, S., Ishino, F., and Surani, M. Peg1/Mest imprinted gene on chromosome 6 identified by cDNA subtraction hybridization. *Nature Genet.* **11**: 52–9 (1995).

52. Mizuno, Y. Sotomaru, Y., Katsuzawa, Y., Kono, T., Meguro, M., Oshimura, M., Kawai, J., Tomaru, Y., Kiyosawa, H., Nikaido, I., Amanuma, H., Hayashizaki, Y., and Okazaki, Y. Asb4, Ata3, and Dcn are novel imprinted genes identified by high-throughput screening using RIKEN cDNA microarray. *Biochem. Biophys. Res. Commun.* **290**: 1499–505 (2002).

53. Nikaido, I., Saito, C., Mizuno, Y., Meguro, M., Bono, H., Kadomura, M., Kono, T., Morris, G., Lyons, P., Oshimura, M., Hayashizaki, Y., and Okazaki, Y. Discovery of imprinted transcripts in the mouse transcriptome using large-scale expression profiling. *Genome Res.* **13**: 1402–9 (2003).

54. Ke, X., Thomas, N. S., Robinson, D. O., and Collins, A. The distinguishing sequence characteristics of mouse imprinted genes. *Mamm. Genome* **13**: 639–45 (2002).

55. Greally, J. M. Short interspersed transposable elements (SINEs) are excluded from imprinted regions in the human genome. *Proc. Natl. Acad. Sci. U. S. A.* **99**: 327–32 (2002).

56. Ke, X., Thomas, N. S., Robinson, D. O., and Collins, A. A novel approach for identifying candidate imprinted genes through sequence analysis of imprinted and control genes. *Hum. Genet.* **111**: 511–20 (2002).

57. Pernis, B., Chiappino, G., Kelus, A. S., and Gell, P. G. Cellular localization of immunoglobulins with different allotypic specificities in rabbit lymphoid tissues. *J. Exp. Med.* **122**: 853–76 (1965).

58. Rajewsky, K. Clonal selection and learning in the antibody system. *Nature* **381**: 751–8 (1996).

59. Held, W., Roland, J., and Raulet, D. H. Allelic exclusion of Ly49-family genes encoding class I MHC-specific receptors on NK cells. *Nature* **376**: 355–8 (1995).

60. Chess, A., Simon, I., Cedar, H., and Axel, R. Allelic inactivation regulates olfactory receptor gene expression. *Cell* **78**: 823–34 (1994).

61. Rodriguez, I., Feinstein, P., and Mombaerts, P. Variable patterns of axonal projections of sensory neurons in the mouse vomeronasal system. *Cell* **97**: 199–208 (1999).

62. Malissen, M., Trucy, J., Jouvin-Marche, E., Cazenave, P. A., Scollay, R., and Malissen, B. Regulation of TCR alpha and beta gene allelic exclusion during T-cell development. *Immunol. Today* **13**: 315–22 (1992).

63. Liang, H.-E., Hsu, L.-Y., Cado, D., and Schlissel, M. S. Variegated transcriptional activation of the immunoglobulin k locus in pre-B cells contributes to the allelic exclusion of light-chain expression. *Cell* **118**: 19–29 (2004).

64. Lewcock, J. W., and Reed, R. R. Neuroscience. ORs rule the roost in the olfactory system. *Science* **302**: 2078–9 (2003).

65. Pastinen, T., Sladek, R., Gurd, S., Sammak, A., Ge, B., Lepage, P., Lavergne, K., Villeneuve, A., Gaudin, T., Brändström, H., Beck, A., Verner, A., Kingsley, J., Harmsen, E., Labuda, D., Morgan, K., Vohl, M., Naumova, A., Sinnett, D., and Hudson, T. A survey of genetic and epigenetic variation affecting human gene expression. *Physiological Genomics* **16**: 184–93 (2003).

66. Hollick, J. B., Dorweiler, J. E., and Chandler, V. L. Paramutation and related allelic interactions. *Trends Genet.* **13**: 302–8 (1997).

67. Rakyan, V. K., Blewitt, M. E., Druker, R., Preis, J. I., and Whitelaw, E. Metastable epialleles in mammals. *Trends Genet.* **18**: 348–51 (2002).

68. Morgan, H. D., Sutherland, H. G., Martin, D. I., and Whitelaw, E. Epigenetic inheritance at the agouti locus in the mouse. *Nature Genet.* **23**: 314–8 (1999).

69. Wolff, G. L. Influence of maternal phenotype on metabolic differentiation of agouti locus mutants in the mouse. *Genetics* **88**: 529–39 (1978).

70. Duhl, D. M., Vrieling, H., Miller, K. A., Wolff, G. L., and Barsh, G. S. Neomorphic agouti mutations in obese yellow mice. *Nature Genet.* **8**: 59–65 (1994).

71. Belyaev, D. K., Ruvinsky, A. O., and Borodin, P. M. Inheritance of alternative states of the fused gene in mice. *J. Heredity* **72**: 107–12 (1981).

72. Rakyan, V. K., Chong, S., Champ, M., Cuthbert, P., Morgan, H., Luu, K., and Whitelaw, E. Transgenerational inheritance of epigenetic states at the murine *Axin*^{Fu} allele occurs after maternal and paternal transmission. *Proc. Natl. Acad. Sci. U. S. A.* **100**: 2538–43 (2003).

73. Rakyan, V. K., Preis, J., Morgan, H. D., and Whitelaw, E. The marks, mechanisms and memory of epigenetic states in mammals. *Biochem. J.* **356**: 1–10 (2001).

74. Wilkins, J. F. and Haig, D. What good is genomic imprinting: the function of parent-specific gene expression. *Nature Rev. Genet.* **4**: 1–10 (2003).

75. Gilligan, A. and Solter, D. The role of imprinting in early mammalian development, in *Genomic Imprinting: Causes and Consequences*, Ohlsson, R., Hall, K., and Ritzen, M. (eds.), Cambridge University Press, Cambridge (1995).

76. Alleman, M. and Doctor, J. Genomic imprinting in plants: observations and evolutionary implications. *Plant Mol. Biol.* **43**: 147–61 (2000).

77. Haig, D. Parental antagonism, relatedness asymmetries, and genomic imprinting. *Proc. R. Soc. Lond. B Biol. Sci.* **264**: 1657–62 (1997).

78. Hamilton, W. D. The genetical evolution of social behaviour. *J. Theor. Biol.* **7**: 1–52 (1964).

79. Killian, J. K., Byrd, J., Jirlte, J., Munday, B., Stoskopf, M., MacDonald, R., and Jirtle, R. M6P/IGF2R imprinting evolution in mammals. *Mol. Cell* **5**: 707–16 (2000).

80. Killian, J. K., Nolan, C., Stwart, N., Munday, B., Andersen, N., Nicol, S., Jirtle, R. Monotreme IGF2 expression and ancestral origin of genomic imprinting. *J. Exp. Zool.* **291**: 205–12 (2001).

81. Varmuza, S. and Mann, M. Genomic imprinting – defusing the ovarian time bomb. *Trends Genet.* **10**: 118–23 (1994).

82. Solter, D. Differential imprinting and expression of maternal and paternal genomes. *Annu. Rev. Genet.* **22**: 127–46 (1988).

83. Kono, T., Obata, Y., Wu, Q., Niwa, K., Ono, Y., Yamamoto, Y., Sung Park, E., Seo, J-S., and Ogawa, H. Birth of parthenogenetic mice that can develop to adulthood. *Nature* **428**: 860–4 (2004).

84. Beaudet, A. L. and Jiang, Y. H. A rheostat model for a rapid and reversible form of imprinting-dependent evolution. *Am. J. Hum. Genet.* **70**: 1389–97 (2002).

85. Hessler, A. Y. A study of parental modification of variegated position effects. *Genetics* **46**: 463–84 (1961).

86. Cohen, J. Position-effect variegation at several closely linked loci in *Drosophila melanogaster*. *Gerontol Clin. (Basel)* **47**: 647–59 (1962).

87. Lyko, F., Brenton, J. D., Surani, M. A., and Paro, R. An imprinting element from the mouse H19 locus functions as a silencer in *Drosophila*. *Nature Genet.* **16**: 171–3 (1997).

88. Lyko, F., Buiting, K., Horsthemke, B., and Paro, R. Identification of a silencing element in the human 15q11-q13 imprinting center by using transgenic *Drosophila*. *Proc. Natl. Acad. Sci. U. S. A.* **95**: 1698–702 (1998).

89. de la Casa-Esperon, E. and Sapienza, C. Natural selection and the evolution of genome imprinting. *Annu. Rev. Genet.* **37**: 349–70 (2003).

90. Pardo-Manuel de Villena, F., de la Casa-Esperon, E., and Sapienza, C. Natural selection and the function of genome imprinting: beyond the silenced minority. *Trends Genet.* **16**: 573–9 (2000).

Epilogue

1. Csete, M. and Doyle, J. Reverse engineering of biological complexity. *Science* **295**: 1664–9 (2002).

2. Jin, C., Wei, D., Low, S. H., Bunn, J., Choe, H. D., Doyle, J. C., et al. FAST TCP: From theory to experiments. *IEEE Network* **19**(1): 4–11 (2005).

3. Willinger, W. and Doyle, J. Robustness and the Internet: Design and evolution, in E. Jen (ed.), *Robust Design: A Repertoire of Biological, Ecological, and Engineering Case Studies*, Santa Fe Institute Studies on the Sciences of Complexity, pp. 231–71, Oxford University Press (2005).

4. Caporale, L. *Darwin in the Genome*. McGraw-Hill, New York, NY (2003).

5. Collis, C. M. and Hall, R. M. Comparison of the structure-activity relationships of the integron-associated recombination sites attI3 and attI1 reveals common features. *Microbiology* **150**: 1591–601(2004).

6. Papachristodoulou, A., Li, L., and Doyle, J. C. Methodological frameworks for large-scale network analysis and design, *Computer Communication Rev.* **34**: 7–20 (2004).

7. Li, L., Alderson, D., Doyle, J., and Willinger, W. A first-principles approach to understanding the Internet's router-level topology. Proceedings of the 2004 conference on Applications, technologies, architectures, and protocols for computer communications, *Computer Communication Rev.* **34**: 3–14 (2004).

8. Yi T.-M., Huang, Y., Simon, M.I., and Doyle, J. Robust perfect adaptation in bacterial chemotaxis through integral feedback control. *Proc. Natl. Acad. Sci. U. S. A.* **97**: 4649–53 (2000).

9. Morohashi, M., Winn, A. E., Borisuk, M. T., Bolouri, H., Doyle, J., and Kitano, H. Robustness as a measure of plausibility in models of biochemical networks, *J. Theor. Biol.* **216**(1): 19–30 (2002).

10. Hucka, M., Finney, A., Sauro, H. M., Bolouri, H., and Doyle, J.C. The systems biology markup language (SBML): a medium for representation and exchange of biochemical network models. *Bioinformatics* **19**(4): 524–31(2003).

11. El-Samad, H., Kurata, H., Doyle, J. C., Gross, C. A., and Khammash M. Surviving heat shock: Control strategies for robustness and performance. *Proc. Natl. Acad. Sci. U. S. A.* **102**(8): 2736–41 (2005).

12. Tanaka, R. Scale-rich metabolic networks. *Phys. Rev. Lett.* **94**: 168101 (2005).

13. Csete, M. E. and Doyle, J. C. Bow ties, metabolism, and disease. *Trends Biotechnol.* **22**(9): 446–50 (2004).

14. Gaspar, H. B., Gilmour, K. C., and Jones, A. M. Severe combined immunodeficiency molecular pathogenesis and diagnosis. *Arch. Dis. Child.* **84**: 169–73 (2001).

15. Carlson, J.M. and Doyle, J. Complexity and robustness. *Proc Natl. Acad. Sci. U. S. A.* **99**(Suppl. 1): 2538–45 (2002).

16. Doyle, J., Csete, M. The sincerest form of flattery. *Nature* **431**: 908–9, 2004.

17. Stelling, J., Sauer, U., Szallasi, Z., Doyle III, F. J., and Doyle, J. Robustness of cellular functions, *Cell* 118: 675–85, 2004.

18. Kitano, H., Oda, K., Kimura, T., Matsuoka, Y., Csete, M., Doyle, J., and Muramatsu, M. Metabolic syndrome and robustness tradeoffs, *Diabetes* **53**(Suppl. 3): S6–15 (2004).

List of Acronyms

ADAR	adenosine deaminases active on RNA
AID	activation-induced deaminase
Alu	a primate SINE
ARP	APOBEC-1 related proteins
BCR	B-cell receptor
bp	base pair(s)
C region	immunoglobulin constant region
CBS	chromosome breakage sequence
CDAR	sytidine deaminases active on RNA
CDR	complementarity-determining region
CS	closely spaced repeat(s)
CSR	class switch recombination
DSB	double-strand break
ESE	exonic splicing enhancer
ESS	exonic splicing silencer
GC	germinal center
gRNA	guide RNA
IES	internal eliminated sequences
Ig	immunoglobulin
IgH	immunoglobulin heavy chain locus
IgL	immunoglobulin light chain locus
IP	internet protocol
ISE	intron splice enhancer
ISS	intron splice suppressor
J region	immunoglobulin "joining" region
kb	kilobases (1000-base long sequence of DNA)
LCR	low copy repeat
LINE	long interspersed element
LTR	long terminal repeat
MDS	macronuclear-destined sequences
MIIC	MHC Class II compartment

MMR	mismatch repair
mRNA	messenger RNA
MRS	mismatch repair system
MULEs	mutator-like transposable elements
mya	million years ago
ORF	open reading frame
PCNA	proliferating cell nuclear antigen
PTGS	post-transcriptional gene silencing
PV	phase variation
QTL	quantitative trait locus
RAG	recombination-activating gene
rDNA	DNA that encodes ribosomal RNA
RISC	RNA-induced silencing complex
R-M	restriction-modification
RMS	root mean square (square root of the mean of the sum of the squares)
RNAi	RNA interference
RS	recombination signal
S region	immunoglobulin heavy chain switch region
SHM	somatic hypermutation
SINE	short interspersed nuclear element (a variety of transposable DNA sequence)
siRNA	short interfering RNAs of ~21–24 nt.
SSCL	simple sequence contingency locus.
SSR	simple sequence repeat
SU-IES	small unique sequences – internal eliminated sequences
TBE	telomere bearing element
TCP	transmission control protocol
TdT	terminal deoxynucleotide transferase
TEs	transposable elements
TCR	T-cell receptor
TIR	terminal inverted repeat
TLS	translesion synthesis
Tm	melting temperature
TPRT	target primed reverse transcription
TR	tandem repeat(s)
TSD	target site duplication
T<>T	thymidine dimer
UDG	uracil DNA glycosylase
USS	uptake signal sequences
UTR	untranslated region
UV	ultraviolet
V region	immunoglobulin "variable" region
VSG	variant surface glycoprotein

Index

Page numbers in italic, e.g. *215*, refer to figures. Page numbers in bold, e.g. **123**, denote tables.

3′ transduction 154–155, *155*
7SL RNA 141
12/23 rule. *See* recombination
    signal sequences
59-base element *130*, 131

A tracts
  biological examples
    effect of MMR on slippage rate 71;
    exclusion from nucleosomes 79;
    Leishmania tarentole kinetoplast
    body DNA curvature due to 28; and
    protein binding 37; as terminus of
    SINEs and LINEs 141
  structure of
    B DNA: differences from and buckling
    at boundary with 29–32; curvature of,
    and its dependence on phase of
    repeat 28–29; fiber diffraction
    studies 28
abasic sites in DNA, consequences
  of formation of and repair/replication
  across 43, 178, 180–182, *181*, 182,
  184, 200
Ac 141, 156. *See also* class II transposons
Acanthomoeba, RNA editing in 249, 261
Acinetobacter variation of *mutS* activity
  in through insertions and
  deletions 118
Actin, gene, scrambled in
  ciliates 239, *240*
adaptation
  to fluctuating environment 74–75
  role of gene duplication in
    67, 188, 150

adaptation (*continued*)
  role of mutable sequences in 63, 67–68,
    88–89, 117–120 (*see also*
    second-order selection;
    contingency loci)
  in populations 56, 58, 121, 127, 136
  in trypanosomes, to host
    immunity 92, 96
adaptive landscape. *See* fitness
  landscape
ADAR *250*, 257–260, *258*, 259
adenovirus, in original discovery
  of introns 267
adhesins
  CEACAM family of cell surface
    proteins, binding to 65
  contingency loci in 63, *64*, 65
  niche tropism and 119
  role in attachment to host cells 65
affinity maturation 178, 188
aging, premature 205
Agrobacterium tumefaciens 134
AID
  cytidine deamination in the
    immune system by 180–186,
    181, 200–201, 262–264
  deficiency of 262
  discovery of 51, 179
  non-Ig substrates 51, 261–262
  RNA editing enzyme, homology
    to 179, *258*
  selectivity of 183, 201, 262
allelic exclusion. *See* monoalleleic expression
  of loci in diploid organisms
Alport syndrome 153

alternative splicing 266–281, *272*.
    See also implied information
    alternative *trans*-splicing *276*, 275–277
    auxiliary splicing factors, role of
        in defining splice sites 277
    as a developmental switch 273–275, *274*
    role in diversifying proteome
        153, 266 (*see also* Dscam; slo)
    in immune system 201
    prevalence of 271–273
    role of RNA editing in 257
Alu 15, *142*. *See also* SINEs
    chromosome rearrangements,
        role in 148–149
    in duplication of gene-rich segments
    in genome evolution 150
    exonization of 15, 153–154, *154*
    human disease, involvement in 153
    in human genome, prevalence
        of 144–145, 150
    microsatellites, association with 85
    in origin of low-copy repeats 149–150
    presumptive RNA editing sites
        within 259
Alx-4 82
amino acid repeats 8, 78–82, 90
    effect on biochemical properties 80
    in "intrinsically unstructured
        proteins" 85
    phenotypic effects in eukaryotes 82
    in *Ubx* 90
amitotic division 227, 241
ampicillin 118
Anaplasma marginale 101–102, *101*
androgen receptor 79, 82–83
angiosperms, imprinting in 289
antagonistic pleiotropy 107
antibiotic resistance. *See also*
        ampicillin; vancomycin
    carried on integrons 14, 121, 129, 133
antibodies. *See* immunoglobulins
antigenic variation. *See also Borrelia spp.*;
        contingency loci; *Neisseria*;
        trypanosomes, antigenic variation in
    common themes among diverse
        antigen variation systems
        7, 63–64, 91, 93–94
    role of gene expression in 64
    role of genome size in mechanism
        of 105
    pseudogenes in *101*
    role of serotype-converting phage
        in 135
    stochastic nature 104, 108
    in viruses 106

antigen receptor. *See* immunoglobulins
antisense transcripts. *See also*
        RNAi; dsRNA
    gene silencing by transposon promoter
        expression generating dsRNA
        16, 146–148, *147*, 158
    in silencing of genes and spreading of
        DNA methylation at
        KCNq1 locus 285
apoB, mRNA editing of *250*, **251**, 253
APOBEC1 179–180, *250*, 254, 257
    homologs of 261–263
    antagonism of homolog antiviral
        role by Vif 263
apoptosis 185, 189
Arabadopsis thaliana
    argonaute, homolog of, in ciliated
        protozoa 242, *243*
    endosperm size in 289
    homopolymeric amino acid repeats in 80
    transposable elements in
        145, 156, 158–159
Arber, Werner 120
ARP 257–258
    homologs of 261
    role of accessory proteins and flanking
        sequences on target site
        selectivity 262
Artemis *167*
attI and *attC 130*, 131
autoimmunity, avoidance of 177,
        186, 189–190
Avery, Oswald T. 121, 131

Babesia bovis 101
Bacillus subtilis 111, 131–133
    regulation of competence by
        quorum sensing 133
Bacteroides contingency loci in 64, 127
bacteriophage
    434, repressor binding to DNA 37–38
    inhibitor of uracil DNA glycosylase
        from 182
    role in antigenic variation in
        Shigella 135
    role of bacterial surface contingency
        loci in resistance to 64–65
    role in horizontal transfer of
        genes 14, 135
    role of restriction/modification systems
        in resistance to 67
barley
    microsatellite length correlation
        with ecological zones 88–89
    transposable elements in 85, 145, 151

barn swallow, tetranuleotide
    repeat in 83
base pairs, structures 25–27
B cells 163–165, 179, 185–190, 199–203.
    See also lymphocytes; cancer, B-cell
    lymphoma, c-*MYC* and
B-cell receptors 185–186, 188. *See also*
    immunoglobulins
BCL6 mutation triggered by
    AID 51, 184
BLM helicase 195, 202, 204–206
Bloom syndrome 204, 206
Borrelia burgdorferi 9
    generation of mosaic surface antigens
        100–101, *101*
    loss of plasmids in laboratory 20
Borrelia hermsii
    vmp-encoded antigenic variation
        system in 100
BOX elements 128
branchpoint adenosine
    attack by 2'OH of in RNA splicing
        268–269, *269*
    binding by splicing factor 269–271,
        270, 280
Brenner, Sydney 183
Buchnera, gene loss in 117, 122

Caenorhabditis elegans
    homopolymeric amino acid repeats
        in 19, 80, 205
    RNA editing in, **251–252**, 256, 260
    trans-spliced leader sequences 277
    transposable elements in
        139, 145, 158–160
C to U deamination. *See also*
    deamination
    causing C to T transition 52
    spontaneous 18, 180
    in immune system 18, 51, 182
    (*see also* AID)
Campylobacter jejuni
    possible role of contingency loci in
        resistance to bacteriophage 65
    repeats in genome **60**
cancer
    B-cell lymphoma, c-*MYC* and
        202–203, 262
    chromosome breakage, possible
        relationship to syntenic boundaries
        of 17, 21
    transposable elements and 161
cat 217
cattle 82, 217
    Trypanosoma vivax infection 94

CDAR 256–257
centroblasts 186, *187*, 189
centrocytes 186, *187*, 189
centromeres as coldspots of meiotic
    recombination 217–218, 220
    change in phase 144
Chlamydiacae, breaking of
    chromosomal symmetry around
    replication origin and
    terminus *126*
Chlamydophila pneumonia 124, *124*
chicken 175, 179, *180*, 182–183, 204
chimpanzee 222
chloroplasts, RNA editing in 252
chromatids *209*
    failure to separate at metaphase due
        to intermolecular G4
        DNA *194*, 195
chromatin 159, 166, 214–220, 222
    immunoprecipitation assay of histone
        modifications 284
    structure
        role in gene silencing 284–285; role in
        macronucleus formation in ciliates
        10, 226, 228 241–242, *243;* open,
        acetylation of histones in 284
chromodomain proteins 242, *243*
chromosomes. *See also* genome
    landscape, unevenness of
    of ciliated protozoa
        chromosome breakage sequences 231, *232;*
        chromosome fragmentation sites 229,
        244, *245;* macronuclear 230–233, *245;*
        micronuclear 231; nature of nuclear
        duality in 226–228
    embryonic lethality, due to maternal or
        paternal disomy at imprinted
        loci 286
    homologous, parental marks on,
        16, 36, 292–293
    megabase (*see Trypanosoma brucei*)
    in meiosis 208–212, *209*, 217, 219,
        223, 292–293
    minichromosomes (*see*
        Trypanosoma brucei)
    neighborhoods in, and
        imprinting 16
    non disjunction *194*, 195, 210
    polytene, in ciliated protozoa
        230, 240–241
    differential replication of 246;
        See also selection
    translocations 203, 217
chromosomal mosaicism, role of
    conjugative plasmids in 133

ciliated protozoa 10, 225–247. *See also*
    DNA, unscrambling of; chromosomes
conjugation in *227*, 228
diploidy of micronucleus *227*
distinction of developmental DNA
    changes in from mating type
    switch and immunoglobulin gene
    rearrangements 226
evolution of 225–226, 228, 241
IES 229, *230*, 233–236, *234*
    proposed mechanism of excision
    of *238*, 238, 241–244, *243*,
    244, *245*, 246–247
lifecycle of 227–228, *227*, *230*, 246
macronucleus, degradation of *227*.
    See also chromosomes, of ciliated
    protozoa, macronuclear;
    chromosomes, of ciliated protozoa,
    micronuclear
phylogeny of 229
SU-IES *230*, 247, 233, *234*,
    235, 237, *238*
    proposed mechanisms of excision of
    239, *245*, 247
timing of DNA replication in 230
zygote *227* 228
CIT1, role in yeast adaptation to
    glucose limitation 148
class I retroelements. *See also*
    transposable elements, LTR
    retrotransposons;
    LINEs; SINEs
IES in ciliates that are
    retrotransposon-like 235
increase in genome size by 141
mechanism of transposition 139
methylation and regulation 160–161
silencing of host genes
    by 160–161
structure and mechanism of
    transposition 139, *140*, 141
class II transposons. *See also*
    transposable elements
MITEs 151–152
structure and mechanism of
    transposition 139–140, *140*
clonal selection 177
closely spaced repeats 9, 107–120
    biological effects of
    role in amplification of β-lactamase
    in adaptation to ampicillin 118;
    generation of variation through
    recombination between 112–113, 117;
    loss and regain of MMR activity
    through recombination 116, *116*

closely spaced repeats *(continued)*
    examples of distribution in "stress"
    genes *115*
gene loss due to lack of recovery when
    horizontal transfer is not
    available 117
recombination between, factors affecting
    rate of 111, 113
code, definitions of 294–295
codon
    and closely spaced repeats 119
    constraints on choices among
    "synonymous" 9, 20–21, 84,
    277–279 (*see also* implied
    information; marked DNA)
    due to overlapping messages 9
    and homopolymeric amino acid repeats
    and trinucleotide repeats 80, 95
combinatorial expansion of possibilities
    13. *See also* exon shuffling;
    mosaic genes; mosaic plasmids
    in antiviral DNA editing 263
    at bacterial contingency loci 73
combinatorial control in alternative
    splicing 279–280, *280*
    immunoglobulin locus 6, 165, 168, 172–174
    trypansome surface antigens
    91, 102–103
competence
    comA 128
    regulation of 19, 121–122, 131–133
complementarity-determining sequences
    170, *173*, 173–174, 188–189.
    See also hypermutable sequences,
    in immune response
computational challenges *See also*
    context; relationships between
    sequences; implied information;
    mosaic genes; genes, gene fragments
    and segments; pseudogenes
    mRNA not complementary to DNA
    from which it was transcribed
    251–252 *(see also* RNA editing)
conjugative plasmids 14, 129, 133–136.
    See also F plasmid; Ti plasmids
    increased transfer when *E. coli* leaves
    exponential growth phase 110
    role of plasmid-encoded proteins
    in conjugation 133
conjugative transposons 128–129,
    133–134, 136
    as distinguished from integrons *130*
consensus sequences. *See also* recombination
    signal sequences; region-specific
    recombination

consensus sequences *See also* recombination
signal sequences; region-specific
recombination *(continued)*
examples of challenge of identifying
171, 235, *236*
amino group in minor groove key 38; in
ciliate DNA sequences 231–236, 241;
imprinted genes 286; mRNA not
complementary to DNA due
to RNA editing 5 (*see also* RNA
editing); palindrome of H bonds as
for P element insertion in
Drosophila 11
in intron splice sites 268, *268*
in mammalian editing factors *258*
context
cellular, and nuclear duality in ciliates 228, 237
"meaning" of a DNA sequence, dependence
on 4–5, 15, 20–21, 280–282 (*see also*
implied information; computational
challenges)
chromosome context, effect of 16, 17, 287
DNA sequence context effects on mutability
10, 12, 19, 52, 62, 71, 113, 183–184,
193, 208 (*see also* second-order
selection)
DNA structure, effect on 11, 27, 38
role of flanking sequences in
MDS 232, 233, *234*
transposable elements, role in creating
context 6
contingency loci 54–76
biological effects of
and alterations in tissue tropism *64, 65;*
bacterial capsule 13, 64; in generation
of antigenic and phenotypic
variants 63–68, 73–74, 92; in genes
involved in interactions with the
environment 63–67; possible
involvement of loci affecting
metabolism 58, 63; role in resistance
to bacteriophage 65
contrast with global mutator phenotype
58, *61,* 73, 92
in eukaryotes 65 (*see also* diversity,
mechanisms of generation of)
evolution of 58, 61, 72
mechanisms of hypermutability at *59*
mutation rate, dependence on *cis* and
trans factors 68, *70,* 70–72
cryptogenes 5, 249, 260–261, *250,* **251**
cytidine deamination.
See deamination; CDAR
cytosine, methylation of in CpG and genomic
imprinting 283

Dam
methylation of promoters and effect
on bacterial gene expression 67
lack of effect on rate of phase variation 71
Darwin, Charles 13
DDM1 159–160
deamination *43*
causing alteration in DNA template 42
enzymatic (*see also* ADAR; CDAR)
appearance of A to I editing in land
animals (speculation) 259; appearance
of C to U editing in land
plants 255–256; of deoxycytidine
in the immune system 177
(*see also* AID)
in RNA editing *250,* **251–252,** 254
repair of 42, 178–179
spontaneous 178
dicer 158, 242, *243,* 260
Dictyostelium
homopolymeric amino acid repeats 80
percent of nucleotide repeats in genome 78
dinB 45–46, *46,* 113. *See also* DNA
polymerases, translesion synthesis
diversity. *See also* contingency loci; antigenic
variation; mosaic genes; second-order
selection
in activity levels generated by recombination
between closely spaced repeats 117–118
balanced diversity in trypanosome
infection 94–95
as a form of fitness 56, 107
(*see also* fitness)
implied 6, 177
mechanisms of generation of
generation of multiple common alleles
by repeat sequences 78–79, 197;
increased by loss of MMR 19 (*see also*
mismatch repair); independent
assortment and 13; junctional diversity
in recombination at RSS 165, *167,*
168, *170,* 172–173; role of mutable
repeat sequences in eukaryotes 77–90;
phenotypic consequences of generation
of multiple common alleles 81;
polytene chromosomes and amitotic
division in ciliates 241; selection
for mechanisms that generate
12, 13, 73; sources of diversity in
bacteria and their increase
during stress 109–110
in a population descended from an
individual 8, 13, 21
preemptive nature of antigenic variation
in trypanosomes 105

diversity (D) regions 6, 164–168, *165*,
    171–173, 175–176
  read in multiple reading frames 6, *170*
  causing variation in length 6, 166–168,
    167, 172–174
DNA amplification
  in ciliates 230, *230*, 226 239–240
  differential, and selection 240– 241
  in dysregulation of c-*Myc* 212
  reversible 12, 59, 68, 113, 118
  of SINEs by LINEs 141
  during stress, of MITEs 161
  by transposable elements 156
DNA, damage to
  proximity to replication fork and probability
    of repair prior to replication 45
  stall of replication fork 45, 81
    and cell death 47; induction of
    protective response in *E. coli* 108
  by ultraviolet light 45, 50, 52
DNA damage tolerance 44, 5–51
DNA, deletion of. *See also* DNA
    amplification, reversible
  in ciliated protozoa 226, 229, *230*, 233,
    240 (*see also* ciliated protozoa IES;
    ciliated protozoa SU-IES)
  deletional recombination in vertebrate
    immune system *165*, *169*, *180*, 200,
    200 (*see also* closely spaced repeats)
  scan RNA model for in *Tetrahymena*
    242, *243*, *244*
DNA fingerprinting 78
DNA framework 13 (*see also* implied
    information; protocols; site-specific
    recombination; metagenome)
  DNA cassettes 14, 129–130
    (*see also* trypanosomes; integrons)
  horizontal transfer, efficient exploration
    through 123
  efficient information storage 93, 96
  rules for, in vertebrate immune system
    and pathogens 6
  variation on a conserved framework 55, 137, 190
DNA glycosylases. *See also* UNG
  in hypermutation in the immune system
    178, 180, *181*, 182
  in repair of deaminated bases 42–43
DNA inversions 125
  at chromosome breakpoint 146–148, *147*, 158
  at *hsdS* operon (restriction/modification)
    in *S. pneumoniae* 127, *127*
  in immunoglobulin gene rearrangements
    168, *169*, *170*
  IS elements, role in 129
  in phase variation *59*, 72

repeats, role in 109, **111**
  symmetric, around origin of replication 125
  in trypanosome gene arrays, role of *ingi* 96
DNA ligase *167*
DNA methyltransferase enzymes, requirement
    of for parent-of-origin imprinting
    283, 285–286
DNA mutations. *See* mutation
DNA-PK *167*, 171
DNA polymerases
  DNA polymerase alpha gene, scrambling
    of in ciliates 239
  fidelity, crystal structure-based
    understanding of 45–47, *47*
  high-fidelity 41–42, *47*
  increase frequency of closely spaced
    repeats in 113
  specialized 49 (*see also* DNA polymerases,
    translesion synthesis)
  template switching 51
  translesion synthesis 44–50, 52,
    178, *180*, 182
    DNA polymerase eta 48–49, *49*, *181*, 184;
    error proneness of 183
DNA rearrangements, global nature of in
    ciliates 225–226
DNA sequence as "snapshot in time" 137.
    See also species genome; metagenome
DNA sequence analysis, challenges. *See*
    computational challenges;
    implied information
DNA sequence context, effects on mutability
    10. *See also* mutable repeats; context
DNA sequence recognition and protein
    binding to. *See also* DNA, spacer sequences
    in; DNA structure
  bacteriophage 434 repressor 37–38
  BOX element binding to DNA secondary
    structure required for competence 128
  direct and indirect sequence effects
    36–38, 79–80
  effect of backbone geometry on 37
  E2 regulatory protein from human
    papilloma virus 31, 38
  hydrogen bonding pattern and consensus
    recognition sequence 11
  integration host factor 37, 68
  RAG 166
  transcription activators 28
  and twist flexibility 34
DNA, single-stranded, regulated accessibility
    to variation-generating mechanisms 184
DNA, spacer sequences in
  effects on immunoglobulin gene
    assembly 10, 166, 169, 171

DNA, spacer sequences in *(continued)*
effect of length on relative orientation of
sequences in space 10, 33
effects on level of transcription 10, 68, 80
effects of noncontacted bases on protein
binding 37
effects on protein affinity for DNA 10
DNA structure 23–38. *See also* DNA,
spacer sequences in
bendable DNA 10, 28 (*see also* A tracts)
AT-rich sequences at IES and SU-IES
boundaries in ciliated protozoa 236, 247;
and transcription activation 28
"canonical" and "non-canonical"
(*see also* G4 DNA)
A-form *24;* B-form 23, *24,* 27, 37
(BI and BII), 191, 193, 204;
Z-DNA 79
multi-chain 25, 81 (*see also* G4 DNA)
parameters to describe (tilt, roll, twist,
rise, etc.) 25–26, *26*
properties of
antiparallel 23, 193; base pair steps 27, 30;
convention for writing 23;
dimensions 11, 23; logic of replication 23;
mechanical properties 25, 79;
persistence length 32–34; supercoiling
184–185; superhelical 25
sequence-dependent variations
curvature 28–29; flexibility of 32;
information in 10–11 23, 25; minor
groove 11, 29, 31, 38, 193; not
straightforward to calculate due to effect
of context on 11, 38; effect on probability
of meiotic recombination 216; sequence
context effects 32, 79; (*see also* A tracts)
twist flexiblity, effect of GC content on 35;
thermal stability, effect of GC content
on 24–25
x-ray crystallography 27, 29
DNA synthesis. *See also* DNA
polymerases
mechanism and fidelity 41–42, *50*
mutations due to misalignment during
43–44, *44,* 83–84
lagging strand and tetranucleotide
repeat slippage 71; lagging strand
and expansion of G-rich
region 194–195
DNA, unscrambling of, during macronuclear
development in ciliates
5, 239–240, *240*
DNA uptake 19. *See also* uptake
signal sequences
dog-1 205

dogs 217
effect of repeat number variation on
skeletal structure of 12–13, 82, 88–89
double-strand breaks
in immunoglobulin gene rearrangement
164–166, *167*
in meiotic recombination 17, 210, *211,*
212, 214, 220
in transposition 139, 144
Down syndrome 209
Drosophila
alternative splicing in
Dscam > 38,016 splice forms 273, *275,*
275; estimate of percentage of alternatively
spliced genes 273; *mod(mdg4)*
and alternative *trans*-splicing
275–277, *276;* role of in sex
determination 273–275, *274*
antennapedia homeotic protein 79
chromodomain proteins 242, *243*
·genome, smaller than micronuclear genome
of ciliated protozoa 246
parental imprinting in 283, 291–292
per gene repeat and temperature compensation
of circadian rhythm 82, 87, 89–90
piwi, homolog of, in *Tetrahymena* 242, *243*
RNA editing in, **251–252,** 259–260
sterility when lacking BLM homolog 206
transposable elements in 139, 145, 159
microsatellite initiating mobile element 85;
P element 11; transposable
element-mediated chromosome
inversion 146–148, *147,* 158
Ubx 90
Ds 141, 144. *See also* class II transposons
Dscam 4, 273, 275, *275,* 280–281
dsRNA 6, 16, 260
in ciliates 235,*243,* 244
dsRNA-activated protein kinase 80
and post-transcriptional gene
silencing 158
in RNA editing site recognition 257

Ebola, RNA editing in **251**
ectopic recombination 15, 95
and *Alu* elements 148–149
and G-rich DNA 206
and repetitive DNA that forms hotspots
of meiotic recombination 216
and segmental duplications 149
and *Ty* elements in yeast adaptation to
glucose limitation 148
editosome *250,* 260
Ehrlich. Paul 177
elasmobranchs 164, *175*

embryo, establishment of imprinting and
   silencing at metastable epialleles in 287
endonucleases
   AP endonuclease 180, *181* (*see also*
      mutation spectra)
   encoded by inteins 128
   Mus81/Mms4p in cleavage of meiotic
      strand invasion structure 212
   Spo11 topoisomerase 17, 214–215, *215*
   in LINEs *140* (*see also* target-primed
      reverse transcription)
   in LINE and SINE amplification 141
   RAG, in immune system variable region
      recombination 164, 166 (*see also* RAG)
   in trypanosomes, hypothesized role in
      gene conversion 99
Enterobacteriaceae 125, 133. *See also*
   E. coli; Shigella flexneri; Yersinia spp.
Enterococci pheromone-induced plasmids 134
environmental DNA samples 131
environmental effects on genetic alterations 19
   on class switch recombination 201
   in meiosis 220–221, 223–224
epidemics
   high phase variation rates in 72
epigenetic mechanisms 228.
   See also silencing
   as consequence of transposable element
      insertion 156–162
   role of old macronucleus in ciliates 244
   transposable elements, regulation of and
      by 145, 160–161
   silencing in parental imprinting
      283–286, *284*
   silencing at metastable epialleles 287–288
Escherichia coli
   chi sequences 58, 215
   contingency loci possible role of in resistance
      to bacteriophage 65
   fimbrial protein
      effect of temperature on recombination
         of 19; phase variation in 72
   horizontally transferred sequences 122
      conjugative DNA transfer in 128–133, *132;*
      extent of genome comprised of increased
      transfer of conjugative plasmids when
      leaving exponential growth phase 110;
      integration of plasmids into chromosome
      134; plasmid transfer of shiga toxin genes
      and virulence 135; virulence traits on
      plasmids, in pathogenic
      strains of 133
   mutator phenotype upon expression of AID
      and/or deletion of UDG in 180, 262
   RecQ, helicases homologous to in 204–205

repeat sequences, prevalence in
   60, 111–112, 119, 122
   closely spaced repeats in, deletions
   between 115; factors affecting
   recombination rates between closely spaced
   repeats 113–114; effects on gene
   activity following recombination
   between closely spaced repeats in *rpoS* 118;
   mutants with increased rates of slippage
   at repeats 71
Euglenozoa 260
Euplotes 229–237, *230, 232, 234, 238*, 244, *246*
evasins, contingency loci in 63, 65
evolution 189, 295. *See also* natural selection
   of the ability to evolve
      (*see also* second-order selection)
   Evolution Canyon 88
   "foresight" in by evolving response to
      repeated challenges to lineage 12
      (*see also* second-order selection)
   of immunoglobulin locus 174
   modern synthesis 12, 86, 121
   modularity, role in 9, 90, 295
   robustness to repeated challenges in 12, 296
exon shuffling. *See also* recombination, in the
   vertebrate immune system; immunoglobulin
   class switch recombination; ectopic
   recombination
   G-rich DNA structures and 14, *200*, 206
   transposons and 154–156, *155, 157*
exons 13, 14, 267, *267, 268, 269*
   alternative *trans*-splicing 276
   constitutive 153
   exon splice enhancers
      10, 277–279, *278, 280*
   exon splice silencers 279, *280*
   ligation of 269, *269*
   mutually exclusive *272, 275*, 280–282
   self-perpetuating splicing pattern as a
      developmental switch *274*
   spliceosome component binding to *270*, 277–281
exonucleases 42, 46, 180, *181*, 183–184,
      197, 203, 210
explicit genome 54, 60

FEN-1 199
fitness
   diversity as 13, 54, 74, 75 (*see also*
      diversity, as a form of fitness)
   reproductive rate as a measure of 56
fitness landscape 65, 66, 86,
   variable nature of *57*, 107
   parallel exploration avoids trapping in
      local minima 189
FMR1 197

Foldback group of transposable elements 146
F plasmid *132*, 133
　rolling circle replication of, in conjugative
　　transfer 133
fragile X 78, *196*, 197
frameshifts
　by co-translational RNA editing **251**
　in cryptogenes 249
　by mutations (*see also* contingency loci)
　effect on ATG start codon 68, *69;*
　reversible when due to mutable repeats 80
　during replication 43, *44*
　in trypanosome pseudogenes 101
functional hemizygosity 282

G4 DNA 14, 19, 191–207
　formation of during transcription and
　　replication 10–11, 18, 193–195,
　　194, 198
　role of C-rich strand template strand 198
　and genome instability 18 (*see also*
　　Saccharomyces cerevisiae,
　　Sgs1 helicase)
　intracellular, evidence for 198–199, *198*
　and region-specific recombination in the
　　immunoglobulin class switch 10–11,
　　17, *200*, 200–202
　structure of 191–193, *192*
　telomeric repeat in vertebrates 193
G4 RNA 192–193
gag 140, 141
Galileo, chromosome rearrangement by
　146–148, *147*
gametes 223, *209*. See also oocytes
GC-rich sequences
　and gene-rich regions 149
　and meiosis 14, 216–217
　and recombination hotspots 14, 216–217
　as untemplated N regions 168
GCN4 transcription factor 216, 221
genes. *See also* alternative splicing;
　immunoglobulins, locus
　assembly from fragments in two different
　　orientations along DNA helix
　alternative *trans*-splicing 275–277, *276;*
　unscrambling in ciliates 239, *240*
　defined as module recognized through
　　rules 296
　duplication and amplification of
　role of *Alu* in duplication of gene-rich
　　segments 150; in APOBEC family
　　evolution 263; gene families 14;
　　mechanism 118, 124; at tandemly
　　arrayed hypermutable loci 59; example
　　of at *tyrP 124*

gene fragments and segments 6, 91, 103,
　164, 173 (*see also* immunoglobulin,
　locus; diversity regions)
　hidden (*see* haploid expression in diploid
　　organisms, due to imprinting)
gene conversion
　as a mechanism of localized
　　hypermutability 59
　in meiotic recombination *211*, 212,
　　213, 219, 222–224
　in trypanosomes 99
　and variation of pathogen surface
　　antigens 7, 63–64, *101*
　in vertebrate immune system
　　179, *180*, 182
gene expression
　effect of transposable elements on 16
　　(*see also* transposable elements)
　silencing of by complementary RNA 11
　　(*see also* dsRNA; RNAi; silencing)
　variation due to contingency
　　loci 67, 68, *69* (*see also* promoter)
　variation due to repeat sequences in
　　eukaryotes 79–82
genetic anticipation 78–79
genetic code, degeneracy of. *See* codon
genetic "intelligence" 55
genome
　multiple forms of information in 248
　　(*see also* implied information)
genome instability 11
　at G-rich repeats 10, 195, 204–205
　induction of through activation of DNA
　　editing enzymes 264
genome landscape, unevenness of 12, 17, 56,
　191. *See also* hypermutable sequences;
　　meiotic recombination, coldspots
G-rich regions 195, 207
　effect of sequence changes on probability
　　of recombination 218–219
　sequence conservation higher near origin
　　of replication in prokaryotes 17
　syntenic region boundaries
　　boundary between early and late
　　replication 17–18; tumor cell chromosome
　　breaks at 21; repetitive sequences
　　at 216–217
genome organization 6, 15, 296–298
　in ciliates 244
　in trypanosomes
　　maxicircles and minicircles 249; silent
　　archive usage 96, 100, 105
genome rearrangments. *See* ciliated protozoa;
　　genome "shock"
genome "shock" 19, 143, 152, 161

genomic context effects on "meaning" of DNA
sequences 13, 15, 125. *See also* context
genomic islands 14, 136–137
germinal centers 186–189, *187*, 262
germline 158, 175–176, 177, 197, 225, 228, 285
germline transcript 201
glucose limitation, yeast adaptation to 148
glutamate-gated calcium channel, RNA
editing of 249, *250*, **241–252**
gradient of polarity in meiotic mismatch
repair 214
8-oxo-dGTP 41
guanine N-7 position
dimethyl sulfate, use in location of sites 193
and G-4 DNA 192–193, *192*
guide RNA 5, 249, *250*, **251**, 260–261, 264

Haemophilus influenza
competence of 131
contingency loci in 57, 61, 62, 64
and restriction/modification 67; and capsule
locus duplication 67; and different start
codons 68, *69;* tetranucleotide repeat
number and slippage rate 71; effect of
MMR mutants on slippage rates 71
pilin dinucleotide repeat mutation rate *61*, 124
uptake signal sequences in 10, 116–117
haploid expression in diploid organisms
16, 209, *209*, 226. S*ee also* monoallelic
expression of loci in diploid organisms
due to imprinting 282
male mealybug, due to heterochromatization
of entire paternal chromosome set 292
heat-shock response 108
Helicobacter pylori 67
heptamer/nonamer. *See* recombination
signal sequences
heterochromatin 6, 15, 158–160. *See also*
transcription, bidirectional, in meiotic
micronucleus of ciliates
association with matrix attachment regions 285
methylation of histone H3 lysine 9 in 284
parent-of-origin specific expression following
gene movement to certain regions
of 291, *292*
heteroduplex DNA
in ciliate circular IES excision products *238*
in meiosis 13, 210, *211*, 212, *213*, 214, 219
removal by mismatch repair 109, 219
histones 159–160, 220
acetylation of 225, 284
"histone code", importance of ciliates
in the discovery of 225, 245
methylation of 225, 242, *243*, 284
role in gene silencing 284–285

HIV
HIV-1 gp120 104
DNA editing of as host defense 262–263
viral infectivity factor, antagonism of host
antiviral DNA deamination by 263
Holliday junctions 204, 210, *211*
homopolymeric amino acid repeats.
See amino acid repeats
homopolymeric nucleotide run. *See* A tracts;
DNA synthesis, mutations due to
misalignment during; second-order
selection and evolution of simple
sequence repeats.
HOP-1 meiosis-specific protein, binds to G4 206
horizontal transfer of DNA 13, 121–137.
See also metagenome; plasmids;
bacteriophage
and adaptation 58, 109
extent of 136
facilitated by shared "protocols" 297
increase during stress, mechanisms of 110
integrons and 129–130
mechanisms of 131, *132*
recombination-mediated integration
into genome 131–132
and recovery of mismatch repair activity
116, *116*
restriction/modification systems, effect
of on 67
host–pathogen interactions. *See also*
contingency loci; trypanosomes
acute infection 55
adaptation to different niches 13, 66
complement, role in 13, 66, 164, 186
diversity generated by immune system
against 177
role of LPS biosynthetic enzyme variation
in resistance to immunity and 66
pathoadaptive mutations 123
pathogen surface antigens 12
need for stochastic element (i.e.,
non predictability) in responses 108
shared strategy of trypanosomes and
immune system to generate
variation 103, 106
unsettled landscape nature of 13, 55, *57*
hotspots. *See also* hypermutable sequences
alpha hotspots 215–216, 218–219, 220–222
beta hotspots 215–217, 222
of double-strand breaks, inheritance of 214
in evolution of contingency loci 58, 60–61
gamma hotspots 215–218
hotspot paradox 221–223
HOX-2.6 79
human endogenous retroviruses 160–161

human genome
  estimate of number of alternatively
    spliced genes in 273
  percent that encodes proteins 4
  repeat sequences in
    homopolymeric amino acid repeats in 80;
    low-copy repeats in 149; prevalence of 78;
    effect of repeat-number variants on
    transcription 81; repeat sequences
    and syntenic boundaries 217
  extent of tolerance of sequence
    divergence during recombination 219
  transposable elements in 139, 145, 148–150
human papilloma virus, E2 regulatory
  protein from 31, 38
human skeletal malformations 82
Huntington's disease 78
hydrogen bonds
  bifurcated, in A tracts in DNA 30
  between N7 and exocyclic amine in G4
    DNA 192–193, *192*
  role in sequence-dependent stabilization
    of DNA 24
hypermutable sequences 14, 57. *See also*
  genome landscape; mutation spectra;
  contingency loci
  in immune response 58, 174, 183, 190
hypermutation in immune system variable
  regions *181*, 182–185, *187*

IgH. *See* immunoglobulin, locus
IgL. *See* immunoglobulin, locus
immunity. *See also* immunoglobulins
  acquired 55
    vertebrate 163, 174, 177, 185–200
  herd immunity 93
  innate 55
    natural killer cells 287
  role of contingency loci in escape from 64,
    92 *(see also* contingency loci, biological
    effects of; antigenic variation)
  evasins 65
immunoglobulin 163–164
  locus 164, *165*, 166, *167*, *170*, *175*
    (see also immunoglobulin class switch
    recombination; diversity (D) regions;
    recombination signal sequences)
  constant region 164, *165*, *173*, *175*,
    177, 179, *180*, *200*, 200–201;
  evolution of 164, 174–176, *175;*
  hypermutation of V regions 14,
    180–185, *181;* immunoglobulin
    fold 172–173, *173;* J regions
    164–168, *169*, *170*, 171–176, *175;*
    N regions 168, *170;* P region 166,

167, *170*, 175; switch regions,
  G-richness of 195–196, *196*,
  200, 200–202; variable (V) regions
  6, 14, 164, *165*, *170*, *173*, *175*,
  177, 179, *180*
  monoallelic expression of 287
  structure 165, 173
immunoglobulin class switch recombination
  14, *180*, *181*, 191, 199–202, *200*, 205
  contrast to alternative splicing 273
  initiated by deamination 179, 262–263
  regulated by transcription 17
implied information 10, 15, 21, 54, 297.
  See also codon, constraints on
  choices among "synonymous";
  relationships between sequences; context
  and alternative splice forms additional
    information 4–5, 266–268, 271,
    273, 277, *278*, *280*, 281
  and ciliate genomes 5, 247
    ciliates distinguish introns from
      IES 235
  closely spaced repeats in MMR genes 19, 119
  and diversity 6, 75, 81
    pathogen surface antigen diversity
      6, 55, 91, 99; vertebrate immune
      system 6, 163–164, 166, 189–190
  and efficient information storage
    6, 91, 93, 103, 172, 248
  evolutionary information 12, 21 *(see also*
    second-order selection)
  internet, analogy to 297
  and mutable sequences 6, 90
  and RNA editing 5, 259–260, 264
  and scrambled gene fragment or exon
    order 239, *240*, 275–277, *276*
  and *trans*-splicing 277
imprinting centers 292, *292*
imprinting, parent-of-origin 282–293
  distinguishing human maternally and
    paternally imprinted genes 287
  laboratory methods to identify imprinted
    genes 286
  timing of establishment of parental marks 16, 285
inosine
  editing of A to in RNA transcripts
    250, **251–252**, 253
  in mRNA, translation as if it
    were G 253, 257
insects, origin of, and *Ubx* 90
insertion elements 55. *See also* integrons;
  transposable elements
insertion sequences 128–129, 134–136
insulator, role in gene silencing 284
intI 130, 131

integration host factor
   influence on phase variation by promoter
      inversion 72
   sequence-dependent effects on DNA
      structure indirect readout and 37
integrons 14, 109, 129–131, *297*. See also
   horizontal transfer of DNA
inteins 127–128
introns 267, *267*, 268. See also alternative
   splicing
   huge, in some human genes 277
   immunoglobulin heavy chains
      18, 200–201, *200*
   intronic splicing enhancers and
      silencers 279, *280*
   lariat *269*
   mutable repeats in 80
   retained, *272*
   RNA splicing in removal of 233
   role in editing of RNA transcripts 257
   spliceosome binding to *270*
ion channel mRNA, editing of **251–252**, 259
iron-acquisition proteins, contingency
   loci in 63–64, 66

"junk" DNA, not 88, 143

Kepler transposable element 146–147,
   147, 160–161
kinetoplastids, RNA editing in.
   See cryptogenes
Ku *167*, 171

L1 *140*, 145, 152, 154–155, *155*.
   See also class I retroelements
lac operon 12, 36
lag phase 108
lariat intron *269*, 269
lateral gene transfer. *See* horizontal transfer
Leishmania tarentole kinetoplast
      body DNA
   curvature due to A tracts repeated in
      helical phase 28
Lepidoptera 85
leucine-responsive regulatory protein 72
lineages, successful. *See* evolution; natural
   selection; second-order selection
LINEs *140*, 145. *See also* class I
   retroelements; target-primed
   reverse transcription
low copy repeats
   in human genome 149–150
LPS biosynthetic enzymes
   contingency loci in 63–64, 66
   role of in biological adaptation 66

LTR retrotransposons *140*, 141, 160–161
lymph node 186, *187*, 262–263
lymphocytes 163–164, 166, 175, 182, 186.
   See also B cells; T cells

maize 138–139, 155–156, 289
malaria. *See Plasmodium*
   falciparum
mammals 164
   avpr1a 82
   metastable epialleles in 287
   Nova proteins as intron splice enhancers
      and silencers 279
   parental imprinting in 282
   RNA editing in, **251–252**, 252, 256
marked DNA
   exons 10, 277–279
   uptake signal sequences 9, 116–117
   use of in protocols 298
Markov chain analysis of repeat
   overrepresentation in *N. meningitidis* 62
marsupials
   parental imprinting in 282, 290
   RNA editing in 256
matrix attachment regions, association
   with imprinted genes 285
mealybugs 283, 292
memory cells 186, *187*
metagenome 13, 21, 131
McClintock, Barbara 19, 138–139,
   141, 143–144, 152, 156, 161
meiosis 6,13,14, *209*. *See also* meiotic
   recombination
   in ciliates 227, 228, *230*, 242, *243*
   in mealy bugs, loss of paternal contribution
      to male genome 292
   pairing of homologous chromosomes
      206, 208, 223, 292
   repeats with high meiotic instability 83
meiotic recombination 208–224
   anti-recombination role of mismatch
      repair, role of in hybrid sterility 219
   coldspots of *215*, 217–218, 220
   double-strand breaks in 17, 210, *211*
   environment, effects of external and
      internal on 220–221, 223
   hotspots of 214–220, 215
      alpha hotspots 215–216, 218–219,
         220–222; beta hotspots 215–217, 222;
         gamma hotspots 215–218
   effect of repeat sequences on rate 215–218
   effect of transcription factors on
      (*see* hotspots, alpha hotspots)
   metabolic state, effects on sites of meiotic
      recombination 220–221

metazoans
    high conservation of splice recognition
        proteins in 278
    relative distances among 229
methylation. *See also* DNA methyltransferase
        enzymes; histones
    differentially-methylated domain
        283–285, *284*
    epigenetic inheritance of 156
    and susceptibility to double-strand breaks
        in meiosis 220
    and transposable elements 156, 158–161
MHC 27, 186, 188
mice 179, 182–184, 205, 217
    establishment of parental marks of
        imprinted genes in 285–286
    metastable epialleles, effect on coat
        color 287–288
    parthenogenetic birth 290
microarrays
    in analysis of mechanism of
        heterochromatin formation 159
    in analysis of numbers of alternatively
        spliced genes in humans and in
        Drosophila 273
    in analysis of recombination hotspots and
        coldspots in yeast 20, 214, *215*, 216–217
    and study of yeast genome adaptation in
        chemostat 148
microsatellites. *See also* simple sequence
        contingency loci; mutable repeats
    at contingency loci *59*
    definition 62, 78, 124
    origin of 84, **111**
Milstein, Cesar 183
miniature inverted-repeat microsatellite
        elements 151–152
minisatellites
    definition 62, 78
    origin of 84, **111**
    increased meiotic recombination at 216
mismatch repair
    antagonism of homologous recombination
        by 109, 219
    correction of DNA synthesis errors 42
    deficiency of and increased acceptance of
        sequence divergence in
        recombination 9, 109
    down-regulation at onset of stationary
        phase 110
    increased frequency of closely spaced
        repeats in MMR genes 113
    role in immunoglobulin variable region
        hypermutation 180, *181*, 183–184
    during meiosis 212–214, *213*

role as modifier of mutation rate at
        contingency loci 70–72, 100, 120
    mosaicism in MMR proteins, due to
        repeated loss and regain of
        segments 123
mitochondria
    symbiosis, analogy made of nuclear
        duality in ciliates to 247
    RNA, editing in 249, 252, 255, 260–261,
        264 (*see also* kinetoplastids;
        cryptogenes)
MMP3 83, 88
mod(mdg4) 275–277, *276*
modifier genes and rate of phase variation
        70. See also mismatch repair
monoallelic expression of loci in diploid
        organisms 13, 16, 282–283, 287–288
    facilitation of selection by 16
    in trypanosomes 16, 99
    of *var* genes 104
morphology
    effect of repeat sequence variation on 82
mosaic genes *101. See also* pseudogenes
        and mosaic gene formation
    Borrelia 100
    mismatch repair protein mosaicism 123
    N. gonorrhoea pilin 100
mosaic plasmids 135
    mPing 151–152
MS1 *196, 197*
MSH2/6 *181*, 183–184, 202
mutable repeats. *See also* simple sequence repeats
    in eukaryotes 8, 77–95
        factors that affect instability at 17, 84, 111;
        G-rich *196*, 197–198 (*see also* G4 DNA)
        in generation of somatic variation 81;
        and human disease 78, 197; instability
        of 7–8, 43–44, *44*, 60, 83, 84, 87;
        meiotic instability 83, 197;
        MS1 (locus D1S7) *196*, 197; premutation
        78–79; selection of, evidence
        for 7–8, 85–90
    in prokaryotes 7–8, **60**
    reversibility of mutations in 80, 84
mutation. *See also* mutable repeats; DNA
        damage tolerance; mutation spectra
    role of damaged nucleotide
        precursors in 41
    dependence on biochemical activity 52
    editing, mutational, of DNA 248, 261–264
    high-fidelity polymerases and 41
    during repair 44 (*see also* DNA polymerases,
        translesion synthesis)
    transitions 52, 178, 180, *181*, 182
    transversions 41, 178, *181*, 182–183

mutational load
  in populations, lower due to mutable repeats
    78 (*see also* second order selection)
  in xeroderma pigmentosum 47
mutation spectra 51–52, 87
  alteration in MMR mutants 72, 81, 92, 123
  bias G/C to T/A 223
  factors affecting 51–52, 184
  genome-wide changes in mutation rate
    and decreased mismatch repair activity 19;
    and inactivation of *mutT* 41; and loss of
    helicases that interact with G4 DNA 19
  leading and lagging strand, difference
    between 52
  natural selection and 53
  regulated alterations in 18
  result of balance of many activities 51–52,
    178, *181*, 182, 184, 215, 261
"mutations of small effect" 86
Mutator-like transposable elements 156
mutator phenotype 41, 58, 114–115
  in epidemics 72
  due to expression of AID in *E. coli* 180
mutL 72, 115–116, *115*
mutS 72, 114–115, *115*, 118–119, 183.
  See also MSH2/6
mutT 41
MVSG coat of trypanosomes 95–96, 99, 103, 106
  arrangement in *T. brucei* genome *97*
  transmission stage 95, 103
  monocistronic 99
c-*MYC* 202–203
Mycobacterium leprae 117
Mycobacterium tuberculosis 55
Mycoplasma 63, 67–68, 112, 117, 119, 124

Natural selection. *See also* second-order selection
  action on mismatch repair and DNA
    polymerase 55
  evolution of response to classes of challenges
    that are repeated in a lineage 12
    (*see also* diversity, in a population
    descended from an individual)
  neutral mutations 122
Neisseria 19
  capsule, phase variation in 65
  competence of 131
  deletion of closely spaced repeats in 115
  phase variation in
    iron acquisition proteins 66;
    pilin 64, 100–101, *101;* effect of
    MMR mutants on phase variation rate 72;
    nadA promoter 68; effect of repeat
    number on phase variation rate 71

repeats overrepresented in genome of
  N. meningitidis **60**, 62, 63
  sporadic virulence of 75
  uptake signal sequences in 117, 133
network organization
  evolvability of, facilitated by protocols
    295–298
Neurospora crassa 161–162
neurotransmitter, mutable repeat in intron
  of rate-limiting biosynthetic enzyme 80
non-Mendelian segregation 212–214, *213*, 217
non-protein coding RNA 11, 21, 284. *See also*
  RNAi; siRNA; dsRNA; relationships
  between sequences; introns
non-random genetic change. *See also* implied
  information; context DNA sequence
  context effects on mutability; second-order
  selection; AID
  with respect to DNA sequence 12, 223
  with respect to function 12 (*see also*
    contingency loci)
  recombination hot and cold spots 208, 222
  selection operating on 18, 121, 207,
    222, 223–224
nuclear duality in ciliates 5, 226–227, 242, *243*
  information transfer from old to new
    macronucleus 242–244, *243*
  mac-destined sequences 230, 232, *232*,
    237, 239, *240*, 241, *245*, 246
  macronucleus 225–231, *227*, *230*
  mic-limited sequences 230, 242, *243*
  micronucleus 226–230, *227*, *230*
  micronuclear genome 233 236 239 246
nucleolus 197, 204
nucleosome assembly
  absence from regions of "open"
    chromatin 284
  sequences affecting 10, 35–36, 79
nucleotide excision repair 47. *See also*
  deamination, repair of
nucleotide mooring sequence *250*, 255, *258*, 262
nucleotide structures *43*
nucleotides, modification of (*see also*
  deamination)
  in nucleotide pool 41
  in RNA 254–256 (*see also* RNA editing)

olfactory receptors, monoallelic expression
  of 16, 287
oocytes 174
  cytosine methylation and imprinting
    in 285–286
Opa 64–65. *See also* adhesins
Oxytricha 229, 231, 233, 237, *238*, 239, 241

pack-MULEs 156, *157*
palindromes. *See* relationships between
    sequences, biological importance of, in
    DNA, hairpins
Paramecium 229–230, 232–236, *234*
    241, 244, 246
paramyxoviruses, RNA editing of **251**, 261
Pasturella 117
pathogenicity islands 58, 134–137
PCNA 42, 50, *50*, 205
per hexanucleotide repeat in and temperature
    compensation in *Drosophila* 82
PfEMP1 104
Phase variation 57. *See also* contingency loci
    in alteration of gene expression through
        altering start codon 68, *69*, 127
    environmental signals in regulation of
        temperature, affected by, at *E. coli* fimbrial
        locus 19, 72
    at *hsdS* operon (restriction-modification
        system) in *Streptococcus* 127, *127*
    by inversion of promoter 72
    models of role of in generation of
        diversity 74
    of pilin 64, 124, 127
    effect of repeat number on rate 71
    in resistance to bacteriophage 65
    Salmonella hin flagella 127
phylogenetic analysis
    of CDARs 256–257
    inconsistency, in Gram negative bacteria,
        of repair and housekeeping genes 123
    of shark immunoglobulin locus germline
        recombination events 175
Physarum
    RNA editing in 249, **251**, 256, 261
pig 217
PilE 133
pilin
    phase variation in 64, 124, 127
    mutation rates of *61*
pheromone-induced plasmids 134
plants
    C to U RNA editing appearance in land plants 255
    parental imprinting in 282
    metastable epialleles in 287–288
    RNA, editing in 249, **251**, 252
plasmids 129. *See also* conjugative plasmids
    exchange of DNA with other plasmids and
        plasmid fusion 134–135
    integration into host chromosome 134
    leading to beta meiotic hotspots and their
        tendency to be lost as a result
        216, 222–223

Plasmodium falciparum 65, 104
Pneumococci 121
Pneumocystis carinii 100
pol 140, 141
Pol I
    complex with WRN accelerates rDNA
        transcription 205
    deletion and tetranucleotide repeat
        slippage 71
Pol V 47. *See also* DNA polymerases,
    Y family
polyploidy 161, 227, 230, 289
position-effect variegation 291
post-meiotic segregation 212, *213*
post-transcriptional gene silencing 158.
    See also RNAi
Prader–Willi syndrome 149, 286
prairie vole
    alleles of vasopressin receptor affect
        expression and male social behavior 82, 87
promoter. *See also* antisense transcripts;
    tuning knobs
    differential methylation upstream of in *H19*
        and parent-of-origin
        imprinting 283–284
    effect of length of spacer on expression
        8, 10, 15, 28, 68, 80, 124
    mechanisms that vary expression at
        contingency loci *59*, *69*
    pol III promoter in SINEs 141
    role of proximity to in probability
        of hypermutation by AID 184
    role in immunoglobulin class switch
        recombination 201
proteome, expansion of. *See also* antigenic
    variation; combinatorial expansion of
    probabilities; *Dscam*; *slo*
    by alternatively splicing cassette exons
        271, *272*
    diversification by alternative splicing
        266, 271, *272*, 273, 275, *275*
    diversification by RNA editing 248
    diversification by RNA splicing in
        trans 275–277, *276*
    by haploid expression and metastable
        epialleles 291, 293
    by information that expands the coding
        capacity of DNA 248–249
protocols 21, 295. *See also* DNA
    framework, efficient information
    storage, rules
    definition of 294
    control, coordination, and integration
        facilitated by 296

pseudogenes
  APOBEC homologs on chromosome,
    in cluster of 262, 263
  and chicken immunoglobulins 15, 179
  and gene evolution 15
  and mosaic gene formation *101*, 102, 105
  processed, from mRNA via TPRT 141
  and templated repair in the immune system *181*
  and trypanosome surface antigens 7, 15,
    91, 96–98, *97*, 100–103, *101*, 105
Pseudomonas 115, 118, 123
pufferfish 78
pyrimidine dimers 47–48, 51

quantitative trait loci 78, 81–83
quorum sensing 133. See *also* competence

rabbits *180*
RAD6-BRE1 220
RAD51
  and gene conversion of laboratory-adapted
    trypanosomes 99
  and homologous recombination in the
    maintenance of yeast telomeres 203
RAG
  apparent derivation from transposon 15, 174
  activity in germline 15, 174–176
  in immunoglobulin formation 166–168,
    199 (see *also* recombination
    signal sequences)
RAP1 218
rat 217
  mRNA editing in, **251–252**
rDNA 195, 197–198, 204, 230
RecA 109–110
RecBC 204
RecQ family helicases 195, 203–206
recombination 13, 14, 17. See *also*
  region-specific recombination; site-specific
    recombination; closely spaced repeats;
    meiotic recombination; recombination
    signal sequences
  between closely spaced repeats 109
  in generation of diversity 109, 129
  homologous 110
  hyperrecombinogenic phenotype of
    chicken cell line DT40 182
  increased in subtelomeric regions 17, 100
  overrepresentation of closely spaced repeats
    in genes involved in 113
  RecA-independent 109, 111
    role in lengthening tandem repeats 109
  in the vertebrate immune system
    14, *165*, *167*, *169*, *170*
recombination activating gene. See RAG

recombination signal sequences 10, 14, 164,
  166–168, *167*, 169, *169*, *170*, 171–176
region-specific recombination
  regulation through transcription 16–18, 216
  role of G-rich regions in 17, 200–202
    (see *also* G4)
  in meiosis 214–215, 220, 223
regulation, novel approaches to
  by generation of stop codons in alternatively
    spliced exons 273–275, *274*
  hierarchy of 15
  by layering of interacting protocols 295
  of resultant protein's activity by RNA
    editing 252, 253, 259
  by varying ratios of protein activity 82
relationships between sequences, biological
    importance of
  in DNA 11, 113
    hairpins 10–11, 71, 80, 166, *167*,
      169, 175, 237
  and recognition of IES in ciliates 242, *243*
  in RNA 11, 20 (see *also* RNA secondary
    structure)
repeat-induced point mutation 161–162
repeat sequences. See *also* mutable repeats
  at ciliate telomeres 231, *234*
  effects on DNA structure 25, 28
    (see *also* A tracts)
  in generation of diversity, role of **111**
  numbers in bacterial genomes **60**
  effects of variation in on rate of
    transcription 8, 81–82
  in wheat, change in length exposure to
    head blight pathogen 19, 81
replication protein A 185
reptiles 164
restriction-modification systems
  contingency loci in 63, 66–67
  effect on uptake of external DNA 67
  inversion in *Streptococcus hsdS*
    operon 127, *127*
retrons 127–128
retrotransfer (DNA "pickpocketing") 134
retrotransposons. See Class I retroelements
retroviruses. See *also* HIV
  resemblance of LTR retrotransposons
    to 141, 154
Rev1 50, *50*. See *also* DNA polymerases,
    translesion synthesis
reverse transcriptase. See *also* pol
  component of telomerase 231
reversible mutation 8, 12. See *also*
    phase variation
  amplification in tandem repeats 118
  contingency loci 57

reversible mutation *(continued)*
  effect of codon choice 20
  recovery of mismatch repair activity
    through 19, 116, *116*
  at repeat sequences 78, 80
reverse transcription 95, 128, 141, 203, 231,
  262. *See also* target-primed reverse
  transcription
rheostats
  and metastable epialleles 9, 291
  proposed dosage sensitivity of
    imprinted genes 291
  RNA editing, effect of on activity of ion
    channel complex 7, 259
  transposable elements, effect on
    transcription of 9, 152
rice
  selection for growth in temperate
    climate 19, 150–152
  transposable elements in 150–152, 156
Rickettsia 117
RNA/DNA hybrid, stability of 198–199, *198*
RNA editing 248–264, **251–252**, *250*
  (*see also* implied information)
  definition, as distinguished from other
    nucleotide modifications 254
  diversification of the proteome by 248
  enzymes, specificity of 254–256
  evolution of 254–261, *258*
  extent of in human genome, estimated 259
  of ion channel RNA 5, 252–253,
    251–252, *250*
  effect on channel kinetics 253, 259
  orphan editing activities 255
RNAi 15, 158–160, 235, 242, *243*
  244, *246*, 260
RNA-induced silencing complex 158, 242, *243*
RNA polymerase 68, *69*
  slippage of and insertional editing 261
RNA secondary structure. *See also* relationships
  between sequences
  alteration of by editing 259
  of CTG repeat structures in UTRs 80
RNase H 199
  mutants of and rate of tetranucleotide
    repeat slippage 71
RNA splicing. *See also* branchpoint adenosine;
  alternative splicing
  chemistry of 268–269, *269*
  discovery of 225
  pyrimidine tract 268, *268*, *270*, 271
RNA transcripts. *See also* RNA editing
  alternative splicing of 248–249
RNA world 254
rpoS 115, 118

Rtel 205
RUP elements 128
Runx-2 82

S-phase asynchrony. *See also* timing
  of replication
  and trypanosome chromosomes 96
Saccharomyces cerevisiae. *See also* RNA editing
  adaptation to glucose limitation 148
  conjugative DNA transfer 133
  homopolymeric amino acid runs in 80
  meiotic recombination in 212, 220–221
    (*see also* meiotic recombination,
    hotspots; meiotic recombination,
    coldspots)
  mismatches, extent of tolerance of,
    in recombination 219
  repeats, mutability and distribution of
    simple sequence repeats in 83, 85
  RNAi in 158
  Sgs1 helicase 14, 204–205
    and chromosome stability 195, 206;
    and telomere maintenance 203
  THO/TREX deficiency and genomic
    instability 199
  transposable elements in 139, 145, 148, 159
Salmonella 115, 135
Schizosaccharomyces pombe 160, 204,
  216, 218, 242, 246
segmental duplications. *See* low
  copy repeats
selection. *See also* natural selection;
  second-order selection
  in immune system during somatic
    evolution 177–178, 185–190, 199–200
  of immune system in evolution 14
  and imprinting 293
  "inclusive fitness" and kinship 289
  among polyploid "alleles" upon amitotic
    division in ciliates 241, 246
  and RNA editing 260–261
second-order selection
  and balance of sequence constraints 58
  and closely spaced repeats in "stress"
    genes 118–120
  and evolution of simple sequence
    repeats 86–87, 89–90
  of hypermutability, and its conservation
    of at certain sites 296
  and location of *Ty* elements 148
  mathematical models of 73–75
  and MMR genes 72
  of protocol-based organization 297
  and repeats of G/C in *N. meningitidis* 62
  and RNA editing 260

second-order selection *(continued)*
    and sites of action of enzymes that
        modify DNA sequences 17
    and uptake signal sequences 119
    and varied bacterial mutation rates 56
sequence context. *See also*
        implied information
    DNA polymerase fidelity, effect of on 12
        (see also DNA synthesis)
    importance of considering in
        experimental design 20
    and region-specific hotspots in meiosis 223
        (see also meiotic recombination,
        hotspots)
    importance of in understanding function 293
    nucleotide sequence recognition *(see also*
        DNA sequence recognition and protein
        binding to)
    in RNA splice sites 268
serotonin transporter 83
SET1 220
shark immunoglobulin locus 164, 166–168,
        170, 175–176
Shigella flexneri 123, 134–135, 137
sialic acid 66, 72
silencing. *See also* antisense transcripts;
        monoallelic expression of loci; RNAi
    in ciliates 241, 246–247, *243*
    in imprinting 283–285
    transposable elements and 145, 156,
        159–161, 246
siRNA 6, 158, 260
simple sequence contingency loci. *See also*
        phase variation; contingency loci
    mutations at microsatellites 62
    overrepresentation in obligate bacterial
        pathogens 63, 112
simple sequence repeats. *See also* simple
        sequence contingency loci; repeat
        sequences; contingency loci)
    computer programs for analysis of 90
    conservation of position suggests functional
        role 81 *(see also* second-order
        selection)
    cryptic 112
    definition 77, 112
    functional consequences of 77–78, 80–81
        (see also phase variation)
    in ORFs 78
    overrepresentation in genomes 77
    overrepresentation in mismatch repair
        proteins 81 *(see also* mutS)
    quantitative effects on gene function 78
    in regulatory regions 78, 80
    at terminus of SINEs and LINEs 141

SINEs. *See also* target-primed reverse
        transcription; *Alu*; class I retroelements
    association with microsatellites 85
    structure and mechanism of transposition
        140, 141, *142*
site-specific recombination. *See also*
        transposable elements
    at contingency loci 59, *61*
    FimB and FimE effects on PV of *E. coli*
        type I fimbriae 72
    and immunoglobulin gene segments 199
        (see also recombination signal sequences)
    and insertion sequences 128–129
    and integron cassettes *130*, 131
    and pathogen surface antigens 7, *59*
    and transposition 125
    variable rate of, effect of prioritizing
        expression
    in trypanosome VSG 99–100; in
        vertebrate immune system 172
slo
    >500 alternative splice forms in
        vertebrates 273
Sm proteins 269
snRNAs in spliceosome, 269–271, *270*
snRNP in spliceosome, 269–271, *270*
solo LTRs 141, 145
Somatic genetic variation
    black spots on red pigs 81
    hypermutation in the immune system
        51, 174, 177, 178–188, 199–200, 264
    immunoglobulin gene rearrangements 163–174
    mutable repeats 81
SopE 135
SOS response
    activated by stress 19
    effect on slippage rates at repeats 71
Southern analysis of meiotic hotspots 214, 216
spaced interspersed repeats 112
speciation 216–217, 219
species genome 8, 19, 21, 54
spliceosome 269–271
    assembly of *270*
    as interpreter of "splicing code" 279–280
split genes. *See* alternative splicing
Spirotrichea 229–230, 232–233, 235,
        240, 244, 246
sporulation 212, 219–221
Spm/dSpm 141, 156. S*ee also* class II
        transposons
SR proteins, role in recognizing splice sites 278
Staphylococcus aureus 55
starvation, effect on sites of meiotic
        recombination 220–221
Streptococcus 115, 127, 131

stress. *See also* starvation
  deterministic responses to frequent
    sources of 108
  diversification responses 19, 108, 120,
    145, 148, 293
  prophage induction by 19, 110, 135
  role of repeat sequences in
    overcoming 107–120
  "stress" genes 113, **114**, *115*
Sturtevant, Alfred 90
Stylonychia 229–230, 239
symbiosis, ciliate nuclear duality analogy to 247
synonymous codons. *See also* implied
    information
  importance of studying 20
  constrained by exon splice enhancers
    and exon splice silencers 277

tandem duplication
  gene amplification 118
  generation of 124
  at hypermutable loci in bacteria *59*, 67,
    111, **111**, 118
  in immune response evolution 174
target-primed reverse transcription 15, 141,
    142, *142*, 154–155
target site duplication
  upon transposable element insertion
    139, *157*
  at boundaries of ciliate internally eliminated
    sequences *234*, 236
T cells 163, 185–186, 188–189, 262.
    See also lymphocytes; germinal centers
T cell receptor 163–166, 171–172, 174
  monoallelic expression of 287
TCP/IP 295
Tc1/mariner superfamily of transposable
    elements 158, 233, 237
Tec elements
  two families of transposons in ciliates
    233, *234*, 236, 237, *245*
telomerase 197, 203–205, 225, 231
telomeres. *See also* telomerase
  ALT pathway 203
  in ciliated protozoa 225–226, 229, *230*,
    232, 233
  as coldspots of meiotic recombination
    217–218, 220
  G richness of in vertebrates 193, 195–197, *196*
  maintenance and regulated formation of
    G4 at 203–204
  subtelomeres
    in antigenic variation in trypanosomes 96, 98;
    ectopic recombination increased in 17, 105
  telomere bearing elements 233, 237

telomere position effect 105
  unusual nucleic acid structures at 193
terminal deoxynucleotidyl transferase
    166–168, *167*
terminal inverted repeats 16, 139, 156–158, *157*
Tetrahymena 225, 229–233, *230*, 232, *232*,
    234, 235, 237, *238*, 240–244,
    243, 246, 260
tetrapods 164–165, *175*
thymidine dimers 48
tilapia, effect of prolactin promoter
    polymorphism on growth of 82
timing of replication 17. *See also* mutation spectra
    in ciliates
  independence of two nuclei in 227;
    and distinguishing retained and eliminated
    DNA sequences 241; of DNA
    unscrambling and deletion 230, *230;*
  heterochromatin in ciliates 241, 242,
    243, 246–247
  imprinting and 17, 285
  role of in pre-meiotic replication and locations
    of meiotic double-strand breaks 220
Ti plasmids 134
T_m 24–25
toxin genes, cone snails 12
trans-acting factors
  effects on repeat mutation rate 71–72
transcription
  bidirectional (*see also*
    antisense transcripts)
  in meiotic micronucleus of ciliates
    242, *243*, 246, 235, 246, 247
  deletion between closely spaced
    repeats during 113
  effect on hypermutation by AID 185
  nucleotide insertion during **251**
  effect on repeat slippage rate 71
  effect of repeats on rate of 79, 81
  silencing by transposable elements 158
transcription factors, role of in meiotic
  recombination 17. *See also* hotspots,
    alpha hotspots
transposable elements 6, 14, 138–162. *See also*
    class I retroelements; class II transposons
  eukaryotic (*see also* transposable elements
    in cilated protozoa; transposable
    elements in eukaryotic evolution)
  autonomous and nonautonomous 139, *140;*
  discovery of 139; effect on expression of
    "host" genes 15, 150, 156–162; genome
    diversification and restructuring
    by 144–145, 161–162, 174; *ingi* association
    with trypanosome gene cassettes 96, *98;*
  microsatellites, role of in origin of 85;

transposable elements *(continued)*
    prevalence of 138–139; variation of activity
        and of predominant element type among
        genomes 145
    increased movement under stress 19
        (*see also* "genome shock")
    and genome rearrangements in prokaryotes
        125–131, 136–137 (*see also* horizontal
        transfer)
transposable elements in ciliated protozoa
    elimination of during macronuclear
        development in ciliates 237
    role of in origin of IES 237–238
    role of heterochromatin in silencing 246–247
    possible "safe haven" for, in micronuclear
        genome of 239
    telomere bearing elements 233, 237
transposable elements in eukaryotic evolution
    15, 138, 143–146, 162
    and exon shuffling 15, 154–156, *155*, 174
    providing homologous sequences for
        recombination 125
transposons. *See* transposable elements
trinucleotide repeats 78–85 88–90. *See also*
    mutable repeats; simple sequence repeats;
    genetic anticipation
    Fragile X 78, *196*, 197
    Huntington's disease 78
    overrepresentation in eukaryotic genes
        encoding developmental regulatory proteins,
        transcription factors, protein kinases 85
tRNA
    nucleotide modification in 254–255
    sequences in transposable elements 136, 141
Trypanosoma brucei 93. *See also* trypanosomes
    chromosomes of 96
    contrast with developmental regulation
        of ciliate chromosomes 96
    pseudogenes in antigenic variation *101*
    VSG structure *94*
Trypanosoma congolense 96
Trypanosoma cruzi 92. *See also* trypanosomes
Trypanosoma vivax 94, 96
trypanosomes. *See also* cryptogenes;
    Trypanosoma spp. tsetse fly
    allelic exclusion in 16
    antigenic variation in 91–106
    bloodstream expression sites 96–99 *97*
    gene fragments in 91
    pseudogenes in 91, 96–97, 100–102, *101*, 105
    slower antigen switch rate in laboratory 20
    metacyclic (mammalian infective) stage 95
    silent gene archive 95–106, *97*
tsetse fly 95–96
type III secretion system 135

type IV secretion apparatus 133
tyrosine hydroxylase, repeats in intron 80, 83, 89
tuning knobs 7, 77, 82, 87, 89–90
Ty elements 145, 148, 217–218

uridine, insertions and deletions of.
    See cryptogenes
ubiquination 50, 220, 263
UDG 178, 180, *181*, 182–184. *See also* UNG
UNG 200. *See also* UDG
Ubx 90
umuC/D 45–46, *46*. *See also* DNA
    polymerases, translesion synthesis
uptake signal sequences 116
    overrepresentation in certain genomes 116
    overrepresentation in "stress-related"
        genes 117, 119
UTRs
    repeats in 80, 90
    editing of **251–252**

vancomycin
    resistance, pheromone-induced
        plasmids and 134
Var 65, 104

variable surface glycoprotein (VSG)
    evolution of 96
    expression of 104
    rate of switching 93
    silent archive 95–106, *97*, *98*
    smokescreen role of 93
    structure of 94, *94*
vasopressin
    receptor alleles affect expression and
        male social behavior in prairie
        vole 82, 87
    RNA editing in **251**
vertebrates
    G4 DNA in 193, 195, 206
    immunoglobulin genes in 6, 11, 14,
        163–164, 169, 174–177, 189
    RNA editing in **251**
    RNA splicing 273 (*see also* alternative
        splicing)
    simple sequence repeats in 78 (*see also*
        simple sequence repeats)
    syntenic breakpoints in 217
Vibrio cholerae 115, 136–137
viruses
    nucleotide editing and, *250*, **251–252**,
        257, 261–262, 263

Wallace, Alfred Russel 1
Werner syndrome 204–205

wheat
    change in repeat length exposure to
        head blight pathogen 19, 81
    correlation of repeat length with
        ecogeographical parameters 88–89
    synthetic polyploids 161
Wigglesworthia glossinidia 122
Witkin, Evelyn 52

Xenopus
    equivalence of distance among ciliates to
        distance between frogs and rice 229

immunoglobulin switch regions in 201
    RNA editing in *250*, 251–252, 257, 261
Xeroderma pigmentosum 47
XRCC *167*, *181*, 182–183

Yersinia enterolitica 115, 133
Yersinia pestis 133, 134

Z-DNA 79
zebrafish 174, 283
zinc finger protein, binding to affected by
    repeat length 80